A Down-To-Earth Guide to

SDLC Project Management

2nd Edition

Joshua Boyde

Getting your System Development Life Cycle project
successfully across the line using PMBOK® ... adaptively.

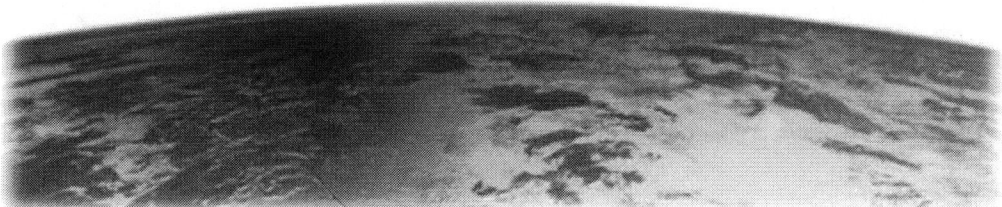

Dedication

As we, lay another to rest in a sea of solemn regrets.

Discovering plans for the future, have all fallen apart.

Decisions made to win, ended up being sink or swim.

May we learn to understand the truth that is blind to management,

The future has already been written in the sand.

This book is dedicated to all those projects that passed away before their time.

Firstly, I wrote this book purely for the enjoyment of writing, and

secondly, to repay all those who have taught me a thing or two during my career.

Hence, I now "pay it forward".

I hope you enjoy my eBook, and consider purchasing it in the physical form.

If you found it useful to your understanding of System / Software Development Life Cycle

(SDLC) project management then please pass this book onto others.

JoshuaBoyde

@Hotmail.com

P.S. This is the book that I wished was available when I started out on my project management career.

Foreword

This book has been crafted for both the project management novice who is ready to confront their first real project, through to the seasoned veteran with several project battle campaigns under their belt.

This book is based on many years of "real-world" System Development Life Cycle (SDLC) project management, as well as the Project Management Body Of Knowledge (PMBOK®), the blending of the useful elements from other management practices & principles, and the incorporation of the past experiences & the lessons learnt from the various industrial backgrounds of those persons who graciously contributed to this book's creation.

Described within is the practical application of field-tested project management techniques to actual situations and prevailing circumstances where the realities of commercial necessities have to be given serious consideration. Additionally, this book does cover some topics and ugly truths that are often not acknowledged in academic textbooks on project management.

Please be aware that this book does espouse some ideas and concepts that are not purely PMBOK® based; hence, this book may not be entirely suitable as a study reference for obtaining Project Management Professional (PMP®) Certification.

Thank You

Thank you for opening my book, "**A Down-To-Earth Guide to SDLC Project Management**", and thank you to those who purchased the 1st Edition of this book, it is much appreciated.

I would also like to thank the following contributors.

SME – Subject Matter Expert

Critique & "jack of all trades" SME: Steven Lipke

Project & program management SME: Michael McAuliffe

Allan Pedersen

Agile SME: Andrew Suduk

Procurement, production, packaging & dispatch SME: Peter Kmita

Management inspiration: Tony Whiteley

Gavin McOnie

Merrenna French

Notes   Warnings

This book is not intended to be like your usual textbook; hence, the use of cartoons, language, colloquialisms, 3rd person comments & war stories, and a turn of phrase that best articulates the actual situations and the prevailing circumstances being described in the various "as-it-happened" example cases.

This book is rather detailed, covering a broad range of topics that may or may not be relevant to your current project and/or your employed industry; however, at some point in your project management career these other topics may become of great concern to you.

Please forgive the abundant use of ampersands "&" as a binder of interlinked thoughts and concepts; *e.g. "needs & wants", "monitoring & control", and "processes & procedures".*

References, Credits & Copyrights

This book is based on the practical knowledge and real-world experiences of the author, the reviewers, and those persons who happened to chip in their thoughts & opinions along the way. ... As well as those work colleagues who took the opportunity to express their frustration with their past & present management.

This book has used the PMBOK® Guide as the primary reference source, public domain materials (thank you Wikipedia), and specific textbooks where referenced.

The model of project management as a process flow and the many associated diagrams were developed by the author while contemplating various project management situations and scenarios.

The cartoon artworks used were mashed-up from Microsoft® provided clipart & characters freely available from their Office® Online image collection. The cover's artwork was composed from NASA public domain images.   *B. McCandless free-flying space venture, 1984.*

A couple of graphs used within this book were extensively modified from those depicted in the PMBOK® Guide and these have been referenced where used.

Every effort has been made to give credit where credit is due and to provide references where possible.

If you decide to use any of the contents of this book in your own works or studies then please give credit to whom credit is due.

Any caricatures to persons you know is purely coincidental,

any similarities to your past projects is "experience",

any uncanny similarity to your current project is ... DEJA VU

Directory

Introduction – why do some projects succeed while other projects fail, and what are the determining factors for declaring a project as a success or as a failure.

Theory – a basic explanation of the concepts and rationales behind project management.

Initiating Phase – discusses getting that proposed project authorized and defining what are the customer "needs", "wants", and "expectations".

Planning Phase – discusses establishing the direction for the project and more specifically, mapping out a path via which the project will be executed and concluded.

Executing Phase – discusses undertaking the project's implementation and specifically getting the work done.

Monitoring & Control Phase – discusses the importance of keeping a watchful eye on the ongoing activities of the project over the entire life of the project.

> *As depicted on the following page, Monitoring & Control consumes a significant portion of this book, and it will also consume a significant portion of the project manager's day.*

Project Rescue & Recovery – discusses saving that runaway project and salvaging that derailed project.

Closing Phase – discusses finishing each major release of the project, confirming that everything was done successfully, concluding the project as an entirety, and learning from the experience of undertaking the project.

Project Integration & Information Management – discusses incorporating the above listed aspects into an amalgamation of methodologies that could work for your particular situation & circumstances, and also the management of all of the information generated during the life of the project.

The End – discusses the quandary of whether project management is currently the right move for you and your career when various factors (both work & personal) are considered.

Table of Contents

AUTHOR'S NOTE

This 2nd Edition of the textbook is NOT a revolutionary revision of the 1st Edition. Rather, this 2nd Edition is an evolution, a tweaking of some points, a sharpening of focus on specific topics, and the inclusion of several real-world example cases & lessons learnt that have been experienced by the book's various contributors.

Thus, if you have already purchased the 1st Edition then (while it would be appreciated to buy the newer version) it is not necessary to update to the 2nd Edition.

Thank you to those who graciously contributed.

1. INTRODUCTION

1.1. Prelude ... why this book?

I wrote this book due to the number of times that I have been engaged to manage a **System / Software Development Life Cycle (SDLC)** project after it has already "run off the rails" and relationships with the customer and internally have become fatigued. Often this unfortunate situation has arisen because the previous person in charge of the project appears to have had limited "relevant" understanding of project management concepts & practices, and thus was baffled with why the project was in trouble.

This book is based on several years of practical project management experience, the **Project Management Body Of Knowledge (PMBOK®)**, and a blending of the useful elements from other project methodologies. That is, this book contains a merging of the concepts, principles, practical experiences, and the lessons learnt from the various industrial backgrounds of those persons who graciously contributed to this book's creation. Additionally, this book has been targeted at the practical application of field-tested project management techniques to actual situations and prevailing circumstances where the realities of commercial necessities have to be given serious consideration.

Please be aware that this book does espouse some ideas & concepts that are not purely PMBOK® based; hence, this book may not be entirely suitable as a study reference for obtaining **Project Management Professional (PMP®) Certification**.

In summary, this book aims to provide you with the knowledge, understanding, and techniques required for guiding that SDLC project to a successful conclusion.

AUTHOR'S NOTE

For more information on obtaining PMP® Certification and the PMBOK® Guide, please refer to the Project Management Institute http://www.pmi.org/

1.2.   What Is A Project?

What is a project?

1) A project is a **limited duration unique endeavour** that produces a **one-off set of deliverables** that are not brought about by continually ongoing repetitive operations; i.e. **not "Business As Usual" (BAU).**

> *For example; developing a new software application is a project, but the daily customer support of this software product once it has been publically released is not a project but rather an ongoing operation.*

2) A project has either a **definitive beginning and/or a definitive end** by when a **specific collection of objectives** will have been **achieved to the satisfaction of the project's stakeholders**, or it is decided that these objectives cannot be effectively achieved, or these objectives are no longer applicable and thus the project is not required anymore.

What is the difference between a project, a program, and a portfolio?

❖ **Project** – is **single**, whereas

❖ **Program** – is **multiple projects** that are **coordinated to contribute towards achieving a common business objective**.

❖ **Portfolio** – is a **collection of projects and programs** that are coordinated to achieving the organization's "strategic goals" (i.e. greater business objectives).

I hope this is not going to be like the usual academic textbook, as I have read too many of those to get this far in my career.

Nope, this ain't like the usual textbook ... promise ... and, you could always use the physical book to beat someone annoying over the head ... I have =)

1.3.   Why Do Projects Fail?

Why do projects fail?

This is a very simple question but unfortunately, this question has numerous possible answers.

However, the first question should in fact be ...

What are the deciding factors for declaring a project as a success or as a failure?

The **traditional determinants of a project's success or failure are**:

1) **TIME** – was the project's output delivered to the customer (i.e. the persons requesting the product, service, or goods) when it was agreed to be delivered?
   That is, **did the project deliver to its date milestones?**

2) **COST** – does the cost of the project not exceed the **budget** that was allocated for undertaking the project?
   That is, **was the project a profitable endeavour?**

3) **SCOPE** – does the project produce the expected results and does the resultant **deliverables contain the agreed features & functionality?**

4) **QUALITY** – does the resultant **deliverables meet the agreed Acceptance Criteria?**

NOTE   ❐ An important point with the above listed determinants of a project's success or failure is the inclusion of the words <u>expected</u>, <u>agreed</u>, and <u>allocated</u>.
These three words are very important because, **the determination of whether a project is deemed a success or a failure is based entirely on the expectations and perspectives of the people involved with the project; i.e. in the opinions of the project's primary stakeholders.**

Consider that, long after the project has concluded, the senior management of the **performing organization** (i.e. **the organization that implemented /executed the project**) may still bemoan how over budget [Cost] and/or how late [Time] the project was delivered. However, it is the **customer organization** (i.e. **the organization that sponsored / paid for the project**) or more specifically the customer's end-users who will be continuously exposed to the features & functionality [Scope] that was & was not delivered, and what was the fitness-for-use [Quality] of those deliverables.

NOTE ☐ The **combination of [Scope] and [Quality]** can be thought of as "**customer satisfaction**", whereas the **combination of [Time] and [Cost]** can be thought of as the project's **"performance measures"**.

I find it intriguing how (Time + Cost) and (Scope + Quality) have a closer affinity with each other than with the other alternate arrangements.

Though, what about the people involved with the project's undertaking and the material resources that were utilized during the life of the project? IMHO, these also affect whether the project is a success or a failure.

1.4. Project Constraints

Previously, **the determinants of a project's success or failure** were stated as being **related to [Scope], [Time], [Cost], and [Quality].**

The combination of these determinants is often referred to in project management as the "**Triple Constraints**" and this is depicted by a triangular diagram; as illustrated below in [Figure 1].

However, if you do an internet search on "Project Triple Constraints" you will also find variations of this diagram with the inclusion of [Resources] replacing the [Cost] aspect, and [Quality] combined with the [Scope] aspect.

Figure 1: The Project "Triple Constraints" Triangle.

Personally, I do not like the idea of combining [Quality] with [Scope] and I especially do not like it when [People] are lumped in with [Resources].

Lessons Learnt: Different definitions for project constraints.

In my humble opinion, the grouping of [Quality] with [Scope], and the grouping of [Resources] with [Cost] underplays the importance and independence of [Quality], [People], and inanimate material [Resources].

[Quality] should not be grouped together with [Scope] ...

A project could comply with all of the scope (i.e. the agreed requirements) but the project's output (i.e. the deliverable) may not fulfil the customer's real-world needs. That is, there is a mismatch with the customer's expectations of the product's durability, reliability, stability, usability, maintainability, and hence the "perception of quality".

[Quality] is today's issue not tomorrow's problem ...

If [Quality] is not considered as an individual project constraint then often the project's current circumstances and situational priorities will push the quality activities to the end of the project life cycle. This ordering of priorities is not due to a deliberate decision or dictate, but rather due to [Time] – [Cost] – [Scope] being judged today whereas [Quality] ends up becoming tomorrow's problem when the deliverable is about to be handed over to the customer.

[People] should not be grouped together with [Resources] ...

A successful project requires the right number of people with the appropriate skill sets. Consider that, people can be trained to improve their productivity, whereas machines can be upgraded. However, once the project commences the machine's performance will stay at approximately the same level, while the humans will vary in performance due to the person's ongoing experiences, attitudes, believes, understandings, expectations, morale, and situations & circumstances outside of the project domain.

Hence, the characteristics of the [People] performance aspect will change dynamically throughout the life of the project, whereas the characteristics of the inanimate [Resource] performance aspect will remain relatively constant.

1.5. Project Variables

As observed previously, there appears to be interchangeable project "variables" that can be used to construct different versions of the **Project Constraints Triangle**; see [Figure 2].

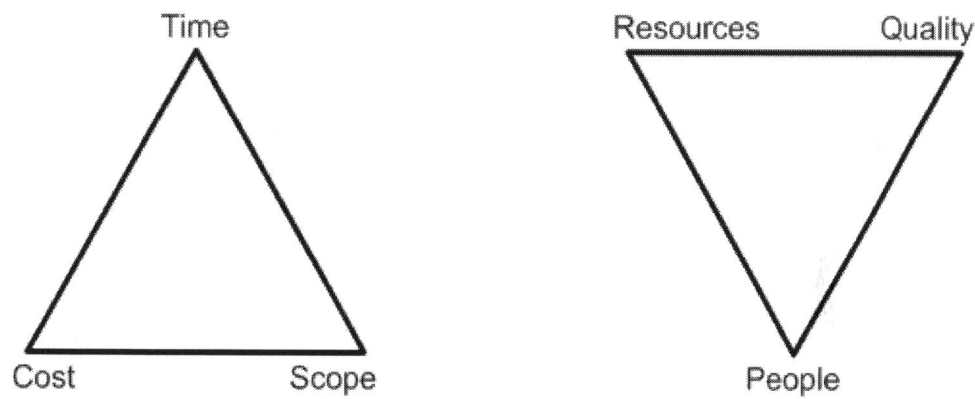

Figure 2: Project Variables.

Let's combine these different project variables together into a single representation; i.e. a "**Project Variable Star**".

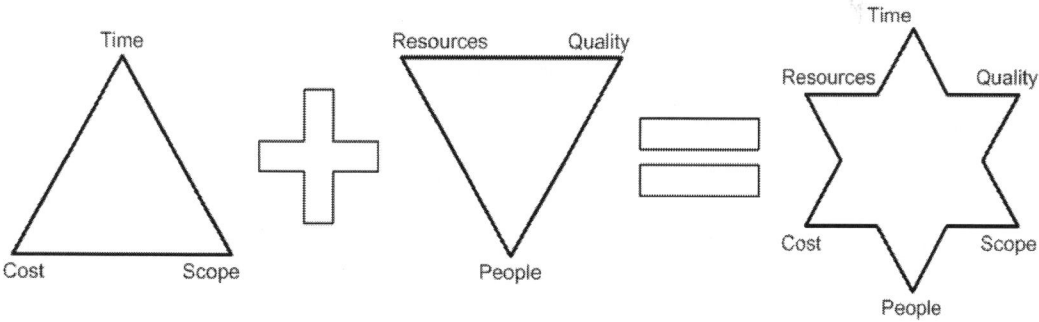

Figure 3: Project Variable Star.

NOTE ❒ The fourth edition onwards of the PMBOK® Guide has wiped out all mention of the "**Triple Constraints**". However, the term "Triple Constraints" still lives on in today's project management vernacular.

Based on the resultant "Project Variable Star" model presented previously in [Figure 3], there are a total of **six (6) project variables / constraints that have to be managed.**

These are; **[SCOPE], [TIME], [COST], [QUALITY], [PEOPLE], [RESOURCES].**

Notice that **<RISK>** has **NOT** been included as **a project variable / constraint.**

What is the difference between project variables and project constraints?

❖ **Project Variable** – is an aspect of the project that can be <u>adjusted</u> in response to changes to the project's current situation and the prevailing circumstances.

> *For example; features [Scope] could be added or removed from the project's deliverables, the delivery date [Time] could be extended or reduced, the Project Budget [Cost] could be increased or slashed, the workers [People] could be increased or decreased in numbers, equipment and tools [Resources] could be reassigned or additionally allocated, and more stringent testing [Quality] could be undertaken on the resultant product.*

❖ **Project Constraint** – is a project variable that has been <u>fixed</u>.
That is, **set to a specific value that will not be changed in the foreseeable future without prior authorization** (via some formalized Baseline Change Control).

> *For example; the project's primary stakeholders have "set-in-stone" the delivery date [Time], they have signed-off on the contract-price [Cost], they have authorized the budget [Cost], they have agreed on the features & functionality to be delivered [Scope], they have allocated and acquired the [Resources] & [People] to be used for the project, and they have agreed on the acceptance criteria [Quality].*

NOTE ☐ A **Project Constraint is also known as a project "Baseline",**
i.e. some agreed constant against which the project's current progress is measured & compared, and by which the project's success / failure is judged.

1.6.   Project Risks

During the life of the project, **deliberate or consequential changes** could be **made to each project variable**. One of these changes **could result in a ripple effect that changes any of the other project variables / constraints.** Subsequently, the ripple effect interaction of these project variables and constraints **creates potential risk to the project's success.**

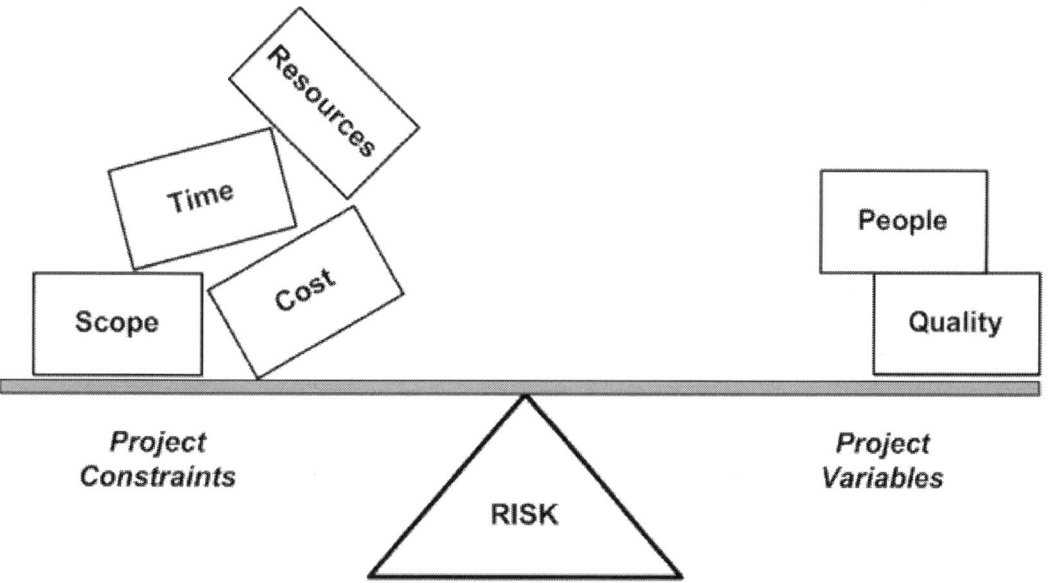

Figure 4: Relationship of Risks to project variables and project constraints.

If too many of these project variables become **project constraints then there may not be enough flexibility** in the remaining unconstrained project variables to **manoeuvre when problems and issues confront the project**.

> *For example; if the delivery date [Time] cannot be changed, but additional features [Scope] have to be added to make the output truly usable, but no additional workers [People] can be added to the project then the project runs the risk of going over budget [Cost] due to the need for overtime work, or the project runs the risk that the delivered product is not as specified [Quality] because corners had to be cut to get everything completed by the mandated delivery date.*

⊠ **In essence, project management entails balancing those project variables and project constraints.**

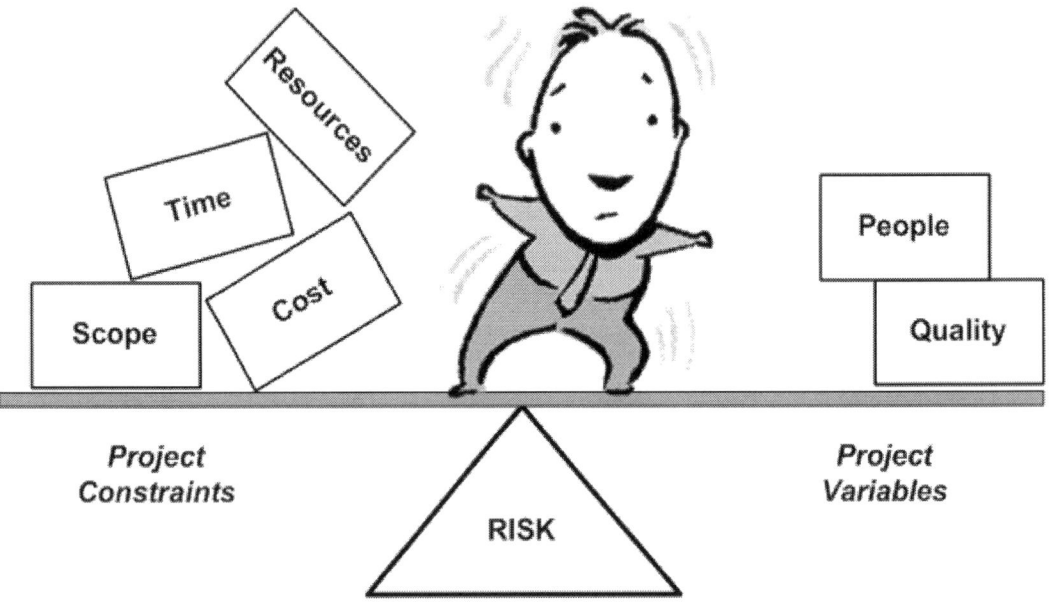

NOTE ❑ **RISK IS NOT A PROJECT VARIABLE, NOR IS RISK A PROJECT CONSTRAINT.**

This is because; the project's management cannot deliberately adjust Risk in order to control the project. Rather, **management adjust the project variables and project constraints to react & adapt to the Risks confronting the project**. However, some of these Risks will have resulted from previous adjustments to the project variables and project constraints.

2nd Edition – general change throughout the entire textbook to emphasize the sequence of; [Scope], then [Time], then [Cost]. Also, emphasize that it is the "approved" Baselines, "agreed" Scope, "agreed" Acceptance Criteria, and the "agreed to level" of Quality.

1.7. Why Really Do Projects Fail?

Projects fail because during the project's life **too many of these project variables become constraints, and hence there is no flexibility left for adjustments when the project's current situation and the prevailing circumstances change**.

Therefore, as additional pressure is exerted onto the project variables and onto the project constraints then the result is that one or more of the project variables "explodes" or "collapses". Where an explosion or collapse can take the form of; a budget overrun [Cost], a schedule blowout [Time], features forgotten or incorrectly implemented [Scope], lack of [Quality] such as high defect counts, fall in morale or [People] resigning, shortfalls in material [Resources] or equipment failures.

If to counter the current failing, the problem project variable is itself "restrained" (i.e. set as a project constraint) **but pressure is still being exerted on the project then one or more of the other project variables or even an existing project constraint could also fail**.

> *For example; if the project must be delivered at date X [Time], for Y dollars [Cost], with Z features [Scope] then the project variables are [Resources],[People], and [Quality]. These three **unrestrained project variables will have to be continually adjusted so that the project can be manoeuvred towards the desired outcome**.*
>
> *If no more [People] and [Resources] can be allocated but additional features are being added to the project then the [Quality] of the project will most probably be the failure point. However, as steps are taken to improve the [Quality] then it will result in blowouts in the project's [Cost] and/or delivery date [Time].*

TOO MANY PROJECT CONSTRAINTS = PROJECT RESTRAINTS ON SUCCESS

2nd Edition – emphasis on the current situation and the prevailing circumstances, where the **current situation** is the project's relationship to the **organizational surrounds**, and **prevailing circumstances** is the project's relationship to the **socio-economic conditions**.

The project manager needs to **maintain control over the balance between these project variables and project constraints, so that as one project variable changes the other project variables (and project constraints) do not change catastrophically**.

Think of it as being like a child's play toy, if you push on one of the points then one or more of the other points pop out. Hence, the aim of the game is to stop any of the points from completely dislodging.

Where the risk is, that one or more will end up on the floor.

Example Case: Too many project constraints and not enough project variables.

Once upon a time, there was a small software engineering company (i.e. performing organization) with a limited number of developers. Due to poor yearly revenue to date, senior management (without consultation with the development team) agreed to a fixed-price contract to deliver an industrial plant simulator by a certain date.

The customer's requirements were for an MS-Windows PC based simulator that "exactly" replicated their new computer controlled plant. The contractual stipulation being that, the simulator had to have an identical user interface to that of the real control system, and the simulator had to duplicate the operational characteristics of the physical plant. These conditions were deemed acceptable by senior management because they hypothesized that the development team could "just merge and modify" their existing simulators (that happen to run on different operating systems).

Therefore, even before the project's implementation had commenced the following project constraints had already been established:

1) the limited number of workers available [People],

2) the fixed delivery date / project duration [Time],

3) the fixed contract price [Cost],

4) the accurate replication of the plant & user interface [Scope], and

5) the exact reproduction of the user interface & operational characteristics [Quality].

Unfortunately, as the development team proceeded with the implementation it became apparent that constructing the new simulator from the older Ms-Windows and UNIX simulators was resulting in a "Frankenstein" software application that was very unstable and convoluted to code.

Secondly, there were a lot more features of the plant and its user interface to be replicated than was guesstimated during the contract negotiations.

Thirdly, due to the limited time remaining to complete the project, there were too few [People] available for work still outstanding.

This project was becoming an absolute failure for this performing organization because to deliver the required full suite of user interface features and the complete simulation of the industrial plant [Scope] was taking longer than was agreed [Time] due to much of the simulator software [Resources] having to be rewritten from scratch.

As a result, the project required significantly more work [Cost] then was originally planned and budgeted for. Consequently, this unplanned work meant that these workers [People] were not available for other projects.

Thus, the performing organization had to decide on how to **"re-baseline"** the project by:

- Engaging more developers; i.e. increase [People] and [Cost].

- Re-writing the software from a clean sheet; i.e. increase [Quality] but also increase in [Time] and [Cost].

- Negotiating for reductions in the coverage of the plant to be simulated; i.e. decrease in [Scope] ... **"de-scoping"**.

- Negotiating for extensions on the delivery date; i.e. increase in [Time] but also increase in [Cost].

- Negotiating for contract deviations; i.e. change in [Scope] and [Cost].

The decision was taken to combine these above listed options.

Thus, the project's deliverables were eventually handed over to the customer with the re-baselined [Scope] and with the re-agreed level of satisfactory [Quality]. However, from the performing organization's perspective, the project could only be deemed a failure due to its excessively long [Time] duration and its significantly inflated [Costs].

For the performing organization, the only positive outcome of conducting this project was the establishment of a new product-line by which to pursue future opportunities.

THE DETERMINATION OF WHETHER A PROJECT IS DEEMED A SUCCESS OR A FAILURE IS BASED ENTIRELY ON THE EXPECTATIONS, PERSPECTIVES, AND OPINIONS OF THE PROJECT'S PRIMARY STAKEHOLDERS.

DO NOT UNDERESTIMATE THE IMPACT THAT INSUFFICIENT END-USER TRAINING CAN HAVE ON THE PERCEPTION OF THE PROJECT DELIVERABLES AS BROKEN, AND HENCE THE CUSTOMER MAY CONSIDER THE PROJECT TO BE A FAILURE. ... EVEN THOUGH, WHAT WAS DELIVERED WAS EXACTLY WHAT WAS COMMISSIONED.

ALL APPLICABLE KNOWLEDGE CAN BE
DISTILLED DOWN TO THE ESSENTIAL
INFORMATION THAT IS NECESSARY TO
UNDERSTAND THE CURRENT SITUATION
AND TO COPE WITH THE PREVAILING
CIRCUMSTANCES.

2. THEORY

2.1. Only Two Types of Projects

All projects irrespective of their size and complexity can be rationalized down into one of **two generic types**:

❖ <u>Start-Date</u> – is where the **project has a specified starting date from when the project work will commence** and at some time in the future it will finish.

Start-Date projects are scheduled forwards from the specified start date towards a yet to be determined end date.

> *For example; development of a new technology or a **bespoke** product.*

❖ <u>End-Date</u> – is where the **project has a specified unmovable "drop-dead" date that the project absolutely has to be completed by, or ELSE!**

End-Date projects are scheduled backwards from the specified end date back towards a date when the work must be commenced by.

> *For example; building a stadium for an upcoming international sporting event, or building a new product to be revealed at a pending international technologies convention.*

NOTE   ❏ **If a Start-Date project is taking too long to deliver then eventually,** the customer organization's representatives or the performing organization's senior management will demand that the project be **transformed into an End-Date project.**

Subsequently, what will be the "End Game" plan to bring this project home by the newly agreed [Time], within the newly agreed [Cost] Budget, with the newly agreed [Scope] of deliverables, and with the newly agreed level of [Quality] ?

2.2.   Project Life Cycle

❖   <u>Project Life Cycle</u> – is a systematic way to get from the beginning to the end of the project.   *... In a determinate amount of [Time] and/or [Cost].*

There are five phases to project life cycles:

Initiating	• What do **THEY WANT**?
	• What will **WE GIVE** them?
	• What will it **COST US TO GIVE** it to them?
	• What will **WE GET IN RETURN** from them?
	• What is **AT STAKE (make & brake)?** *break*
Planning	• **WE NEED WHAT WHEN** and **HOW MUCH?**
	• How will **WE KNOW WE GOT IT RIGHT?**
	• What do **WE THINK CAN GO WRONG?**
Executing	• How exactly **WILL** the **TEAM DO IT?**
	• The **TEAM** is **DOING IT.**
	• The **TEAM** is **CHECKING IT.**
	• Is the **TEAM DONE YET?**
Closing	• Did **WE GET IT RIGHT?**
	• Does **EVERYONE AGREE** it is **ALL DONE?**
	• Let us say **GOOD BYE**... as friends.
	• What did **WE LEARN** from this?

Monitoring & Control

• Keep one eye on the road, keep the other eye on the gauges, and both hands on the steering wheel.

There are **two generic project life cycles** that form the basis for all other variants that you will encounter:

(1) **Linear / Waterfall**, and

(2) **Iterative.** *& Agile.*

2.2.1. Life Cycle: Linear-Waterfall

Linear-Waterfall is a project life cycle **where each phase sequentially follows the completion of the previous phase**; like water flowing down a stream and toppling over waterfalls along the way, see [Figure 5] below.

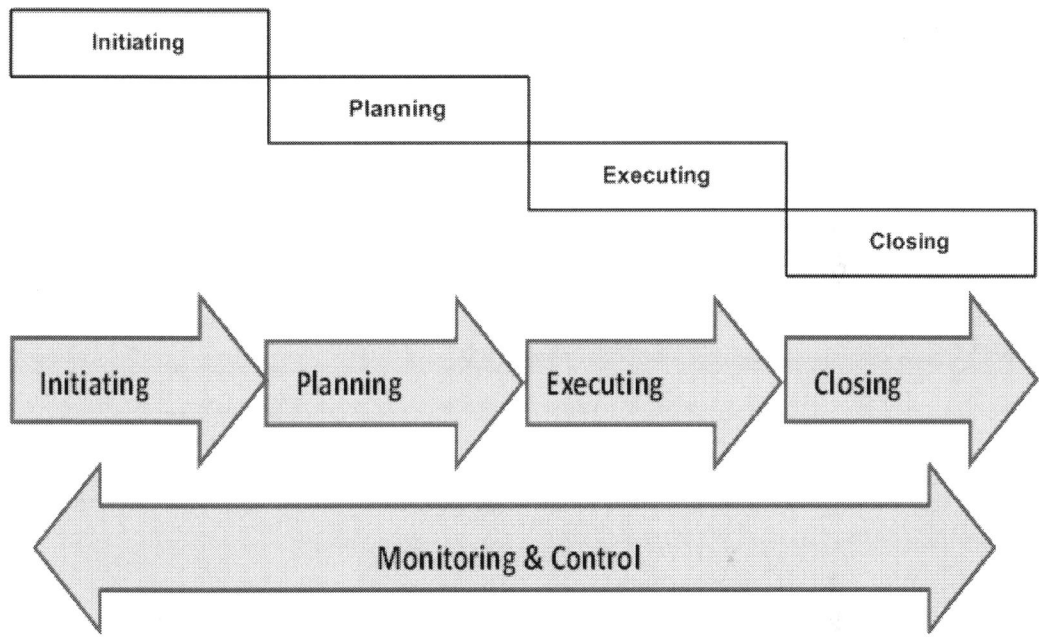

Figure 5: Project life cycle phases outlined as a Waterfall model.

Some organizations will **conclude one phase** (or even sub-phases) **with a formal review "Gate" to reassess the project and to give permission to either; proceed, change, or terminate the project**. These points can also be used to formalize **the "handoff"** (i.e. handover) **of one phase to the next phase**.

 For more information on the **Phase Completion Review,** please refer to [Section 3.6].

With the Linear-Waterfall life cycle model, each phase is finished before advancing onto the next phase. However, in reality, work will often start on the next phase prior to the current phase being completely finished; i.e. these phases will "**overlap**" as illustrated below in [Figure 6]. Phases may also start sooner, finish later, and plateau for some time; *e.g., this Executing Phase has a second plateau for bug fixing.*

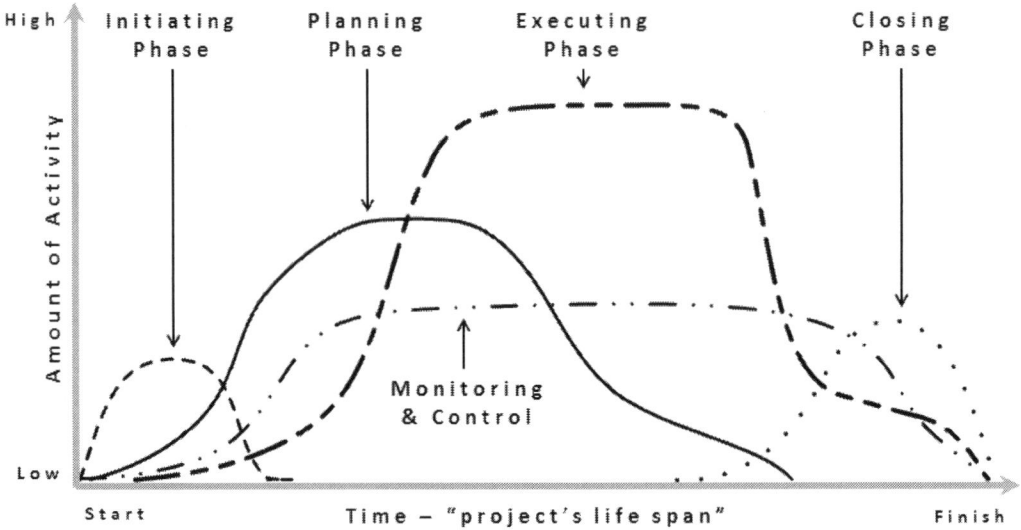

Figure 6: Overlap of life cycle phases.[1]

Advantages of the Linear-Waterfall life cycle

- It is easy to define the project **milestones** because the end of each phase is an **obvious achievement point**.

- It is easy to **silo** each phase of the project as a **clearly defined body of work** and hence it is easy to **assign responsibility for each phase**.

- **Dedicated teams** can be easily assigned to **work on a particular phase** of the project; *e.g. sales & senior management (Initiating), project manager & system architects (Planning), system developers (Executing), and system testers (Closing).*

---

[1] Based on PMI 2008, "*A Guide to the Project Management Body of Knowledge (PMBOK® Guide), 4th Edition*".

Disadvantages of the Linear-Waterfall life cycle

- It is possible for the project to be almost finished before it is realized that the project's **[Scope] of deliverables does not align correctly with the customer's expectations of [Quality]**.

- It is possible to realize excessively late that some **feature / functionality** has been **completely forgotten** about, and hence to incorporate this missing item of [Scope] could require a **major rework** of the project.

- **Lacks flexibility to adjust** the [Scope] and the direction of implementation if the project's current situation and/or the prevailing circumstances change.

- There is a potential for a **lack of ownership** due to the long duration that one team works on their phase before "**throwing it over the fence**" to the next phase's team.

    For example; the system architects (Planning) pass it onto the system developers (Executing) who in turn pass it onto the system testers (Closing).

- Because of the siloing of the project's work, **things can fall between the gaps** when one phase passes to the next phase.

- Because of the long time before the customer receives proof of something usable, there could be **niggling doubts about** the project team's performance and subsequently lack of faith in the project's **potential for success**.

- Because of the long time before the customer can touch & feel something usable, it can be **harder to engage the customer's participation** and subsequently takes longer to establish the customer's **personal investment** in the project's outcome.

- Because of the inherent lack of customer participation, customer feedback, and at hand proof of the project's potential for success, it is therefore **harder to negotiate project extensions and contract deviations** when the current situation and the prevailing circumstances change.

2.2.2. Life Cycle: Iterative *& Agile*

Iterative *& Agile* is a project life cycle **where a limited number of features & functionality (of the total collection) are completed during each cycle** through the project; i.e. each cycle builds upon the previous cycle, see [Figure 7] below.

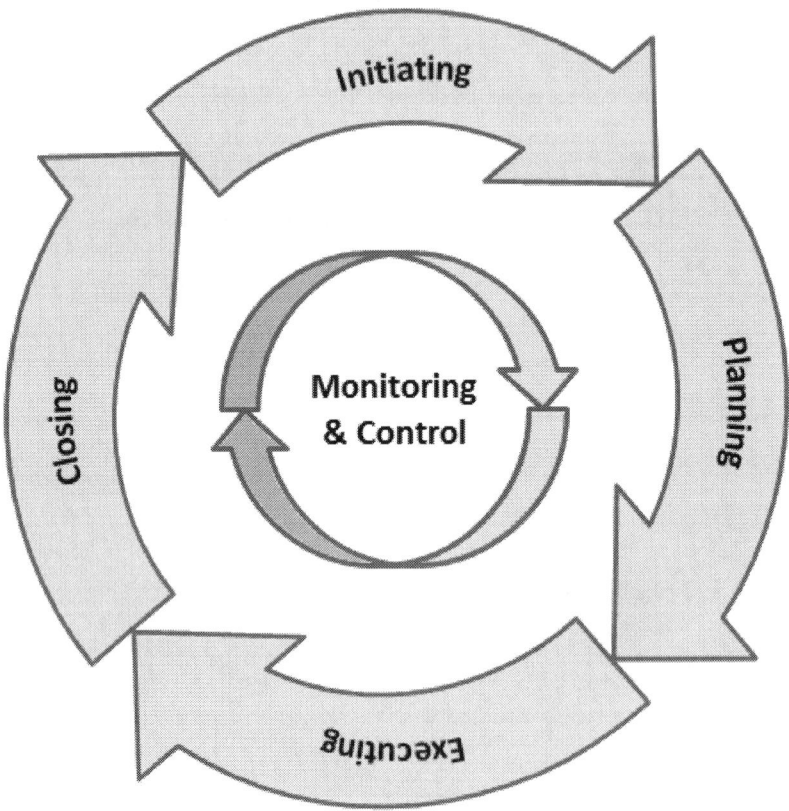

Figure 7: Project life cycle outlined as an iterative model.

The most noteworthy thing about the **iterative** *& Agile* **life cycle is** that **at semi-regular intervals the life cycle returns to the beginning**, where the customer and the performing organization can re-evaluate the project's situation & circumstances; i.e. change driven.

At the start of the next cycle the project's deliverables can **be refined or completely** redefined, and re-prioritized in the next Initiating Phase; i.e. **"Progressive Elaboration"**.

Advantages of the Iterative & Agile life cycles

- It breaks the project down into smaller deliverable packages of work and thus it is possible to **obtain something usable a lot earlier** and thereby **learn quicker from the outcomes & mistakes made during that cycle**.

- This earlier partial delivery enables the project to be **re-evaluated sooner and more often** and thus by progressive elaboration provides more **flexibility to change** the project's **direction** when the situation & circumstances deem it necessary to do so.

- This earlier partial delivery enables the customer to be engaged sooner in the testing and evaluation of the deliverables. This customer engagement thereby results in **timely feedback** that improves the chances of the project deliverables meeting the customer's expectations, and enables the **earlier detection of missing or misunderstood features & functionality**.

- This earlier partial delivery improves the customer's belief in the project's potential for success because they have **physical proof of progress**.

- It is **easier to negotiate project extensions and contract deviations** because the customer can be involved sooner with the ongoing life of the project and hence the customer would be functioning more as a project participant rather than as a project combatant; i.e. the customer develops a personal investment in the project's outcome.

Disadvantages of the Iterative & Agile life cycles

- It is **harder to determine when** the project is **finished**, especially if after each cycle the project's scope is further evolved.

- Similarly, it is **harder to determine the cost** of the project especially when the project's duration keeps changing **due to further "Scope Elaboration"**.

- The **project could keep growing** as more features & functionality is added (which could be a good thing depending on how the project is being paid for).

Suggested rules for the Iterative & *Agile* life cycles

- Each cycle should only contain a **limited number of features & functionality**. Where, a feature / functionality is an individual section or bullet point in the ~~Customer Requirements~~ approved Detailed Specifications (Functional Specifications). Hence, someone can finger-point to exactly where that specific feature / functionality is required and defined.

- The **end of each cycle is a release point** either internally or potentially externally to the customer.

- **Not every cycle** needs to **produce** a customer **releasable deliverable**.

    For example; one cycle may be only a proof-of-concept.

- **Multiple cycles** can be undertaken **at the same time**.

    For example; one cycle could be for feature-set 'A', while concurrently another cycle for feature-set 'B' could be worked on. When both cycles 'A' and 'B' are completed then the next iterative cycle 'C' could be the merging of 'A' and 'B'.

- A **larger cycle** can be **composed of smaller cycles**.

- It is best to **do the project's "make-or-break"** (high priority) **features as early as possible in the order of cycles**. That is, without this feature / functionality being successfully implemented then there is no point doing the rest of the project.

    Hence, if your project is dependent on some evolving technology or some strategically important feature then try to confirm that this is going to work successfully before spending a lot of time on implementing peripheral features & functionality.

    For example; with a television set's development if the TV cannot receive the signal, display the picture, and play the audio then it is meaningless whether the TV utilizes a remote control, has on-screen menus, or includes a subwoofer.

2.2.3. Life Cycle: Linear & Iterative as a couple

In reality, many projects are implemented by using both linear and iterative methods at the same time. From a practical perspective, the **conclusion of each iterative cycle** could be declared a **milestone**. Therefore, the project would consist of a sequence of iterative milestones; see [Figure 8] below.

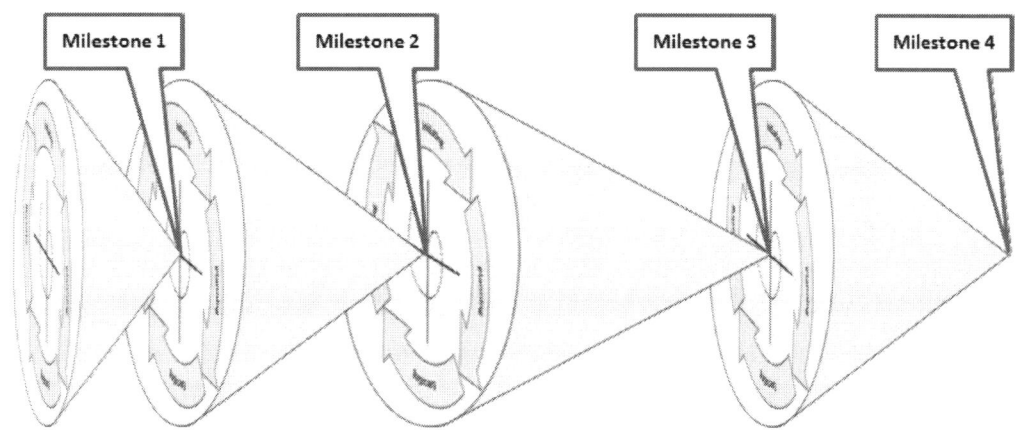

Figure 8: Iterative cycles delivered as linear milestones.

Alternatively, **each iterative & *Agile* cycle could be thought of as a mini waterfall project**; see [Figure 9] below.

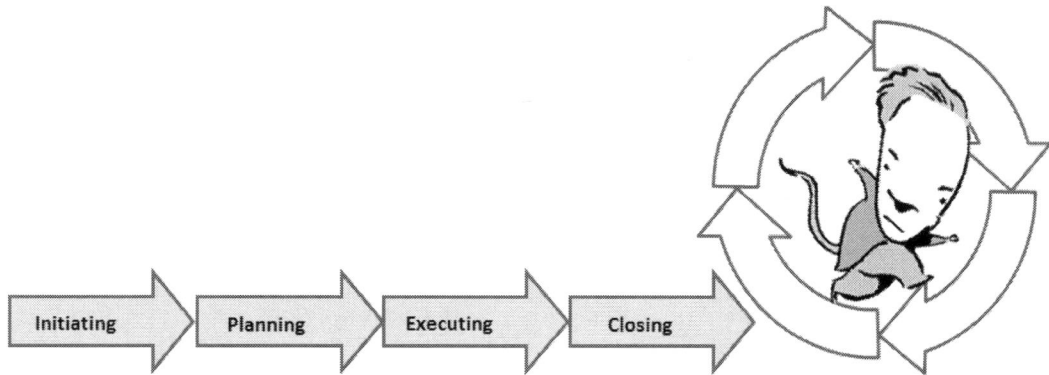

Figure 9: An iterative & *Agile* life cycle as a mini waterfall project.

2.2.4.   Life Cycle: Iterative & Agile as siblings

The iterative life cycle described so far in [Figure 8], does not define or limit the duration of
each cyclic pass hence it could be interpreted as breaking the waterfall project up into
"meaningful chunks of functionality" that is handed over to the customer at variable
durations depending on what has to be delivered for each milestone.

Consequently, a cycle could be days, weeks, months, or years in duration.

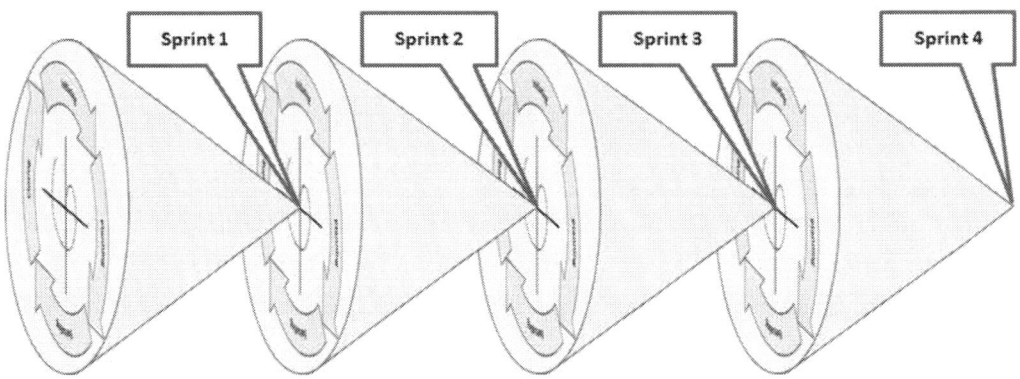

Figure 10: Iterative cycles delivered as fixed duration sprints.

Alternatively, **what if the maximum duration for any single cycle was set to a consistent
period of only 3, 4, or 6 weeks?** As illustrated above in [Figure 10].

Because of these short duration iterative cycles, the project would be **sprint**ing along
at a **constant beat**, **rapidly turning out useful components** that could be quickly evaluated
by both the performing organization and the customer's representative to verify that the
deliverables were going to meet a selected portion of the agreed **Acceptance Criteria**.

If this evaluation found that a deliverable was "non-conforming" or it was realized that
the project was heading down the wrong path then it would not be a major tragedy to halt
the project, throw away the last cycle or cycles, and then restart in the right direction.

Well there may be some level of costs and heartbreak involved in this "do-over",
however this would be significantly cheaper and more time efficient then getting all the

way to the end of a waterfall (or long duration iterative release) only to discover that the envisioned project was all wrong.

Additionally, **these "regular as clockwork" iterative cycles** would have the deliverable functionality for each cycle being defined, designed, developed, tested, and delivered in a relatively short period of time. This in affect **would limit the window of opportunity for changes to be made to that cycle's functionality**.

Consequently, these "Sprint" releases would provide:

- a **measurable proof of progress**,

- the opportunity to obtain **timely feedback**,

- **earlier detection of potential issues** sooner rather than later, and

- gives the project a **great amount of flexibility and agility** to change direction and adapt to the current situation and the prevailing circumstances.

Agile techniques are especially useful when the project's situation and the prevailing circumstances are changing rapidly, and/or when the project's requirements / Scope are not clearly defined; i.e. the project stakeholders will understand it better once something is presented.

AUTHOR'S NOTE

This book will NOT be delving deeply into agile methodologies, as this would be a book inside a book. However, this book will be borrowing ideas from these methodologies.

Hence, the agile approach described in this book is more akin to an iterative process, with a sprinkling of the techniques from true agile methodologies, and a degree of the discipline of the waterfall process thrown in for good measure.

Why you may ask; well in your career you will encounter organizations that are far to the left (i.e. waterfall) or far to the right (i.e. agile), hence a balanced middle perspective will provide you with the capabilities to easily adapt to either doctrine.
Secondly, a more formalized approach to agile makes it a very formidable tool for getting your System Development Life Cycle project successfully across the line.

For more information on agile techniques, please refer to [Section 7.4].

2.2.5. Life Cycle: Hybrid of Waterfall & Agile

I have heard it bemoaned that; *"one of the problems with this project is that the customer thinks it should be implemented using that old-fashioned archaic waterfall process"*, whereas the performing organization's project team believe *"it should be implemented using modern agile methods"* such as Scrum or Extreme Programming.

Can a project be both agile and waterfall?

"Yes, Virginia"... agile and waterfall life cycles can coexist and be mutually inclusive.

Example Case: A mad suicidal rush to deliver via a waterfall methodology.

Once upon a time, there was a customer who had rigorous project governance (maybe to the point of being bureaucratic) where every aspect of the project had to be reviewed, approved, and signed-off prior to advancing to the next phase of the project.

However, this customer had business obligations that necessitated that the project be delivered at a soon to be approaching fixed deadline date, yet the complete set of project requirements and Detailed Specifications had not been determined let alone agreed to nor signed-off.

Consequently, the options available to the performing organization were to;

1) Wait for the requirements and specifications to be solidified and

   then fast track the project; i.e. *"make a mad suicidal rush to deliver"*.

OR

2) Immediately start the project's implementation based on those requirements and

   specifications that had been agreed to so far.

3) Look for some exit strategy that will result in the least possible
   financial and reputational damage to the performing organization.

A compromise solution to the problem outlined in this example case would be to **break the project**'s remaining time (i.e. the duration for the phases of Planning, Executing, and Closing) **up into constant fixed durations of 3, 4, or 6 week iterative cycles called "Sprints"**; as illustrated below in [Figure 11].

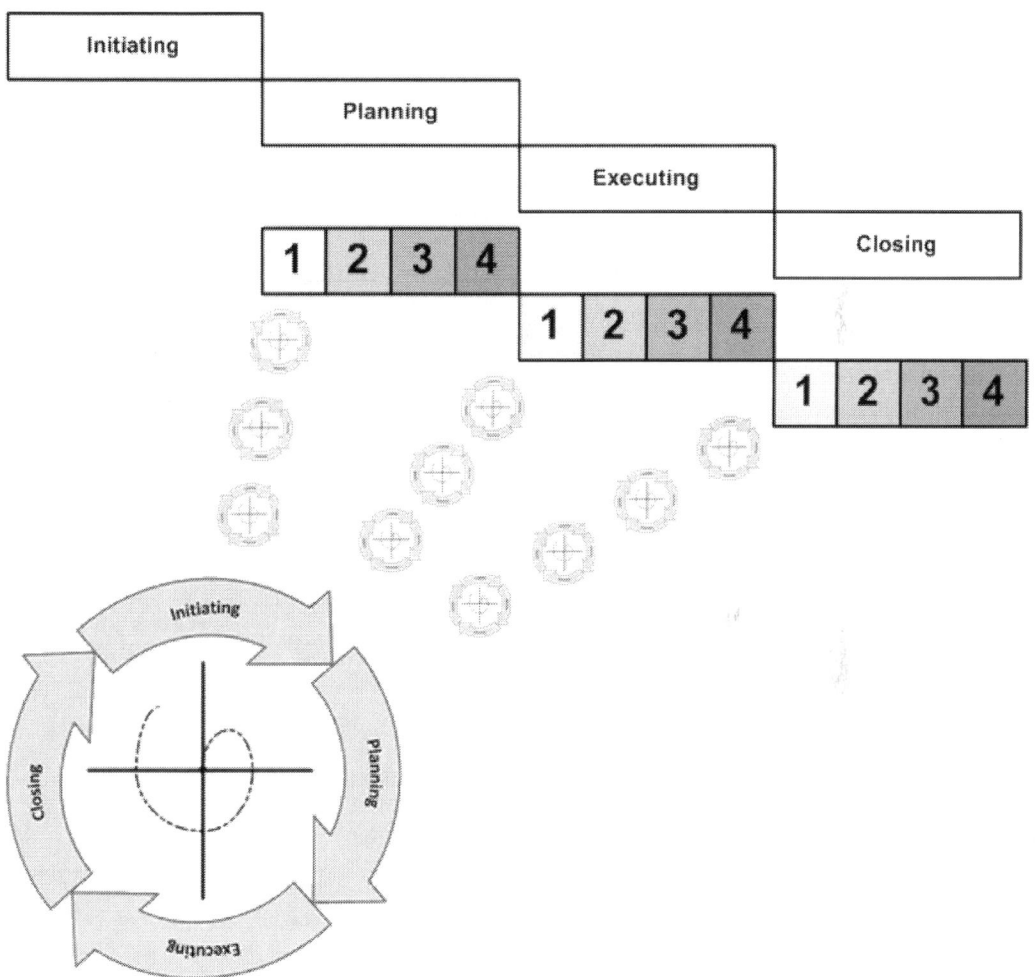

Figure 11: A waterfall project implemented via iterative sprint cycles.

What this solution would do in effect is **break the project up into multiple short duration waterfall projects**; i.e. **"modularizing"** the project into mini-projects.

Consequently, only the requirements & specifications for the current iterative sprint would have to be reviewed, approved, and signed-off.

Once the current iterative sprint's requirements & specifications are ratified (in the Planning Phase), then the development (in the Executing Phase) and the testing (in the Closing Phase) can commence.

While the current sprint's development & testing are underway then simultaneously the next sprint's requirements & specifications can be established.

This **hybrid of waterfall and agile methodologies** does provide some useful **benefits**:

- It does **kick into gear the project's management** (both from the customer and performing organizations) because the **end date** for the next mini-waterfall project is **a lot closer to today** than was the situation with the larger single waterfall project. Subsequently, the project's management are **forced** to **focus on the relevant requirements for the pending release** and thus the project's management is **impelled to come to a mutual agreement on the relevant specifications** for this release.

- Each mini-waterfall project can be tailored to produce a **usable deliverable** that can be **evaluated** and thus be used to **judge** the project's **overall progress**.

- Enables the **requirements & specifications to evolve based on practical evaluation and feedback**, instead of the project being delayed while every possible scenario is considered and the resultant mammoth "cover-all" specifications is thoroughly reviewed & reworked to perfection before eventually being approved.

NOTE ❒ The Initiating Phase was not considered in this blending of waterfall and iterative methods, because as detailed in [Chapter 3] the project cannot officially commence until the Initiating Phase is concluded with the project's formal authorization via the sign-off of the Project Charter.

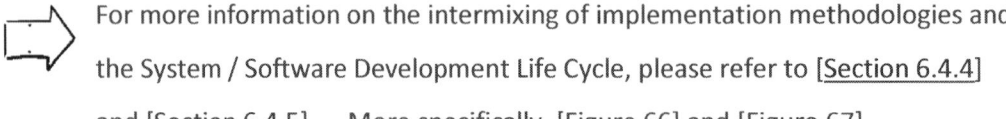

For more information on the intermixing of implementation methodologies and the System / Software Development Life Cycle, please refer to [Section 6.4.4] and [Section 6.4.5]. ... More specifically, [Figure 66] and [Figure 67].

2.2.6. Life Cycle: System / Software Development Life Cycle (SDLC)

❖ **System / Software Development Life Cycle (SDLC)** – is a project life cycle that **is tailored specifically towards the creation, alteration, and maintenance of software applications, hardware platforms, and information technology systems.**

SDLC is also an abbreviation for Solution Delivery Life Cycle.

The most visual difference between the project life cycles previously outlined in [Figure 5] and [Figure 7] and the SDLC presented in [Figure 12] and [Figure 13] is that instead of there being five project phases (of Initiating, Planning, Executing, Closing, and Monitoring & Control) these phases are instead described as:

1. **Requirements / Specification**   DEFINE

2. **Design**   DESIGN

3. **Development / Implementation**   DEVELOP

4. **Verification / Testing**   TEST

5. **Release, Maintenance, and Support**   RELEASE

Alternatively, these SDLC phases could be broken into **multiple sub-phases**:

1. **Initiation**
   - Identifying a need or opportunity for the "end-product" (i.e. application and/or system to be delivered).

2. **Conceptualization**
   - Defining the objectives and the boundaries of the project.

3. **Work Approval**
   - Authorization to commence work in earnest.

4. **Requirements Gathering & Specifications**
   - Defining the [Scope], requirements, functionality, and features.

5.	**Planning**	• Defining the needed [Scope], [Time], [Cost], [Quality], [People], and [Resources].
6.	**Design**	• Defining how the specifications will be transformed into reality during implementation.
7.	**Development**	• Transforming the designs into reality.
8.	**Debugging**	• Correcting the implementation to ensure conformance with the specifications.
9.	**Integration**	• Merging different components of the implementation together and achieving operability with samples of other related end-products.
10.	**Testing**	• Internal confirmation that the end-product conforms to the specifications.
11.	**Verification**	• End-user perspective confirmation of the end-product's usability and conformity to expectations.
12.	**Release / Rollout**	• Connecting the end-product to its operational environment and training of the end-users.
13.	**Operations**	• Ongoing end-user usage of the end-product during its in-service life.
14.	**Maintenance & In-Service Support**	• Keeping the end-product operational, resolving and correcting issues encountered during end-user usage.
15.	**Decommissioning**	• Disconnecting the end-product from its operational environment.
16.	**Disposal**	• Getting rid of the retired end-product.

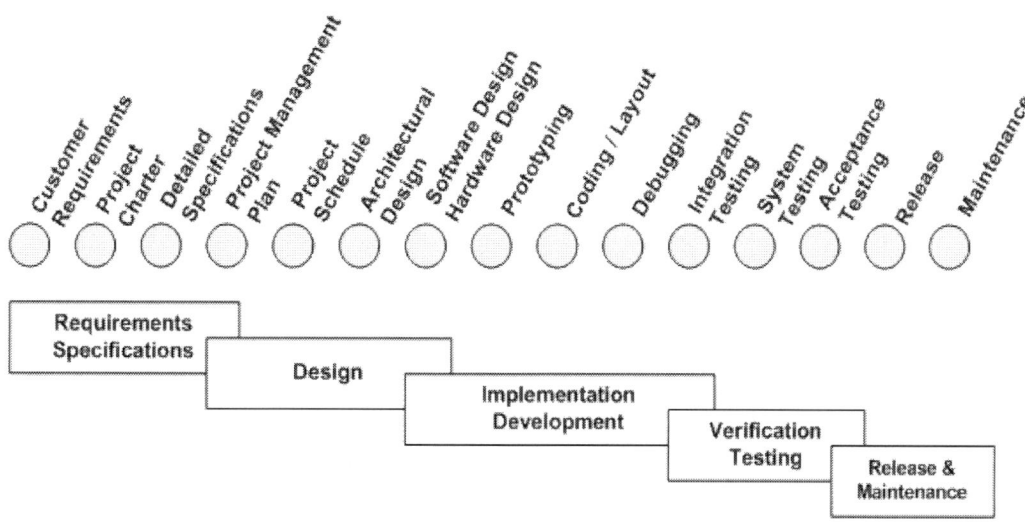

Figure 12: SDLC related to a waterfall life cycle.

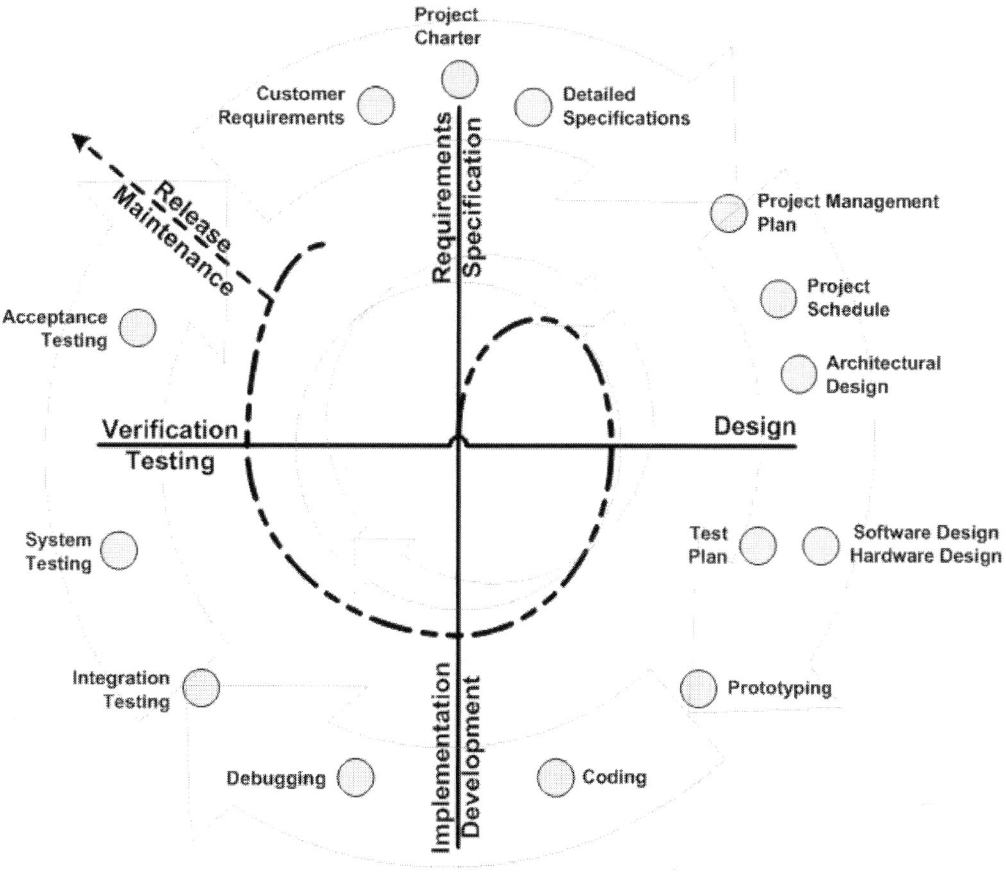

Figure 13: SDLC related to an iterative life cycle.

2.3. Project Layering and Slicing … even Dicing

For many development projects, a choice can be made to either break up the project's implementation based on "layering" or "slicing".

What is project "layering" and "slicing"?

The answer is easier to explain by using a hypothetical development project.

Consider a productivity suite (*e.g. Microsoft Office*) that is to be composed of multiple applications (*e.g. Word, Excel, PowerPoint, Publisher*) that will have a common user interface appearance, similar editing functionality, underlying data structures, and utilize similar file storage techniques.

> *For this example case, consider the layers as:*
>
> + *Layer 1: each application's user interface,*
>
> + *Layer 2: each application's editing engine,*
>
> + *Layer 3: each application's data structure, and*
>
> + *Layer 4: each application's file storage processing.*

Now one option would be to break the project's implementation up based on "layering" as illustrated in [Figure 14], whereas the other option would be to break the project's implementation up based on "slicing" as in [Figure 15].

With "layering", a specialist group could concentrate on the same layer for all of the applications; *e.g. an interface team, an editor engine team, a data structure team.* Using this approach, **all of the layers of the applications have to be completed together before product evaluation and delivery can occur.**

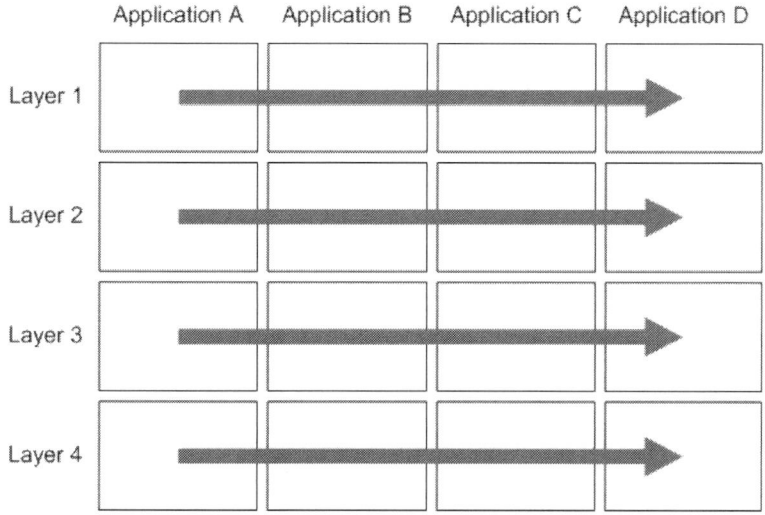

Figure 14: Project broken up via "layering".

With "slicing", individual applications are assigned to individual teams.

Using this approach, **it is possible for evaluation and delivery to occur once one of the applications is completed** instead of having to wait for all of the applications to be finished.

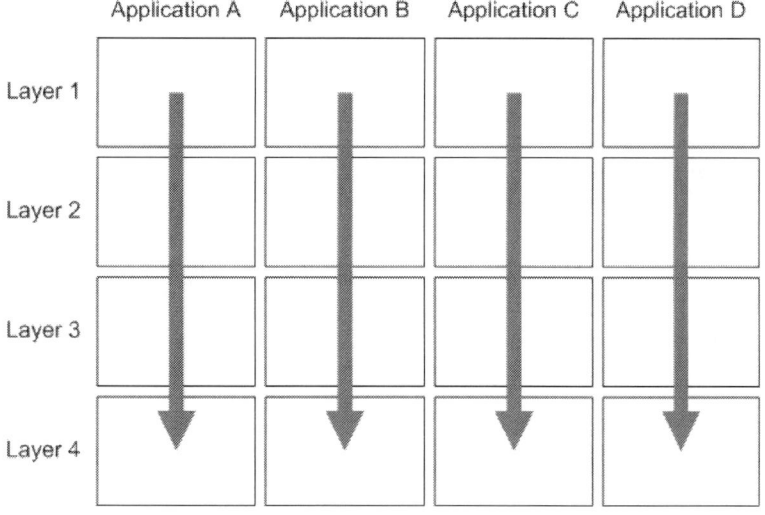

Figure 15: Project broken up via "slicing".

The problem with **"layering"** is that, if while implementing one application it is found that it will be **necessary to change a pair of interconnected layers** then it is very probable that **complementary changes could ripple across to all of the other applications' matching layers**. Accordingly, this could result in a **significant amount of rework** for each application.

With **"slicing"**, it is possible to concentrate entirely on the completion of one application as a proof-of-concept (i.e. prototype) and to work out all of the problems with the design and with the implementation process before commencing work on all of the other applications. This would mean that **necessitated changes could be inherited from one completed application's development to the next**.

However, the most important benefit of **"slicing"** is that the **customer evaluation can occur very early in the project's life** (i.e. evaluation of the proof-of-concept). This will help to ensure that the deliverables do comply with the customer's expectations and therefore **necessary scope changes** could be **made at an early stage** before a significant proportion of the development is underway.

Why would one ever choose the "layering" approach instead of the "slicing" approach?

Well, the **"layering" implementation approach is tailored for a <u>matrix</u> organizational structure** where specialist people concentrate on specific aspects of the project that they are expert at; *e.g. user interfaces, databases, file systems*.

Whereas, the **"slicing" implementation approach is tailored towards a <u>functional</u> organizational structure** where a dedicated team works on one specific application / module that is siloed from the other parts of the project.

The functional structure is more akin to those days of traditional Waterfall methodology, whereas the matrix structure has come into prevalence with the adaptation of Iterative and Agile methodologies.

Alternatively as illustrated below in [Figure 16], one could use a **hybrid combination of "layering" and "slicing" to "dice" up the project's implementation**.

For example; there could be a project team per application, but Subject Matter Experts (SME) would be brought in to work on a specific area of the application to ensure consistency across all of the applications being developed.

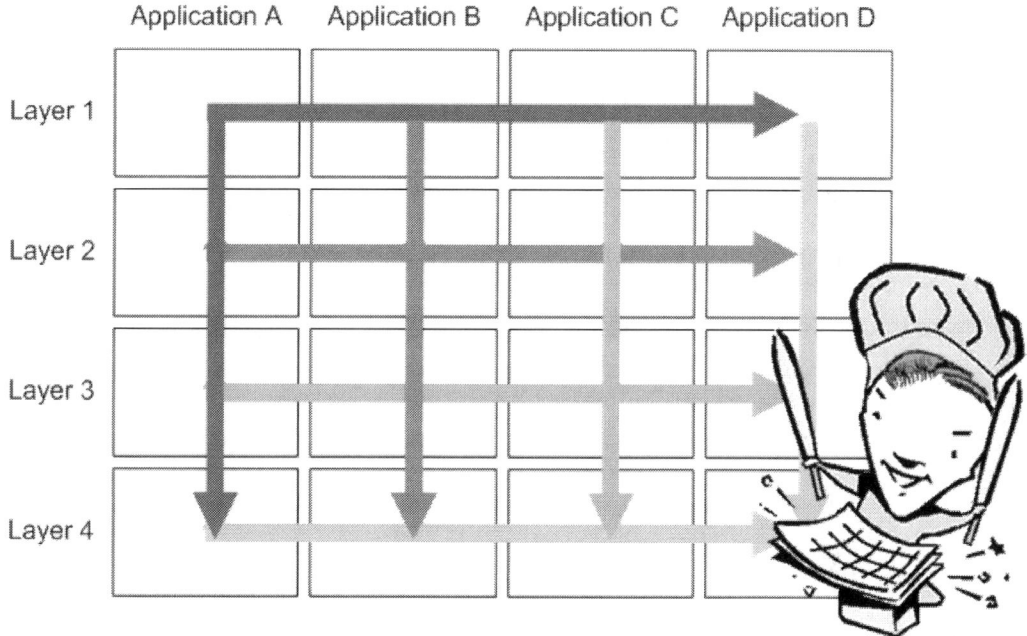

Figure 16: Project broken up via "dicing".

This **"dicing"** of the project's implementation will **occur with large projects**, with **large organizations where economies of scale and specialization** will result in **productivity savings**, and/or with highly structured organizations where there are **clearly defined demarcations on roles and responsibilities**.

This "dicing" can be combined with **production-line processes** where portions of the project are sequentially slotted into position in a conveyer-belt like **work-stream** with other projects to be executed by dedicated groups that are responsible for specific activities; *e.g. graphic arts, verification testing, end-user platform integration.*

2.4.   Project Stakeholders

2.4.1.   Stakeholders: A Short Introduction

As stated back in [Section 1.3]:

> *"The determination of whether a project is deemed a success or a failure is based entirely on the expectations and perspectives of the people involved with the project; i.e. in the opinions of the project's primary stakeholders".*

One way to ensure that the project is perceived as a success is to **know what the stakeholders expect, keeping them informed, providing them with what they need to know when they need to know it, and making sure things are done by when they need them to be done.**

Now you literally can read entire textbooks on Stakeholder Management, but the basic concepts are that there are **two categories of stakeholders**[2] [Figure 17]:

1) **Primary Stakeholders** - who directly affect the project's outcome, and

2) **Secondary Stakeholders** - who influence the project's outcome and/or are affected by its outcome.

You may read other Stakeholder Management theory and find that they describe **three types of stakeholders**[3] [Figure 18]:

a) **Core Stakeholders** – that are essential to the project's survival.

b) **Strategic Stakeholders** – that are vital to the project at a particular time.

c) **Environmental Stakeholders** – that form the backdrop and surrounding environment to the project.

---

[2] Cleland D. 1998, "Stakeholder Management", In J. Pinto (Ed.), "Project Management Handbook", NY, USA.
[3] Clarkson, MBE, 1999, "Principle of Stakeholder Management", University Of Toronto Press, Toronto, Canada.

⊠ **An important purpose of project management is to carefully manage the stakeholders' expectations, perspectives, and opinions.**

I cannot stress this enough; **the determination of whether a project is deemed a success or a failure is based entirely on the expectations, perspectives, and opinions of the project's primary stakeholders.** ... *And, not necessarily on the factuality.*

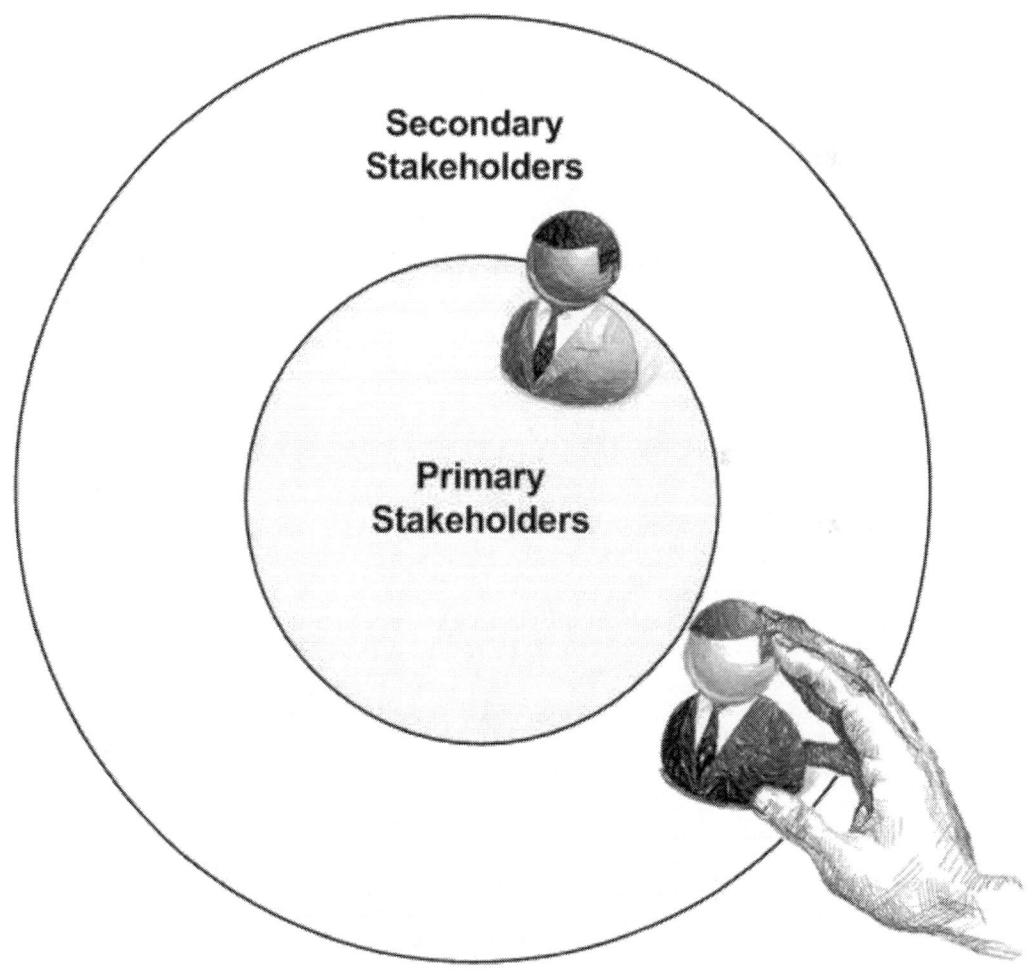

Figure 17: Stakeholder primary-secondary categories model.

NOTE ☐ A **"customer"** can be either external or internal to that organization who is performing the project work; i.e. for **a client** or for **fellow staff**.

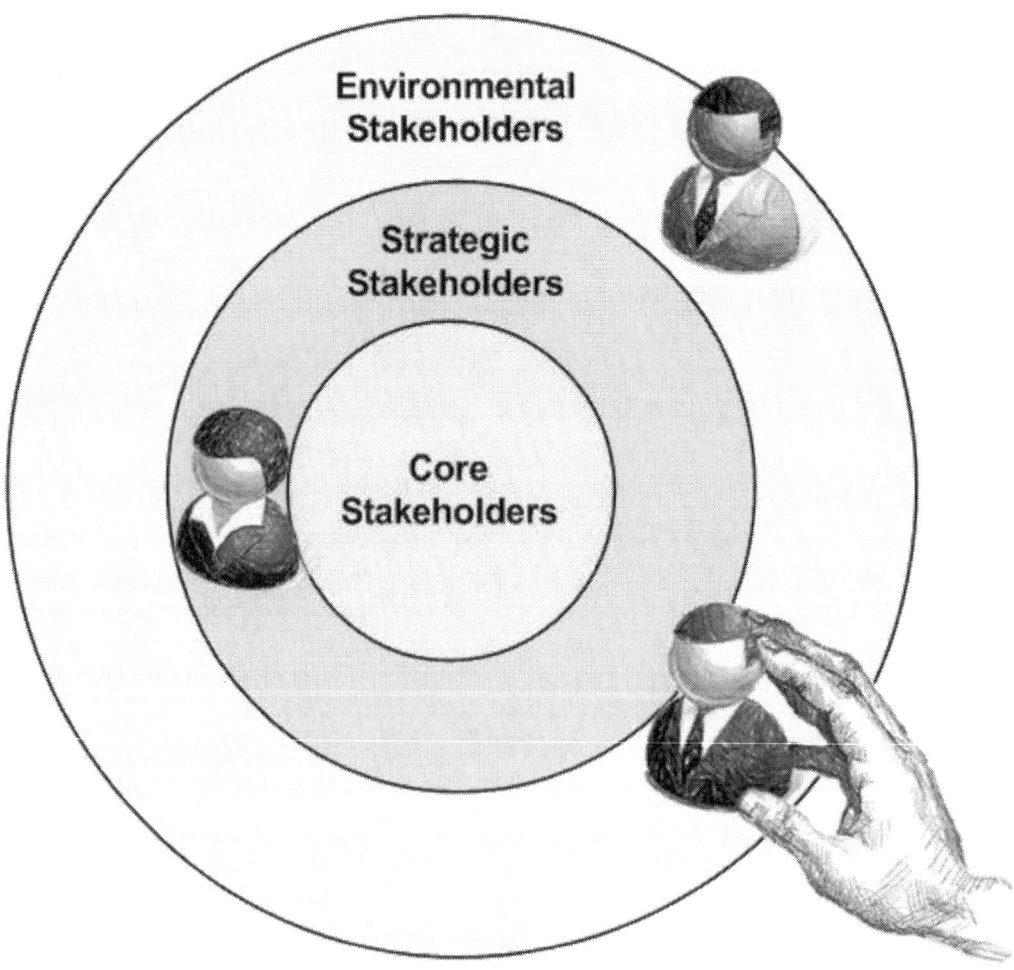

Figure 18: Stakeholder core-strategic-environmental groupings model.

2.4.2. Stakeholders: Relationship to the SDLC

Consider your typical SDLC project and identify the key people and groups that would be involved with that project.

This list would probably contain; the customer's representative, the customer's management, sales people and senior management, subject matter experts, system architects, system builders/developers, system testers, the customer's testers, the project manager, and many others (including the end-users).

With the **Waterfall SDLC**, these **stakeholders** would be **laid out along the life cycle** as illustrated below in [Figure 19].

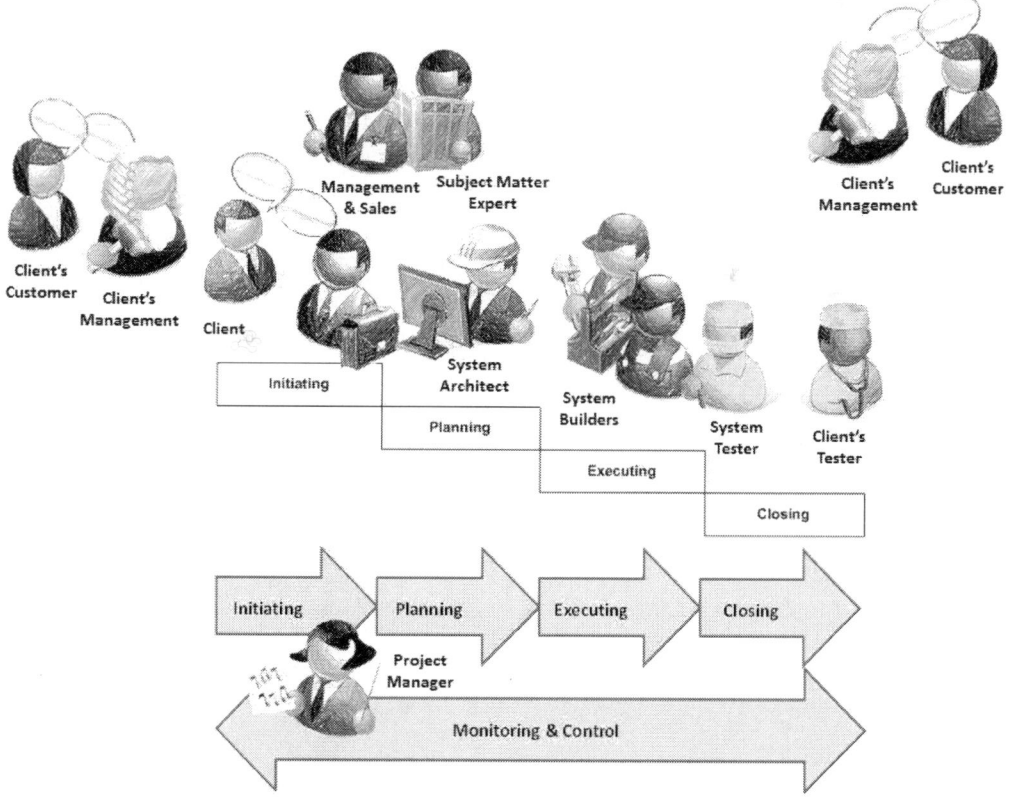

Figure 19: Stakeholders common to a waterfall project life cycle.

With the **Iterative SDLC**, the **stakeholders** would be **laid around the life cycle** as illustrated

below in [Figure 20].

 For more information on stakeholder identification and management, please

refer to [Section 10.3].

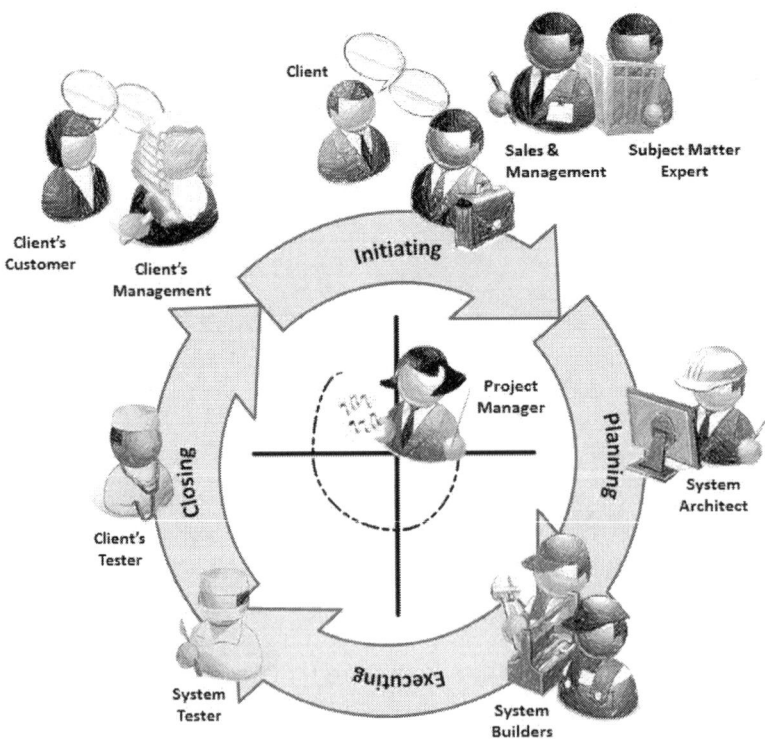

Figure 20: Stakeholders common to an iterative project life cycle.

2.4.3. Stakeholders: Influence on the SDLC

If you work through the stakeholder analysis as outlined in [Section 10.3] you will probably find that the stakeholders who have the greatest potential influence on the project being successful are those standouts in [Figure 21] and [Figure 22].

Interestingly, these standout influential stakeholders relate to the Initiating Phase.

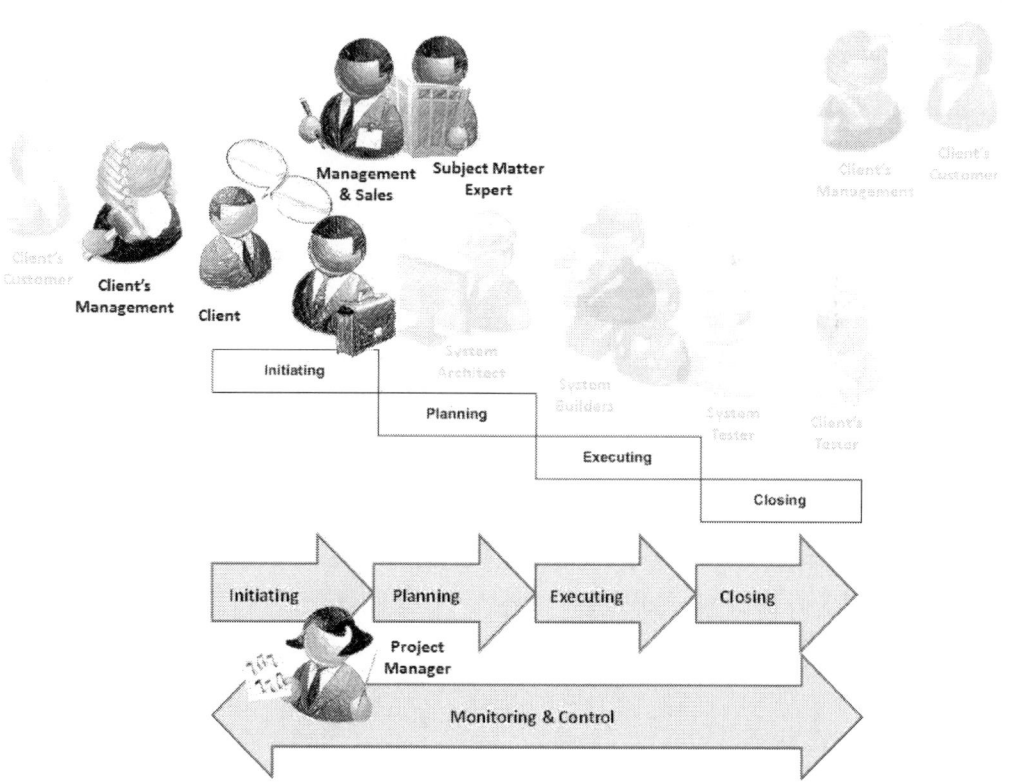

Figure 21: Stakeholder influence on a waterfall project life cycle.

NOTE ❑ Due to these stakeholders' significant influence on the project's success,

it is wise to manage these stakeholders right from the start of the project's life;

i.e. at the Initiating Phase.

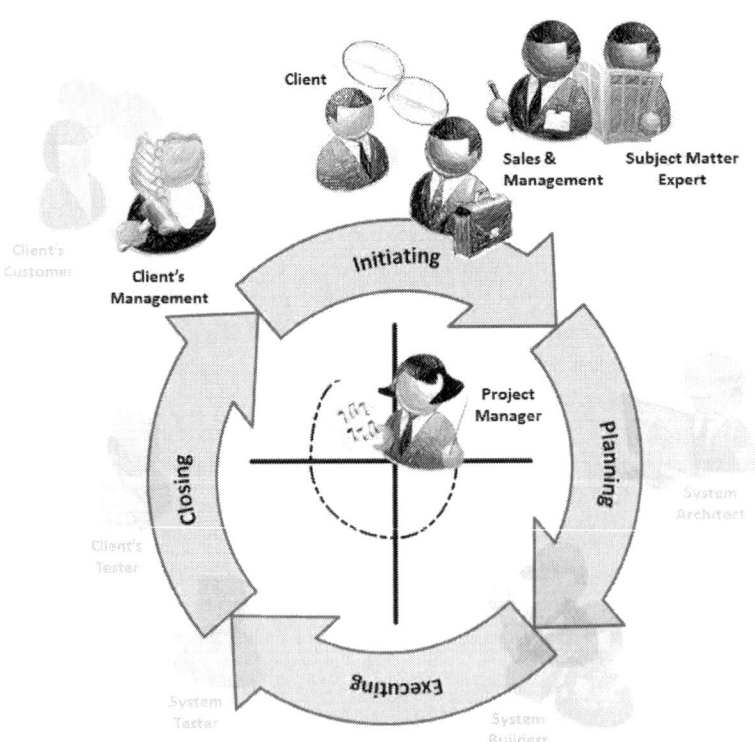

Figure 22: Stakeholder influence on an iterative project life cycle.

2.5.   Project Change Management

Of all of the phases of the project's life cycle, the Executing Phase usually takes the longest [Time] to complete and usually requires the greatest allocation of [People] & [Resources]. Hence, as illustrated in the top half of [Figure 23] on the following page, this **Executing Phase is usually the most [Cost] expensive phase to undertake.**

However, experience finds that this **Executing Phase does not have the greatest influence on whether the project will be a success or a failure.** Rather, as illustrated in the bottom half of [Figure 23], **a greater influence on the project being a failure are changes made later on in the project's life when these changes are more costly to undertake.**

> *For example; changes due to misinterpretation of requirements, additional features being added, features forgotten about entirely, incorrect designs, inappropriate architecture, incorrect data, and missing information.*

Thus, due to how costly late stage changes are for the project, **it is cheaper in the long-run to ensure that due-diligence is observed when performing the Initiating Phase and the Planning Phase.** Unfortunately, these two phases are occasionally underdone due to the desire to rush into the project's implementation and then onto the getting paid part.

NOTE   ❑ It is these **early phases of Initiating and Planning when the stakeholders have the greatest influence on ensuring the success of the project**; see [Figure 21], [Figure 22], and [Figure 23].

❑ It is the **project manager's responsibility to ensure that the Initiating and Planning Phases were and are done properly**, because this is when the project is most likely to start "running-off the rails".

Many a doomed project was destine for failure long before the implementers got involved. Yet, these implementers (as if expected to perform miracles) often received a disproportionate amount of the blame for the project's failure. While those who long ago initiated the project found greener pastures.

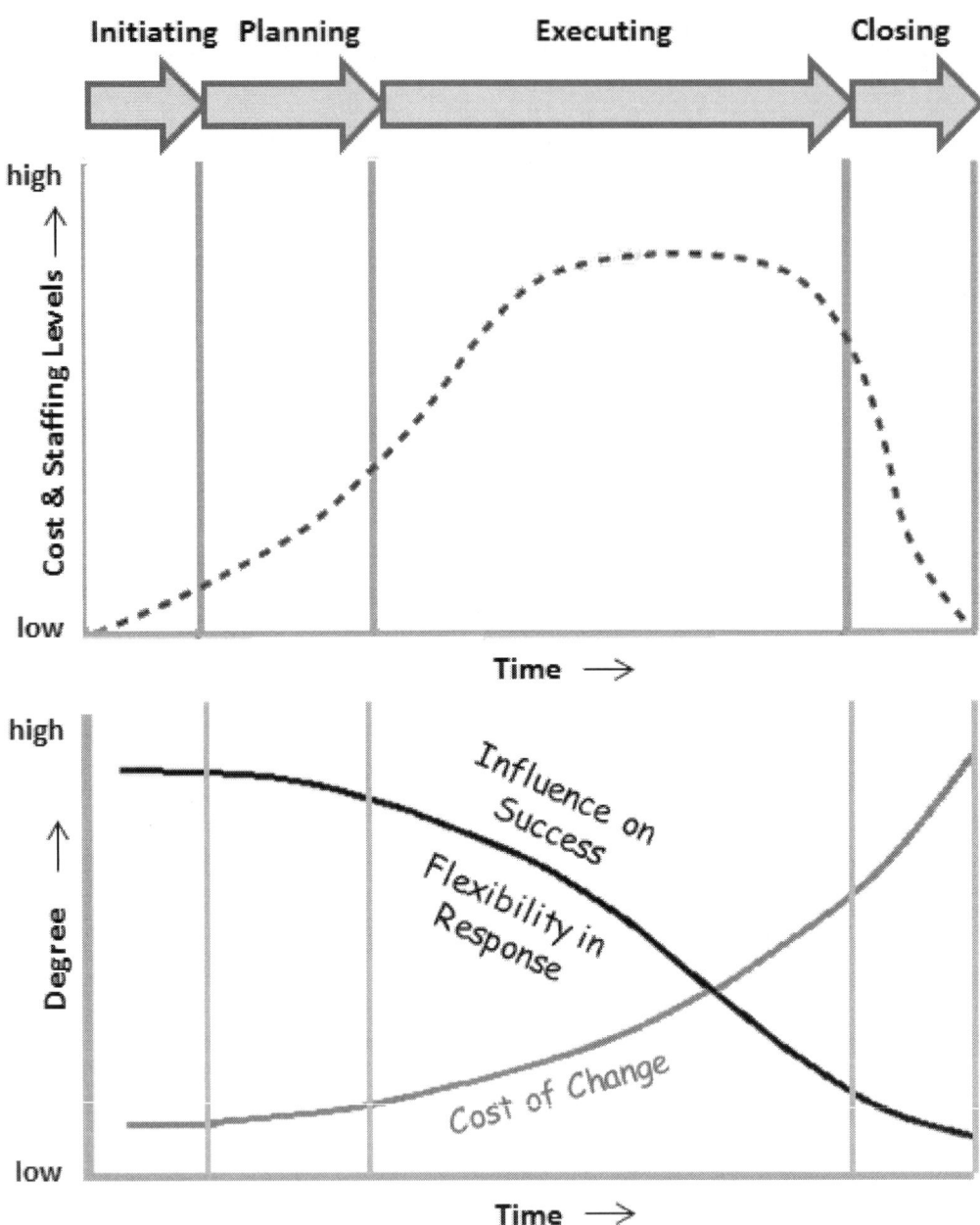

Figure 23: Typical cost and staffing levels during the project's life cycle, and the comparison of the Cost of Change verses the Influence on Success.[4]

---

[4] Based on PMI 2008, "*A Guide to the Project Management Body of Knowledge (PMBOK® Guide),4th Edition*".

2.6. Project Monitoring & Control

2.6.1. Project Monitoring

A method for project monitoring (& control) is to use a problem-solving process of

PLAN – DO – CHECK – ACT [ref: Deming / Shewhart], as shown in [Figure 24] below.

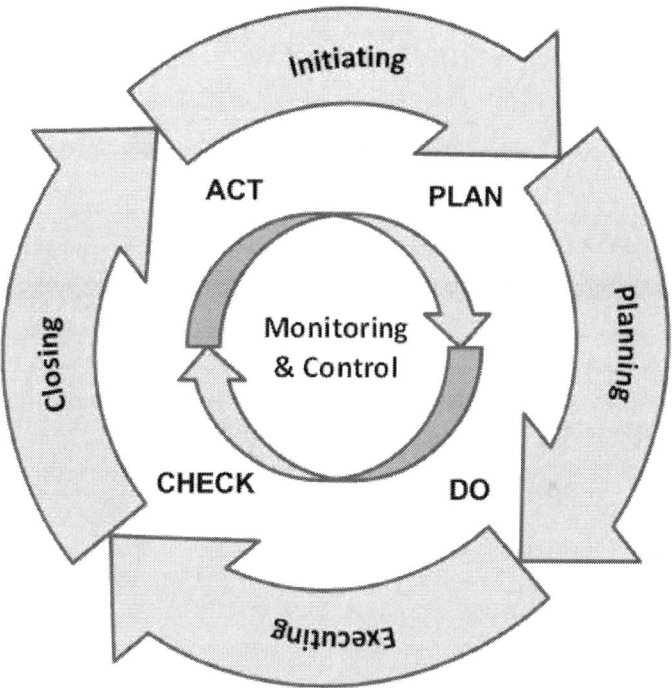

Figure 24: PLAN – DO – CHECK – ACT relationship with the SDLC.

1. **PLAN** – decide **what** has **to** be **done** and **what** are the **expected** results.

2. **DO** – **implement** what was decided as needing to be done.

3. **CHECK /** *Study* – **compare** the **actual** results **with** the **expected** results.

4. **ACT** – determine the **cause of** any **differences** between the actual results and the expected results, decide whether **improvements** have to be made, and decide whether it is necessary to **go around** the cycle **again** ... *(and again, and again)*.

2.6.2.   Project Control

As stated back in [Section 1.7], the reasons why projects fail are because;

> *"During the project's life too many of these project variables become*
> *project constraints and hence there is no flexibility left for adjustments when the*
> *project's current situation and the prevailing circumstances change"* ...

Moreover, because of this inflexibility the project's management are unable to maintain,

> *"Control over the balance between these project variables and project constraints,*
> *so that as one project variable changes the other project variables (or project*
> *constraints) do not change catastrophically".*

Therefore, **the objective of project control** (and hence that of the project manager) **is
to maintain the balance between these project variables and project constraints**.

As illustrated below in [Figure 25], a key element to achieving this control is
via "Communications".

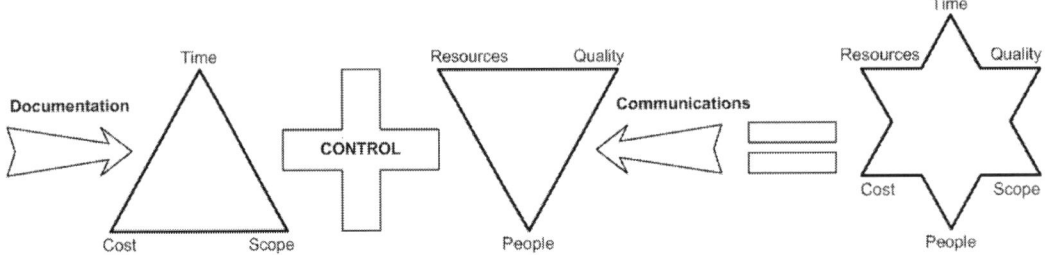

Figure 25: Project control over the project variables and project constraints.

⊠ Poor communications is a project killer.

*During the entire life of the project (especially when the project has just
begun), the project manager needs to instill trust that they know what
they are doing and that they have things under control. An easy way to
achieve this is by effective communications with the project's participants.*

 A project manager who is **NOT** prepared to spend their time communicating with all of the project's stakeholders is a project manager who is destine to lose control of the project.

This is because, **without continuous communication flows and feedback,** the project manager and the project's stakeholders will not know and understand what is relevant to them, not know what is going on, and not know what is not going on (as they thought it was).

For more information on communications, please read [Section 4.3.4].

⊠ **A project manager will spend 80-90% of their time communicating in both written and oral forms.**

2.7. Project Documentation

2.7.1. Documentation: Quality Not Quantity

If one was to follow a formalized project management methodology (such as outlined in the PMBOK® Guide) to the nth degree then one could literally be buried in documentation.

For example; Project Management Plan, Scope Management Plan, Cost Management Plan, Schedule Management Plan, Risk Management Plan, Procurement Management Plan, Staffing Management Plan, Communications Management Plan, Quality Management Plan, Integration Management Plan, as well as ...

Functional Specifications, architectural documentation, design documents, system and integration test plans, acceptance test plans, product release plans, maintenance schedules ... etc.

NOTE   ❏ **The quantity and quality (i.e. the detail) of documentation applied to a project should be proportional to the size, scale, and significance of the project**. That is, the larger the project, the more complex the project, the more strategically important / linchpin is the project, and/or the more costly the project then the more detailed documentation will be required.

     ❏ Sometimes **certain documentation will be specified as a mandatory project deliverable and necessary for the purpose of project conformance** to processes, procedures, standards, and/or certification.

     ❏ **The level of detail** that is included in each document **is dependent on the "needs & wants" of the stakeholders** that will use and contribute to that documentation.

I would recommend that ... **different document authors** write or contribute to appropriate sections of the documentation. This way, more specialist knowledge and skills (i.e. subject matter expertise) can be **concentrated on specific topics.** Thereby, **generating different perspectives** of the project and hence **catching potential issues before they arise.**

 I would also recommend that ... these documents be **peer reviewed** to take full advantage of the performing organization's existing knowledge and experience base.

In conclusion, the **purpose of project documentation** is:

1) As a **medium of communications.**

2) To catch things that could be forgotten, under-examined, or misinterpreted; i.e. the documentation has to be written to a sufficient level of detail to **prevent misunderstandings, misconceptions, and forgetfulness.**

3) To orchestrate the **coordination** of people, resources, requirements, objectives, time, and finances involved with the project.

4) To **comply with the agreed Acceptance Criteria** for the project's deliverables, for the Quality Assurance & Quality Control, and for quality auditing.

NOTE ☐ **The production of documentation should contribute to the project's success and not be a burden to its progress nor hinder its successful delivery.**

How demoralizing it is to spend hours / days / weeks writing a document knowing full well that it will never be read nor used, but is solely there as a bureaucratic tick-a-box checklist item.

And yes, some of those stories are true about "mistakenly pasted text" finding its way into a document that was supposed to be reviewed. E.g. a shopping list and even cooking recipes.

2.7.2. Documentation: DO NOT Forgo

When a project is running late, one of the most obvious ways to save time is to leave out those activities that are not required as a project deliverable. Alas, one of the first casualties is often documentation. Some documentation gets glossed over; such as the Project Management Plan and even the Detailed Specifications, while other documentation such as architectural designs, software designs, source code comments, and test plans could receive even less attention or not done at all.

 **DO NOT forgo documentation due to the need to speed up the project, as this will only result in nightmares of misunderstandings, misinterpretations, and rework.**

2.7.3. Documentation: is NOT static

Project documentation is NOT static, but rather a "living and breathing" chronicle of the project, whose purpose needs to alter over time to meet the requirements of the project stakeholders. Therefore, the documentation has to be continually updated with the latest information.

NOTE ❒ **Documentation is vital for communications, and especially as a record of interactions and proof of what was & was not agreed to.**

Documentation is also a very effective shield when the finger of blame starts pointing.

Plus, documentation is a very useful source of starting information and self-plagiarism materials for other projects.

2.8. PMBOK® – Project Management Body Of Knowledge

How does all of this relate to PMBOK? [Project Management Body of Knowledge]

Well, the **Project Management Body Of Knowledge (PMBOK®)** is divided up into **ten knowledge areas** as outlined in the PMBOK® Guide and listed below.

These knowledge areas happen to entail the handling of those six project variables / constraints that form part of the underlying basis of this book.

1) **Project Scope Management** – purpose is to ensure that only the agreed work activities are undertaken, and that only the mutually agreed functionality & artefacts are delivered by the project. Hence, the delivery of nothing more or less than what was agreed to and approved (signed off) by both the representatives of the customer organization and the performing organization.

2) **Project Time Management** – purpose is to ensure that the project is completed within the agreed timeframe, and that each of the agreed milestone dates are achieved.

3) **Project Cost Management** – purpose is to ensure that the project is completed within the agreed & approved budget.

4) **Project Quality Management** – purpose is to ensure that the project's deliverables conform to the agreed Acceptance Criteria, and to ensure that the project team members are following the relevant quality processes & procedures.

5) **Project Human Resource Management** – purpose is to ensure that the People aspects of the project are handled in a humane manner, and to ensure that the project team members are utilized effectively & efficiently.

6) **Project Communications Management** – purpose is to ensure the timely and relevant bi-directional exchange of project information, and to ensure that records of such stakeholders' interactions are kept for future reference.

7) **Project Risk Management** – purpose is to ensure that the project is not derailed by risks & issues that confront the project.

8) **Project Procurement Management** – purpose is to ensure that the inanimate <u>Resources</u> aspects of the project are handled appropriately, and to ensure that these resources are utilized effectively & efficiently.

9) **Project Stakeholder Management** – purposes is to ensure that the project stakeholders' needs, wants, expectations, perceptions and concerns are handled appropriately. … *Prior to the 5ᵗʰ Edition of the PMBOK Guide, this was included as part of communications management.*

10) **Project Integration Management** – relates to the coordinated integration of all of those previously listed knowledge areas of project management.

NOTE ❑ These knowledge areas take the massive topic of project management and break it down into specialized areas of concern. The book that you are currently reading puts an SDLC centric spin on these knowledge areas and depicts them in a sequence of process flows.

The model in [Figure 26] on the following page represents the complex interrelationship between the various aspects of project management:

- the 10 PMBOK® knowledge areas,
- the 6 project variables / constraints,
- the 5 phases of a project's life cycle,
- the 3 different facets to project Risk,
- and 1 project manager.

AUTHOR'S NOTE

For more information on obtaining PMP® Certification and the PMBOK® Guide, please refer to the Project Management Institute http://www.pmi.org/

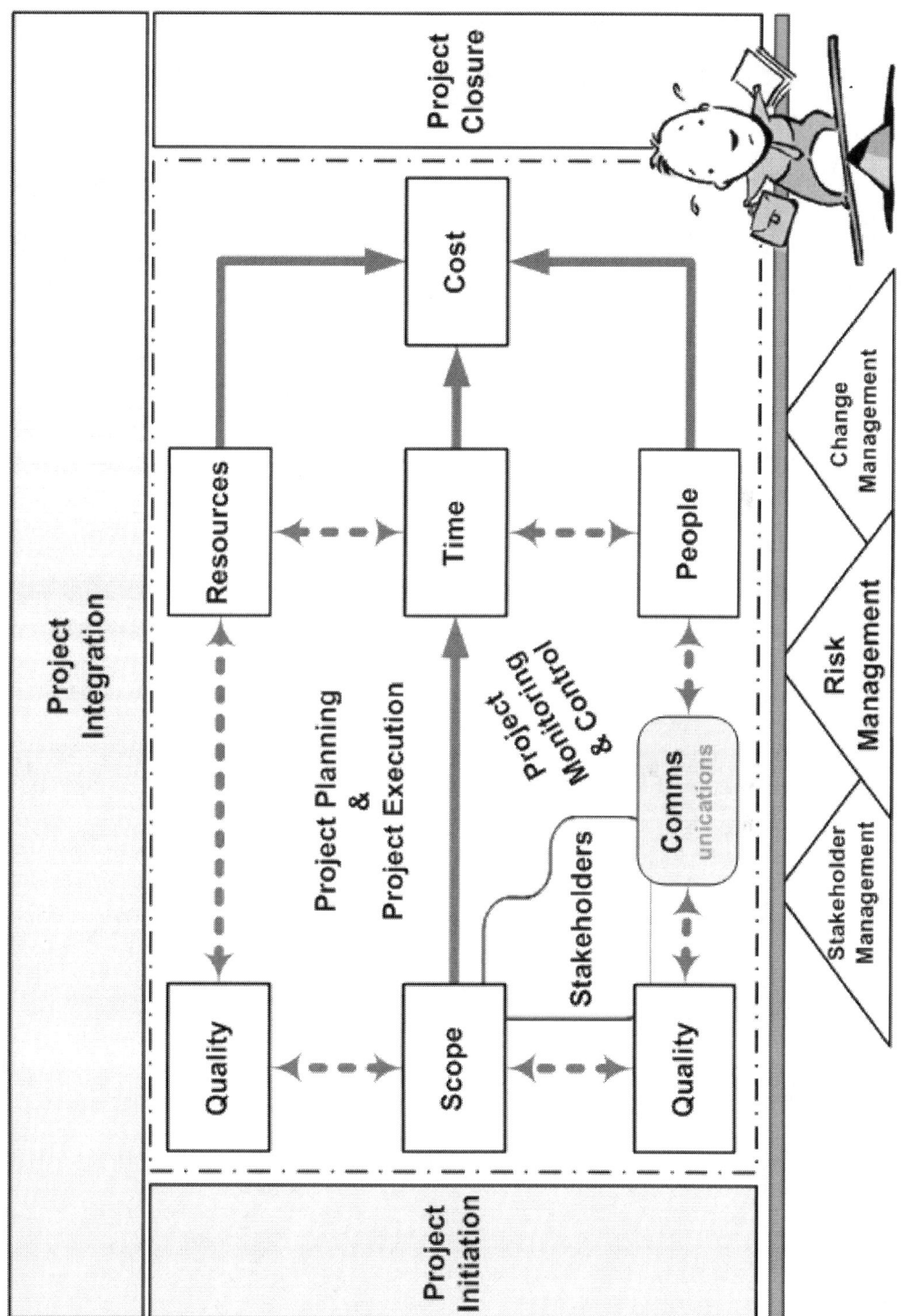

Figure 26: PMBOK® as a model of project constraints & project variables.

AN IMPORTANT PURPOSE OF PROJECT MANAGEMENT IS TO CAREFULLY MANAGE THE STAKEHOLDERS' EXPECTATIONS, PERSPECTIVES, AND OPINIONS ... THEREBY, INFLUENCING THESE STAKEHOLDERS' INTERPRETATION OF THE PROJECT AS A SUCCESS.

3. INITIATING Phase

3.1. Overview

As illustrated below in [Figure 27], the Initiating Phase is concerned with the darker shaded area of the Project Management Process model.

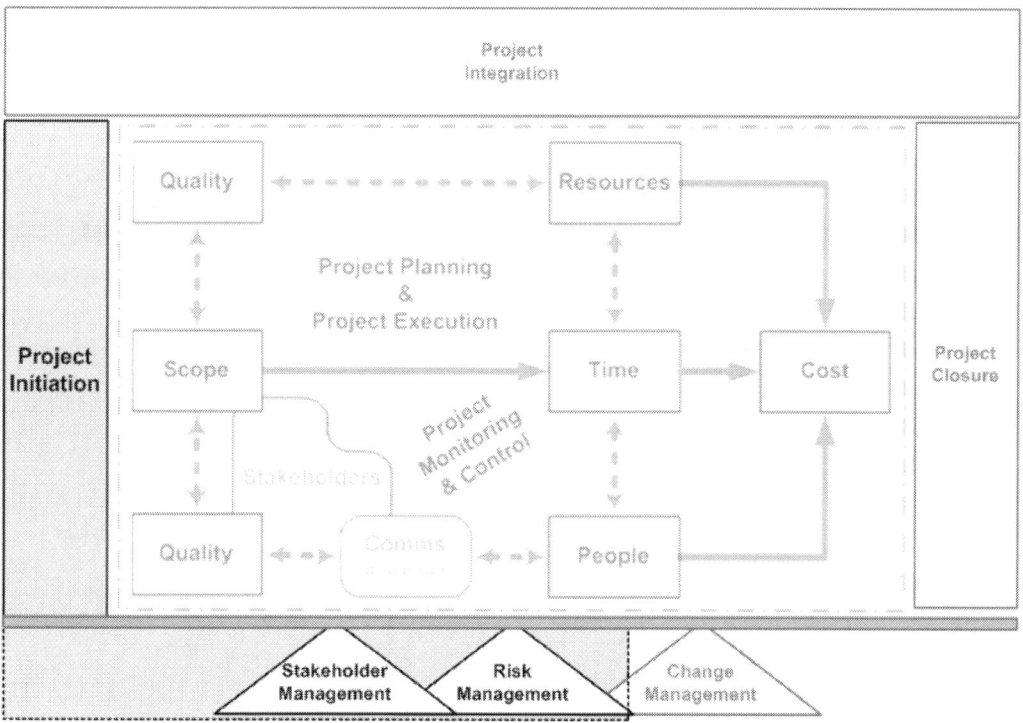

Figure 27: The Initiating Phase of the Project Management Process model.

NOTICE how in [Figure 27] the Initiating Phase also involves aspects of Stakeholder Management and Risk Management.

3.2.   Purpose

The purpose of the Initiating Phase is to define a potential new project or a proposed new phase to an existing project **based on the "needs & wants" of the customer,** and to then **obtain formal approval & authorization to commence the project**. This official **project authorization is via the sign-off of the Project Charter.**

The Project Charter can also be known as a Tasking Statement, and some organizations may refer to it as a Statement Of Work (SOW).

The process to create the Project Charter is illustrated below in [Figure 28].

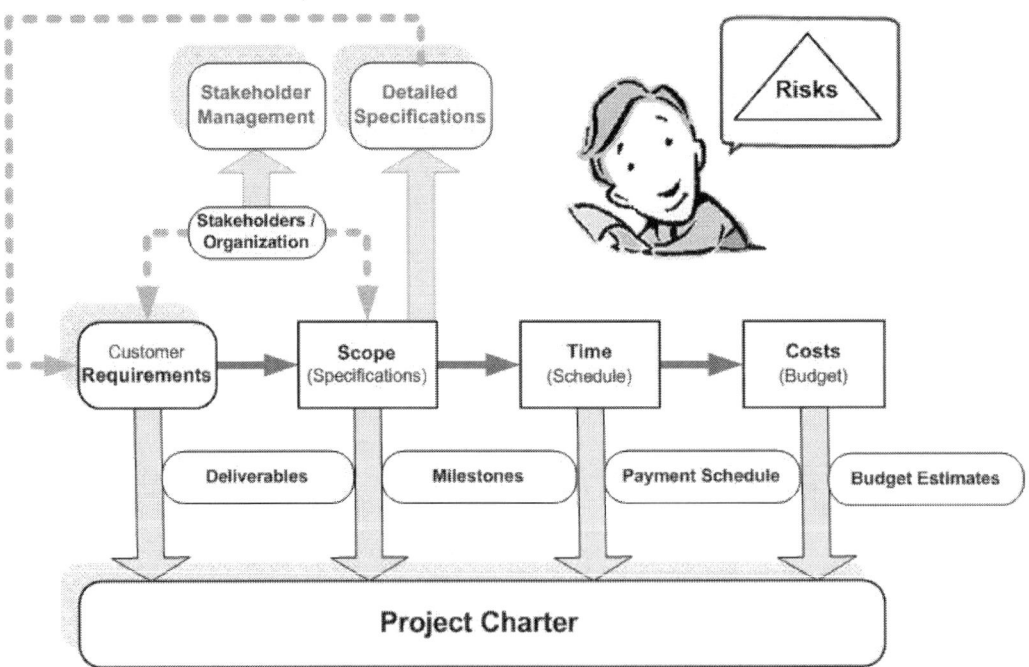

Figure 28: The Initiating Phase as a process to create the Project Charter.

NOTE   ☐ **Before any substantial work can commence** on the project,
           the Project Charter must have been approved and signed off by the
           representatives from **both the performing and customer organizations.**

3.3.   The Big Questions

Sometime you will hear it bemoaned;

"in hindsight this project was a failure before it even started."

Though, this statement does raise a rather interesting question.

How can a project fail before it even starts?

The reasons that **the foundations for project failure can be laid so early in the project's life is due** to the project's **primary-core stakeholders not answering** the following questions **in enough depth**. And, the performing organization needs to **answer these questions (to themselves) as honestly and truthfully** as they can:

1.   What do **THEY WANT**?

2.   What will **WE GIVE** them? ... *And, what will we not be giving to them?*

3.   What will it **COST US TO GIVE** it to them?

4.   What will **WE GET IN RETURN** from them?

5.   What is **AT STAKE (make and brake)**? ... *Break*

The performing organization must carefully **answer these questions right now**, so that a **responsible "due-diligent" decision can be made to undertake, halt, delay, or abort** the project **before significant** amounts of [Time], [Cost], [People] & [Resources] are **expended**.

⊠ The Initiating Phase is the last opportunity to terminate the project without incurring significant financial penalties and reputational damage.

This opportunity to prudently terminate the project before it is regretted is true for both the performing & customer organizations. So, seriously think it through, cause could this project END IT ALL?

3.3.1.   Question: "What do THEY WANT?"

Who are they?

"They" are the project's stakeholders, and chief among these stakeholders is the **"Project Sponsor"**; *e.g. the customer for external projects or the performing organization's senior management for internal projects.*

That is, the **Project Sponsor is the person who is paying for this project's undertaking**.

How do I know who they are?

To be able to answer this question, one must **first undertake an initial pass at stakeholder identification**, as outlined in [Section 10.3.1].

However, this stakeholder identification does not have to be undertaken in-depth; rather, deep enough to know whom to communicate with in order to obtain reasonable answers to the following question.

What do they want?

The proposed project's primary-core stakeholders should have some ideas as to what they hope to achieve by undertaking this project.

The information about their "needs & wants" can be extracted from; the Contract, Request For Proposal (RFP), Request For Tender (RFT), Request For Quotation (RFQ), Business / Customer Requirements, pre-existing Detailed Specifications, existing Change Requests, work orders, sales & marketing feedback, product requirements & proposals, feasibility studies, business case studies, and by direct communications (both written and oral) with these stakeholders.

Accordingly, the outcome of this *"what do they want"* question is an **overview of the project's [Scope] and some preliminary details about the Customer Requirements**; or as a bare minimum, references to where this information can be obtained.

What if they have no clue as to what they want?

This situation is surprisingly common; they (the customer) know they need something but they don't know what it is, let alone what it needs to do. Hence, the best way to approach this situation is to undertake a **"Top-Down Analysis"** as illustrated below in [Figure 29].

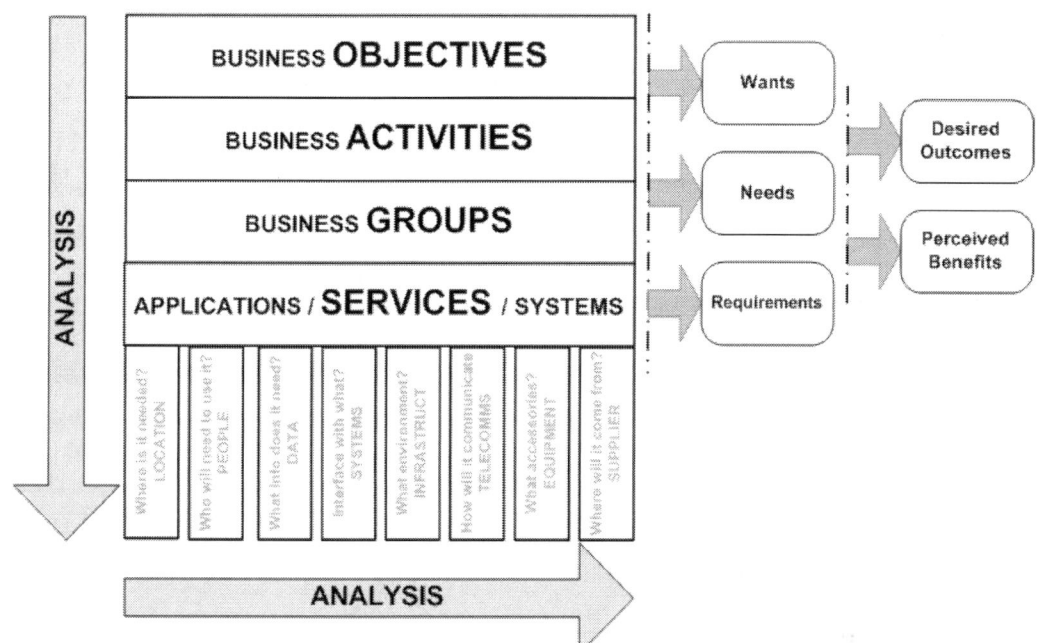

Figure 29: Top Down analysis to determine what they "want", "need", and "require".

The logic of the **top-down analysis** is as follows; the organization (the customer) has specific **business objectives** that they are trying to achieve, they undertakes specific **business activities** to achieve these objectives, these activities are executed by specific **business groups** within their organization, these business groups use specific **services** (applications & systems) when executing their activities.

Buried somewhere within each of these layers are the details of their; **"wants"**, **"needs"**, and **"requirements"**. From this can be derived their **"desired outcomes"**, and what are their **"perceive benefits"** of undertaking the project.

Now, all that is involved is an examination of each of these layers.

How to access those hidden customer "wants", "needs", and "requirements".

1. What are the **business's objectives**?

    For example; sales growth, market expansion, client retention.

    - What is the purpose of each business objective?

2. What **business activities** are undertaken to achieve each objective?

    For example; sales, marketing, customer support, client training, customer billing, development, internal support services.

    - What is the purpose of each business activity?

3. What **business groups** of people / departments / business units undertake these

    activities? *For example; regional sales team, online marketing, R&D, customer*

    support helpline, IT support... etc.

    - What is the purpose of each business group?

4. What **services**/applications/systems does each of these groups use to undertake

    these activities? *For example; CRM, billing system, databases, office suites... email,*

    phone, fax, printers... internet, intranet, workstations, and network file shares... etc.

    - Does one of these services/applications/systems have the problem that
      has to be resolved?

    - Is this new service/application/system required for; new functionality,
      upgrade to existing functionality, or to complement additional functionality?

Now that the target services/applications/systems have been identified, the next step is

to ask specific questions to focus in on the details of their "wants", "needs", and

"requirements".

DO NOT venture too deep into the specifics of the details, as that is the purpose of the

Detailed Specifications ... though, **just enough to establish the foundations**.

NOTE   ❐ for now, **Do not question, "when do they need it by" [Time] and "how much are they prepared to spend" [Cost]**, because for the moment we are only interested in sketching out the basics of the Customer Requirements [Scope] that will be expanded on later. So don't worry about [Time] & [Cost] as these will be dealt with soon enough.

Having completed the **top-down analysis of the customer's business** then hopefully the performing organization has a **better understanding of the customer's "wants"**, **"needs", and "requirements"**.

Thereby, enabling the **determination of the customer's "desired outcomes"** and obtaining an **insight into what are the customer's "perceived benefits" from undertaking this project**.

⊠ Misunderstandings and misinterpretations are a project killer.

"Assumption is the mother of all screw ups."

"SIMILAR" IS NOT CLOSE ENOUGH TO THE "SAME", AS IT WILL ONLY END IN FLAMES AND BLAME.

3.3.2.   Question: "What will WE GIVE them?"

Who are we?

"We" are the **performing organization** ... your company, your business unit, your department, your team, or just plain Me-Myself-I. That is, "We" are the other primary-core stakeholder **directly opposite the project sponsor**.

What will we give them?

The answer to this question does not need to go into great levels of detail, though **the answer should be broad enough to enable an understanding of the performing organization's response to the customer's requirements and to provide a generalized indication of how it is proposed that the project will be implemented**. Depending on the organizations involved and their industrial backgrounds, this answer may be;

a **Tender Response** if responding to a government sector's **Request For Tender (RFT)**,

a **Bid Response** if responding to a private sector's **Request For Proposal (RFP)**, or

in some cases of close organizational relationships this **may be the Detailed Specifications**.

DO NOT include detailed architectural designs, design concepts, or technical descriptions in the response to the customer.

Unless, these have been **specifically requested as** a **mandatory inclusion in the response**; i.e. a stipulation in the request document.

The reason for being "hazy" with this technical information is that; **if too great a level of detail is included** with the response and the project is signed-off with this information in tow **then potentially the project could be hamstrung by these** when **in retrospect** these were not **the best options** for the successful execution of the project.

 I would recommend that ... a few senior implementers be brought together for a **quick whiteboard brainstorming session to verify that what is proposed is technically feasible.** ... *Any risks involved?*

I would recommend that ... **the response** to the customer **only include the level of technical detail** which would be **found in a glossy sales & marketing brochure**. Thereby giving the project room to manoeuvre if need be during the Executing Phase.

The answer to this *"what will WE GIVE them"* question (which could be **incorporated into the Project Charter** / *Tasking Statement* / *Statement Of Work*) would be the following:

1) **The <u>overview</u> for this project.** This would include; an **introduction** to the project, a **background** to the project, the **goals** of the project, the reasons for **need**ing this project, the perceived **benefits** of this project, the **desired outcomes** of this project, and a description of the project's **measurable & tangible objectives**. Some of this information will be a reiteration of what the project sponsor as already stated, but all of this would be in the performing organization's own words as an indication that the performing organization understands what the customer is trying to achieve.

    Once the performing organization understands the customer's "needs & wants" then **the requirements can be collected in detail**. This will help in defining the [Scope] of the project and thereby aid in detailing the project's assumptions, constraints, scope boundaries, deliverables, milestones, and in turn establish what is the [Time] scale and [Cost] size of the project.

2) **The <u>assumptions</u> made about this project**; i.e. a description of those **things that are believed to be true**. Such as, the availability of key personnel, equipment, material resources, technologies, data, skillsets, environmental conditions... etc.

3) **The project <u>constraints</u>**; *e.g. budget limitations, unmovable delivery dates, compulsory quality standards, mandatory conditions... etc.*
    Additionally, the **<u>priorities</u>** for these project constraints may be specified.

4) **The <u>scope boundaries</u> for this project**; i.e. a generalized description of **those things that will be within the project's domain**, and **those things that will be "out-of-scope"**.

5) **A generalized list of the project's high-level <u>requirements</u>** and references to any documents that detail these requirements in-depth.

6) **A list of** the project's major / **high-level <u>deliverables</u>**.

These deliverables would be those things that will be handed over to the customer's representative(s) by the end of the project and possibly at various release points during the project's life. The receipt and acceptance of these deliverables by the customer would form some of the "tick-a-box" information accumulated to decide whether the performing organization qualifies to receive **the financial remunerations associated with each particular deliverable and release milestone.**

7) **A list of** the major / **high-level dates for the project <u>milestones</u>**.

Each project milestone should be targeted at providing the customer with functional deliverables that can be used to evaluate the project's progress towards the agreed objectives (and include any associated partial / progress payment points).

NOTE ❑ For the project to have a chance of being successful, then **both the customer's and the performing organization's understanding and interpretations** of the requirements, deliverables, milestones, and the scale of the project **must align.**

Intellectual Property Rights

Ah, don't they make a lovely couple.

Yeah but, at some point in the future this relationship will come to an end.

It is now during the Initiating Phase when the foundations for an amicable separation need to be established, else a bitter divorce could very well be in the making.

That is, when this project finishes and the relationship between the performing organization and the customer organization concludes then **"who exactly owns what"**?

This is not referring to the specified project deliverables (i.e. end products), but rather to those **intangible assets of Intellectual Property (IP)** that were derived during the project's life and that IP which is now entangled within the project's deliverables. Also, what about the **ownership of the "knowledge learnt"** during the creation of those deliverables?

> *For example; patents, copyrights, registered & unregistered designs, registered & unregistered trademarks ... inventions, proprietary information, data, source code, circuit layouts, algorithms ... the know-how.*

Depending on the laws of the state / country where the project's contractual agreement pertains then ownership of such intellectual property could reside with the performing organization, it could be vested with the customer organization, or a combination thereof. Unless that is, **the Terms & Conditions of the project's signed contract clearly specifies what are the Intellectual Property Rights of each party involved with the project**. This would not only **include the customer organization and the performing organization,** but **also any third parties such as sub-contractors who will be engaged to participate in the project**.

 **BEWARE the complications involved with the ownership and access rights to Intellectual Property**, given that just because you created it does not necessarily mean (that after the project has finished) you'll legally own it or have a right to reuse it.

For SDLC projects, it is common for both the performing and customer organizations **to bring their own pre-existing intellectual property into the project** as components of the deliverables or as parts of the processes & procedures used to produce those deliverables.

Additionally, for SDLC projects it is almost certain that during the course of the project, new **intellectual property will come into being** either due to the collaboration of those involved parties, or as a result of the experience of undertaking the project.

So, who exactly owns the project's intellectual property?

Well, just as with a marriage, each person will bring into the relationship some assets they possessed before the relationship commenced. In intellectual property terms, this is known as "Background IP". Whereas, those assets which the couple acquired during the relationship is known in intellectual property terms as "Foreground IP".

❖ **Background IP** – is that **intellectual property which existed prior** to the contracted relationship commencing, or is independent of the contract.

❖ **Foreground IP** – is that **intellectual property which resulted from or is generated pursuant** to the contracted relationship.   *... Oh "pursuant", legal speak.*

That all sounds simple, but what if the resultant Foreground IP is so intertwined with the Background IP that the Foreground IP cannot feasibly & reasonably be divorced from the Background IP, then who owns this IP?

Sounds like a prenuptial agreement is required.

The answers to these questions of IP ownership need to be clearly understood (by all involved parties) during the Initiating Phase and not be left to be argued out during the Executing Phase, and especially not to be fought over during the Closing Phase.

Therefore, during the Initiating Phase the representatives from the performing organization and the customer organization (as well as any involved third parties) need to **negotiate the rights each party has with respect to this Foreground IP and Background IP**.

For example; it may be agreed that:

- *Neither party gives the other party/parties rights or title to its Background IP, other than to the limited use of the Background IP for the purpose of producing the project deliverables and the resultant use of those project deliverables for the sole purpose for which the deliverable was intended.*

- *All rights or titles to all Foreground IP shall be the sole and exclusive property of the performing organization, excluding that Foreground IP related specifically to the project deliverable. That Foreground IP related specifically to the project deliverable shall be immediately vested in and become the property of the customer organization upon delivery of the deliverable.*

Wonder whose contract Terms & Conditions these are from?

Yes well, and did you read all of those clauses and subclauses of legal mumbo-jumbo ... obfuscation speak be proud.

One thing for sure, **it is essential that the senior management (and legal representatives) of all parties involved with the project understand the contract's Terms & Conditions**, and especially with respect to their intellectual property rights. As, **a lack of due-diligence here could result in an expensive IP dispute, potentially ongoing litigation, and the forced revealing of information & things that were not intended to be seen by others**.

When using Open Source, need to understand the Terms & Conditions of the associated license, because these T&Cs could require that the produced source code be made available to the public domain for inspection & distribution, and thereby possibly contradicting the IP agreements between the performing and customer organizations.

 I would strongly recommend that ... an **Intellectual Property Register** be kept to keep track of who brought what particular IP into the project, and who owns specific IP that is generated during the course of the project.

3.3.3.   Question: "What will it COST US TO GIVE it to them?"

Based on past experiences and historical data make a reasonable **estimate** of the project's; **approximate [Time] duration**, the **approximate [Cost]** to implement, and the **approximate [People] & [Resources]** that will be required to undertake the project; see [Figure 30].

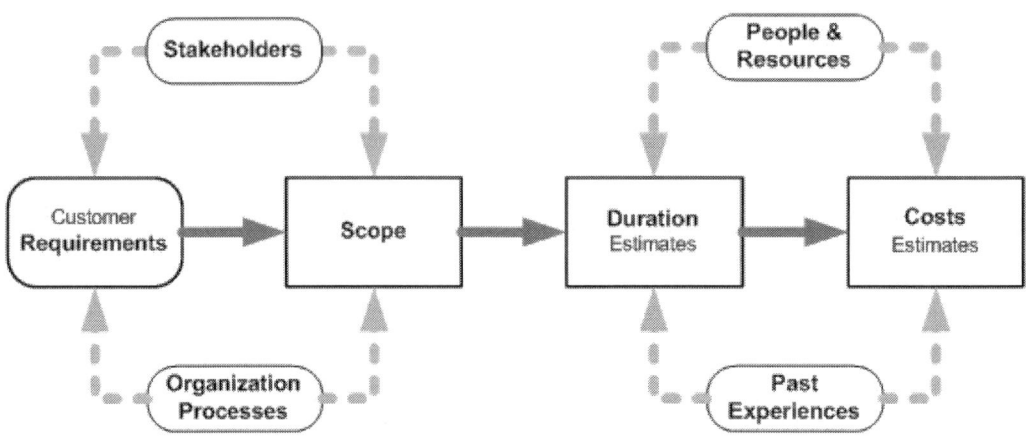

Figure 30: Process for the Initiating Phase estimates.

Now these estimations need only be "within the ballpark" of that which would be produced by a detailed budget and schedule analysis; i.e. a **"Rough Order of Magnitude" (ROM)**. This ROM should at least distinguish between $1 - 3 - 6 - 12 - 18$ month's duration projects, multi thousand – million budget projects, and $5 - 10 - 50 - 100$ people projects. If one is **unable to make logical distinctions between these scales for the project then this project is destine for failure.**   ... *ROM is non-committal info on what is involved.*

> *For example; a sales manager who signs the business up to a six months delivery contract without realizing that the requirements would necessitate a major rework of their existing product, and hence the project would need at least 12 months for development. Consequently, the project will be a failure due to incorrect scaling.*

Sometimes the ROM is sent in advance, followed by a detailed proposal response that contains the Contract Master Schedule and Cost Breakdown.

3.3.4. Question: "What will WE GET IN RETURN from them?"

Now that we have a reasonable understanding of the customer's requirements and subsequently and overview of the project's [Scope], as well as estimates for the project's [Time] duration, [People] & [Resources] utilization, and a basic [Cost] for undertaking the project, then the next step is to **calculate the marked-up "Sell Price"** and also **factor into the quotation the appropriate "Safety Margins"** (i.e. additional expenses that based on past experiences are likely to be incurred by the project). This cost safety margin would include such things as; **Cost Escalation** (i.e. exchange rates, inflation), **"Cost Contingency"** (i.e. coverage for the potential expense of providing **warranty support** and the **repair of latent defects**), and a **percentage for Project Risk**. Thereby, ensuring that this project will most likely be a **profitable endeavour** for the performing organization.

What will we get in return from them?

The answer to this question would be; the contract's value, the payment methods, and the breakdown of partial / progress **Payment Milestones**.

> *For example; 5% at contract sign-off,*
> *5% at completion of the Detailed Specifications &*
> *Project Management Plan, 5% at completion of the*
> *architectural designs, 35% at completion of the Site*
> *Acceptance Tests, 20% at completion of the User*
> *Acceptance Tests, 10% at the project closure sign-off,*
> *and the remaining 20% at the conclusion of the*
> *warranty period.*

 I would recommend, **tailoring the Payment Milestones to result in positive cash flow** / revenue (for the performing organization) where the accumulated amount of payments up to that point exceeds the accumulated expense of conducting the project to that same point.

3.3.5. Question: "What is AT STAKE ... make & brake?"

The performing organization's answer to this question is a **feasibility study, business case, and/or cost-benefit analysis** considering such things as:

- What is the **Return On Investment (ROI)** ? *Payback Period, Net Present Value ?*

    ROI = (Gain From Investment – Cost Of Investment) / Cost Of Investment

- What **strategic value** will this project provide to the performing organization?

- What are the project's **intangible benefits** to the performing organization?

- What are the **benefits to other** projects or other groups in the organization?

- Does this project **suit the performing organization's processes,** its **culture,** its **strategic objectives & plans,** and its **existing capabilities & capacities**?

- What are the prevailing **socio-economic circumstances**? *For example; the global economy, financial funding, statutory-standards compliance, and ecological impact.*

- What is the performing organization's **history with this customer**?

- What is the performing organization's **history with this type of project**?

- What is the **likelihood of success** for this (type of) project?

- Does a **Performance Bond** need to be paid as **guarantee** of performance?

- Are **Liquidated Damages** involved if the performing organization does not deliver on time? What are the **Terms & Conditions** related to these liquidated damages?

- What are the **risks involved** with or with not undertaking the project?

- What are the **alternatives available** to not undertaking this project?

- *Would the failure of this project possibly be the straw that broke the performing organization's back?*

⊠ **Does this proposed project meet the organization's selection criteria when compared to the alternatives?**

⊠ **Does undertaking this project make sense by providing value to the business and its owners / shareholders?**

[#] *Are we making a rational decision or an emotional one?*

There are several reasons that it may be decided to undertake a project:

- A **customer request** or an **outstanding** contract **obligation**.

- There is a **market demand** or a trend towards an **emerging market**.

- **Statutory, certification, legal requirements** to have certain procedures and/or processes in place; *e.g. quality, safety, ecological.*

- **Competitive advantage** or just maintaining competitiveness in the market.

- **Productivity / process improvement** to internal operations.

Then **sometimes projects are undertaken for no obvious financial benefit** to the business.

For example;

- *The undertaking of a predestine loss-making project for the sole purpose of* **building / rebuilding the working relationship with a key customer**.

- *A project undertaken purely for the* **status and promotional image** *that it would provide to the organization (as a* **technical innovator**) *even though the business would be making a loss on each unit produced.*

- *An interim development project intended to* **keep the essential personnel busy** *while awaiting the outcome of negotiations for a larger pending contract.*

- *An underquoted contract accepted because it would* **keep the business ticking over**, *as there is no other contract work available at that time.*

A Project's Face Value Versus it's Business Value

A project's "face value" (i.e. cost of implementation) **is by no means a determinant of** the project's strategic / **Business Value** to the customer organization (and to the performing organization), **especially those "keystone" projects** that are internal to the organization.

> *For example; this I.T. systems upgrade project may only have a price tag of $100k, yet it results in an essential part of the continued and effective operations of the organization's multi-million dollar business. If this "cheap" project was to perform badly, then the business could very well find its operations gravely affected.*

> *For example; this relatively low budget development project is the first steppingstone to a greater program of work. Where the customer organization is in essence using this project as a litmus-test of the performing organization's capabilities and competencies. If this "low budget" project was to perform badly and fail, then the business could find itself scrounging for other work.*

> *For example; this low budget compliance project needs to be completed and rolled out by a specific date, else the performing organization would lose it certification / license, and subsequently not be able to legally operate in its primary market.*

A PROJECT'S WORTH IS NOT ALWAYS PROPORTIONAL TO ITS COST OF IMPLEMENTATION.

3.4.   Outputs

3.4.1.   Output: Documents & Deliverables

Now that the big questions have been answered, the next step is to ensure that these answers have been documented. The resultant documentation from the Initiating Phase is highlighted below in [Figure 31], and the inter-relationship of this documentation to other documentation produced during the project's life is illustrated in [Figure 32].

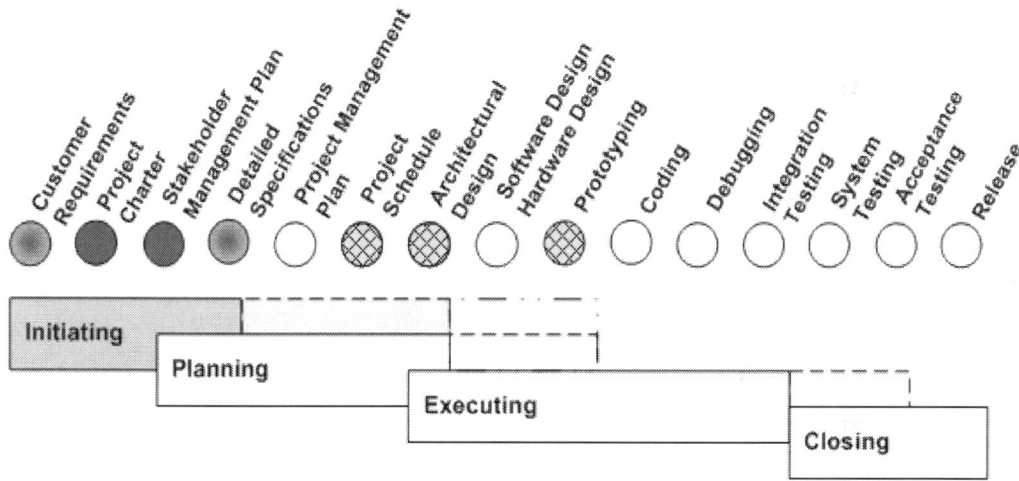

Figure 31: Documents & deliverables output of the Initiating Phase.

Notice that in [Figure 31] there is **potential for crossover between the Initiating Phase and the Planning Phase**, and maybe **even some crossover with the Executing Phase**. These crossovers are due to the senior management of the performing organization requiring some preliminary planning and prototyping to be implemented so that (based on the resultant outcomes) they may obtain a better understanding of what is involved with the project. Thereby, with clearer insight into the definition of the project, the senior management may be able to make a more responsible decision as to whether the project should OR should not be undertaken.

Hence, for some proposed projects this may require a more in-depth understanding of the alternate architectural designs and even the creation of "quick & dirty" proof-of-concept prototypes to determine which alternatives are possible, technically feasible, and financially & strategically viable.

NOTE   ❐ The important thing is to **provide enough detail and insight to enable a responsible decision on whether to select the proposed project and NOT to covertly accelerate the project's implementation.**

ICARUS'S PROBLEM WAS NOT THAT HE TRIED TO FLY,
BUT RATHER THAT HE IGNORED
THE RECOMMENDATIONS OF DAEDALUS
THE MASTER CRAFTSMAN, AND
TRIED TO FLY TOO CLOSE TO THE SUN.

3.4.2. Output: PMBOK-SDLC Map

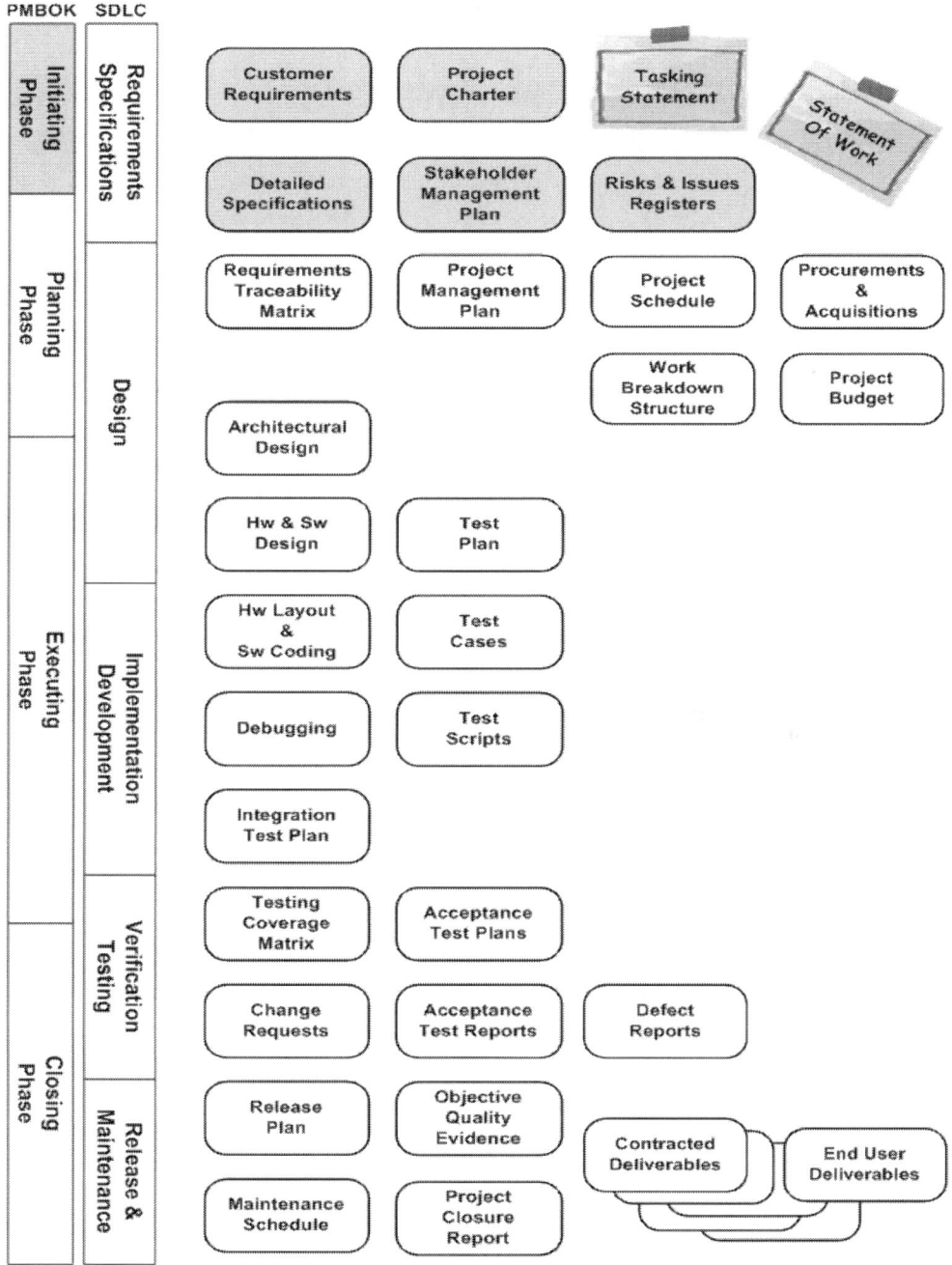

Figure 32: Relationship of PMBOK - SDLC and the documents & deliverables produced for the Initiating Phase.

3.4.3.　Output: Customer Requirements

By this point in the Initiating Phase, a better understanding of what the customer really "needs & wants" should have become known.

Hence, due to the interactions of the customer's representatives and the performing organization, it is **common for the Customer Requirements to evolve during the Initiating Phase.**

3.4.4.　Output: Project Charter

The major output of the Initiating Phase is to obtain official authorization to commence the project; i.e. a signed-off Project Charter. */ Tasking Statement / Statement Of Work.* Unfortunately, the creation of the Project Charter does not always involve the participation of the project manager. In some situations, the **Project Sponsor** (or their representative) hands over the Project Charter to the project manager.

With all of the information collected during the Initiating Phase, the subsequent Project Charter would potentially contain the following sections:

1. **Project Identification** – that is the project's ID Code and official title.

    Where will the electronic version of the Project Charter document reside in the document repository (or where will it be located on the file server if a Document Control System is not going to be used)?

2. **Project Authorization** – a list of the persons who approved the undertaking of this project. Also, include each person's job title, (optionally their contact details), and which organization or business group they represent.

3. **Project Manager** – the name of the project manager assigned to manage this project, and their **authority level**; e.*g. budgetary approval (**Cost Account Authorization**), and commitment of resources & personnel.*

4. **Key Stakeholders** – list the primary-core and strategic stakeholders for this project. Include each person's position title, and which organization or business group they represent. Optionally include their contact details (**Points Of Contact**), their function / role, and their responsibilities on this project (**Responsibilities Assignment Matrix**).

5. **Project Overview** – see [Section 3.3.2], include the project introduction, background, needs, goals, objectives, perceived benefits & outcomes.

6. **Project Scope Boundaries** – see [Section 3.3.2]. … *In Scope and Out Of Scope.*

7. **Project Assumptions** – see [Section 3.3.2].

8. **Project Constraints** – see [Section 3.3.2].

9. **Project Priorities** – see [Section 3.3.2].

10. **High-Level Project Risks** – see [Section 3.3.5] … *execution, operational, external.*

11. **High-Level Project Requirements** – see [Section 3.3.2].

12. **High-Level Project Deliverables** – see [Section 3.3.2].

13. **Intellectual Property Rights** – see [Section 3.3.2].

14. **High-Level Project Milestones** – see [Section 3.3.2].

15. **Contract Master Schedule** – "rolled up" **Summation Gantt Chart**.

16. **Cost Estimates** – see [Section 3.3.4].

17. **Payment Milestones / Payment Schedule** – see [Section 3.3.4].

18. **Tangible Measures** - list the tangible measures to be used to gauge whether the project has achieved its objectives.

19. **Project Acceptance Criteria** – state the determinants for accepting the project's deliverables and the associated Objective Quality Evidence that is to be presented.

20. **Project Closure & Termination** – outline the processes involved with concluding the project. Also, consider the worst-case scenario of the project being terminated and outline any associated costs & penalties that would be incurred.

21. **Additional Information** – include any extra information deemed necessary by the customer's representative or the senior management of the performing organization to underpin the agreement.

NOTE   ☐ **The Project Charter should not have to be changed once the project is authorized**; i.e. the Project Charter should not evolve during the project's life.

 I would recommend that ... notice should be taken of how much "toing & froing" occurs in getting the Project Charter signed-off, as this will provide insight into what can be expected for those other inter-organizational project documents that have to be reviewed & approved.

3.4.5.   Output: Risks & Issues Register

While answering [Section 3.3.5] question of "What is **AT STAKE (make and brake)**" a few apparent risks & issues confronting the project came to light, these should be recorded in the Risks & Issues Register to be dealt with as the project progresses.

⇨ For more information on Risk Management, please refer to [Section 4.3.5] and [Section 10.1].

3.4.6.   Output: Detailed Specifications

Depending on the circumstances of the project and the relationship between the customer and the performing organization, a Detailed Specifications (*e.g. Functional Specifications – FS, Detailed Business Requirements Specifications – DBRS*) maybe an output of the Initiating Phase or alternatively an early output of the Planning Phase. This Detailed Specifications would contain a feature-by-feature list with appropriate explanations of what will be delivered to the customer. See [Section 4.3.1].

NOTE   ☐ The **Customer Requirements** are a **"wish list"**, whereas
        the **Detailed Specifications** is a **"what will be delivered list"**.

3.4.7. Output: Stakeholder Management Plan

Once the Project Charter is signed-off then the next step is to **complete the stakeholder analysis** and thereby determine who all of the stakeholders are, and specify how to **manage these stakeholders' expectations**; i.e. the Stakeholder Management Plan.

 For more information on Stakeholder Management, refer to [Section 10.3].

Though honestly, rarely have I witness a stakeholder analysis being done because often it is a "Them" (the customer) & "Us" (the provider) affair.

3.5. Things to Watch Out for

When undertaking the Initiating Phase there are a few things that have to be watched out for, these are:

✖ **DO NOT commit to a project without exercising appropriate levels of managerial and technical due diligence.** That is, **do not allow one part of the organization to make commitments that other parts of the organization cannot realistically deliver.** As this will only result in unrealistic customer expectations that cannot be met without significant impact on the performing organization; i.e. a burden on [Cost] & [Time], possible financial penalties, and definitely reputational damage.

> *For example; Sales makes commitments on price and/or delivery dates without obtaining input from the development group on the technical viability of the project with respect to time and the resources that would be available.*

✖ **DO NOT base [Cost] & [Time] quotations on a principle of "whatever values are necessary to win the business" instead of those quotations being derived by sensible & logical estimates.** Because, if a customer is allowed to put their faith in these ridiculous prices and unrealistic delivery dates then **the project will most probably end in a catastrophic failure due to project blow-outs.** ... *It is a foolish strategy to plan to lose on the project, but win on the follow-on work.*

✖ **DO NOT assume that the [Cost] & [Time] of an under-estimated "win at all costs"
quotation can be recoup with contract deviations, time extensions, and future
maintenance agreements.** Because, eventually some party will feel that they are
being taken unfair advantage of, and hence this losing strategy will **end up in a
disgruntled and combative relationship for the parties involved with the project.**

✖ **Watch out for copious amounts of verbal exchanges** when there is **minimal written
correspondence to summarize the outcomes of the discussions.**
This minimalist documentation will result in future disagreements over;
what was said, what was agreed to, and whether such-n-such is a scope change.

**Being lazy during the Initiating Phase can result in a badly defined project and an
inappropriately selected project.** This will have a significant impact on the success of the
project and more specifically on the performing organization's bottom-line. **This impact
will be felt irrespective of how much effort is put into the later phases of the project.**

✓ **Clearly list the project's deliverables.** For a small / inexpensive project, this may be as
simple as a bullet-point list of the Customer Requirements with notes outlining the
details of the specifications.

✓ **Clearly define in written format** what **the purpose and measurable & tangible
objectives** of the project are, and what the **specifics of the project deliverables** are.

DON'T PLACE SHORT TERM GAINS AHEAD OF LONG TERM VIABILITY.

3.6.   Phase Completion Review

Now that the work related to the Initiating Phase has concluded, then some form of "**Phase Completion Review**" meeting would be held to **determine whether the project is in an acceptable state to advance onto the Planning Phase**.

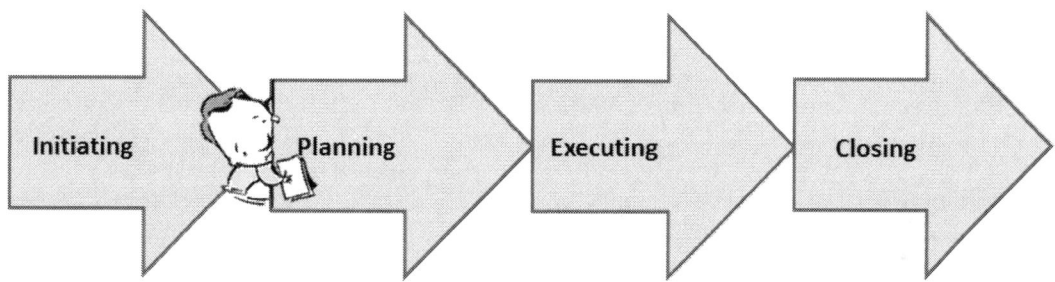

Would the outcome of this **"Gate" review** be to **re-envision** the entire project or would the resultant decision be that the project should be **terminated** before it gets underway. That is, based on the resultant outputs of the Initiating Phase [Section 3.4], and when all things are taken into consideration [Section 3.5] will the decision be a **"GO" or "NO GO"** for the project?

NOTE   ❑ Due to the importance of the resulting decisions from this meeting, it is **essential that the Phase Completion Review involve representatives from the senior management of the performing organization, representatives of the customer organization, possibly any other concerned primary-core stakeholders, and the project manager** (if known).

In conclusion, if the resultant decision from this Phase Completion Review meeting were **permission to proceed**, then the **Project Charter would be signed-off by the authorized representatives from the customer organization, the performing organization, and any other contractually obligated primary-core stakeholders**.

3.7.   Phase Completion Check List

1) Have you identified **who** the project **stakeholders** are?	✓
2) Do you understand what they **"need" and "want"** from this project?	✓
3) Do you understand their **"desired outcomes" and "perceived benefits"** from undertaking this project?	✓
4) Do you understand their **requirements**?	✓
5) Do you understand the project's **deliverables** and **milestones**? ... *IP Rights.*	✓
6) Do you understand the **scope** of this project and its **measurable & tangible objectives**?	✓
7) Do you understand the **scale** of this project with respect to its duration, resourcing levels, and costs?	✓
8) Do you understand the **constraints imposed** on this project?	✓
9) Do you understand the **risks involved** with undertaking this project?	✓
10) Do you understand **why it makes sense** for the performing organization to undertake this project, and what will be their **return on investment** for undertaking this project?	✓
11) Do you believe this project is an **appropriate choice** for the performing organization given its; circumstances, alternatives, capabilities & capacities, processes & procedures, history, and culture?	✓
12) Do you believe that this project has a **chance of succeeding**?	✓
13) Do you have a documented list of **Customer Requirements**?	✓
14) Do you have a **signed-off Project Charter**?	✓
15) Do you know **who the project manager** will be?	✓

Example Case: The Project Initiated To Certain Failure.

Unfortunately, at some point in your project management career, you possibly will be given or will be involved with what colloquially is described as a *"hospital pass"* project, or a *"shit sandwich"* project. That is, the agreed project constraints of [Scope] – [Time] – [Cost] – [Quality] are set in stone, but these were incorrectly established during the Initiating Phase, and hence if the project was to proceed as these project constraints currently are then the project is destine to end in abject failure. … And, there is almost nothing that you as the project manager and the Project Implementation Team can realistically do during the Executing Phase of the project's life cycle, to prevent this failure from occurring (irrespective of how much effort you all put in).

This project was *"born-to-fail"* because it was ill conceived during the Initiation Phase due to; misunderstandings & misinterpretations of the customer's needs – wants – requirements – expectations, and/or the [Scope] – [Time] – [Cost] – [Quality] project constraints do not align with what is realistically achievable given the project variables and the other project constraints, and/or project constraints were wrongly evaluated.

> *For example; during the contract negotiations the sales group agreed with the customer to a certain fixed price. However, when the implementation group planned the project, it was then apparent that there was a significant difference between their calculated Budget and the much smaller Sales Price.*

Thus, it is extremely important that, during the Planning Phase the project manager diligently determine (with the assistance of the Project Implementation Team) what is realistically achievable, what are realistic values for the project's constraints, and when misalignments are found then these need to be highlighted to the performing organization's senior management to resolve with the customer's representatives, asap.

Consider the following situation where, during the contract negotiation a customer was given a fixed price [Cost] of $800,000 to deliver a product with specific functionality.

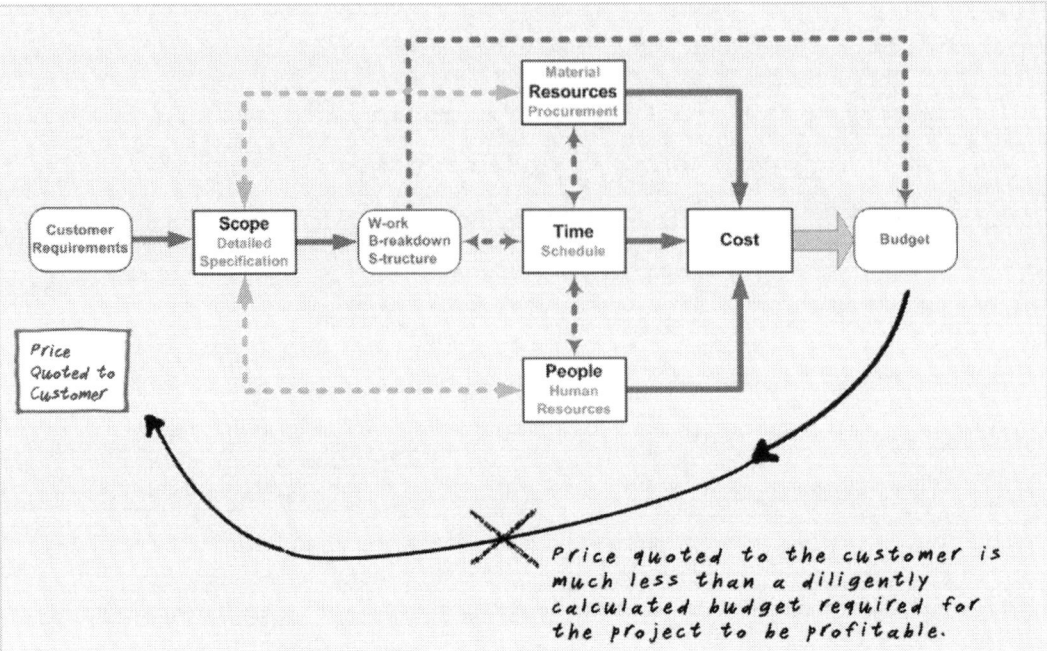

Price quoted to the customer is much less than a diligently calculated budget required for the project to be profitable.

However, during the Planning Phase; when implementation of the agreed Detailed Specifications [Scope] was broken down into a Work Breakdown Structure and all of the project's involved tasks were [Time] scheduled based on those [People] & [Resources] that would be utilized & available, then the project manager discovered that the project would [Cost] something closer to $1,300,000 to deliver, i.e. a 62.5% difference in [Cost].

Firstly, the project manager needs to diligently determine why there is such a significant difference in [Costs]. Was this difference due to; a whole heap of activities not being included when the project was conceived (*e.g. bespoke hardware drivers will have to be developed for the hardware platform which is different to that used previously*), were activities underestimated in complexity and duration (*e.g. packaging each unit will take 30 minutes and not 10 minutes as thought, so for 200 units it will take 100 hours = $15k and not 34 hours = $5k*), were 3rd party activities and costings forgotten or underestimated in complexity and price (*e.g. environmental testing by an authorized & statutory recognised test-house will cost $150k and not $15k as guesstimated*).

Secondly, the project manager needs to speak up, before the Executing Phase starts.

War Story … writing cheques your body can't cash.

It is unbelievable, the number of times during my career that I have either seen or been the new project manager who was "dropped into a shitfight"; all because, the project's Initiating Phase was evidently conceived, contracts negotiated, and strategic plans made "on the back of a few used beer coasters during an executive luncheon".

It was all nice and friendly during the job interview, with a closing comment of "the first project should be a milk run for you".

However, after the pleasantries of the first few days, once you got the lay-of-the-land and your first independent project status report to the customer was due; the realization that, this project was in fact a disaster zone and it was going to be excruciatingly difficult to extract the project out of this mess.

The individual members of the project team are demoralized with working such long hours just to keep this project afloat. Some members of the project team are openly talking about "hating this job" while others are mumbling about resigning, team cohesion is non-existent, there must be a roster in effect for sick leave, and their bitterness towards senior management and especially towards the Executives (who signed-off on the project's approval) is vitriol.

The Head Of Development is at loggerheads with the sales & marketing departments.

The executive management are frustrated at the lack of progress, as they feel the effects of this project vampire sucking the life out of the organization.

How could this project have ended up in such a predicament?

Unfortunately, you can't ask the project's previous manager as they have moved on after having "pushed this project uphill for so long" … according to the rumours and hearsay.

As for, the sales person who signed off on the contract for this project? Well, after they collected their commission for this vapour-ware, they either got promoted into an untouchable position of authority or received a golden-handshake while heading out the door.

And there is the project manager, left completely exposed while holding this ticking time-bomb of a disaster, having been made entirely responsible for the project's outcome.

Reviewer's Comment ... Yep, I have recently seen that one.

This nice guy was hired as a project manager and had the misfortune to be assigned this "completely screwed" project. By the end of the 2nd week, the General Manager was on this guy's case due to the poor state of the project. But here's the kicker, the General Manager had for the last few months, been the acting project manager for that particular project.

DON'T ALLOW ONE PART OF THE ORGANIZATION TO MAKE COMMITMENTS THAT OTHER PARTS OF THE ORGANIZATION CANNOT REALISTICALLY DELIVER.

War Story ... Cutting one's own throat.

My other favourite is when, they undercut the profitability margin to unsustainable levels (i.e. none existent, a.k.a. loss maker), in the hope of winning over all of that customer's future business. However, what they have effectively done is establish a price precedent expectation for that customer.

Hence, when the subsequent work eventuates and the customer is quoted the realistic profit margin price (not a greedy price), then the customer gets argumentative and strongly negotiates for a price at the previous loss maker levels.

Alternately, the follow on business does not eventuate, or does not eventuate at the quantities that the original loss maker strategy was based on. Thus, the company is left in the RED, or even DEAD.

HISTORY HAS AN UNCANNY PROPENSITY TO REPEAT ITSELF.

WHILE THE INITIATING PHASE IS THE LAST OPPORTUNITY TO TERMINATE THE PROJECT WITHOUT INCURRING SIGNIFICANT FINANCIAL PENALTIES AND REPUTATIONAL DAMAGE ...

THE PLANNING PHASE IS THE MOST OPPORTUNE MOMENT TO CORRECT ANY OVERSIGHTS WITH THE INITIATING PHASE AND LAY THE FOUNDATIONS FOR THE PROJECT'S SUCCESS.

WITH SIGNED-OFF
PROJECT CHARTER
IN HAND, AND A
DRAFT DETAILED
SPECIFICATIONS THAT
I UNDERSTAND,
I AM NOW OFF
TO DISCOVER THE
INTRICACIES OF
THE PROMISED LAND

4. PLANNING Phase

4.1. Overview

As illustrated below in [Figure 33], the Planning Phase is concerned with the darker shaded area of the Project Management Process model.

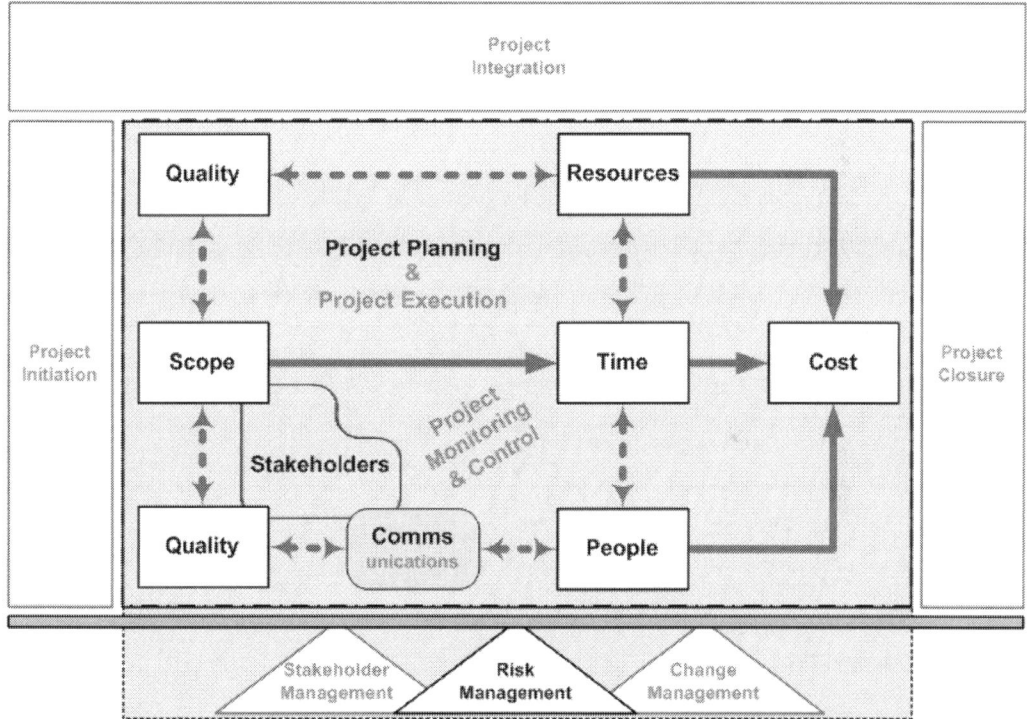

Figure 33: The Planning Phase of the Project Management Process model.

NOTICE how in [Figure 33] the Planning Phase also involves Risk Management as well as Stakeholder Management and Change Management.

⊠ **The Planning Phase is the most opportune moment to correct any oversights with the Initiating Phase and lay the foundations for the project's success.**

4.2. Purpose

The purpose of the Planning Phase is to juggle all of the possibilities, prioritizing the probables, recording them for posterity, and thereby;

1) **Refining** the project's objectives, requirements, and scope boundary.

2) **Defining** exactly what has to be done, and how it will all be coordinated.

3) **Communicating** the plans.

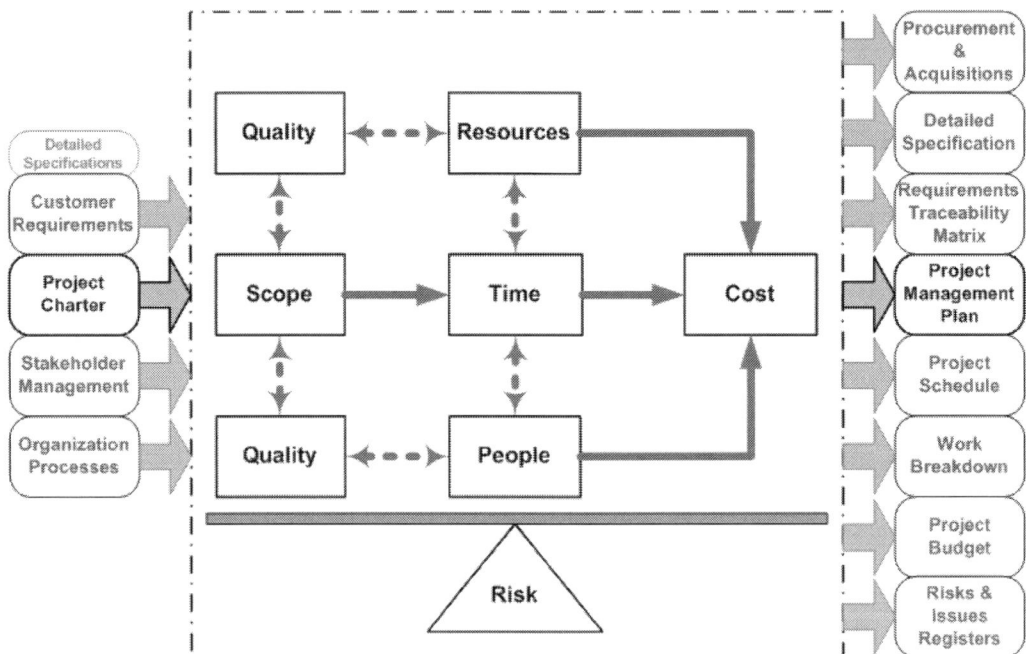

Figure 34: The Planning Phase as a process model of project variables/constraints and documentation inputs & outputs.

NOTICE that the representation of the Planning Phase depicted above in [Figure 34] has been modified from [Figure 26], which modelled the relationship between PMBOK® and the project variables & project constraints.

4.3.   The Big Questions

Given that the project manager may not have been directly involved with the Initiating Phase; consequently, **the Planning Phase** becomes the **most opportune moment** for the project manager **to correct any oversights with the Initiating Phase** and to **lay the foundations for** the **success**ful execution of the project.

⊠ **Insufficient planning is a project killer, by derailing the project's chances of ever being successfully executed.**

For the Planning Phase to contribute to the project's success, then the performing organization has to answer the following questions honestly and realistically:

1. **WE NEED WHAT WHEN** and **HOW MUCH**?

2. How will **WE KNOW WE GOT IT RIGHT**?

3. What do **WE THINK CAN GO WRONG**?

PLANNING IS A DOCUMENTED THOUGHT EXERCISE TO WALK THROUGH THE IMPLEMENTATION OF THE ENTIRE PROJECT FROM START TO FINISH.

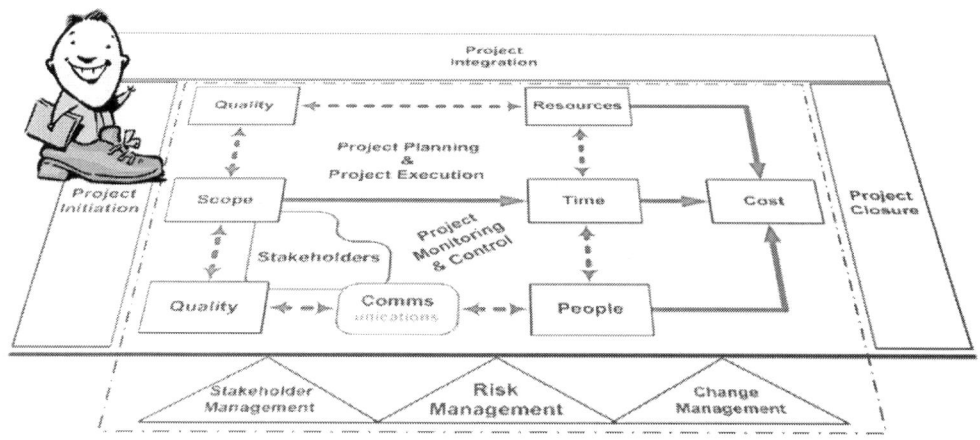

4.3.1. Question: "WE NEED – WHAT – when and how much?"

What is needed?

The answer to this sub-question is concerned with **defining the project's [Scope]**, and this is built upon the response given to the Initiating Phase's question of *"what do they want"*; i.e. those agreed to things listed in the Customer Requirements.

While the **Initiating Phase's answer** to this particular question may have been **very general**, the **Planning Phase's answer** has to be **a lot more specific with its details**.

4.3.1.1. Need: Detailed Specifications

To answer this "need WHAT" question the following activities have to be undertaken:

1. **Accumulate** all of the **Requirements** from the customer and from within the performing organization (taking into consideration their own strategic interests).

2. Interact with the project's stakeholders and **involve Subject Matter Experts (SME)** to enable the **expansion of the detailed descriptions** of these requirements.

3. Examine the business's **organizational processes**; i.e. procedures, policies, structures ... the way they go about doing business.
   Would this be compatible with the underlying basis of the project?

4. Define the **deliverables** and state the **Acceptance Criteria** for these deliverables.

5. Refine the understanding of what the project **stakeholders** are **expect**ing.

6. Record the resultant information in a version-controlled **Detailed Specifications**; *also known as a Functional Specifications, or Scope Statement*.

 Think of the **Customer Requirements** as a "wish list" and the **Detailed Specifications** as a "what will be **delivered** list".

NOTE ☐ The **Detailed Specifications** needs to be one of the first **deliverable milestones** that must be **signed-off** (by both the representatives of the customer and the performing organizations) **as an agreement on what functionality & features are to be delivered**; i.e. **Detailed Specifications is the [Scope Baseline].**

The **creation of the Detailed Specifications** should be thought of as a **mini iterative project**, as illustrated below in [Figure 35]. *... Were any new risks identified?*

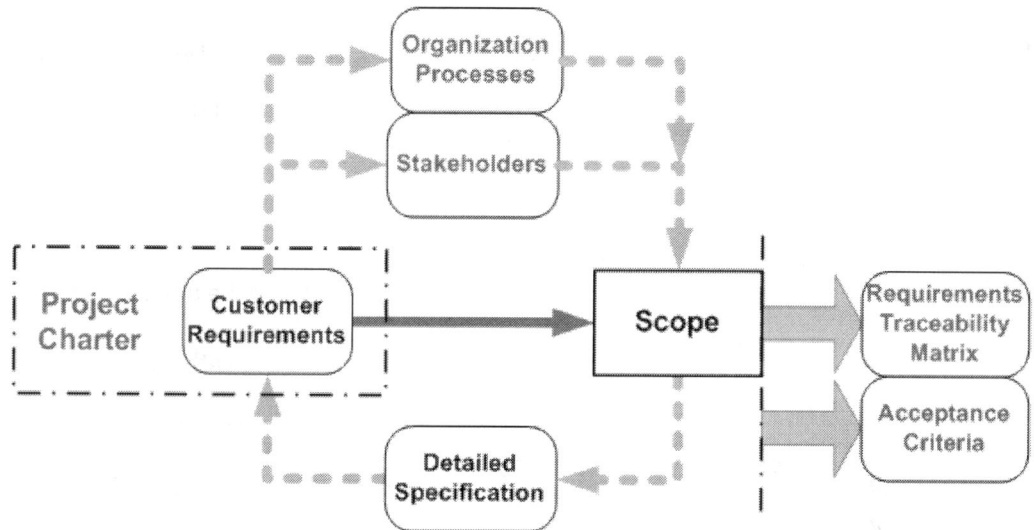

Figure 35: Process to refine the project's [Scope] and the Detailed Specifications.

 **The Detailed Specifications is about the [Scope] of the requirements and NOT about the solution.** Hence, it **DOES NOT contain the design.**

Rather, the Detailed Specification should contain the information upon which the design and architecture are built.

This non-inclusion is because the technical solution may have to change/evolve to achieve the agreed requirements, but when the technical solution is included in the Detailed Specifications then this could potentially be interpreted as a locked-in deliverable. In hindsight, this locked-in technical solution may be found to be inappropriate or detrimental to a successful outcome of the project.

The Detailed Specifications should contain sufficient explanation information to enable all of the project's stakeholders to understand what is and is not going to be delivered.

> *For example include appendices for such things as; overview of the proposed architecture, use-cases, mock-up of the user-interfaces, the definition for interfaces with other systems / applications, references to technical standards.*

Where appropriate also include comments on; points that have to be clarified, questions about potentially missing functionality, and highlight issues that have to be resolved prior to the sign-off / approval of this Detailed Specifications.

NOTE ☐ This **Detailed Specifications will evolve during the project's life** as features & functionality are added, removed, and/or expanded upon.

☐ The approved **Detailed Specifications has to be "frozen" for each milestone release and subsequently a new edition** of the Detailed Specifications **produced for each proceeding release if there are any agreed changes in [Scope]**.

"WALKING ON WATER AND DEVELOPING SOFTWARE FROM A SPECIFICATION ARE EASY IF BOTH ARE FROZEN."

EDWARD V. BERARD

4.3.1.2. Need: Scope Coverage

Now that all of the "requirements" (i.e. features & functionality) have been determined, these can be listed in the **Requirements Traceability Matrix (RTM)**; i.e. a spreadsheet **itemizing each individual Customer Requirement against where it appears in the approved Detailed Specifications.** *... Were any new risks identified?*

Thus, the **Requirements Traceability Matrix depicts the "Scope Coverage"** of the Customer Requirements. For more information on the RTM, refer to [Section 12.1.1].

4.3.1.3. Need: Scope Boundary

Any feature or functionality that is not included in the signed-off / **approved Detailed Specifications [Scope Baseline], or any feature or functionality not as described in this approved Detailed Specifications will be deemed as outside the project's [Scope]; i.e. outside the "Scope Boundary".**   *... were any new risks identified?*

 I would recommend that ... an **'Out Of Scope' section** be included **in the Detailed Specifications** for outlining those contentious points.

This would also exclude from the scope boundary, any subsequent after-signature additions to the current release irrespective of whether these changes are necessary for the successful delivery of that release. **To have any additional changes included (or existing features & functionality removed) would constitute a Change Request**, and thereby result in the possible renegotiation of the contract price and potentially extensions to the delivery date or some other form of compensation to the affected stakeholders.

One thing for sure, it is an **absolute certainty that the [Scope] is likely to change during the life of the project**. Hence, the project will **need to implement some form of "Scope Change Control",** as described later on in [Section 10.2].

Scope Creep

"Scope Creep" is where additional features & functionality is snuck into the Detailed Specifications, and/or **is added to the project's implementation work without prior agreement** on what compensation is appropriate for it to be included.

Scope change must be promptly monitored & controlled; see [Section 7.1].

⊠ Scope Creep is a project killer.

Scope Creep will definitely result in the death of a project by causing additional work to be undertaken. This additional work will cause the project's duration to increase.
The increased duration will result in missed delivery milestones and increased costs of implementation. These increased costs will erode away the profit margin for undertaking the project. With the project, being delivered late and over budget then this will result in disgruntled stakeholders from both the performing and customer organizations.

Conversely, watch out for **"Scope Shrinkage" where features & functionality are forgotten or misplaced.**　　*… As this will also upset the project's stakeholders.*

4.3.1.4. Need: Work Breakdown

Now that the project's [Scope] has been defined then the next step is to break this [Scope] into distinct manageable "**Work Packages**"; i.e. **decomposing the project's scope into separate deliverables and work activities that have to be implemented**.

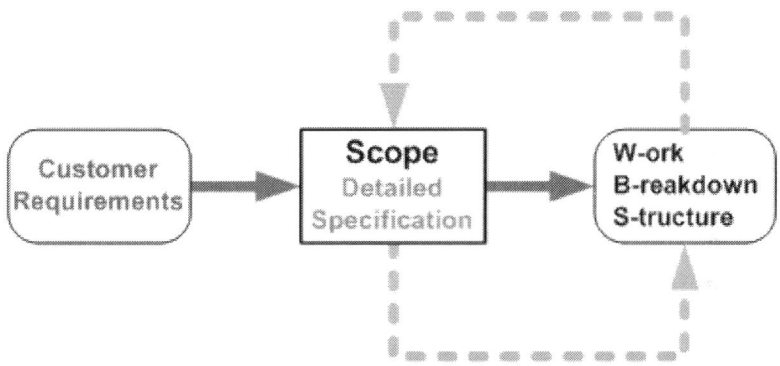

Figure 36: Process to break the project's [Scope] into work packages.

One of the easiest ways to tackle this work breakdown is to take the approved Detailed Specifications, consider each major point (i.e. features or functionality) as a cardboard box, and then imagine throwing each one of these boxes into the corner of a big room. Once all of the boxes are piled up in the corner then start stacking these boxes into topic-related columns. *... Were any new risks identified?*

> *For example; consider the parent fantasy world of 'No-Television-Land'.*
> *The project is to produce a device that will receive differentiated signals transmitted from independent base stations (i.e. channels).*
> *Depending on which channel was selected by the user, the device will then display moving images and play audio sound.*

> **Rolling Wave Planning** – uses the planning techniques described in this chapter except that the planning is done iteratively in waves, where those activities to be performed in the immediate future are planned out now in detail, and those activities to occur in the not so immediate future are only broadly planned for, and these will be planned out in more detail later on. This is **very similar to what happens with Agile – Scrum.**

I will admit that I don't know much about the architecture of a television, other than what I could find via an internet search. However, if the example's sentences were a summation of the approved Detailed Specifications for the television project then the boxes I would be throwing in the corner would be as follows.

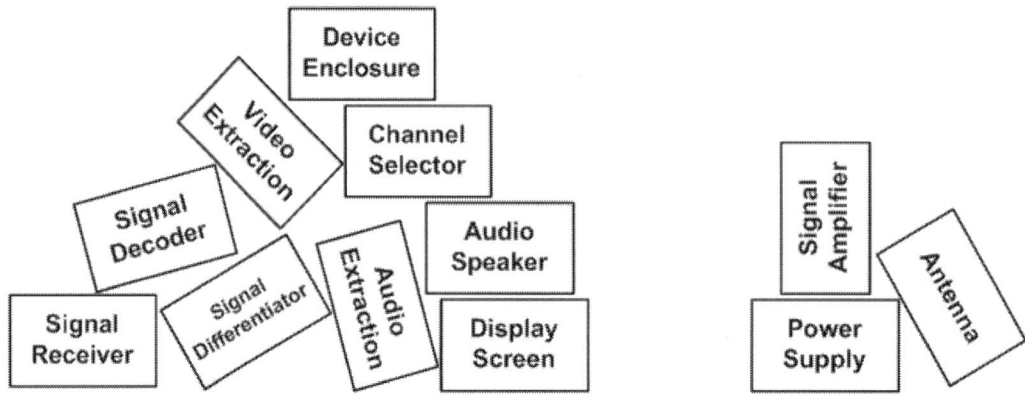

Figure 37: Example construction of a Work Breakdown Structure for a television.

Now it is a given that a television expert would be able to think of a whole lot more boxes that should be added to this diagram. Hence, **the creation of the Work Breakdown Structure (WBS) is the perfect opportunity to engage subject matter experts to advise on this particular field of endeavour**. *... Were any new risks identified?*

With all of the boxes piled up, the next step would be to organize these into hierarchical columns of related boxes, as illustrated on the following page in [Figure 38].

Given that this work breakdown example is **composed of physical componentry** then it could also be referred to as a **Product Breakdown Structure (PBS)**, i.e. a **Bill Of Materials**.

LITTLE BOXES,
LITTLE BOXES,
LITTLE BOXES
PILED EVERYWHERE...
THESE LITTLE BOXES
REQUIRE SORTING
BY SOMEONE
WHO DARES.

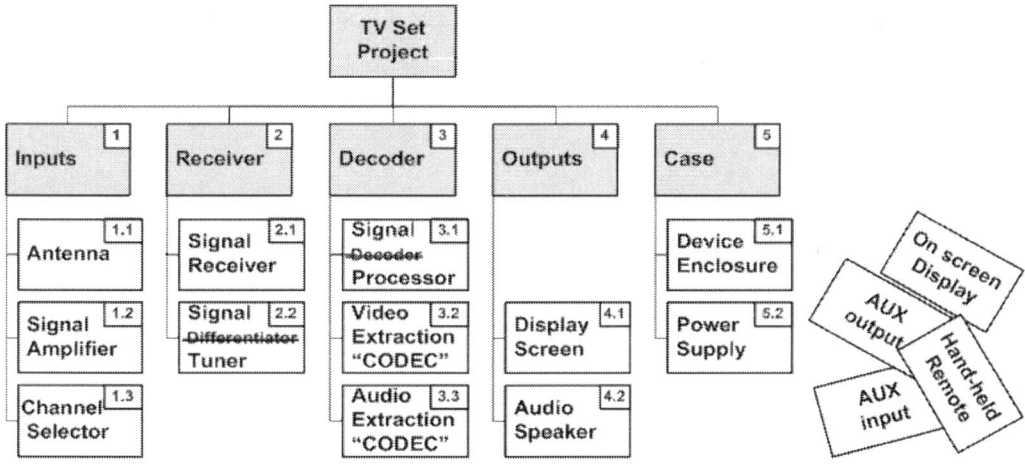

Figure 38: Example refinement of the Work Breakdown Structure for a television.

The trick with the column-izing of these boxes is to use an appropriate scheme.

For example; by the phases of the project, the major deliverables, the people groups that will undertake the work, or by the architectural layout.

You could take the labelling of these boxes a step further, give each one a unique identifying number based on which column it belongs to, and thereby give related boxes similar ID numbers. This activity identifier (i.e. **WBS Code / Code Of Accounts Identifier**) could be coupled with a detailed description of the individual work package and thereby compiled into a **WBS Dictionary**. Where each entry is akin to a mini project charter for that work package; including such information as a description of the work, [Cost] estimates, [Scope] requirements & technical references, acceptance criteria & [Quality] standards, [People] & [Resource] needs, assumptions & constraints, [Time] duration & milestones. *And, the Risks associated with each particular package of work.*

In [Figure 38], there were a few more boxes (i.e. **additional features**) found that weren't mentioned in the example's Detailed Specifications. However, to go ahead and **include these would constitute scope creep**. Hence, it is better to note these down and discuss them with the project's stakeholders as to whether these should be amended to the current set of deliverables with [Time] & [Cost] adjusted accordingly and compensated for, or whether these extras should be included at a later date in some future release.

4.3.1.5.   Need: Understand the Inter-Dependencies

With the project's **Work Breakdown Structure (WBS)** having been constructed the next step would be to start thinking about the **inter-dependencies and the sequencing of these boxes**; i.e. **activity dependencies**.

A **Network Diagram / Precedence Diagram** as pictured below in [Figure 39] would be used to **illustrate these relationships**.

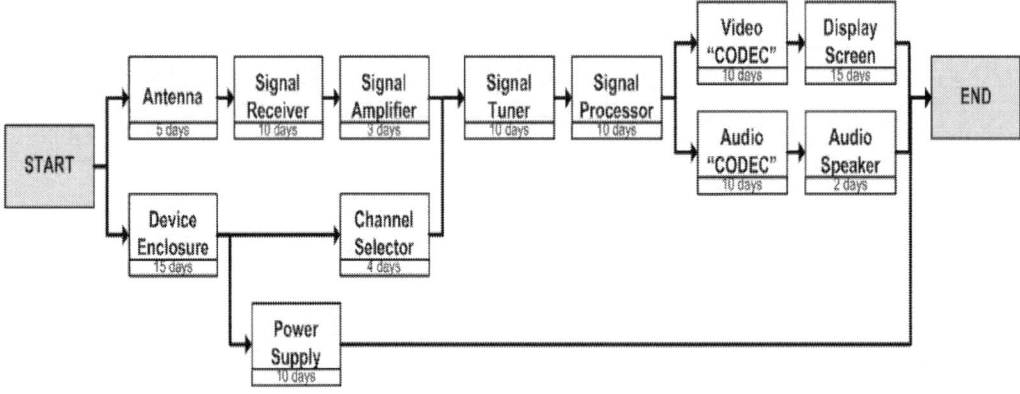

Figure 39: Example of a Network Diagram for a television development.

You could also take this diagram a step further and give each box its own duration value on how long it would take to complete that individual activity; i.e. 'task'.

I personally think of a "task" as an "activity" with an assigned duration, and very soon to be assigned a monetary value.

NOTE   ☐ Project scheduling tools such as MS-Project can easily produce a Network Diagram. Thus, the manual creation of Network Diagrams is often bypassed, instead going straight from the Work Breakdown Structure to the Project Schedule. For this book, Network Diagrams / Precedence Diagrams / *Activity On Node Diagrams* while having been acknowledged don't receive extensive analysis or use.

Using these duration values, you could then traverse each path from left-to-right and sum-up the total task durations to determine the "critical path". Where, **the Critical Path is the sequence of linked tasks that results in the longest duration to complete the project,** and if a task on this path takes longer than planned then the project's duration will increase.

This technique is known as CPM — Critical Path Management.

Looking back at [Figure 39], a task cannot begin until all of its preceding tasks have been completed. If there are two or more paths coming into one task-box then that task cannot start until all of its preceding paths are completed.

> *For example; with the television you cannot start working on the "Signal Tuner" until both the "Signal Amplifier" and the "Channel Selector" are finished.*

Reviewer's Comment ... Diving straight into scheduling.

You would not believe how many times I have seen project managers jump straight from the Detailed Specifications into producing the project's schedule, and completely forget about producing the most simple of WBS. Yet, how often have these same project managers overlooked some feature or functionality in their project's schedule. Personally, I throw together a quick WBS at the same time as producing the draft of the project schedule. This way, these complement each other's development and I am a lot less likely to forget something.

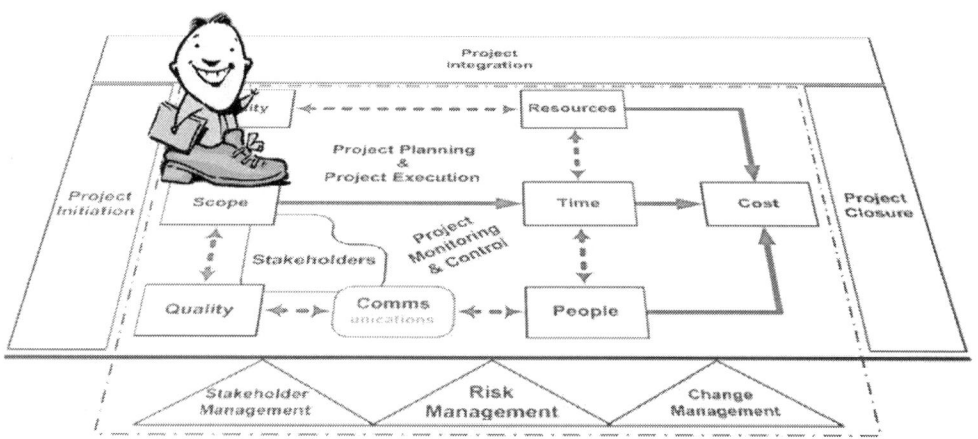

4.3.1.6.   Need: Project Schedule ... 1st Pass

Now that the Work Breakdown Structure (WBS) ~~and network diagram~~ has been completed, the next step is to produce the project's "1st Pass" schedule (**Gantt Chart**); as illustrated below in [Figure 40].

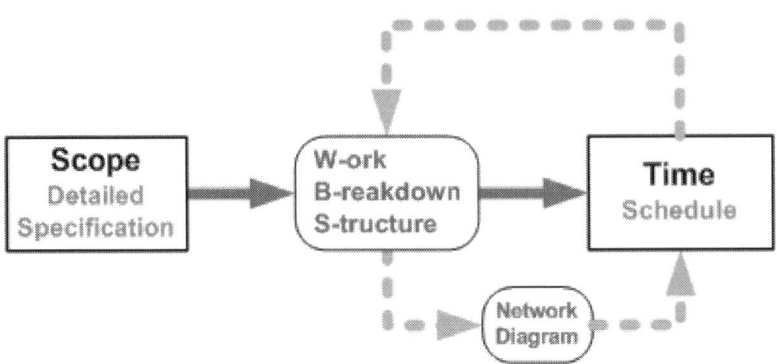

Figure 40: The Planning Process Model to create the "1st Pass" Project Schedule.

1st Pass Schedule Creation Rules

The rules for producing this 1st Pass schedule are as follows:

1)   Ignore whether this project has a specified Start-Date or a stipulated End -Date as in [Section 2.1], instead **make the project's Start-Date as today.**

2)   Ignore the exact number of people that are available for the project, instead imagine that there is **only one "good-worker" who will do the entire project by themselves**; think of that worker as **an "Expert You"**. However, different workers will need to be interviewed to obtain their specialist opinions on what activities the good-worker has to do to complete that relevant section of the work breakdown. Did this "Expert You" discover additional activities not currently thought of?   *... were new risks identified?*

3)   Ignore the fact that some activities can be done in parallel because for this 1st Pass schedule there hypothetically is only one good-worker who is only able to concentrate on one thing at a time, so **all activities have to be completed sequentially.**

(Optionally) Next to each task in the 1st Pass schedule include its WBS Code for later use when determining the project's budget; i.e. that unique activity identifier code attached to that task's box in the WBS diagram shown in [Figure 38].

Once this 1st Pass schedule is completed then incorporate any missing activities back into the Work Breakdown Structure and where necessary re-organize the WBS to determine if anymore activities have been overlooked.

That is, **progressively elaborate the WBS and the 1st Pass schedule until confidence is high that no project activities remain undiscovered**. *... were any more risks identified?*

NOTE   ❑ **For the 1st Pass schedule, we are only concerned with finding all of the activities and NOT with the schedule's duration, milestones, end date, nor with the implementation method to be used.**

     ❑ This 1st Pass Project Schedule will never be publically seen by the customer let alone by the performing organization's senior management.

Though, a Gantt Chart schedule is not always what the primary-core stakeholders want to be presented with, but at this stage of planning the Gantt Chart schedule is a rather useful tool for understanding the complexities of the project and more specifically the [Scope] to [Time] relationship.

> I would recommend that ... it be determined how and in what format the project's primary-core stakeholders expect to have presented the information related to the project's duration and milestone dates.

TO SUCCEED, ONE MUST FIRSTLY
FIGURE OUT WHERE & HOW TO PROCEED
WITH WHO & WITH WHAT IN HAND.

4.3.2.   Question: "WE NEED what – WHEN – and how much?"

When is it needed?

The answer to this sub-question is concerned with; determining the **sequence of each [Scope] activity**, estimating the **[Time] duration of each activity**, estimating the **[Resources] required** for each activity, estimating the **[People] required** for each activity, and developing the "2nd Pass" schedule into a form that is presentable to the stakeholders.

This sub-question's answer is built upon the 1st Pass schedule developed during the *"what"* part of this question; see [Figure 41] below.   *... Were any new risks identified?*

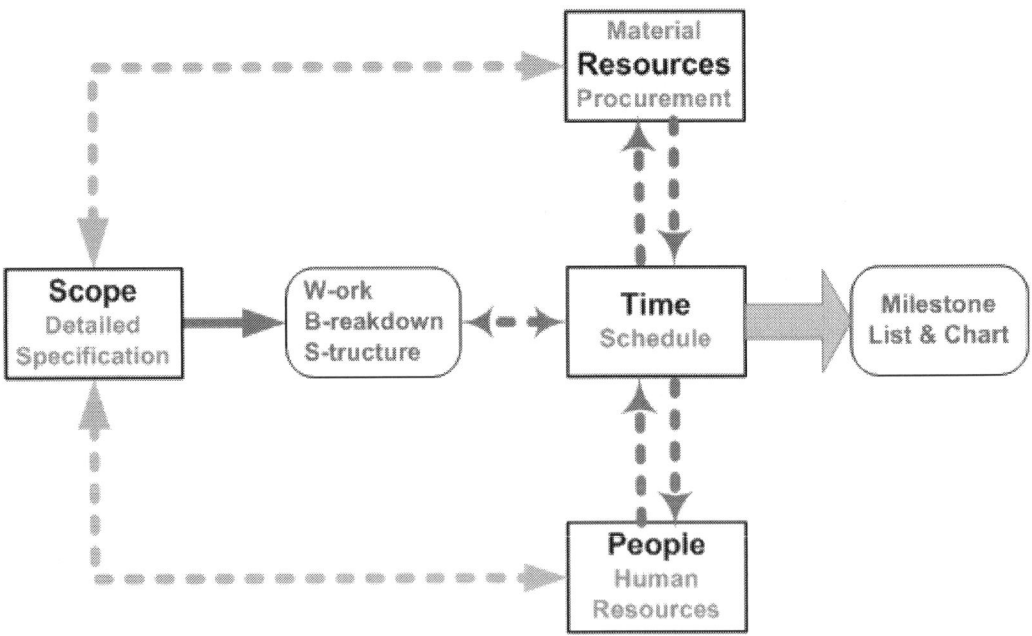

Figure 41: The Planning Process Model to create the "2nd Pass" Project Schedule.

NOTICE how in [Figure 41] the process of estimating the requirements for [Resources] and [People] will have a bi-directional "progressive elaboration" effect on the [Time] estimates, as well as a bi-directional relationship on the project's [Scope].

[Resources] and [People] availability restraint on [Scope]

Prior to establishing the 2nd Pass schedule, consideration must be given to the bi-directional relationship between the project's [Scope] and those [Resources] and [People] that are potentially available to the performing organization.

While the project's [Scope] will mean that certain [Resources] and [People] will be required to produce the project's deliverables, **the [Resources] and/or [People] that are available on the market accessible to the performing organization (as well as to the customer organization) will influence the [Scope] that can be effectively implemented**.

> *For example; if the Customer Requirements were for a new aircraft to be able to travel at speeds of "reduculum^2", then without access to acceptable quantities of high grade "Unobtainium" metal then this requirement is not realistically possible let alone feasible.*

Therefore, **the potential availability of appropriate [Resources] must be taken into consideration when specifying the details of the project's [Scope]**.

Similarly, **the potential availability of the necessary [People] must be taken into consideration when specifying the details of the project's [Scope]**.

> *For example; with the Customer Requirements for the new aircraft to be able to travel at speeds of "reduculum^2", then this would necessitate the hiring of temporal-plasma-fluid-aero dynamists to participate in this project. ... And, currently there ain't too many (if any) of those persons available on the market.*

Hence, the reliable access and timely availability of specific [People] and [Resources] will influence what [Scope] can realistically be achieved.

 I would recommend that ... the project's agreed [Scope] be updated and signed-off to reflect the real-world commercial & practical realities of the availability of the necessary [People] & [Resources], and not allow the project to proceed into a "pipe-dream" disaster.

4.3.2.1. Need: Project Schedule ... 2nd Pass

The 2nd Pass schedule is concerned with;

1) **Sequential and parallel tasks**; i.e. identifying those sequences of tasks that are co-dependent, and identifying those sequences of tasks that are not inter-dependent and could possibly be undertaken simultaneously.

   For example; the coding of the application software that runs on the device is in all probability completely independent of the design of the packaging that with be used to transport the device to the customer. Therefore, these two tasks could be undertaken in parallel.

2) The **estimated [Time] duration** (i.e. **hours of effort**) required to undertake **each task**.

3) **How many [People]** will be available to undertake **each task**?

4) **What skill sets & skill levels** will be required to undertake **each task**?

5) **What [Resources]** will be required to undertake **each task**?

6) Is the project a **Start-Date or End-Date type of project**; i.e. the project must commence by a certain start date or the project must be finished by a specific end date.

When determining those tasks that have to be done sequentially and those tasks that can be done in parallel, there are a few deciding factors to consider:

1) The **logical sequence of construction**; *e.g. Part-C is built on Part-B which is built on Part-A, therefore A is followed by B which is followed by C.*

2) The **availability of the person to undertake the task**; i.e. one person is not able to do multiple tasks at the same time with much proficiency.

3) The **availability of the resource needed to undertake the task**; i.e. a task requiring a specific resource can only take place when that resource is available.

4) The **mandatory order of deliverables** as stipulated in the contract.

When determining the duration of tasks, then need to ask knowledgeable persons for their estimates on the duration / effort, and the [Resources] required to complete those tasks.

✘ **DO NOT accept a long duration / effort estimate for a single task** because when this long duration / effort task is broken down the resultant child-tasks often increase or decrease the task's length to something closer to what will really be required for its implementation. ... *Most people are prone to giving overly optimistic estimates.*

> *For example; if to develop Module Zeta is said to take a month, then break the development of Module Zeta down into 4x one week tasks (e.g. design – develop – test – integrate) and obtain estimates for each of these sub-tasks.*

✘ **DO NOT accept too small a duration / effort estimate for a single task** because too fine a granularity will result in micro managing the [People] and an overly bulky and complex schedule. Instead, **focus on tasks that are apparent steps towards a deliverable** and not just a minor step along the way.

✓ **Set a maximum limit on the estimated duration / effort for any individual task** (*e.g. 80 hours*) **anything quoted as longer than this duration has to be broken down into shorter duration sub-tasks.** Though, if the project is of a relatively short duration say one month then the maximum estimation limit per task maybe set as a couple of days, hence any task longer than that has to be subdivided into smaller tasks.

... were any new risks identified?

ESTIMATION OF TASK DURATION... WHEN SIZE REALLY DOES MATTER.

JUST RIGHT

TOO SMALL

TOO BIG

 **Obtain as accurate as possible estimates for task duration / effort** [Time] and NOT "ball-park" guesstimates because the schedule's [Time] duration / effort directly affects the [Cost] calculations.

✓ **Use 'weighted average' duration / effort** where you obtain from the project team members for each of their tasks an estimate of; (t_o) **optimistic** ... best case, (t_m) **most likely** ... normal case, and (t_p) **pessimistic** ... worst case.

$$Weighted\ Average = \frac{(t_o + 4t_m + t_p)}{6}$$

This technique provides more realistic estimates, i.e. bit more pessimistic. Have a wiki at, PERT – Project (Program) Evaluation Review Technique.

✓ **Set realistic work durations per workday per person**, because not every project team member is going to be able to do a full 7 to 8+ hours worked per day on his or her assigned tasks. While they may do a full day's work, there will be meetings to attend, workmates to assist, other competing projects requiring attention, administrative activities... and short notice leaves of absence (i.e. sick leave, single days annual leave).

Therefore, **plan for** each project team member **to be able to operate at a maximum of ~80% utilization per day on his or her assigned tasks** (i.e. not 100%).

Though, the percentage used in this calculation will be influenced by the work place environment, the industry, the business's culture, the individual's other work allotments, and the individual's work ethic.

ESTIMATION OF TASK DURATION DIRECTLY AFFECTS MY BUDGET ALLOCATION

 Ensure that the project manager's time is accounted for; i.e. **does the project manager's time appear in the schedule as a distinct task** that is charged to the project's Cost Account Code.

When scheduling the project include a task for the project manager such as "manage project" (that is a 'Start-to-Start' with the project's first task and 'Start-to-Finish' with the project's last task) though not at 100% availability but rather that proportion of the typical day that would normally be spent working on the project, *e.g. 15% of total hrs*.

As presented in the next section, the project's [Costs] are derived from the tasks in the schedule; so **if no specific tasks are assigned to the project manager then that project manager will not be included in the project's estimated costing and therefore not budgeted for**.

While on the topic of things to include, **remember to update the project's calendar to reflect those days when project tasks may or may not be worked on**. Also include such things as; **public holidays**, team-members' **approved leave**, team-members' work status (*e.g. **full time or part-time***), and when specific **resources will and will not be available** (*e.g. assigned to other projects, or when their services expire*).

HE HAD THE MOST DETAILED OF PLANS ... BUT
FORGOT TO INCLUDE THE HOLIDAY BREAKS.

Now that the 2nd Pass schedule [Figure 43] is completed, then the following can be produced; the **Milestones List, Milestone Schedule & Summation Schedule** [Figure 44], and the **Overview Schedule** [Figure 42].   *… were any new risks identified?*

This Summation Schedule would provide an overview to the project's implementation over the coming time periods; i.e. used to visually summarize to the project's stakeholders the relationship between the project's activities and the project's calendar.

In the scheduling software application, **these milestones would be tasks assigned to zero day durations and used to indicate the completion of significant events in the project's life**, such as the product delivery to the customer.

NOTE   ☐ **Choose milestones that correspond to those events of notable interest to the project's primary-core stakeholders**; *e.g. when each stage will finish.*

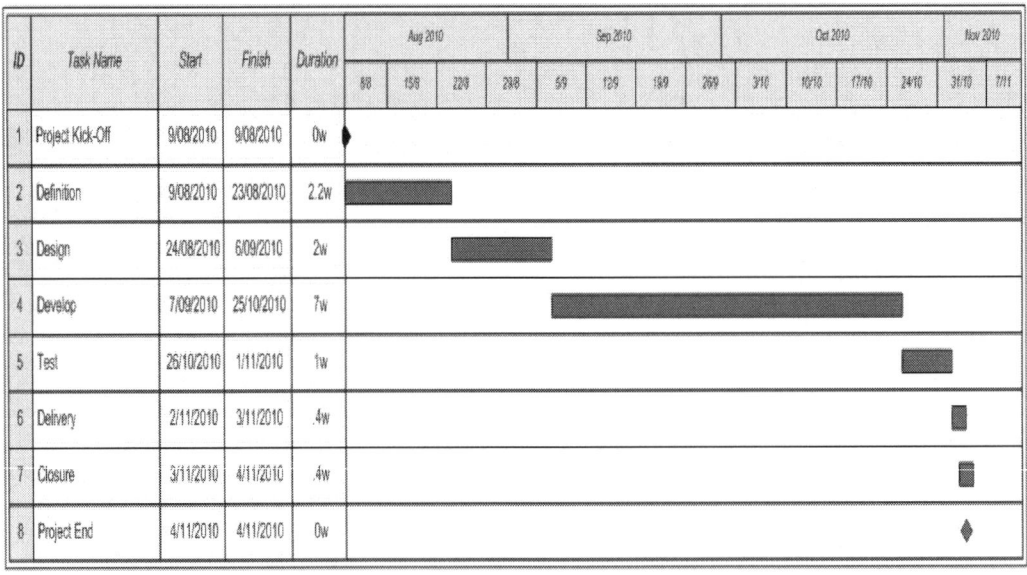

Figure 42: This Overview Schedule was produced with MS-Visio.

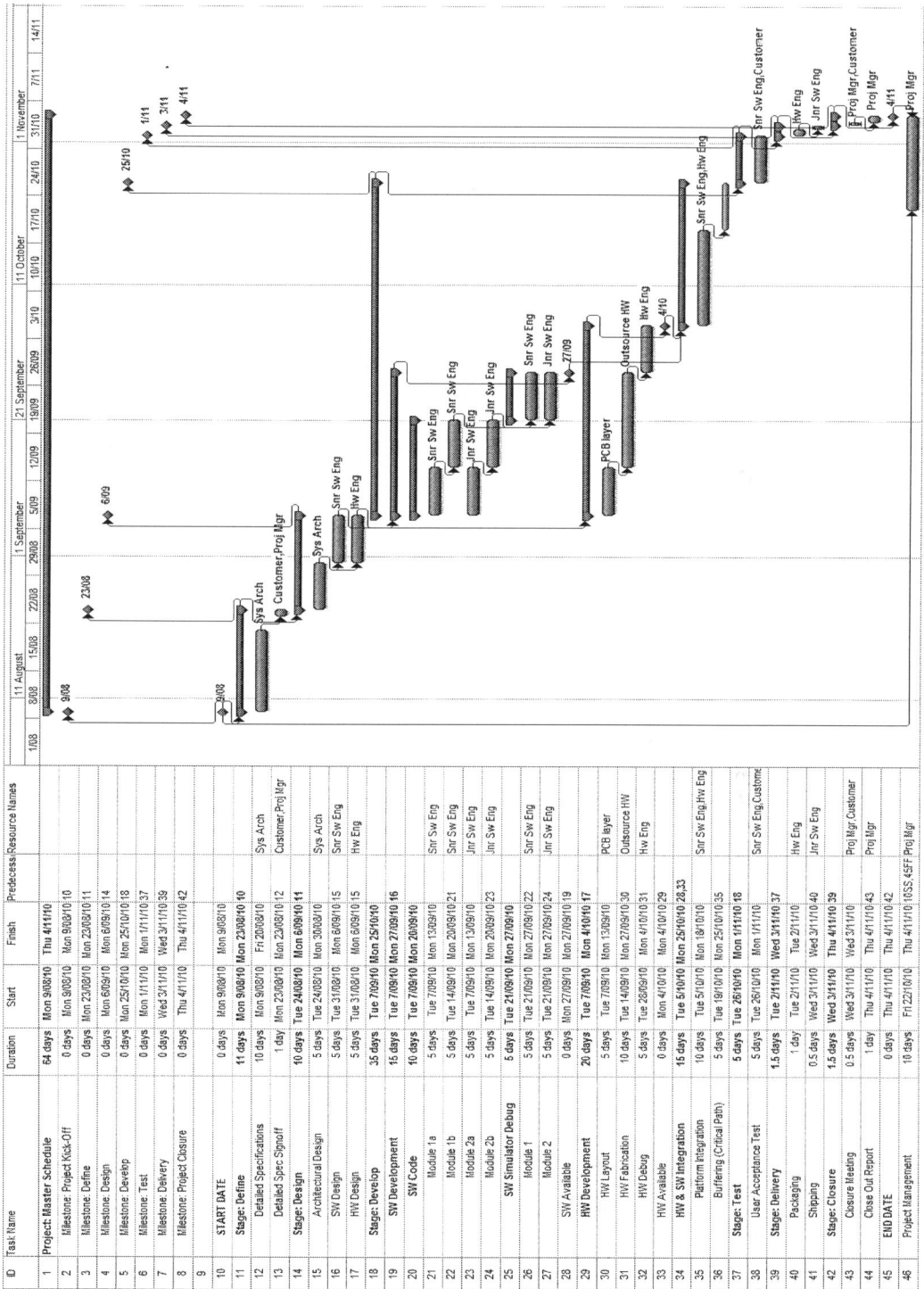

Figure 43: This Detailed Schedule (Gantt Chart) was produced with MS-Project.

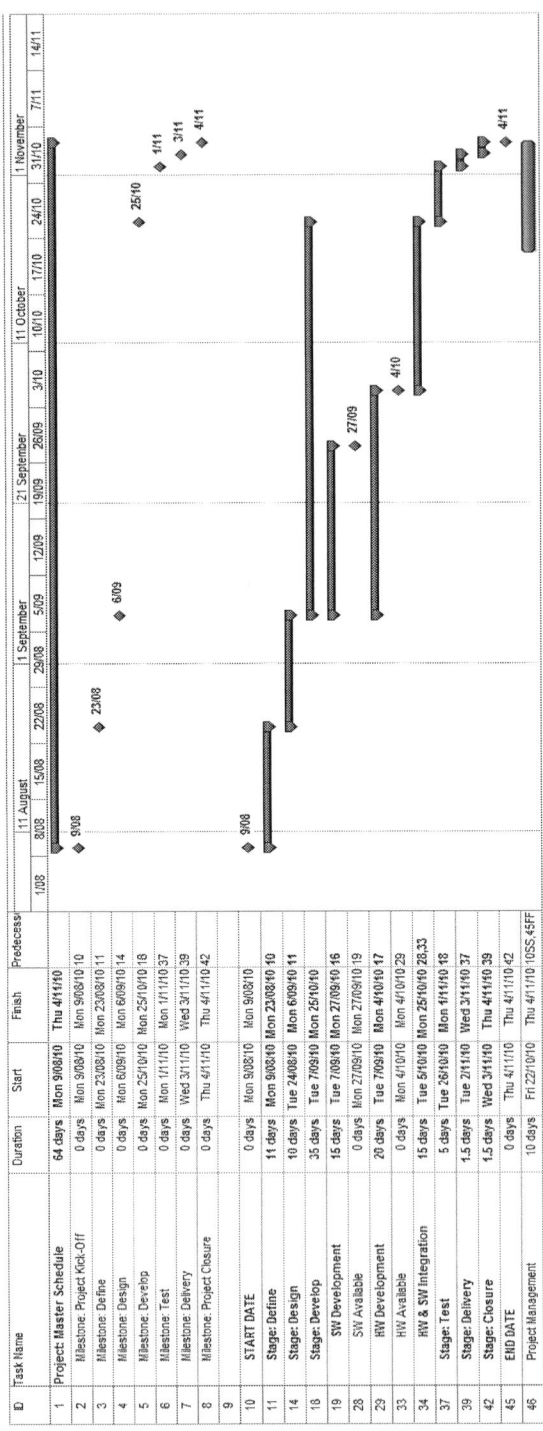

Figure 44: This Milestone Schedule (top half) and the Summation Schedule (bottom) was produced with MS-Project.

A down-to-earth guide for creating the project schedule.

The following is by no means a complete set of scheduling techniques,

but it should be enough to get the project's schedule started (in MS-Project).

1. Create a task of zero days duration and give it the same name as your project

    (also make it a larger bold font so that it will stand out).

    This 'Project:' task is here simply to make apparent what project this schedule is for

    when the Gantt Chart is printed out in a physical form.

		Task Name	Duration	Start	Finish	Predecessors
1		**Project: Stocking-Stuffer**	0 days	Sat 25-12-10	Sat 25-12-10	

2. In the proceeding rows of the schedule, add some tasks of zero days duration for

    each of the project's major milestones / releases.

    Give each task a meaningful name to identify the associated milestone.

    These 'Milestone:' tasks are here to clearly indicate the delivery date for these

    specific releases without the need to trawl through the entire schedule, rolling-up

    and rolling-down tasks to determine this information.

		Task Name	Duration	Start	Finish	Predecessors
1		**Project: Stocking-Stuffer**	0 days	Sat 25-12-10	Sat 25-12-10	
2		Milestone: Release Forwards	0 days	Fri 07-01-11	Fri 07-01-11	6
3		Milestone: Release Backwards	0 days	Mon 10-01-11	Mon 10-01-11	13
4		Milestone: Release Bookends	0 days	Wed 05-01-11	Wed 05-01-11	20

I would indent all of those milestone tasks under the Project Heading task; thereby getting a total duration (in days) for the project, and obtaining what the project's start date and end date will be.

3. The first milestone for this example project will be a 'Start-Date' mini-project containing four tasks to be completed in sequential order (A1 – A2 – A3 – A4):

- Add a title-task to identify clearly this release.

- On the proceeding rows, add each work-task to be completed, and right-indent these tasks to rollup underneath this release's title-task.

- Assign the 'Duration' for each work-task.

- For each work-task set the 'Predecessors' to the row number of the previous row; e.g. row-10's predecessor is row-9, row-9's predecessor is row-8, row-8's predecessor is row-7.

- Now that this release has been scheduled then add the corresponding 'Milestone:' task to be dependent on this release's title-task; i.e. row-2's predecessor is row-6.

	❶	Task Name	Duration	Start	Finish	Predecessors
1		⊟ Project: Stocking-Stuffer	10 days	Mon 27/12/10	Mon 10/01/11	
2		Milestone: Release Forwards	0 days	Fri 7/01/11	Fri 7/01/11	6
5						
6		⊟ Release: Forwards Demo	10 days	Mon 27/12/10	Fri 7/01/11	
7		Task A1	2 days	Mon 27/12/10	Tue 28/12/10	
8		Task A2	3 days	Wed 29/12/10	Fri 31/12/10	7
9		Task A3	2 days	Mon 3/01/11	Tue 4/01/11	8
10		Task A4	3 days	Wed 5/01/11	Fri 7/01/11	9
11		Milestone	0 days	Fri 7/01/11	Fri 7/01/11	10

This example uses a **Finish-to-Start (FS) dependency where the predecessor task has to (finish) prior to the successor task being able to (start)**.

This dependency technique is **used for Start-Date projects**.

This is the most common form of scheduling you will encounter, is the default technique in Ms-Project, and is how the 1st Pass schedule would be created.

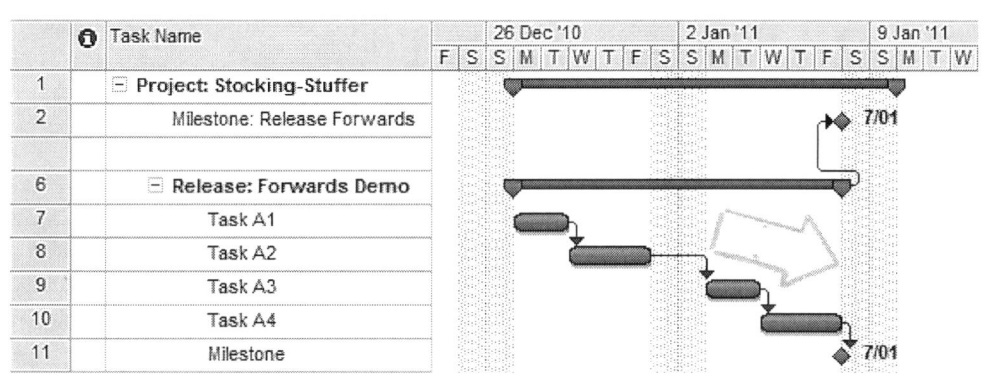

The diagram above has a nice formation of waterfall activities to be completed sequentially.

4. The second milestone for this example project will be an End-Date mini-project containing four tasks that must be completed by a specific 'drop-dead' date (of Monday the 10th of January).

- Add a title-task to identify clearly this release. Though make sure that this title-task is left-indented back to the same level as the title-task and doesn't remain at the default indent level of the previous work-tasks.

- On the proceeding rows, add each work-task to be completed, and right-indent these tasks to rollup underneath this release's title-task.

- Assign the 'Duration' for each work-task.

- For each work-task set the 'Predecessors' to the row number of the next row; e.g. row-16's predecessor is row-17SF, row-15's predecessor is row-16SF, and row-14's predecessor is row-15SF.

- Now that this release has been scheduled then assign the corresponding 'Milestone:' task to be dependent on this release's title-task; i.e. row-3's predecessor is row-13.

	❶	Task Name	Duration	Start	Finish	Predecessors
1		⊟ Project: Stocking-Stuffer	**10 days**	**Mon 27/12/10**	**Mon 10/01/11**	
2		Milestone: Release Forwards	0 days	Fri 7/01/11	Fri 7/01/11	6
3		Milestone: Release Backwards	0 days	Mon 10/01/11	Mon 10/01/11	13
13		⊟ **Release: Backwards Demo**	**10 days**	**Mon 27/12/10**	**Mon 10/01/11**	
14		Task B1	2 days	Mon 27/12/10	Wed 29/12/10	15SF
15		Task B2	3 days	Wed 29/12/10	Mon 3/01/11	16SF
16		Task B3	2 days	Mon 3/01/11	Wed 5/01/11	17SF
17		Task B4	3 days	Wed 5/01/11	Mon 10/01/11	18SF
18	▦	Milestone	0 days	Mon 10/01/11	Mon 10/01/11	

> ▦ This task has a 'Start No Earlier Than' constraint on Mon 10-01-11.

Notice how the milestone-task on row-18 has in the info column a note about it having a 'Start No Earlier Than' constraint; i.e. a specified End Date.

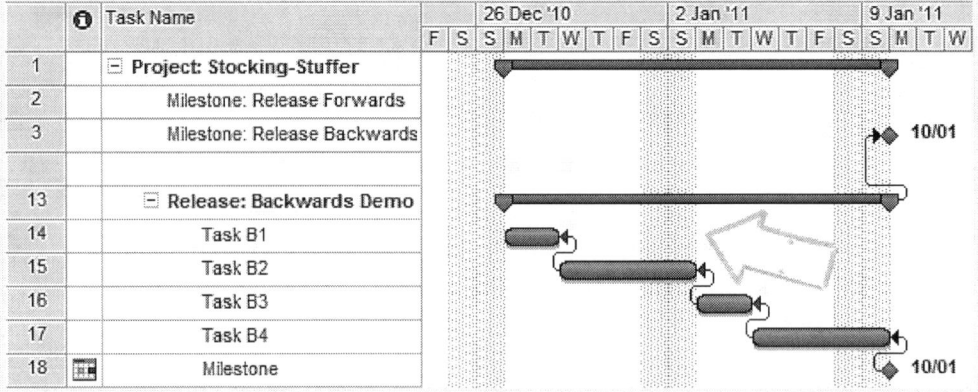

This example uses a **Start-to-Finish (SF) dependency where the predecessor has to have (started) before the successor task can (finish)**.

This dependency technique is **used for End-Date projects**, where the drop-dead end date would be set as a milestone, and the predecessor tasks would be set as dependent on their successor task, and thereby working backwards to determine when each task and the project needs to commence by in order to have a chance to complete the project by the dictated end date.

5. The third milestone for this example project will be a mix of the two remaining dependency techniques; i.e. **Start-to-Start (SS)** and **Finish-to-Finish (FF)**.

I think of (SS) and (FF) as the book-ending of tasks.

- Add a title-task to identify clearly the release.

  Make sure that this title-task is left-indented back to the same level as the title-task and doesn't remain at the default indent of the previous work-tasks.

- On the rows proceeding, add each of the work-tasks to be completed, and right-indent these tasks to rollup underneath this release's title-task.

- Assign the 'Duration' for each work-task.

- Set the predecessor, 'Task C3' on row-23 to start at the same time as 'Task C2' on row-22, hence row-23 'Predecessors' is set to row-22SS.

- Set the predecessor, 'Task C5' on row-25 to finish at the same time as 'Task-C4' on row-24, hence row-25 'Predecessors' is set to row-24FF.

- Now that this release has been scheduled then assign the corresponding 'Milestone:' task to be dependent on this release's title-task; i.e. row-4's predecessor is row-20.

		Task Name	Duration	Start	Finish	Predecessors
1		⊟ Project: Stocking-Stuffer	10 days	Mon 27/12/10	Mon 10/01/11	
2		Milestone: Release Forwards	0 days	Fri 7/01/11	Fri 7/01/11	6
3		Milestone: Release Backwards	0 days	Mon 10/01/11	Mon 10/01/11	13
4		Milestone: Release Bookends	0 days	Wed 5/01/11	Wed 5/01/11	20
20		⊟ Release: Bookends Demo	8 days	Mon 27/12/10	Wed 5/01/11	
21		Task C1	3 days	Mon 27/12/10	Wed 29/12/10	
22		Task C2	2 days	Thu 30/12/10	Fri 31/12/10	21
23		Task C3	3 days	Thu 30/12/10	Mon 3/01/11	22SS
24		Task C4	2 days	Tue 4/01/11	Wed 5/01/11	23
25		Task C5	3 days	Mon 3/01/11	Wed 5/01/11	24FF

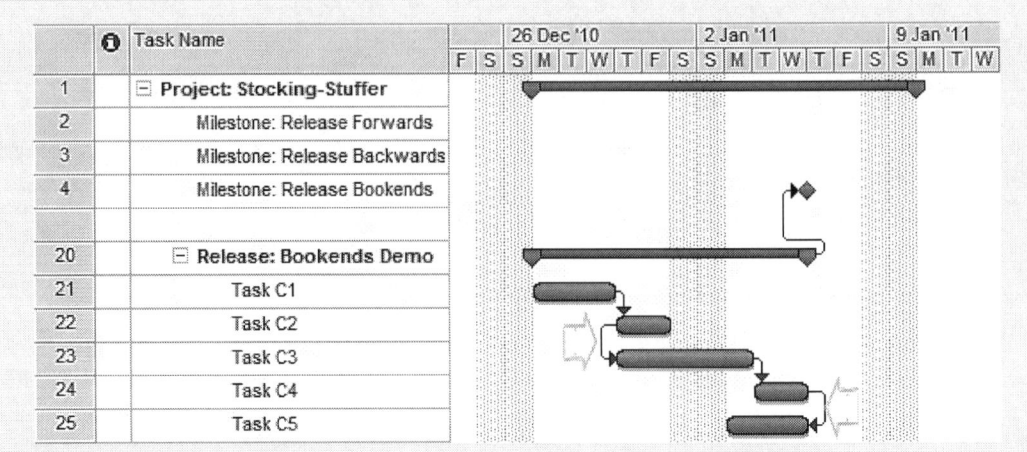

This example uses;

- The **Start-to-Start (SS) dependency is where the current / predecessor task cannot (Start) until its sibling / successor task is also ready to (Start)**.

- The **Finish-to-Finish (FF) dependency is where the current / predecessor task cannot (Finish) until its sibling / successor task is also ready to (Finish)**.

Below is a combined view of each of the scheduling techniques; Finish-To-Start (FS), then Start-To-Finish (SF), and finally Start-To-Start (SS) and Finish-To-Finish (FF).

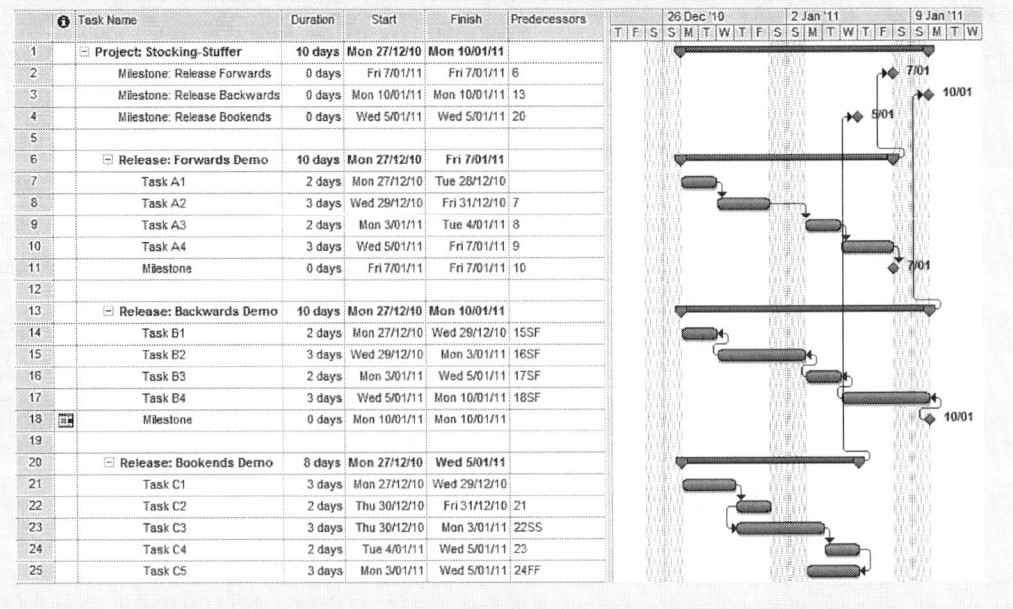

A few other terms should be mentioned at this point:

- **Lead Time** – is a period of **time <u>before</u>, that the successor can start prior to the predecessor finishing**; i.e. the 2nd task can be underway by a certain amount of time, before the 1st task has finished. A dependency of **[FS] – No. days**.

  > *For example; the ordering of the computer server hardware can occur 2 weeks, before the destination server room construction is completed.*

- **Lag Time** – is a period of **time <u>after</u>, that the predecessor task has finished prior to the successor task being allowed to commence**; i.e. the 2nd task has to wait a certain amount of time, after the 1st has finished. A dependency of **[FS] + No. days**.

  > *For example; The concrete floor for the power room has to have settled / dried for 2 weeks, before the generator equipment can be installed.*

- **Float / Slack** – is a period of **time that a task could be delayed or exceed its planned time allocation without resulting in the project / release milestone being delivered late.** *... "Free Float" per task, "Total Float" per milestone.*

- **Critical Path** – is a **sequence of dependent tasks that if one of these tasks exceeded its planned time allocation then its *"knock-on effect"* would result in the project / release milestone being delivered late.** Hence, the **Critical Path has no float / slack.**

NOTE ☐ **A spreadsheet DOES NOT make a credible project schedule** because it doesn't effectively represent the relationship between [Scope] and [Time].

A down-to-earth guide for adding buffering to the project schedule.

This previous project schedule was relatively simple but it didn't include any **contingency buffering / reserve for something going wrong**.

> *For example; some safety margin just in case a group of tasks is a bit more technically difficult than was planned for.*

Steps for adding contingency buffering to a schedule:

1. Make a copy of the existing project schedule.

2. At the end of the sequence of tasks that require a contingency buffer, add a new task, assign the 'Duration' for this buffer task to the desired safety margin, and set the 'Predecessors' to the row number of the previous row in the sequence.

3. Optionally, Right-Click this 'Buffering' task's bar and select 'Format Bar' from the menu, and choose an appropriate pattern and colour to differentiate this buffering from those activity tasks.

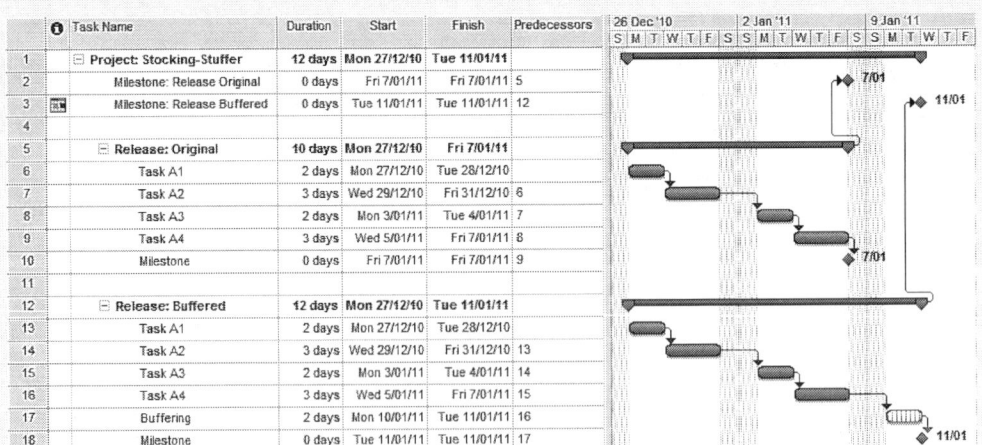

	Task Name	Duration	Start	Finish	Predecessors
1	⊟ Project: Stocking-Stuffer	12 days	Mon 27/12/10	Tue 11/01/11	
2	Milestone: Release Original	0 days	Fri 7/01/11	Fri 7/01/11	5
3	Milestone: Release Buffered	0 days	Tue 11/01/11	Tue 11/01/11	12
4					
5	⊟ Release: Original	10 days	Mon 27/12/10	Fri 7/01/11	
6	Task A1	2 days	Mon 27/12/10	Tue 28/12/10	
7	Task A2	3 days	Wed 29/12/10	Fri 31/12/10	6
8	Task A3	2 days	Mon 3/01/11	Tue 4/01/11	7
9	Task A4	3 days	Wed 5/01/11	Fri 7/01/11	8
10	Milestone	0 days	Fri 7/01/11	Fri 7/01/11	9
11					
12	⊟ Release: Buffered	12 days	Mon 27/12/10	Tue 11/01/11	
13	Task A1	2 days	Mon 27/12/10	Tue 28/12/10	
14	Task A2	3 days	Wed 29/12/10	Fri 31/12/10	13
15	Task A3	2 days	Mon 3/01/11	Tue 4/01/11	14
16	Task A4	3 days	Wed 5/01/11	Fri 7/01/11	15
17	Buffering	2 days	Mon 10/01/11	Tue 11/01/11	16
18	Milestone	0 days	Tue 11/01/11	Tue 11/01/11	17

NOTICE how the Buffered version of the release has a milestone date later than the original version of this release. This is the result of the added buffering.

Note: only add buffering to the "critical path", else it could be taken as permission to slacken off on other paths' tasks.

How much buffering should be added?

One method is to take away (i.e. accumulate) all of the "float / slack" for the tasks on the task path that is to be buffered, accumulating these contingencies together to get a total contingency value, then dividing this total contingency value by two (i.e. 50%), and then adding this resultant amount to the end of the task path as the buffer.

An alternative simplified buffer calculating method would be to; accumulate all of the "most-likely" estimates for the tasks on the path that is to be buffered, then adding onto this accumulated duration an agreed fudge-factor percentage.

> *For example; for certain types of projects the performing organization has a "rule of thumb" that a 20% "safety-margin" duration is added at the end of the project. Hence, a project that is realistically expected to take 100 days to complete is told to the customer as requiring 120 days to deliver.*

NOTE   ❏ Irrespective of what **buffer calculating method** is used, it is essential that this method be **used consistently across all of the performing organization's projects, so that these projects can be compared under a common set of rules.**

       ❏ **Buffering should not be added to every possible path through the project.**

Buffering should be added to the critical path and to those feeder paths where it is expected that some delay is likely to occur or that delays would have significant consequences. However, these feeder buffers should not push out the critical path, and their inclusion should not be incorporated / calculated into the critical path's buffer.

> *For example; the critical path is calculated to be primarily related to the development of the software application (which is estimated to take 4 weeks + 1 week buffer), with the feeder path of delivering the ordered end-user hardware expected to take only 2 weeks. But, what if the hardware delivery takes an additional 3 weeks (which sometimes happens)? Hence, would it not be wise to add in a contingency buffer to this feeder path, just in case history repeats itself.*

I would recommend that ... when creating the schedule you should add **buffering near the end (prior to an important deliverable milestone) and not to each group of tasks or minor milestones.** Clearly mark each buffering component as a distinct project task.

We use a lean agile implementation technique, how is this going to affect things?

This is not a problem, because when it comes time to execute the project's implementation just take the tasks that were determined in the 2nd Pass schedule and mix-n-match these tasks to fit into development "sprints"; i.e. mini iterative cycles with a constant duration of 3, 4, or 6 weeks.

The thing to remember is that the total duration of the project's implementation and its delivery milestones will not change. All that will change is the order and arrangement of the project's tasks that make up each mini iterative cycle (sprint) that compose the delivery milestones; see [Figure 45] and [Figure 46].

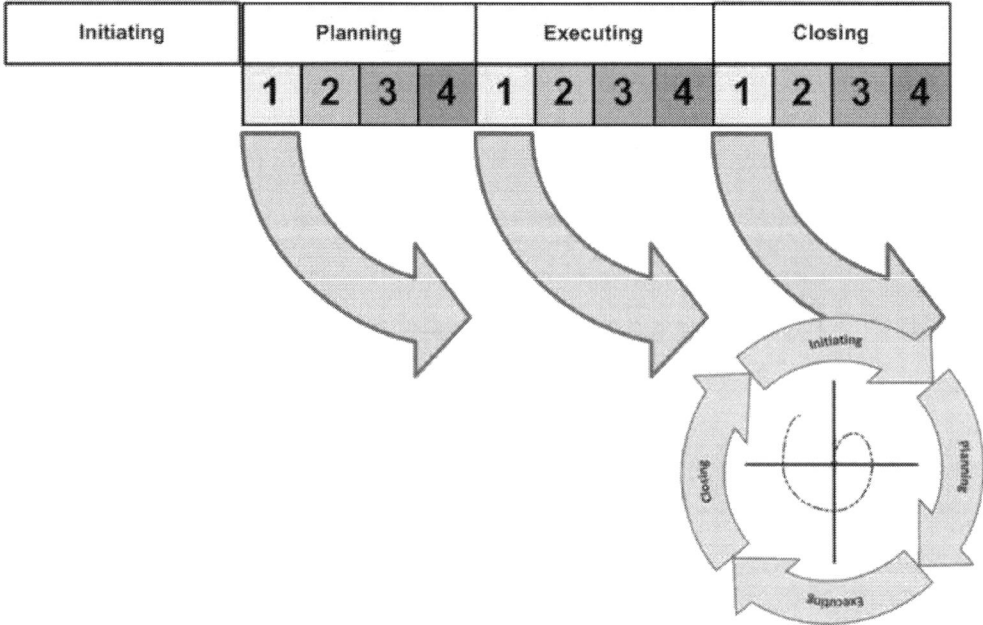

Figure 45: Working agile into waterfall planning.

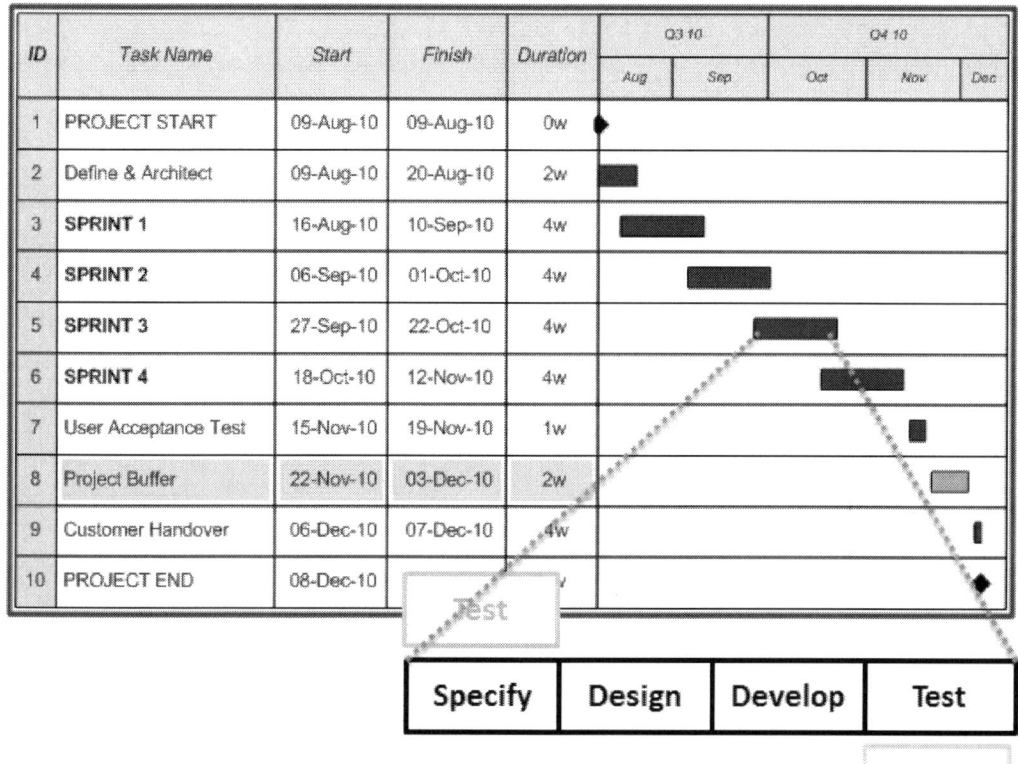

ID	Task Name	Start	Finish	Duration	Q3 10		Q4 10		
					Aug	Sep	Oct	Nov	Dec
1	PROJECT START	09-Aug-10	09-Aug-10	0w					
2	Define & Architect	09-Aug-10	20-Aug-10	2w					
3	SPRINT 1	16-Aug-10	10-Sep-10	4w					
4	SPRINT 2	06-Sep-10	01-Oct-10	4w					
5	SPRINT 3	27-Sep-10	22-Oct-10	4w					
6	SPRINT 4	18-Oct-10	12-Nov-10	4w					
7	User Acceptance Test	15-Nov-10	19-Nov-10	1w					
8	Project Buffer	22-Nov-10	03-Dec-10	2w					
9	Customer Handover	06-Dec-10	07-Dec-10	4w					
10	PROJECT END	08-Dec-10							

Figure 46: A overview schedule for an agile implementation

NOTICE how in [Figure 46] the 'specify' part of one sprint can overlap with the 'test' part of the previous sprint. This is not unusual, as often the contents of one sprint is determined by what was and was not included in the sprints that came before.

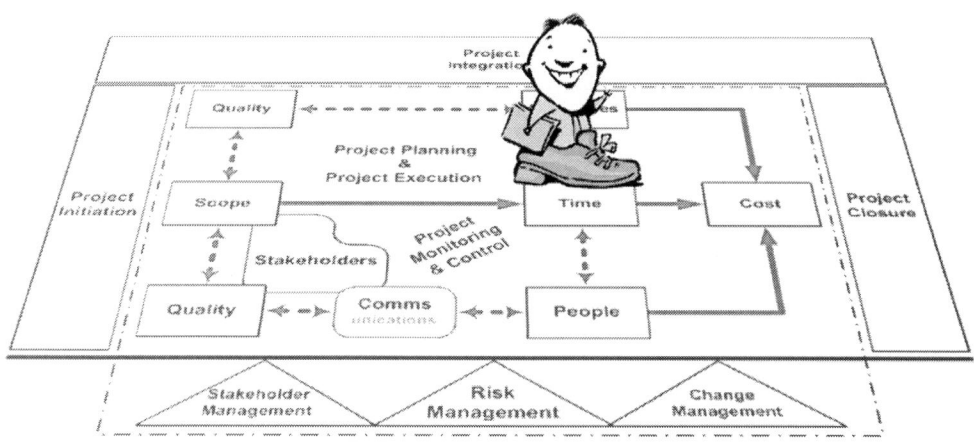

4.3.2.2. Need: People & Resources

When it comes to [People] resourcing, the project's planner will either:

1) **Know the exact individual makeup** of the project team; *e.g. John, Jack, Jim, and Bud.*

2) **Know the general makeup** of the project team; *e.g. 2x leads, 3x seniors, 4x juniors, and 1x graduate.*

3) The team members are **yet to be specified but they will come from a "pool of resources"**.

 I would recommend overlaying the given people onto a copy of the WBS and/or the system architecture diagram, and identify if there are any areas that are not covered. Could also grade each person's technical proficiency in that area and identify any skills weaknesses.

Once the project's team members have been nominated against their tasks thought should then be given to **assigning** them project **roles, responsibilities, authorization, reporting relationships,** and determining any **potential skills deficiencies.**

What training will they require? Did you allocate time for training?

 When obtaining **estimates from those persons who will be undertaking the task, they may subconsciously add in their own safety buffering** per task or group of tasks.

Potentially the person estimating the duration / effort of the task could be inadvertently adding a 20% safety margin **"padding"**; *e.g. a 4 day task stated as requiring 5 days.* However, when it comes to implementing that task that same person will most probably unwittingly see this as the 5-day task and not the 4-day task that they originally thought.

NOTE ☐ A **weighted average duration technique could be used to counter this covert padding of the task durations.**

A down-to-earth guide for adding [People] resources to the project schedule.

For this example, there are two very similar releases with one in the first week and the other in the second week. Both releases consist of four tasks; where the 1st & 2nd tasks are sequentially dependent, the 3rd & 4th tasks are sequentially dependent but are completely independent of the first two tasks. Therefore, if the available resources allowed, then tasks 3 & 4 could be implemented in parallel with tasks 1 & 2.

For this example, two definite people resources have been assigned to the project, and two more people may be available in the second week.

Steps for adding resources to a schedule:

1. Create the project schedule using (the default) 'Finish-to-Start' (FS) task dependencies to form a Start-Date project for both Release 1 and Release 2.

	ⓘ	Task Name	Duration	Start	Finish	Prede
1		**Project: People-Placer**	0 days	Sat 08-01-11	Sat 08-01-11	
2		Milestone: Release 1 Persons	0 days	Fri 14-01-11	Fri 14-01-11	5
3		Milestone: Release 2 Persons	0 days	Fri 21-01-11	Fri 21-01-11	11
4						
5		⊟ **Release: 1 Persons Task**	**5 days**	**Mon 10-01-11**	**Fri 14-01-11**	
6		Task A1	2 days	Mon 10-01-11	Tue 11-01-11	
7		Task A2	3 days	Wed 12-01-11	Fri 14-01-11	6
8		Task A3	2 days	Mon 10-01-11	Tue 11-01-11	
9		Task A4	3 days	Wed 12-01-11	Fri 14-01-11	8
10		Milestone	0 days	Fri 14-01-11	Fri 14-01-11	9,7
11		⊟ **Release: 2 Persons Task**	**5 days**	**Mon 17-01-11**	**Fri 21-01-11**	**5**
12		Task B1	2 days	Mon 17-01-11	Tue 18-01-11	
13		Task B2	3 days	Wed 19-01-11	Fri 21-01-11	12
14		Task B3	2 days	Mon 17-01-11	Tue 18-01-11	
15		Task B4	3 days	Wed 19-01-11	Fri 21-01-11	14
16		Milestone	0 days	Fri 21-01-11	Fri 21-01-11	15,13

Note: can make tasks not only dependent on logical progression but also on who & when specific people are available to do the work.

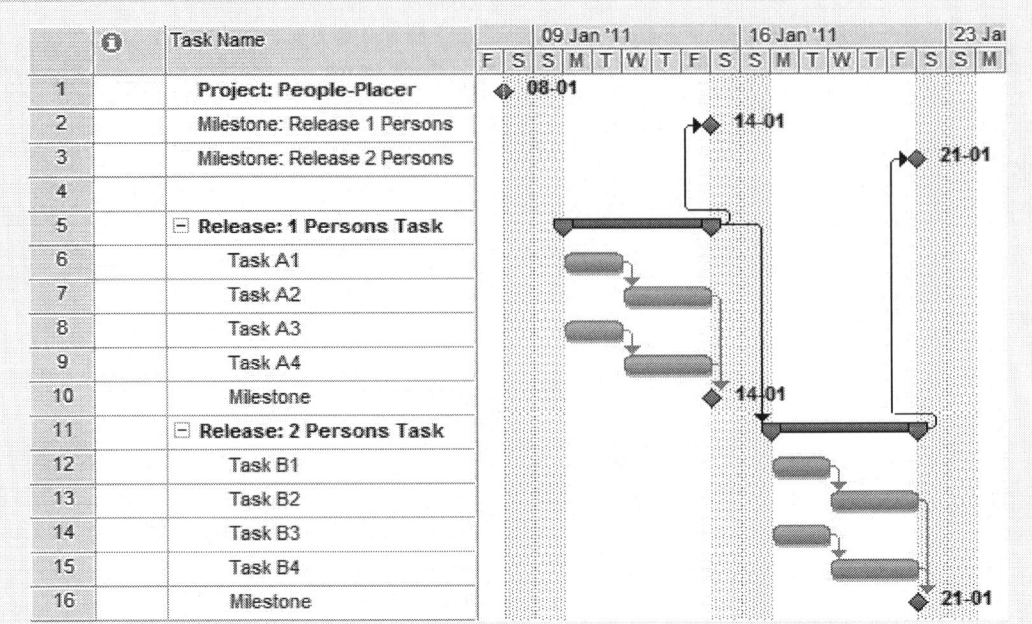

2. Next to the 'Predecessors' column, there may be a 'Resource Names' column. If this column is not present then right-click the last column and choose 'Insert Column... ' and select 'Resource Names' from the list.

3. In this example case there are two [People] resources assigned to the project, so in the 'Resource Names' column enter the person's name against each task.

	ⓘ	Task Name	Duration	Start	Finish	Prede	Resource Names
1		**Project: People-Placer**	0 days	Sat 08-01-11	Sat 08-01-11		
2		Milestone: Release 1 Persons	0 days	Fri 14-01-11	Fri 14-01-11	5	
3		Milestone: Release 2 Persons	0 days	Fri 21-01-11	Fri 21-01-11	11	
4							
5		⊟ **Release: 1 Persons Task**	**5 days**	**Mon 10-01-11**	**Fri 14-01-11**		
6		Task A1	2 days	Mon 10-01-11	Tue 11-01-11		Person 1
7		Task A2	3 days	Wed 12-01-11	Fri 14-01-11	6	Person 1
8		Task A3	2 days	Mon 10-01-11	Tue 11-01-11		Person 2
9		Task A4	3 days	Wed 12-01-11	Fri 14-01-11	8	Person 2
10		Milestone	0 days	Fri 14-01-11	Fri 14-01-11	9,7	
11		⊟ **Release: 2 Persons Task**	**5 days**	**Mon 17-01-11**	**Fri 21-01-11**	5	
12		Task B1	2 days	Mon 17-01-11	Tue 18-01-11		Person 1
13		Task B2	3 days	Wed 19-01-11	Fri 21-01-11	12	Person 1
14		Task B3	2 days	Mon 17-01-11	Tue 18-01-11		Person 2
15		Task B4	3 days	Wed 19-01-11	Fri 21-01-11	14	Person 2
16		Milestone	0 days	Fri 21-01-11	Fri 21-01-11	15,13	

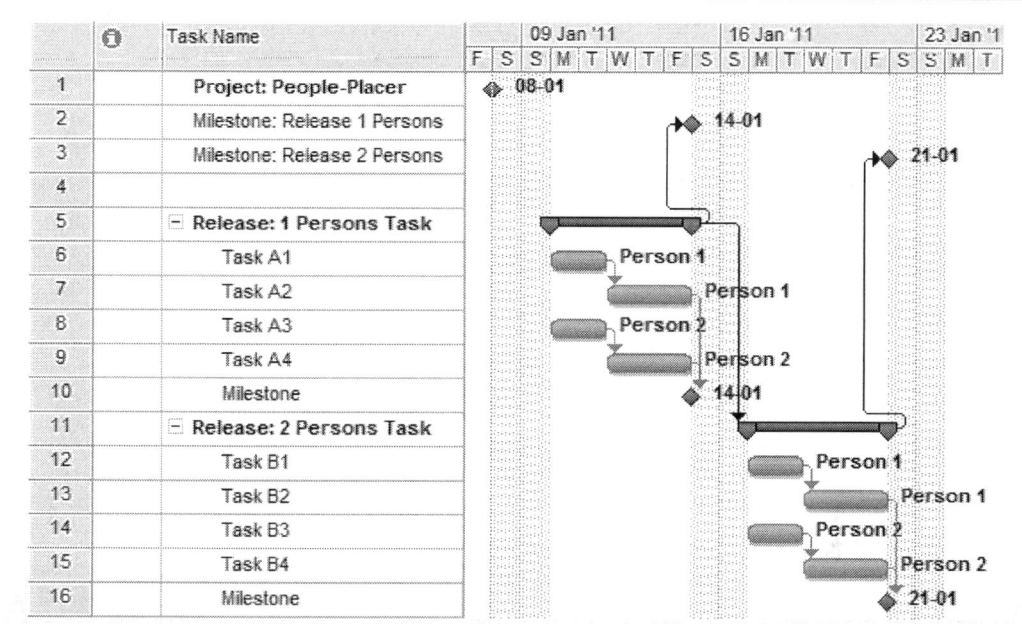

4. Fortunately, the 3rd & 4th persons have appeared for the second week and they have been assigned to work alongside (assist) the 1st & 2nd persons respectively.

   Therefore, in the 'Resource Names' column next to Person_1 put comma Person_3, and next to Person_2 put comma Person_4.

	❶	Task Name	Duration	Start	Finish	Prede	Resource Names
1		**Project: People-Placer**	0 days	Sat 08-01-11	Sat 08-01-11		
2		Milestone: Release 1 Persons	0 days	Fri 14-01-11	Fri 14-01-11	5	
3		Milestone: Release 2 Persons	0 days	Wed 19-01-11	Wed 19-01-11	11	
4							
5		⊟ **Release: 1 Persons Task**	5 days	**Mon 10-01-11**	**Fri 14-01-11**		
6		Task A1	2 days	Mon 10-01-11	Tue 11-01-11		Person 1
7		Task A2	3 days	Wed 12-01-11	Fri 14-01-11	6	Person 1
8		Task A3	2 days	Mon 10-01-11	Tue 11-01-11		Person 2
9		Task A4	3 days	Wed 12-01-11	Fri 14-01-11	8	Person 2
10		Milestone	0 days	Fri 14-01-11	Fri 14-01-11	9,7	
11		⊟ **Release: 2 Persons Task**	2.5 days	**Mon 17-01-11**	**Wed 19-01-11**	5	
12		Task B1	1 day	Mon 17-01-11	Mon 17-01-11		Person 1,Person 3
13		Task B2	1.5 days	Tue 18-01-11	Wed 19-01-11	12	Person 1,Person 3
14		Task B3	1 day	Mon 17-01-11	Mon 17-01-11		Person 2,Person 4
15		Task B4	1.5 days	Tue 18-01-11	Wed 19-01-11	14	Person 2,Person 4
16		Milestone	0 days	Wed 19-01-11	Wed 19-01-11	15,13	

	ⓘ	Task Name	09 Jan '11	16 Jan '11	23 Jan '11
			F S S M T W T F	S S M T W T F	S S M T W
1		Project: People-Placer	◈ 08-01		
2		Milestone: Release 1 Persons	◆ 14-01		
3		Milestone: Release 2 Persons		◆ 19-01	
4					
5		⊟ Release: 1 Persons Task	▽▽		
6		Task A1	Person 1		
7		Task A2	Person 1		
8		Task A3	Person 2		
9		Task A4	Person 2		
10		Milestone	◆ 14-01		
11		⊟ Release: 2 Persons Task	▽▽		
12		Task B1	Person 1,Person 3		
13		Task B2	Person 1,Person 3		
14		Task B3	Person 2,Person 4		
15		Task B4	Person 2,Person 4		
16		Milestone	◆ 19-01		

NOTICE how when the additional person was added to the task, that task's duration was automatically reduced; e.g. tasks B1 & B3 from 2 days to 1 day, and tasks B2 & B4 from 3 days to 1.5 days. However, if for some reason the task has to have a fixed duration then this task's duration will have to be manually increased back to what it should be.

NOTICE how in Row-11 for Release 2, the total duration has shrunk from 5 days down to 2.5 days, and thereby shifting the completion date from Friday (21st) to Wednesday (19th).

This example has demonstrated how to add [People] resources to the project schedule, and how additional [People] resources can reduce the project's [Time] duration. However, "there is no such thing as a free lunch", and as the following costing section will demonstrate, **reducing the [Time] duration by adding [People] resources can also increase the project's [Costs]**.

Note: ensure that people are not overloaded with doing too many tasks at the same time; working 15hr days. ... "Resource Levelling".

4.3.2.3. Need: Procurement of People & Resources

At this point, the schedule has outlined what [People] and what [Resources] will be required and when these have to be deployed to the project. ... *Any new risks identified?* Now, the decision is whether:

1) Should the organization **use** its own **in-house people or** should people **external** to the organization be engaged; *e.g. contracting, or outsourcing*?

2) Should subcomponents of the project be **custom built or** should these be **procured externally**; *e.g. purchasing, leasing, hiring, internal requisitioning, or development*?

The simple reality is, **the performing organization only has a certain number of [People]** with a certain amount of relevant skills, and **a certain amount of material [Resources] at its disposal.** Hence, **would the performing organization be strategically better served by paying to use other organizations capabilities & capacities?**

Example Case: If it is so cheap then why can't we produce it ourselves?

I was at the corner store, picked up this small snack-pack of sesame-seed wafers.

> *"Why the @#$% are we importing these from overseas when I'm sure that a nation such as ours could easily make this product ourselves?"*

Other than the cost of manufacture, the product must also have to include in its price; the costs of shipping, storing at both ends and in between, plus some profit margin for all of the hands that it goes through from manufacture to the point of sale.

Yet, this snack-pack is so cheap.

Hmm, why is this so?

The reasons being that; our nation has relatively more expensive labour rates when compared to the manufacturing nation, and they are supplying to a significantly larger market (i.e. to the whole world) instead of confined to our limited national market.

This means that to manufacture this product locally doesn't make business sense when our limited people and material resources could be better utilized on more high-end products with greater profit margins.

This is the same for projects, one needs to consider;

- **Labour rates**, and
- **Economies of scale.**

Example Case: I could do it myself, but ...

The family car is due for its major service but, you are looking at a bill from the mechanic of $1500 to $2000.

Now, you could order the parts yourself and spend the entire weekend under the bonnet (hood) covered in grease.

But to be honest, you don't know one-end of the dip-stick from the one holding it.

Plus, you got to take the kids to sports on Saturday and you promised your partner that on Sunday you would. ... *Insert nagging here*.

The truth is that, you don't have the relevant knowledge, skills and experience to undertake the task successfully and/or your time would be better spent elsewhere.

And, you ain't got all of the necessary tools to do the job properly.

This is the same for projects, one needs to consider;

- The **skill levels of your internal people**,
- Any **time limitations** imposed by other **competing commitments and obligations.**
- And, wouldn't it be better to **engage a professional to do the job properly** and thereby **mitigate against** the **potential damage** caused by you screwing it up.

Once these decisions have been made about how the [People] and the [Resources] will be procured, then the schedule needs to consider:

- The **availability of each [Resource]**? *... Were any new risks identified?*
  e.g. how many are available and what is the level of competing demand for these.

- **Allow reasonable lead times for the delivery of these [Resources]** as there could be several weeks delay and even potential availability shortages. *... new risks identified?*

  Question: is there a **planned contingency in place** if there are delays or the resource is not available when required? That is, have **alternate arrangements** been considered and **contingency buffering** added to the project schedule?

- **Update the project's calendar** to include; when the [Resource] is and is not available, *e.g., the project is only allocated a specific slot in the work-stream.*
  When does the [Resource] have to be relinquished to another project?
  What about ongoing service & maintenance of the resource?

- Does the organization's own **financial calendar impose a "window of opportunity"** for deploying these externally procured [Resources]; *e.g. the project only has a defined budget for the current financial period... come the End Of the Financial Year then what?*

Once the decision is made to procure [People] and/or [Resources] externally to the performing organization then this opens up additional concerns and considerations.

1) **What are the procurement policies**, rules, and regulations; *e.g. public-sector tendering*?

2) **Who has the authorization** to engage external parties?

3) **What are the legal considerations** for the contracts that will have to be established with the chosen vendors, suppliers, contracting agencies, and consultants?

NOTE   ❐ **Once engaged the external party becomes a project stakeholder** and hence has to be managed as such; see [Section 10.3].

However, **the involvement of each external party is a mini-project unto itself**.

1.	**Initiating**	• Who are the possible external parties and how will opening contact be made with them? *E.g., tenders, bids, quotations, or direct contact.*

• How will they know what we "need" and "want"?

• What is the perceived scope of their deliverable/work?

• How will the price be agreed? *E.g., based on a fixed price, reimbursed costs, or time & materials usage.*

• How will the external party be selected? *E.g., what are the selection criteria?*

2. **Planning**   • When and what exactly will they be required to deliver? *E.g., a Statement Of Work (SOW) outlining the project's scope that is relevant to them.*

• What are the acceptance criteria for their deliverables?

• When will the external party be required to commence work?

• What information will the external party require to do their work?

• How will they be communicated with and what progress reporting will be required?

3. **Executing**   • They are doing the work and producing the deliverable.

4. **Closing**   • How will it be verified that they have delivered what was agreed to?

• How will they be paid and how will the relationship be ended?

5.	**Monitoring & Control**	• Need to baseline their performance, scope coverage, and cost claims.
		• How will their performance be assessed, and if necessary how will it be corrected?
		• How will changes to scope & costs be handled? That is, what change control will be put in place?
		• How will they be managed and their costs monitored?

Reviewer's Comment ... Establishing the right relationships.

When selecting external parties and when hiring new employees, there is more to the selection process than simply choosing who has the required technical skills & qualifications. ... There is also, who has the appropriate interpersonal attitude to work with / integrate into the project team; as engaging the wrong personality could be highly disruptive and even destructive to the team's cohesion & morale.

IT IS THE PEOPLE + THE RESOURCES UTILIZED THAT WILL EVENTUALLY PRODUCE THE DELIVERABLES ... WHILE THE PROJECT MANAGEMENT PLANS WILL ONLY PROVIDE GUIDANCE TO THEIR HANDS.

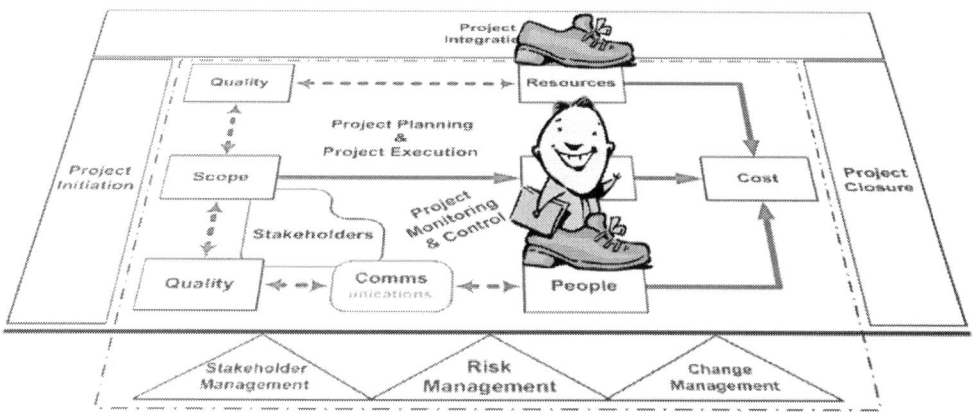

4.3.3. Question: "WE NEED what, when and – HOW MUCH?"

How much ... will it Cost?

The answer to this sub-question is concerned with estimating what is the monetary funding necessary to complete the project.

What budget will be required to undertake each task, work package (i.e. group of tasks), and the project as a whole?

What is the difference between Cost and Budget?

❖ **Cost** – is how much **money will be needed to complete a task**, or

in past-tense terms is how much money was needed to complete the task.

❖ **Budget** – is how much **money remains** to complete a task, or

how much money has been **allocated and approved to spend to complete a task**.

Therefore, this "*how much will it cost*" question is intended to determine **what is the [Cost Baseline] for the project? Once approved the [Cost Baseline] will then define the project's budget**.

The answer to this sub-question's is built upon the 2nd Pass Project Schedule developed during the "*when*" part of this larger question; see [Figure 47].

4.3.3.1. Need: Costs & Budget

The **[Cost] estimate is determined from** the schedule's **[Time] estimates,**
the **[People] to be engaged**, and the **[Resources] needed to be procured**.
This is why it was **so important to obtain reasonably accurate estimations** for the
2nd Pass Project Schedule, **so as to produce a reasonably accurate [Cost Baseline]**.

Hence, the **cost estimate should include; work time, anyone, or anything that would be charged against the project's financial accounting codes.**

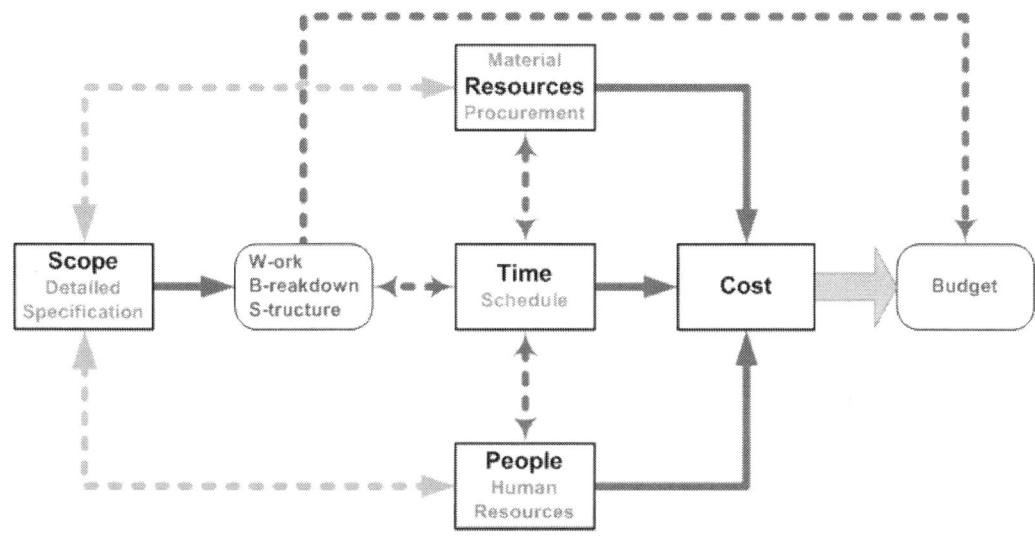

Figure 47: The Planning Process Model to determine the project's cost & budget.

Recall [Section 4.3.1] where unique identifying numbers were assigned in the Work Breakdown Structure (WBS), well these ID numbers can now be used to correspond with unique numbers in the performing organization's accounting system; i.e. **Cost Account Codes**. These ID numbers will then enable the costs to be aggregated based on **WBS Work Packages**. In turn, these work packages could be themselves **aggregated into** larger groupings to obtain **"Accrued Component Costs"**.

How do I go about calculating the Cost?

For the project's [Costs] to be determined from the schedule, need to:

1. Assign [People] and/or [Resources] to every task in the project's schedule.

2. Assign **hourly/daily rates** or a **per-unit-cost** to each [People] and [Resource];

   e.g. daily for permanent staff, hourly for contractors, daily or hourly for hired

   equipment, and per-unit for measurable quantities (e.g. tonnes, litres, square-meters).

3. Alternately assign a **Fixed-Cost** to a task or group of tasks;

   *e.g. outsourcing a task to an external contractor based on a **fixed price quotation**.*

 I would recommend that ... thought should be given to whether it is a better idea **to build, buy, lease, outsource, or share** the costs with other projects or groups.

4. With all of the tasks, now having [Cost] to [Time] values assigned it is now just a case of accumulating these [Costs] and [Time].   *... Were any new risks identified?*

   The greater the monetary units required and the more hours to be worked to complete each task then the greater will be the project's accrued [Costs] and hence the **greater** will be the project's **budgetary requirements**.

5. Once a representation of the project's costing has been established, **examine the monthly and quarterly distribution of these [Costs].**

   - Are there some months when the expenditure that would be incurred are disproportionately higher or lower than that for other months?

   - Would it be possible to rearrange some of the project's tasks so that the [Costs] are more evenly spread across the months?

   However, this rearrangement of tasks will require a rework of the project's schedule [Time], and progressively elaborating the schedule and expenditures to a mutual compromise of [Time] vs. [Cost].

6. Add some **contingency buffering** (i.e. **Contingency Reserve**) to the [Cost] estimate in a similar fashion to the buffering that was added to the schedule's [Time] estimate. Just in case, there are price increases during the project's life, *e.g. freight, labour rates*.

7. This **Cost Contingency Buffering should be clearly indicated as a distinct item in the costings** that are presented to the performing organization's senior management (**NOT to the customer's representatives**). This cost contingency buffering should be calculated using a **predefined methodology** specified by the performing organization's senior management and **used consistently across the organization**'s projects.

This contingency buffer could be established as either:

1) A **predefined percentage** of the total accrued cost estimate; *e.g. 10%.*

2) Assigned as a **monetary value per unit of the time buffer** that was added to the schedule's time estimate; *e.g. each time buffer day is worth $5,000.*

3) A **predefined amount derived from some formula** the performing organization uses for all of its projects; *e.g. X-amount per project team-member, Y-amount per block of resources, Z-amount per financial quarter, or some fudge factor.*

Should not the contingency buffer be hidden?

The final decision on whether the contingency buffer is shown or hidden should be made by the senior management of the performing organization and NOT by the project manager. The reasoning being to ensure that all projects will use exactly the same buffering methodologies, hence all projects could be compared from the same perspective and therefore senior management would be able to make consistent logical decisions about these projects. One of the reasons for showing the buffering is that if the buffering were not clearly shown then assumptions can and will be made.

> For example; "there is no buffering present so I better add some", but unbeknown to this good-samaritan there was already buffering hidden away inside the project. Thus, there is now additional buffering on top of the original project buffering.

Also, without buffering being clearly shown then it would be difficult to assess and compare the [Times] & [Costs] between projects because some projects would include it, while others could include it but calculated a different way, and others may not include it at all. Thus, an accurate comparative determination of the relative success or failure of projects would not be readily possible.

However, the main reason for not making this buffering information visible is to disadvantage the project's other stakeholders who are not privy to what the amount of buffering truly is.

 I would recommend that ... the **contingency buffering be hidden from** some stakeholders (i.e. **the customer** who may view it as room to negotiate downwards) and **made apparent to** other stakeholders (i.e. the performing organization's **senior management** who need to ensure that reasonable profitability margins are retained).

Often the reasons for not showing the buffering is because some project stakeholder "might" do something. But the reality is, by not showing the buffering the stakeholders will definitely make some "assumptions" which potentially could be the wrong assumptions.

Therefore, **the decision on whether or not to show the buffering boils down to a decision on what damage the resultant "mights" or "assumptions" will have on the successful completion of the project**.

BUFFERING IS NO FREE LUNCH,
BUT ... IT IS WORTH A BUNCH
TO THOSE WHO END UP IN A
TIME AND MONEY CRUNCH.

A down-to-earth guide for calculating [Costs] from the project schedule.

This cost calculation example will build on the resource allocation schedule example used previously.

The scenario is that, the project has two very similar releases, one in the first week and the other in the second week. The project definitely has two people assigned.

The project may have two more people available, but if these people were to appear then it would only be in the second week.

Steps for calculating [Costs] from the project schedule:

1. Create / open the project schedule containing both Release 1 and 2.

   At this point, the two additional resources have not appeared and so the schedule only contains Person_1 and Person_2.

	⊙	Task Name	Duration	Start	Finish	Prede	Resource Names
1		Project: People-Placer	0 days	Sat 08-01-11	Sat 08-01-11		
2		Milestone: Release 1 Persons	0 days	Fri 14-01-11	Fri 14-01-11	5	
3		Milestone: Release 2 Persons	0 days	Fri 21-01-11	Fri 21-01-11	11	
4							
5		⊟ Release: 1 Persons Task	5 days	Mon 10-01-11	Fri 14-01-11		
6		Task A1	2 days	Mon 10-01-11	Tue 11-01-11		Person 1
7		Task A2	3 days	Wed 12-01-11	Fri 14-01-11	6	Person 1
8		Task A3	2 days	Mon 10-01-11	Tue 11-01-11		Person 2
9		Task A4	3 days	Wed 12-01-11	Fri 14-01-11	8	Person 2
10		Milestone	0 days	Fri 14-01-11	Fri 14-01-11	9,7	
11		⊟ Release: 2 Persons Task	5 days	Mon 17-01-11	Fri 21-01-11	5	
12		Task B1	2 days	Mon 17-01-11	Tue 18-01-11		Person 1
13		Task B2	3 days	Wed 19-01-11	Fri 21-01-11	12	Person 1
14		Task B3	2 days	Mon 17-01-11	Tue 18-01-11		Person 2
15		Task B4	3 days	Wed 19-01-11	Fri 21-01-11	14	Person 2
16		Milestone	0 days	Fri 21-01-11	Fri 21-01-11	15,13	

2. Next to the 'Resource Names' column there probably is not a 'Cost' column.

   If this column is not present, then right-click the last column and choose to 'Insert Column...' and select 'Cost' from the list.

❶	Task Name	Duration	Start	Finish	Prede	Resource Names	Cost
1	**Project: People-Placer**	0 days	Sat 08-01-11	Sat 08-01-11			$0.00
2	Milestone: Release 1 Persons	0 days	Fri 14-01-11	Fri 14-01-11	5		$0.00
3	Milestone: Release 2 Persons	0 days	Fri 21-01-11	Fri 21-01-11	11		$0.00
4							
5	⊟ **Release: 1 Persons Task**	**5 days**	**Mon 10-01-11**	**Fri 14-01-11**			**$0.00**
6	Task A1	2 days	Mon 10-01-11	Tue 11-01-11		Person 1	$0.00
7	Task A2	3 days	Wed 12-01-11	Fri 14-01-11	6	Person 1	$0.00
8	Task A3	2 days	Mon 10-01-11	Tue 11-01-11		Person 2	$0.00
9	Task A4	3 days	Wed 12-01-11	Fri 14-01-11	8	Person 2	$0.00
10	Milestone	0 days	Fri 14-01-11	Fri 14-01-11	9,7		$0.00
11	⊟ **Release: 2 Persons Task**	**5 days**	**Mon 17-01-11**	**Fri 21-01-11**	5		**$0.00**
12	Task B1	2 days	Mon 17-01-11	Tue 18-01-11		Person 1	$0.00
13	Task B2	3 days	Wed 19-01-11	Fri 21-01-11	12	Person 1	$0.00
14	Task B3	2 days	Mon 17-01-11	Tue 18-01-11		Person 2	$0.00
15	Task B4	3 days	Wed 19-01-11	Fri 21-01-11	14	Person 2	$0.00
16	Milestone	0 days	Fri 21-01-11	Fri 21-01-11	15,13		$0.00

3. These two people resources are yet to have their individual cost assigned.

    Select "Menu > View > Resource Sheet".

❶	Resource Name	Type	Material Label	Initials	Group	Max. Units	Std. Rate	Ovt. Rate	Cost/Use	Accrue At	Base Calendar	Code
1	Person 1	Work		P		100%	$0.00/hr	$0.00/hr	$0.00	Prorated	Standard	
2	Person 2	Work		P		100%	$0.00/hr	$0.00/hr	$0.00	Prorated	Standard	
3	Person 4	Work		P		100%	$0.00/hr	$0.00/hr	$0.00	Prorated	Standard	
4	Person 3	Work		P		100%	$0.00/hr	$0.00/hr	$0.00	Prorated	Standard	

4. Person_1's standard rate is $60.00/hour, whereas Person_2's standard rate is

    $50.00/hour.

❶	Resource Name	Type	Material Label	Initials	Group	Max. Units	Std. Rate	Ovt. Rate	Cost/Use	Accrue At	Base Calendar	Code
1	Person 1	Work		P		100%	$60.00/hr	$0.00/hr	$0.00	Prorated	Standard	
2	Person 2	Work		P		100%	$50.00/hr	$0.00/hr	$0.00	Prorated	Standard	
3	Person 4	Work		P		100%	$0.00/hr	$0.00/hr	$0.00	Prorated	Standard	
4	Person 3	Work		P		100%	$0.00/hr	$0.00/hr	$0.00	Prorated	Standard	

Now that these costs have been included for Person_1 and Person_2 there is an

immediate change to the project schedule's cost column per task.

		Task Name	Duration	Start	Finish	Prede	Resource Names	Cost
1		Project: People-Placer	0 days	Sat 08-01-11	Sat 08-01-11			$0.00
2		Milestone: Release 1 Persons	0 days	Fri 14-01-11	Fri 14-01-11	5		$0.00
3		Milestone: Release 2 Persons	0 days	Fri 21-01-11	Fri 21-01-11	11		$0.00
4								
5		⊟ Release: 1 Persons Task	5 days	Mon 10-01-11	Fri 14-01-11			$4,400.00
6		Task A1	2 days	Mon 10-01-11	Tue 11-01-11		Person 1	$960.00
7		Task A2	3 days	Wed 12-01-11	Fri 14-01-11	6	Person 1	$1,440.00
8		Task A3	2 days	Mon 10-01-11	Tue 11-01-11		Person 2	$800.00
9		Task A4	3 days	Wed 12-01-11	Fri 14-01-11	8	Person 2	$1,200.00
10		Milestone	0 days	Fri 14-01-11	Fri 14-01-11	9,7		$0.00
11		⊟ Release: 2 Persons Task	5 days	Mon 17-01-11	Fri 21-01-11	5		$4,400.00
12		Task B1	2 days	Mon 17-01-11	Tue 18-01-11		Person 1	$960.00
13		Task B2	3 days	Wed 19-01-11	Fri 21-01-11	12	Person 1	$1,440.00
14		Task B3	2 days	Mon 17-01-11	Tue 18-01-11		Person 2	$800.00
15		Task B4	3 days	Wed 19-01-11	Fri 21-01-11	14	Person 2	$1,200.00
16		Milestone	0 days	Fri 21-01-11	Fri 21-01-11	15,13		$0.00

NOTICE how Release 1 and Release 2 have a total cost of $4,400 each.

Hey, why don't you right-indent all of the tasks so that these roll up under the project's name and therefore obtain a total duration and total cost.

		Task Name	Duration	Start	Finish	Prede	Resource Names	Cost
1		⊟ Project: People-Placer	10 days	Mon 10-01-11	Fri 21-01-11			$8,800.00
2		Milestone: Release 1 Persor	0 days	Fri 14-01-11	Fri 14-01-11	5		$0.00
3		Milestone: Release 2 Persor	0 days	Fri 21-01-11	Fri 21-01-11	11		$0.00
4								
5		⊟ Release: 1 Persons Task	5 days	Mon 10-01-11	Fri 14-01-11			$4,400.00
6		Task A1	2 days	Mon 10-01-11	Tue 11-01-11		Person 1	$960.00
7		Task A2	3 days	Wed 12-01-11	Fri 14-01-11	6	Person 1	$1,440.00
8		Task A3	2 days	Mon 10-01-11	Tue 11-01-11		Person 2	$800.00
9		Task A4	3 days	Wed 12-01-11	Fri 14-01-11	8	Person 2	$1,200.00
10		Milestone	0 days	Fri 14-01-11	Fri 14-01-11	9,7		$0.00
11		⊟ Release: 2 Persons Task	5 days	Mon 17-01-11	Fri 21-01-11	5		$4,400.00
12		Task B1	2 days	Mon 17-01-11	Tue 18-01-11		Person 1	$960.00
13		Task B2	3 days	Wed 19-01-11	Fri 21-01-11	12	Person 1	$1,440.00
14		Task B3	2 days	Mon 17-01-11	Tue 18-01-11		Person 2	$800.00
15		Task B4	3 days	Wed 19-01-11	Fri 21-01-11	14	Person 2	$1,200.00
16		Milestone	0 days	Fri 21-01-11	Fri 21-01-11	15,13		$0.00

The project now has a total [Cost] of $8,800, and a total [Time] duration of 10 days.

5. The 3rd & 4th persons have arrived for the second release, but unfortunately they are a bit more expensive at $80.00/hour and $90.00/hour respectively.

	Resource Name	Type	Material Label	Initials	Group	Max. Units	Std. Rate	Ovt. Rate	Cost/Use	Accrue At	Base Calendar	Code
1	Person 1	Work		P		100%	$60.00/hr	$0.00/hr	$0.00	Prorated	Standard	
2	Person 2	Work		P		100%	$50.00/hr	$0.00/hr	$0.00	Prorated	Standard	
3	Person 4	Work		P		100%	$80.00/hr	$0.00/hr	$0.00	Prorated	Standard	
4	Person 3	Work		P		100%	$90.00/hr	$0.00/hr	$0.00	Prorated	Standard	

With these rates for Person_3 and Person_4 included then there is an immediate change to the project schedule's cost column per task.

	Task Name	Duration	Start	Finish	Prede	Resource Names	Cost
1	⊟ Project: People-Placer	7.5 days	Mon 10-01-11	Wed 19-01-11			$10,000.00
2	Milestone: Release 1 Person	0 days	Fri 14-01-11	Fri 14-01-11	5		$0.00
3	Milestone: Release 2 Person	0 days	Wed 19-01-11	Wed 19-01-11	11		$0.00
4							
5	⊟ Release: 1 Persons Task	5 days	Mon 10-01-11	Fri 14-01-11			$4,400.00
6	Task A1	2 days	Mon 10-01-11	Tue 11-01-11		Person 1	$960.00
7	Task A2	3 days	Wed 12-01-11	Fri 14-01-11	6	Person 1	$1,440.00
8	Task A3	2 days	Mon 10-01-11	Tue 11-01-11		Person 2	$800.00
9	Task A4	3 days	Wed 12-01-11	Fri 14-01-11	8	Person 2	$1,200.00
10	Milestone	0 days	Fri 14-01-11	Fri 14-01-11	9,7		$0.00
11	⊟ Release: 2 Persons Task	2.5 days	Mon 17-01-11	Wed 19-01-11	5		$5,600.00
12	Task B1	1 day	Mon 17-01-11	Mon 17-01-11		Person 1,Person 3	$1,200.00
13	Task B2	1.5 days	Tue 18-01-11	Wed 19-01-11	12	Person 1,Person 3	$1,800.00
14	Task B3	1 day	Mon 17-01-11	Mon 17-01-11		Person 2,Person 4	$1,040.00
15	Task B4	1.5 days	Tue 18-01-11	Wed 19-01-11	14	Person 2,Person 4	$1,560.00
16	Milestone	0 days	Wed 19-01-11	Wed 19-01-11	15,13		$0.00

The inclusion of Person_3 and Person_4 has increased the project's total cost ($10,000), though it did shorten the project's duration by 2.5 days to 7.5 days.

But, what if Person_3 and Person_4 cost the same as the first two people?

	Resource Name	Type	Material Label	Initials	Group	Max. Units	Std. Rate	Ovt. Rate	Cost/Use	Accrue At	Base Calendar	Code
1	Person 1	Work		P		100%	$60.00/hr	$0.00/hr	$0.00	Prorated	Standard	
2	Person 2	Work		P		100%	$50.00/hr	$0.00/hr	$0.00	Prorated	Standard	
3	Person 4	Work		P		100%	$60.00/hr	$0.00/hr	$0.00	Prorated	Standard	
4	Person 3	Work		P		100%	$50.00/hr	$0.00/hr	$0.00	Prorated	Standard	

	❶	Task Name	Duration	Start	Finish	Prede	Resource Names	Cost
1		⊟ Project: People-Placer	7.5 days	Mon 10-01-11	Wed 19-01-11			$8,800.00
2		Milestone: Release 1 Person	0 days	Fri 14-01-11	Fri 14-01-11	5		$0.00
3		Milestone: Release 2 Person	0 days	Wed 19-01-11	Wed 19-01-11	11		$0.00
4								
5		⊟ Release: 1 Persons Task	5 days	Mon 10-01-11	Fri 14-01-11			$4,400.00
6		Task A1	2 days	Mon 10-01-11	Tue 11-01-11		Person 1	$960.00
7		Task A2	3 days	Wed 12-01-11	Fri 14-01-11	6	Person 1	$1,440.00
8		Task A3	2 days	Mon 10-01-11	Tue 11-01-11		Person 2	$800.00
9		Task A4	3 days	Wed 12-01-11	Fri 14-01-11	8	Person 2	$1,200.00
10		Milestone	0 days	Fri 14-01-11	Fri 14-01-11	9,7		$0.00
11		⊟ Release: 2 Persons Task	2.5 days	Mon 17-01-11	Wed 19-01-11	5		$4,400.00
12		Task B1	1 day	Mon 17-01-11	Mon 17-01-11		Person 1,Person 3	$880.00
13		Task B2	1.5 days	Tue 18-01-11	Wed 19-01-11	12	Person 1,Person 3	$1,320.00
14		Task B3	1 day	Mon 17-01-11	Mon 17-01-11		Person 2,Person 4	$880.00
15		Task B4	1.5 days	Tue 18-01-11	Wed 19-01-11	14	Person 2,Person 4	$1,320.00
16		Milestone	0 days	Wed 19-01-11	Wed 19-01-11	15,13		$0.00

With the cheaper Person_3 and Person_4 then the total cost of the project ($8800) is the same as the original schedule.

This example has demonstrated that while **adding [People] "resources" to a project can reduce the total [Time] duration of the scheduled tasks, these additional [People] "resources" can also increase the project's total [Cost]**.

Therefore, the decision comes down to what is better for the project; a reduced [Time] duration with a potential increase in the associated [Costs], or keeping project's expenses down while potentially having the project delivered late.

Reviewer's Comment ... Don't calculate cost via the schedule.

WOOH!!! Using MS-Project for cost calculations !!!

This is one situation where you should definitely use a spreadsheet. Because, cost calculations via the schedule will require a constant balancing of getting 100% people utilization which will mean that the project manager will spend more time in MS-Project than they would managing the people working on that project.

Accepted ... however, what I was trying to demonstrate by including the [Costs] calculation in the schedule example was how the [People] and the [Resources] assigned to the tasks will affect the project's [Costs].

IMHO, the project schedule should only be used for what the Gantt Chart is good for; and that is for the tracking of time, the determination of the percentage complete, and the assignment of people & resources to tasks.

I would recommend that ... you should obtain from the organization's senior management or the finance department a project budget calculation spreadsheet and replicate in this spreadsheet the project's tasking (as well as the people and resourcing).

In this spreadsheet, create an individual worksheet for each of the major work packages that were determined during the progressive elaboration of the Work Breakdown Structure (WBS) and schedule. Then once each worksheet has replicated that portion of the project schedule then accumulate this cost information on a "summation" worksheet at the front of the spreadsheet.

DO NOT use MS-Project for cost and budgeting calculations.

Instead, use a spreadsheet that has been tailored for such a purpose.

Only use Ms-Project for [Time], [People] & [Resource] analysis.

How do I present these Cost Estimates?

At this point don't be overly concerned with how this costing information will be presented to the customer as it is reasonable to assume that the performing organization's senior management (and the finance group) will want to modify this [Cost] estimate to include **"Sell Price" mark-up**, **safety margins** (*e.g. Cost Escalation for exchange rates & inflation*, **Cost Contingencies** *for warranty repairs & latent defects*), and a **percentage for project risk**. Thereby ensuring that this project will most likely be a **profitable endeavour**.

 I would recommend that ... the budget presented conform to the "standardized" template (**Cost Account Authorization** spreadsheet) that is used for all of the performing organization's projects, as the profitability margin calculations have probably already been included.

The cost ~~estimate~~ calculations information should include such details as:

- **How** were the cost estimates **determined**?
  What techniques were used to calculate these costs?

- **What** is the perceived **accuracy** / precision of these cost estimates? *E.g. ± 10%.*

- **What** is the **range** of the cost estimates? *E.g. $275K to $300K.*

- **What** is the cost **contingency** buffer (risk factoring) and how was this decided?
  E.g. 5% of the subtotal costs.

- **What assumptions** were made when determining these cost estimates?
  E.g. equipment leasing or purchasing, exchange rates, inflation.

- **What constraints** were imposed on the determination of these cost estimates?
  E.g. expenditure has to be relatively level over every quarter instead of having individual months with disproportionately high capital outlays (i.e. seeking positive cashflow "revenue" into the performing organization over "expenditures").

- **What** is the level of **confidence** with these cost estimates?
  What are the reasons why and why not for this confidence?

NOTE   ☐ The [Cost] calculations have to be presented in the form that was deemed necessary by the performing organization's primary-core stakeholders.

☐ The [Cost Baseline] DOES NOT include any "Management Reserves" for unplanned costs to the project that could not be reasonably predicted.

With the 2nd Pass schedule's [Scope] tasks, [Time] durations, [People], [Resources], and [Costs] determined, then the next step is to seek **official acceptance and approval of the schedule and budget**. Then *"freeze"* this information as the project's **[Time Baseline]** and **[Cost Baseline]**.

What if these Baselines need to be changed?

Once the [Scope Baseline], [Time Baseline], and [Cost Baseline] have been signed-off, **IF changes / deviations are required to any of these authorized baselines, THEN some form of Baseline Change Request (BCR) process will be needed**. This BCR process would involve aspects of Risk Management [Section 10.1], Change Management [Section 10.2], and Stakeholder Management [Section 10.3]. Where representatives from the customer organization and the performing organization (and Subject Matter Experts) would come together in a **Change Control Board (CCB)** [Section 10.2.1] to; (1) determine the necessity of the change, (2) mutually agree on the revisions to the relevant baselines, and (3) authorize the revised baselines. See [Chapter 16] on Project Rescue & Project Recovery.

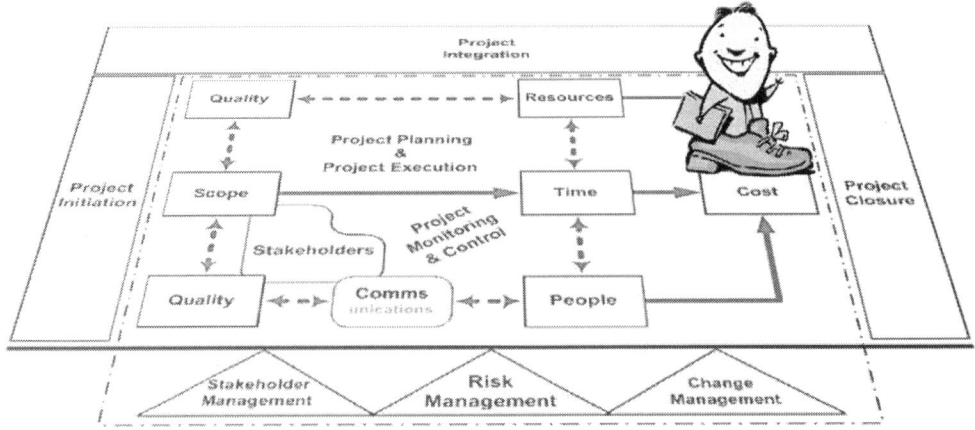

What if there is only a limited Budget available?

When [Cost] is the major delimiting factor for the project (i.e. the prominent project constraint) then the [Scope] of the project should be broken up into work packages with an estimated [Time] duration and an associated price-tag per work package.

The representatives of the customer organization and the performing organization would then come together and figure-out (i.e. negotiate) the **most cost-effective combination of work packages to result in the most beneficial composition of features & functionality**; i.e. **maximizing the "return on investment"**. ... *What is the best Bang for Buck?*

This selected [Scope] would then constitute the scope boundary for the first major milestone release of the project that would be delivered first as a mini project.

As the first major milestone release nears completion (delivery) then the customer's representatives would figure out exactly how much money they have remaining in their budget. Then as previously, the customer in consultancy with the performing organization would determine the next combination of remaining work packages that would provide the most cost-effective composition of desired features & functionality that would be implemented in the next release with the remaining budget.

So-on & so-forth for each subsequent release until either; there is not enough of the budget [Cost] remaining, there is not enough [Time] remaining for another release, or all of the desired [Scope] has been implemented to the satisfaction of the customer's representatives. ... *By which time, the customer may have been able to scrape together the necessary budget to fund the additional development, and in some cases loosening their paymaster's purse strings sufficiently to fund all of the outstanding work. ... But first, they need actual tangible proof of the project's potential for success.*

This is similar to a sprint backlog as used for an agile project, [Section 7.4]. However, this would be for a waterfall or iterative implementation of months / years duration instead of an agile "sprint" of a few weeks duration.

War Story ... The dangers of PDOOMA estimates.

From my experience as an engineer, the common causes for projects being significantly over budget and delivered late are due to:

✗ The deplorable estimations for the "real world" [Time] scales required to undertake the technical & production activities.

✗ The pitiful estimations for the "real world" [Costs] involved with such technical & production activities.

✗ A lack of understanding of the skilled [People] that are required to participate in such technical & production activities.

✗ An oversight on the quantities & speciality of the [Resources] required and utilized by such technical & production activities.

So please, ADD buffering for these technical & production activities. Better still, DO NOT let non-technical people guesstimate (or as one engineer succinctly put it, "butt pluck") the Time, Cost, People, and Resources involved with these technical & production activities as they will only "stuff it up".

PDOOMA
PULLED DIRECTLY OUT OF MY ASS

WHAT MORONS BUTT PLUCKED THESE DEPLORABLE ESTIMATES ?

4.3.4.   Question: "How will WE KNOW WE GOT IT RIGHT?"

How will we know we got it right?

As stated previously in [Section 1.3], the decision on whether a project is deemed a success or a failure is based on the expectations & perspectives of the project's stakeholders. Therefore, to "*Get It Right*" will require that these stakeholders' expectations & perspectives be managed appropriately.

However, the effectiveness & efficiency of the communications with these stakeholders will significantly influence these stakeholders' expectations & perspectives.

4.3.4.1.   Get It Right: Communications

To ensure success, the Planning Phase must consider the communications aspect.

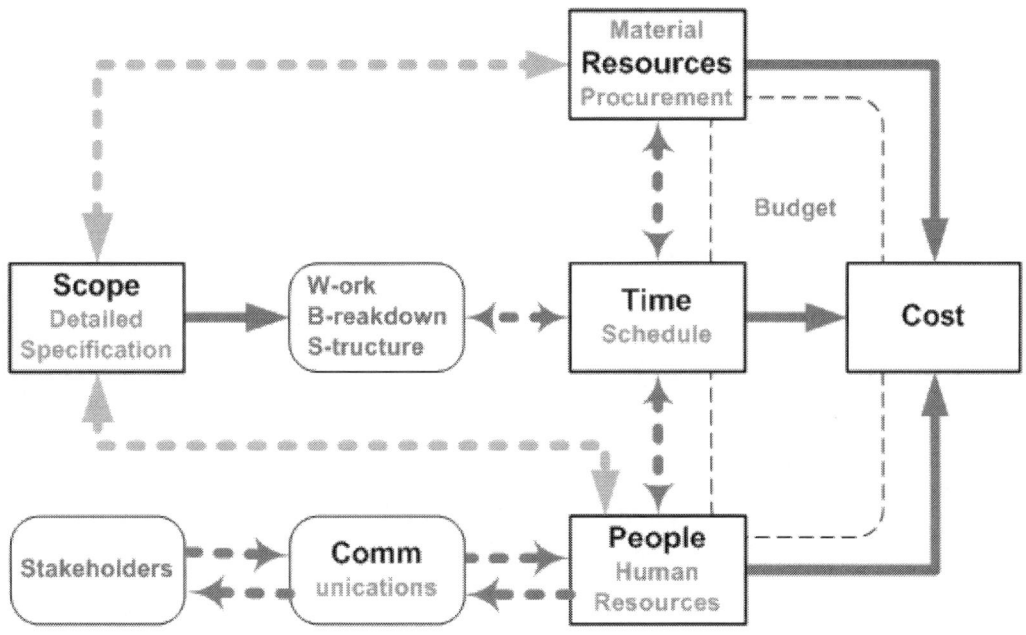

Figure 48: The Planning Process Model including the communications aspect.

From [Figure 48], notice that the **communications are bi-directional** between those [People] working on the project and the project's stakeholders.

How else will you know if you are satisfying their expectations and are in tune with their perspectives unless you communicate with them? "HELL LOWW"

Now, with communications the old adage of *"see no evil, hear no evil, speak no evil"* comes into play. If the people that are working on the project are not providing the project's stakeholders with the information that they require, when they need that information, in a form that they can use, then these stakeholders will not be able to make the appropriate adjustments & corrections to the project in sufficient time when the project's current situation and the prevailing circumstances necessitates.

That is, if they don't know that the bus is coming, if they can't hear the bus, if they don't see the bus, and if they are not told about the bus ... then it should not be a surprise when they get **SPLATTERED** by the bus.

⊠ Poor bi-directional communications is a project killer.

Effective & Efficient Communications

Communications has to be both effective & efficient.

❖ **Effective communications** – are providing the information in the format they need, when they need it.

❖ **Efficient communications** – are providing only the relevant information they need.

Two very similar sounding words, but these have major differences when it comes to satisfying the project's primary-core stakeholders.

The reality is, **the higher up the management hierarchy** then generally **the less quantity of information per project they require but the better the quality of the information that they need.** This is because, senior management don't have time to read or listen to copious amounts of information; rather, they need it in a concise summation of the facts & recommendations so that they can quickly make a rational decision or they can ask for more information on the specific area of concern to them.

Hence, always try to ensure that the latest information is available and provide the most up-to-date (visual) representation of the project's true status, as these primary-core stakeholders' satisfaction is dependent on this information.

Therefore, the project's communications needs to consider such things as:

● **What are the communications methods & mediums** that will be used internally within the project team and externally between the performing organization and the customer's organization?

Additionally, consider the communications with subcontractors, outsourced vendors, and suppliers.

Also, consider communications with environmental stakeholders such as other members of staff and the general public at large.

For example; face-to-face meetings, video conferencing, emails, telephone, bulletin boards, information packs ... etc.

- **What are the communications paths and escalation hierarchy?**

    For example; does all communications have to go up through the performing organization's hierarchy then across to the customer's organization then down its hierarchy and visa-versa, or are certain direct peer-to-peer communications permitted between the performing organization's project team members to their corresponding equivalents in the customer's organization.

- **What are the Points Of Contact (POC)** in the performing organization and the customer's organization (and with subcontractors, vendors, and suppliers)?

    For example; who are the involved parties project managers, subject matter experts, senior management, quality assurance, account managers ... etc.

- **What are the standardized communications protocols** and formats to be used?

    For example; report templates, plan templates, document templates, forms, Risks & Issues Register, presentation materials ... etc.

Whenever possible try to remain consistent with the formats used by the rest of the organization; as there are probably good (work practices) reasons for these current formats.

- **What tools, applications, and facilities will be used for communications, document storage, and archiving?** See [Chapter 18] on Project Information Management.

- **What is the frequency of the communication?**

    For example; monthly customer project updates, weekly project reports, and daily 15 minute "stand-up" Project Implementation Team meetings ... etc.

 I would recommend that ... where possible use the existing communications methods, procedures, and templates that are in common usage within the performing organization. This way, at least those stakeholders within the performing organization will already have a generalized expectation of what will be the forms of communication and what will be the timings for such communications.

Lessons Learnt: Communications, some things to watch out for.

1) **DO NOT have a meeting without some idea of what is the productive purpose of that meeting.**

   - Let us be honest, we have all sat through a meeting or few thinking;
     "well this is a complete waste of my time".
     Just remember that **"time is money", and every bum on a seat in that meeting is burning through money.** Therefore, before requesting the meeting consider **who exactly has to be present for the meeting to be productive.**

   - If a meeting produces no actions other than a lot of *"we will discuss this after the meeting"* then **rather than a meeting, one-on-one discussions** should have been undertaken **with the outcomes summarized and distributed to the other members via email.**

   - **7-8 persons are about the maximum number of attendees for a productive meeting;** any more than this and you will end-up with a few meeting-passengers (i.e. non-contributors) whose time could probably be better utilized elsewhere.

   - An **odd number of attendees are good for** a meeting where **a decision** must be made, as there (subconsciously) is always someone to act as a **tiebreaker.**

   - **When meetings become epidemic in an organization / business unit then productivity & efficiency decline** because the "workers" are constantly delayed waiting for the "decision makers" to come out of the latest meeting.

2) **DO NOT hold a meeting without first having issued an agenda prior to the meeting or at least providing a "heads-up" reason for holding the meeting**, and give the meeting's participants enough advance notice to prepare for the meeting.
   This may require issuing the agenda a few days before hand. ... *Especially when dealing with external parties such as the customer representative and subcontractors ... when some form of resolution is required.*

3) **DO NOT conclude a meeting without summarising the outcomes and reiterating the required actions to be undertaken. If minutes are to be produced then these should be distributed within a day of the conclusion of the meeting.** Also, provide written summation of the outcomes and more specifically the actions, who has been assigned to which actions, and when each action is expected to be completed by.

4) **Verbal communications** with a project's stakeholder that results in a decision or a new / changed action **should be reiterated in a summation written form.**

   For example; an email summarizing what was and was not agreed to.

5) Consider **all emails whether they be internal or external as legal documents;** so watch what you type, as it could come back to haunt you.

6) **Social networking (e.g. Facebook, Twitter, and blogs) is not your friend. Consider anything written about your project** (by anyone internal or external to the project) **as being in the public domain and directly viewable by your customers and your senior management.** ... *Yep, joking about the crappy project ain't good.*

COMMUNICATIONS ... THAT BRIDGE BETWEEN THOSE PEOPLE WHO IMPLEMENT THE PROJECT AND THE STAKEHOLDERS WHO EXPECT QUALITY DELIVERABLES.

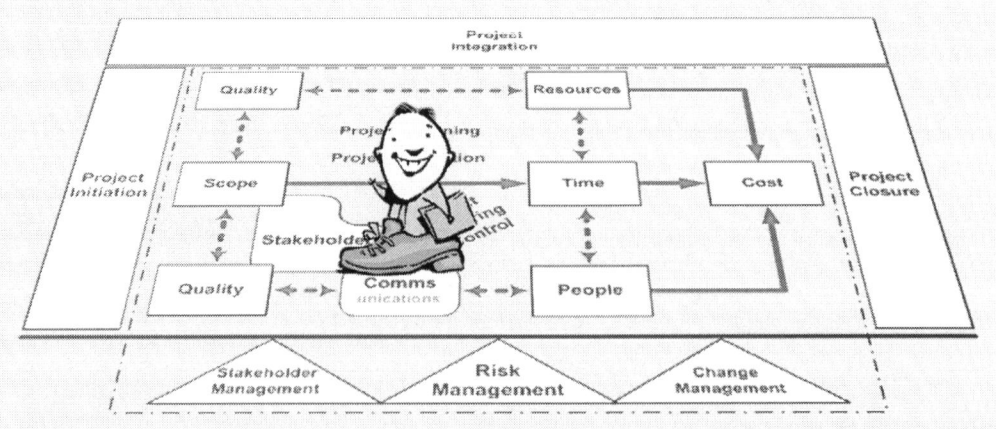

4.3.4.2. Get It Right: Quality

Now that we have established the importance of communications between the project's stakeholders, then the next concern for *"Getting It Right"* is ensuring that the project stakeholders perceive that the deliverables ~~meet the requirements~~ meet the approved specifications and those deliverables are fit for use.

For the customer, the true measure of quality is analogous to, "the taste of the cake and not the processes used to bake the cake".

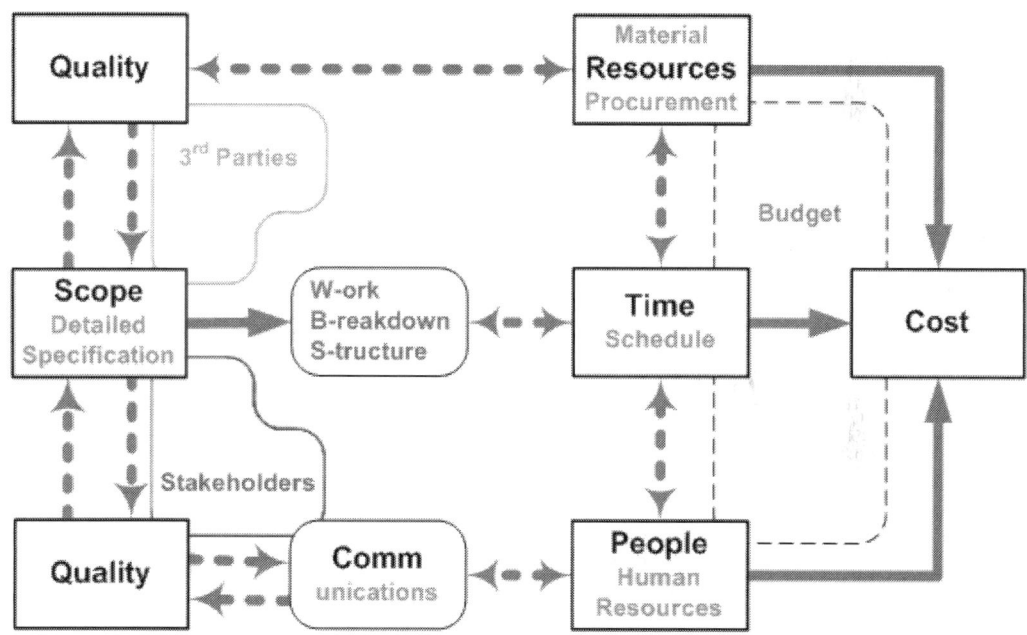

Figure 49: The Planning Process Model including the quality aspect.

However, **from the performing organization's perspective, quality is concerned with the correctness of the project's measurable & tangible deliverables when compared to the agreed [Scope] and the agreed Acceptance Criteria.** See [Figure 50].

Quality does not mean ~~exceeding the Customer Requirements~~ ... exceeding the approved Detailed Specifications; i.e. a "gold plated" output if not specified is thus not a desired level of quality output.

You might think that the purpose of quality processes is to ensure that the project's deliverables satisfy the needs for which they were intended. However, this is not the case because **to produce a quality deliverable is very much dependent on the definition and description of that specific deliverable** (i.e. agreed [Scope]) **and the expectations of the end-users** (i.e. the stakeholders). As depicted previously in [Figure 49], [Quality] has a binding relationship with; the project's [Scope], the project's stakeholders, the [People] implementing the deliverables, and how these persons are communicated with.

Consequently, **poorly defined project [Scope] and/or poor communications with the project's stakeholders will result in their "needs" and expectations not being clearly understood, which can only result in a poor [Quality] deliverable irrespective of the "grade" of the deliverable**. … *What, hare, that went straight over my head.*

What's the difference between Quality and Grade?

❖ **Grade** – is the level or category of the end-product due to **the number of features that it contains, or its characteristics when compared to its competitors and siblings.**

> *For example; each brand-name mobile phone producer has one model that is their premium phone for that year, even though that particular model of phone is manufactured in the same industrial plant (with the same Quality Assurance and Quality Control processes & procedures) as its entry-level sibling models.*

❖ **Quality** – is how well the "end-product" **meets its specifications and is fit for use.**

> *For example; application forced closures (i.e. crashes), the number of defects encountered, the rate of failure / **Mean Time Between Failure (MTBF)** … etc.*

The grade of the deliverable will be affected by increasing or decreasing the project's [Scope]. However, these **increases / decreases in [Scope] will not necessarily affect the [Quality]** of the deliverable.

What will definitely affect the [Quality] of the deliverable are the [People] & [Resources] utilized, and the Quality Assurance & Quality Control processes that are in use.

What's the difference between Quality Assurance and Quality Control?

❖ **Quality Assurance** – is concerned **with the processes & procedures used to produce the deliverable**, so that the resultant deliverable will conform to the **agreed Acceptance Criteria**; i.e. "**doing**".

❖ **Quality Control** – is concerned **with evaluating & verifying that the deliverable does meet the agreed Acceptance Criteria**; i.e. "**checking**".

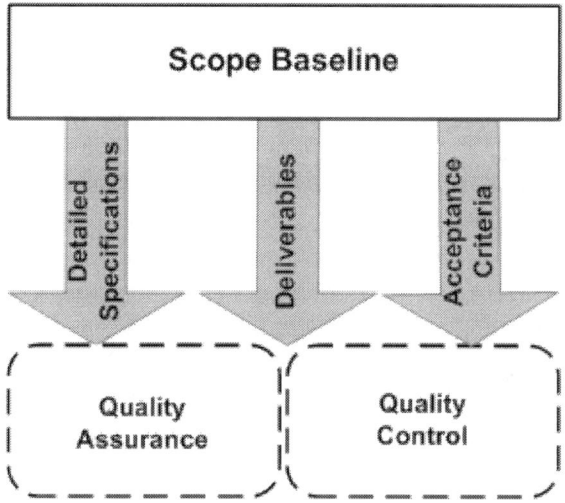

Figure 50: Relationship of Quality Assurance & Quality Control to the Scope Baseline.

With this distinction made, it is common to find projects where there is **an active attempt near the end of the project to inspect quality into the deliverables rather than to have designed & built quality in during the project's life cycle.**

That is, having firstly worried about "getting it done" then secondly worrying about "getting it right". Unfortunately, this approach to quality usually **results in a lot of rework while trying to get the deliverables to meet the agreed Acceptance Criteria.**

⊠ The lack of an appropriate amount of Quality Assurance & Quality Control is a project killer.

> At this point in this book, I would like to advice that Quality Assurance & Quality Control (as well as Process Improvement & Change Control) will make a re-appearance in [Chapter 5] for the Executing Phase, and in [Chapter 6] and [Chapter 8] for the Monitoring & Control Phase.
>
> However, for the moment I don't wish to go into these topics in any more depth, because this would involve covering the "PLAN – DO – CHECK – ACT" process model that those later chapters will examine in detail.

With this basic understanding of quality and its importance, the next step is to plan how to integrate [Quality] into the Executing Phase of the project.

To do this, one **needs to identify those quality requirements that must be complied with**; *e.g. standards, certifications, policies, procedures, workflows, checklists... etc.*

> For more information on [Quality] Monitoring & Control, please refer to [Section 8.1].

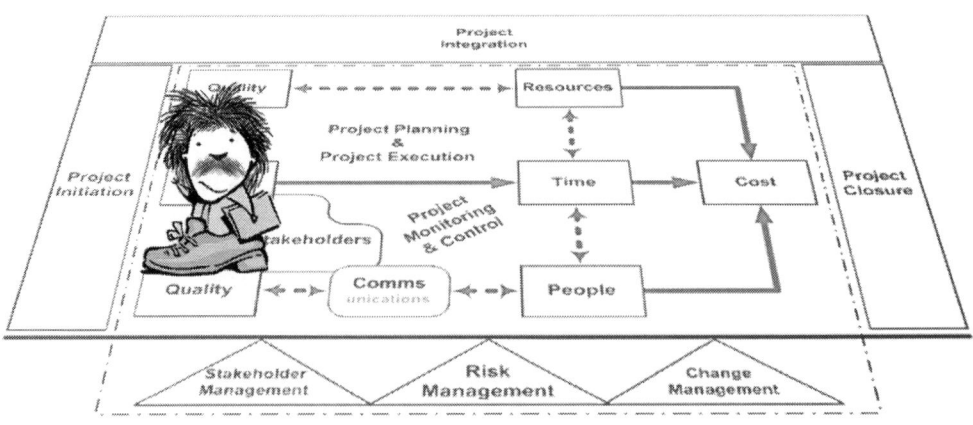

4.3.5. Question: "What do WE THINK CAN GO WRONG?"

Now that we have finished mapping out the process flow relating to the project variables & project constraints then the next thing to consider is.

What do we think can go wrong?

As mentioned back in [Section 1.6], **the <Risk> to the project is the balancing act of controlling the project variables & project constraints**. Hence, the Project Planning Process model will need to consider the risks aspect; see [Figure 51].

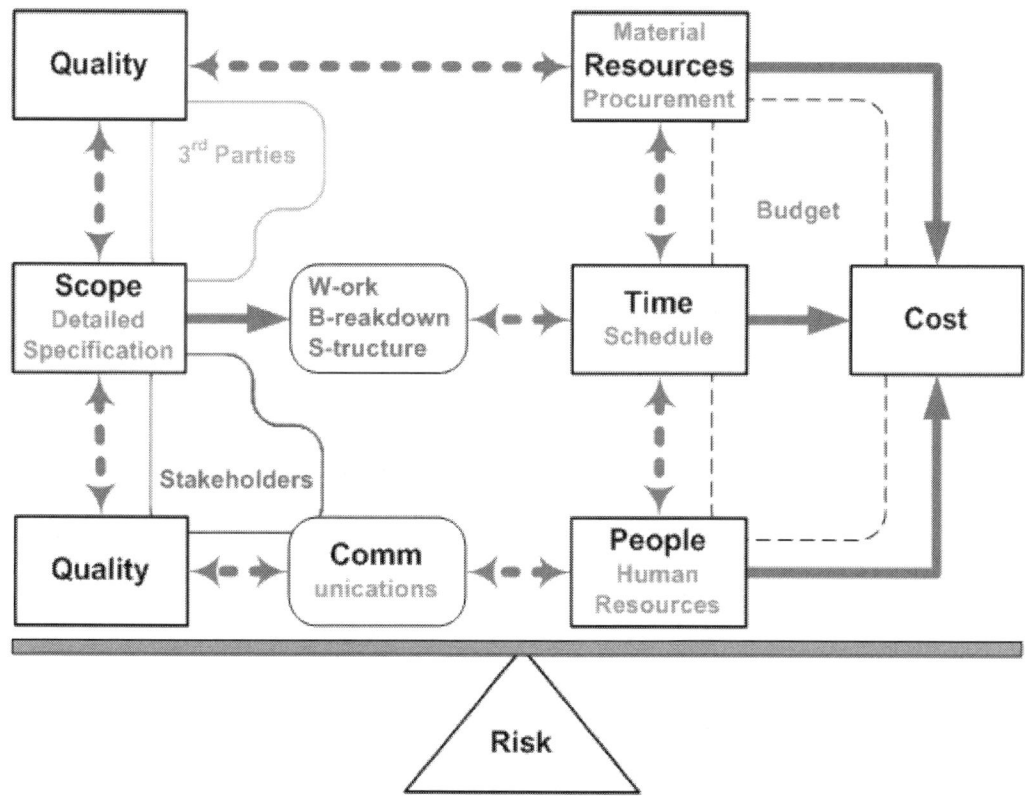

Figure 51: The Planning Process Model with consideration for Risk.

4.3.5.1. Do Not Get It Wrong: Risk

To answer the question of *"what do we think can go wrong"* will require an understanding of what risks are, planning for when risk eventuates, and knowing how to handle the response to a risk once it has evolved into an issue.

What's the difference between a Risk and an Issue?

❖ <u>Risk</u> – **could happen in the future**, whereas

❖ <u>Issue</u> – **has happened, or is happening right now,** or is definitely about to happen real soon from now.

A risk is an uncertain event or situation that could affect one or more of the project constraints and/or project variables and thereby **endanger the achievement of any number of the project's outcomes & deliverables.**

A risk becomes an issue when it prevents the project from delivering the agreed [Scope], by the agreed [Time], within the approved [Cost] budget, and/or with the agreed level of [Quality].

⊠ **Unmanaged Risks and out-of-control Issues are a project killer.**

Hypothetically, the effect of a risk could be either positive or negative; but most people think of the word "risk" as having negative connotations whereas the word "opportunity" is seen in a positive light.

This book will consider **three very broad categories of risks**:

1) **Inherent Risk** - is that risk due to the nature of the project being undertaken. An analogous example; being killed is a risk for a soldier who is being sent to a war-zone.

   For example; a project implemented using a new chip-set for a technology whose international standard has yet to be ratified risks having to be reworked to become standards compliant.

2) **Control & Management Risk** - is that risk directly related to the management and control of the project's constraints & project variables.

   For example; change control [Scope], time management [Time], financial management [Cost], stakeholder management [People], procurement management [Resources] & [People], Q&A management [Quality].

3) **Confrontational Risk** - is that risk which simply comes into being and confronts the project.   *... "Shit happens", so better get used to dealing with it.*

   For example; rolling power blackouts, transport strikes, economic / financial crisis, changing market trends, customer shifts, another project that is dramatically increased in priority.

Scratch those above broad categories of risks; because the truth is there are so many different ways that risks could be categorized with respect to the project. Hence, **choose risk categories that make most sense to the project's stakeholders and contribute best to the management of the project.**

   For example; architectural, software, hardware, user interface, testing, infrastructure, supplier, outsourcing, environmental ... functionality [Scope], budget [Cost], scheduling [Time], human resource [People], procurement [Resources] ...etc.

In addition to those previously mentioned "risk & issues", risks can also involve risks related to the interaction with the project stakeholders (i.e. Stakeholder Management), as well as risks related to changes made to the project's scope (i.e. Change Management).

NOTE ☐ **Risk Management** (aka Risk Monitoring & Control) **needs to start now during the Planning Phase and NOT wait until the Executing Phase.**

This Risk Management also has to be an ongoing iterative activity performed throughout the entire project life cycle.

And, not just when the proverbial hits the fan.

 For more information on Risk Management, please refer to [Section 10.1].

 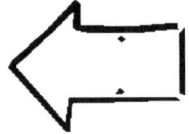

GO ON, TAKE A RISK ... I DARE YOU.

SECTION 10.1

WHAT ARE YOU ... CHICKEN?

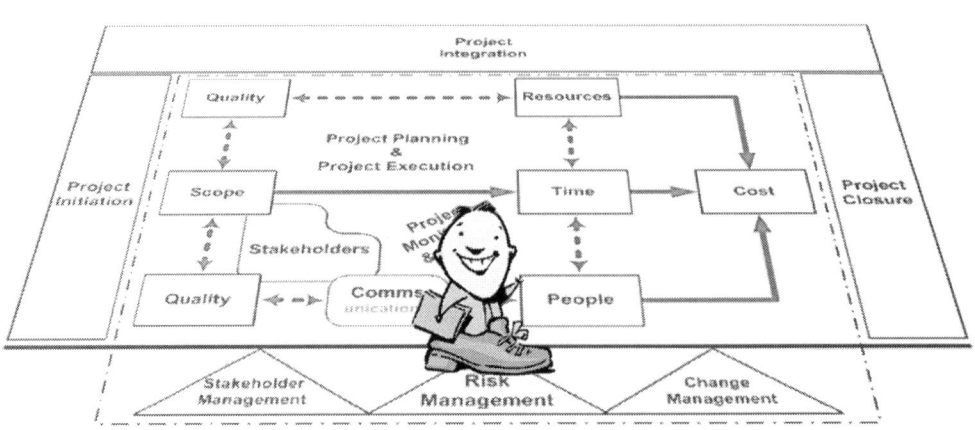

4.4.   Outputs

4.4.1.   Output: Documents & Deliverables

Now that the big questions have been answered, the next step is to ensure that these answers have been documented. The resultant documentation from the Planning Phase is highlighted below in [Figure 52], and the inter-relationship of this documentation to the other documents produced during the project's life is illustrated in [Figure 53].

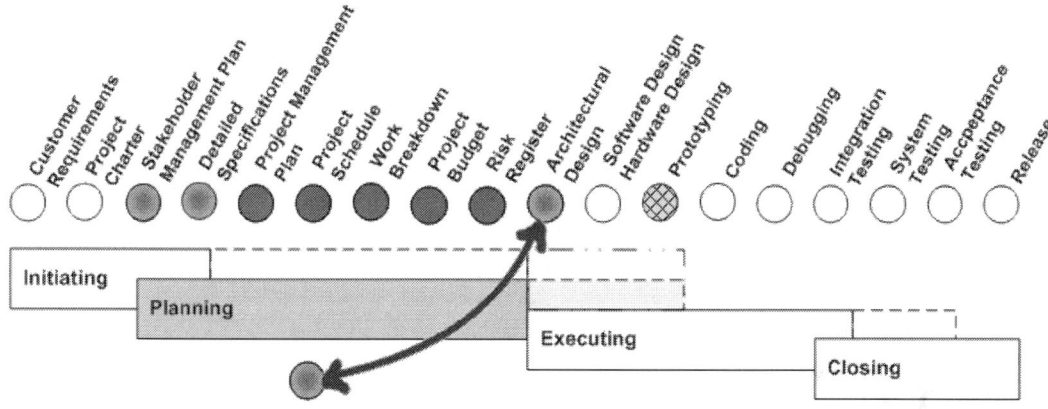

Figure 52: Documents & deliverables output of the Planning Phase.

NOTICE how in [Figure 52] there is potential for crossover between the Initiating Phase, the Planning Phase, as well as with the Executing Phase.

This crossover could occur because for some projects to be planned successfully, one has to better understand the solution's design and the system architecture that will be used. Hence, the production of these documents & deliverables may occur earlier than shown above in [Figure 52].

⊠ **Failure to plan ... ...** ADEQUATELY
is planning to fail ... DRAMATICALLY!

4.4.2. Output: PMBOK-SDLC Map

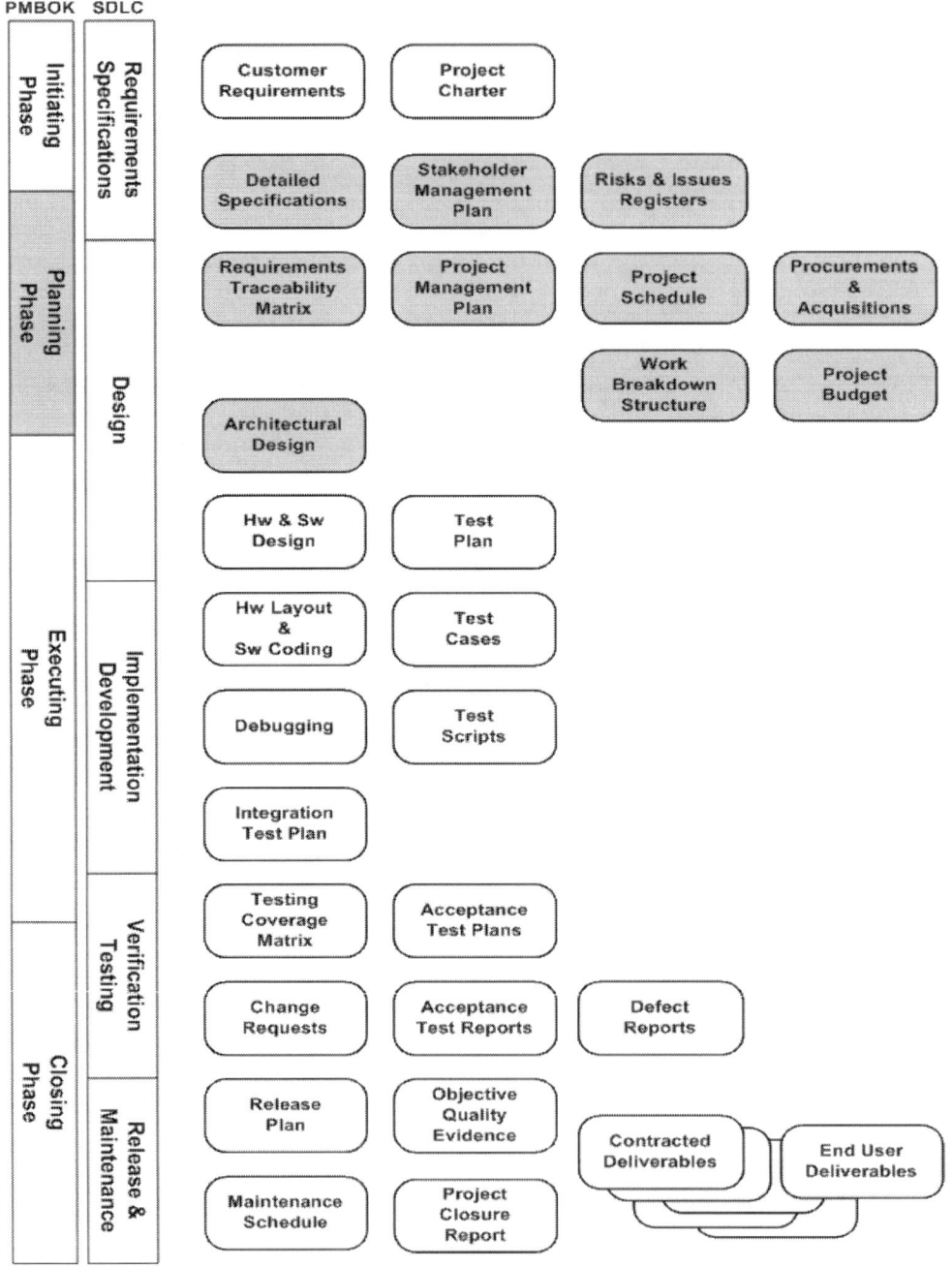

Figure 53: Relationship of PMBOK - SDLC and the documents & deliverables produced for the Planning Phase.

Minimalist Set of Documents & Deliverables

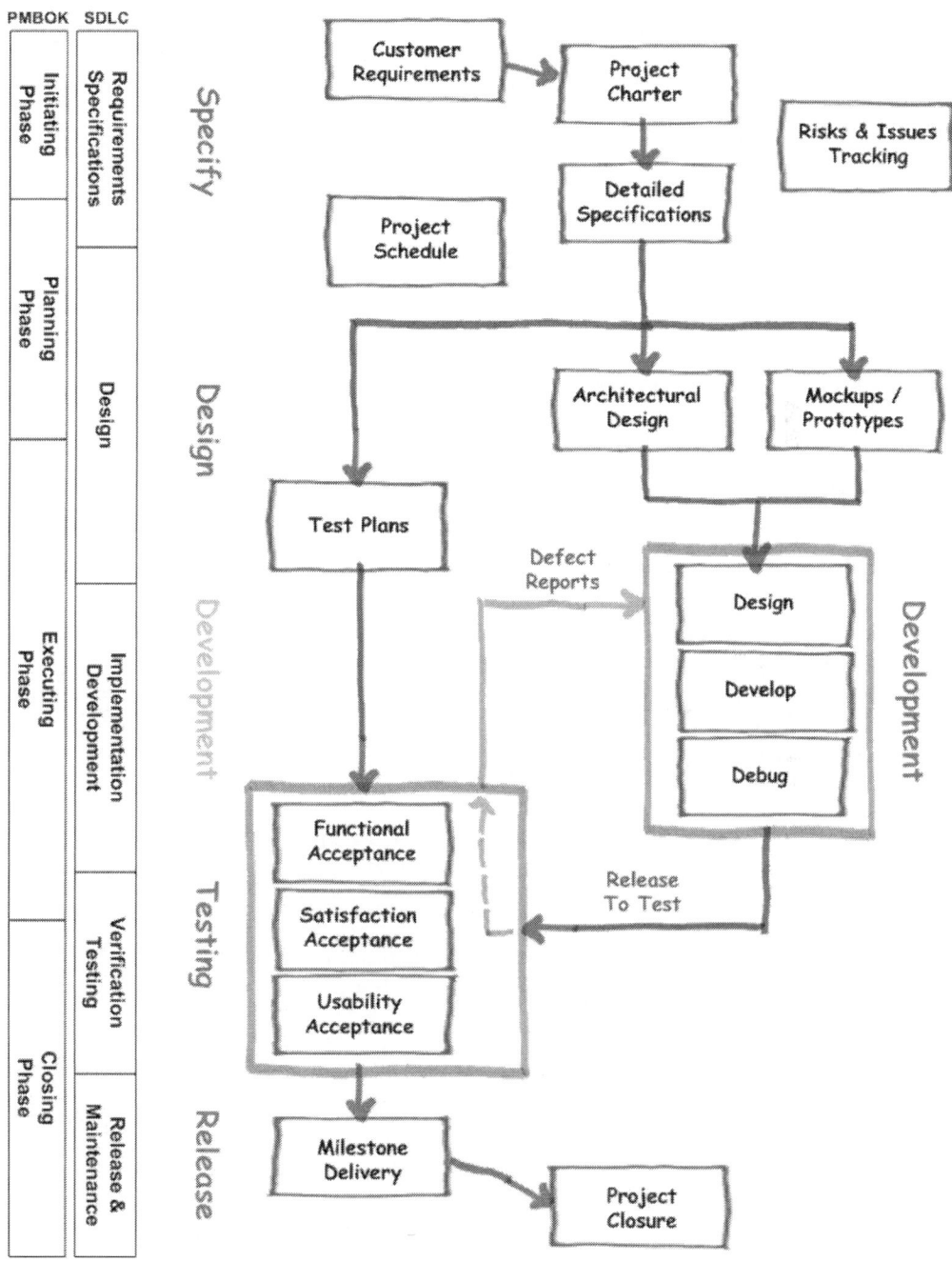

Figure 54: Minimalist set of documents & deliverables necessary to even have a chance of delivering the project successfully.

4.4.3. Output: Project Management Plan

 Just be wary when you're asked by senior management to produce a **"Project Plan"** because sometimes they may simply be **referring to a Project Schedule** where as others **may be referring to** a complete **Project Management Plan.**

For a small organization, or a small project, (or with management that don't have extensive project management training) they may be referring to just a Project Schedule; i.e. Gantt Chart. However, for stakeholder satisfaction (and your perceived worth as a project manager) remember to **give the primary-core stakeholders what they "want" and not what project management guidelines recommend.**

The Project Management Plan (PMP) can be used for a **few different purposes**:

1) **As a list of instructions on how the project**'s execution, monitoring, and closure **will be controlled** and how changes will be handled. Thereby, mitigating ambiguities over how certain situations & scenarios will be dealt with. *.. i.e. The Rules Of Play.*

2) **As a common reference point** for all of the project's stakeholders to come to, and thereby obtain a consistent understanding of how the project will be managed and where are the project baselines defined & detailed. *.. i.e. Guide Book Of Plays.*

3) **As a binder for the project's subsidiary management plans**, see [Table 1]. Where, each subsidiary management plan describes how that relevant section of the project will be managed, integrated, and coordinated.

4) **As a link directory to each of the subsidiary documents which detail** that particular aspect of the project; i.e. analogous to a library index.

Because of these multiple purposes for the Project Management Plan, there are many different formats and structures that you may encounter in use.

QUESTION: Does your organization have a standardized PMP template?

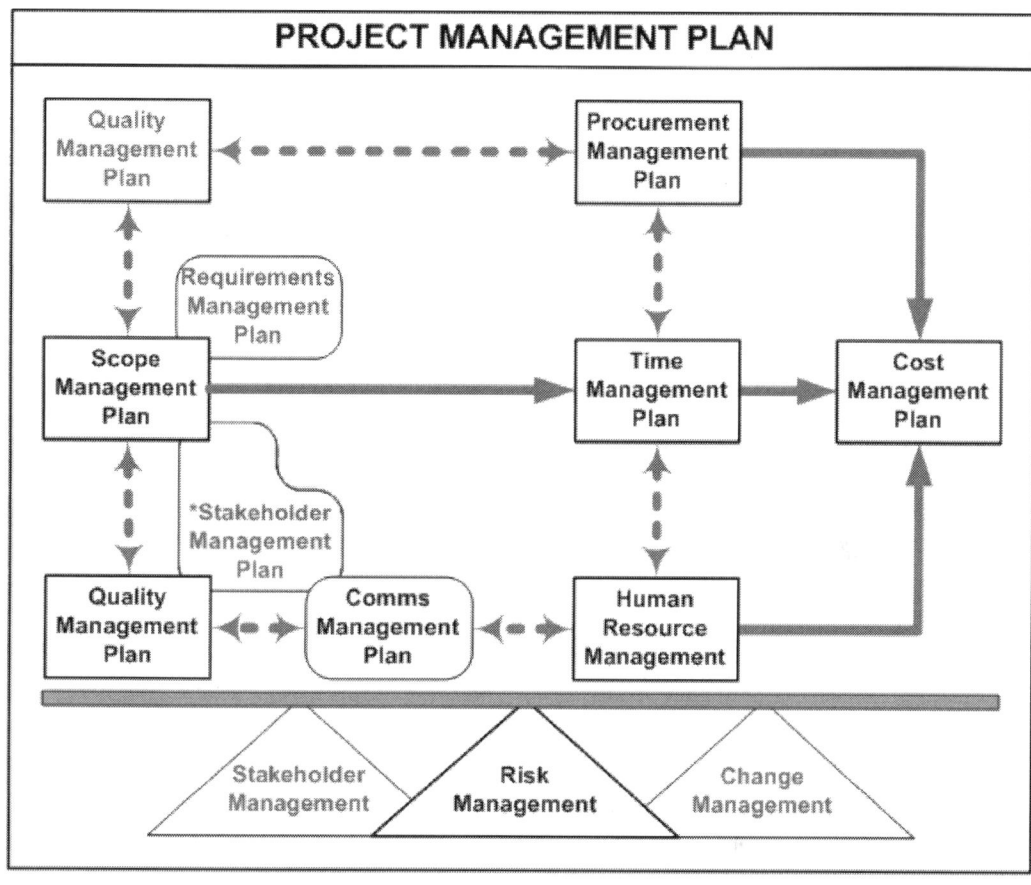

Figure 55: Relationship of the Planning Phase to the Project Management Plan and the subsidiary management plans.

NOTE   ❏ The model in [Figure 55], like the 5th Edition of the PMBOK® Guide, has the Stakeholder Management Plan as an independent part of the Project Management Plan. Whereas, the earlier editions of the PMBOK® Guide incorporated Stakeholder Management into Communications Management.

❏ The **level of detail** contained in the Project Management Plan **will vary depending on the scale & complexity of the project**, as well as the processes & procedures dictated by the primary-core stakeholders. Therefore, **the Project Management Plan could be highly detailed, broadly framed, or variations thereof**. Hence, choose a format that is appropriate for the given project.

Don't blow a noticeable portion of the project's budget on the PMP.

NOTE ❐ The **Project Management Plan will progressively evolve during the project's life until the project is closed out**, thus the circular nature in [Figure 55].

The Project Management Plan organizes all of the planning information and documented subsidiary plans generated during the Planning Phase into one integrated plan.
As outlined in the PMBOK® Guide, the Project Management Plan would be composed of the following subsidiary management plans listed below in [Table 1].

Table 1: Subsidiary Plans of the Project Management Plan.	
Scope Management Plan	• Details the processes to ensure that the project includes all of the agreed features & functionality, and only those authorized features & functionality are included in each corresponding deliverables release (i.e. the scope boundary).
	• Defines how changes to the project's [Scope Baseline] will be managed & controlled.
	• Consider the project's 'Scope Statement'; *e.g. Project Charter / Tasking Statement, Detailed Specifications, and Work Breakdown Structure.*
Requirements Management Plan	• Details the processes to capture, analyse, prioritize, and document the customer's requirements.
	• Defines how changes to the customer's requirements will be managed & controlled.
	• Consider the Customer Requirements, Functional Specifications, and the Requirements Traceability Matrix.

Time Management Plan *Schedule Management Plan*	• Details the processes to ensure the timely completion of the activities required to successfully complete the project. • Defines how changes to the project's schedule [Time Baseline] will be managed & controlled. • Consider the project's schedule / Gantt Chart, and Milestone Chart.
Cost Management Plan	• Details the processes to estimate, budget, and control expenditure, to ensure that the project is successfully completed within the approved budget. • Defines how changes to the project's budget [Cost Baseline] will be managed & controlled. • Consider the project's Budget and Costing spreadsheets.
Procurement Management Plan	• Details the processes to acquire materials, products, services, and personnel from outside the project team to successfully conclude the project. • Defines how the project's [Resource] procurements will be managed & controlled. • Consider the contracts, Statements Of Work, 3rd party agreements, and 3rd party Terms & Conditions

Human Resources Management Plan	• Details the processes to organize, manage, and lead the [People] involved with the project team; including defining roles, responsibilities, required skillsets, training, and reporting hierarchy. • Defines how changes to the project's [People] human resources will be managed & controlled. • Consider the staffing plans, training plans, and resource calendars.
Communications Management Plan	• Details the processes to ensure the timely delivery of appropriate information to the relevant stakeholders. Details how the information will be; generated, collected, consolidated, stored, retrieved, distributed, archived, and disposed of. • Defines how the project's communications paths will be managed & controlled. • Consider emails, agendas, meeting minutes, reports, document repositories & archives.
Stakeholder Management Plan	• Details the processes for identifying the project's stakeholders and managing their needs, wants, concerns, expectations, and perceptions. • Define how these stakeholders are to be communicated with. • Consider organizational hierarchies, interest groups, and organizational politics.

Quality Management Plan	• Details the quality procedures & policies to ensure that the project's deliverables shall satisfy the agreed Acceptance Criteria.
	• Defines how changes to the project's [Quality] will be managed & controlled.
	• Details the Objective Quality Evidence (OQE) that will be produced and presented as proof of the quality of the deliverables.
	• Consider the Quality Assurance & Quality Control procedures and policies, acceptance test plans, test records & reports.
Risk Management Plan	• Details the processes for identifying the risks to the project, how these risks will be analysed, planning the responses to these risks, how these risks will be monitored & controlled.
	• Consider the Risks & Issues Register, risk response strategies, problem reports.
Project Baselines	• Details the agreed & approved progress measures for [Scope], [Time], and [Costs] against which the project's real-world progress will be compared to determine if corrective actions are required.
Appendices	• Any additional information that would be highly beneficial, if this information was attached to the Project Management Plan or any of the subsidiary plans.

As can be seen by the number of subsidiary plans listed in [Table 1], the complete Project Management Plan can be a sizable document; but then this would be the perfect opportunity to **break the Project Management Plan's production into those smaller subsidiary plans and have different people (i.e. "experts") work on each one.**

What about the Change Management Plan, Configuration Management Plan, Operational Readiness Plan, Test Plans, and Training Plans?

 I would recommend that ... there be one overarching "standardized" Project Management Plan that is common for all projects executed by the performing organization, and that project specific deviations away from this base document are attached as an Addendum. ... Similar to how governments / states / nations have a standardized "Contract For The Sale Of Land / Property", and any deviation clauses from this norm are attached as "Special Conditions" at the end of the document.

NOTE ☐ The **Project Management Plan should be one of the first project deliverable milestones** (after or in parallel with the Detailed Specifications) **that must be signed-off by representatives of both the performing and customer organizations as an agreement on how the project will be managed and how changes will be controlled.**

Once the Project Management Plan has been signed off it should be "frozen" (i.e. baselined) for the current release cycle of the project.

Any changes to the Project Management Plan will subsequently need to go through a **Change Management Process.**

4.4.4. Output: Detailed Specifications

Review, Approve, and Sign-off

The "draft version" of the **Detailed Specifications** was discussed previously during the Initiating Phase in [Section 3.4.6].

The "final version" of the Detailed Specifications is a very important output of the Planning Phase as it defines the [Scope Baseline], and hence this document **must be signed-off by representatives from both the performing and customer organizations before the project can proceed into the Executing Phase.**

To head into the Executing Phase without the Detailed Specifications having been agreed to (signed-off) is just asking for trouble; with potential for having to rework implemented features & functionality that has either been redefined or misunderstood, and the resultant disputes over compensation for work that was or was not "in scope".

I agree wholeheartedly, you definitely don't want to find your project in the untenable position of entering Acceptance Testing without the Detailed Specifications having already been agreed to and signed off by the relevant parties.

⊠ **An unsigned and uncontrolled Detailed Specifications is a project killer.**

Document Contents

Interestingly, often a sizeable proportion of the Detailed Specifications will be quickly determined and in-principle agreed to. However, the remaining portion either becomes a drawn out process or is overlooked.

That is, the Detailed Specifications either is over or under cooked:

- With neither side being confident to sign on the dotted line, and hence the document is alternately reviewed and reworked (i.e. **"document tennis"**) with both sides trying to ensure that their respective interests and details are absolutely correct.

- Both parties are **"overconfidently blinded"** with the simplicity of the project and/or the ease of establishing the specifications and hence "game-changing" features & functionality are overlooked, understated, or incorrectly specified.

A practical solution to this "document tennis" and "overconfident blindness" is to **release the Detailed Specifications as a series of approved (signed off) versions that correspond to the cycles of delivery milestones**; i.e. a **"controlled version"** of the document **for each major iterative release** of the project.

Therefore, only the information that is relevant to the current release would be reviewed and subsequently signed off. Any other details for features & functionality outside the scope of the current release would be put into an appendix of "functionality & features for a future release", or that section of the Detailed Specifications would be marked as "Out-Of-Scope" for the current release.

Additional [Time] buffering for document preparation is especially necessary for the situation when **someone has to sign on the bottom-line** and hence they may be hesitant to take legal and/or financial responsibility for the outcomes that result because of that approved document.

With respect to the review & approval process for the Detailed Specifications, you should ask yourself:

1) How much "toing & froing" occurred in getting agreement on this document and then obtaining the final sign-off?

2) Was this a drawn out process or was this quick & painless?

3) Was the experience with the Detailed Specifications similar to the difficulty or ease of getting the Project Charter and the Project Management Plan signed-off?

If all of these signatory documents were similar experiences then you had better rethink how long it is going to take to get the project's other reviewed & approved documents signed-off.

I would thus recommend that; you **adjust the project's schedule accordingly to ensure that you have provided enough duration buffering for each document that requires review & approval.**

Only the Details and Not Solidifying the Solution

NOTE ❐ The **Detailed Specifications** should be used to itemize the performing organization's **"what will be provided" response** to the customer organization's "wish list" of requirements, and hence **the Detailed Specifications should NOT be used to solidify the solution.**

This is because, when the Detailed Specifications is approved then these architectures & designs that are embedded in the document could be interpreted as locked-in deliverables. Subsequently, during the Executing Phase these architectures & designs may be found to be inappropriate or detrimental to a successful implementation.

THE DETAILED SPECIFICATIONS SHOULD NOT BE USED TO SOLIDIFY THE SOLUTION.

Document Writing

When it comes to writing the Detailed Specifications, it must be remember that the **Detailed Specifications is NOT a work of literature**; i.e. paragraph after descriptive paragraph only introduces complexity and the potential for discrepancies.

For the **Detailed Specifications, the must-haves** are; **bullet-points, numbered-lists, tables, section numbers, diagrams, and mock-ups**. Though, do not go to the extreme of having section numbers that look like latitude & longitude coordinate systems, but rather to a depth where an individual requirement can easily be referred to.

Look out for **ambiguities in the Detailed Specifications** as these grey areas can be a ticking time bomb to the project's success as these can **result in rework, scope creep, and be the catalyst for a disgruntled relationship** between the customer and the performing organisation.

DO NOT reiterate specific details in multiple locations in the Detailed Specifications (and associated documents) as this can **result in conflicting requirements** if the specifics are changed in some locations but forgotten to be updated in other sections & documents.

If any specifics must be reiterated for clarity of another requirement then make sure that any "hard-coded" details are removed from that repeated requirement.

For example; unread messages will be removed from the queue after the specified default period ~~of 30 days~~.

I would recommend that ... a decree be issued that when conflicting specifications are found to exist between the Detailed Specifications and the project's other documentation then the signed-off and approved Detailed Specifications related to that specific release is to be used as the deciding reference. ... *Also, get them to ask for clarification so that rectification can be made to other docs.*

What about the Customer Requirements?

Remember that, the **Customer Requirements are a "wish list"** of the functionality that the customer thinks that they "want" implemented.

Whereas, **the Detailed Specifications is a "what will be delivered list"** outlining exactly what functionality the performing organization will be providing; i.e. the project's [Scope].

Therefore, **the creation of the Detailed Specifications is the perfect opportunity to write into the project exactly what the performing organization wants to deliver and conversely, what the performing organization does not want to deliver**.

> *For example; if a feature / functionality was interpreted a certain way then the subsequent implementation of that feature / functionality maybe beneficially reused on other projects.*

> *For example; if one specific customer requirement could be interpreted a couple of different ways where one of these ways would be relatively easy to implement whereas the other way would be a real pain-in-the-neck. Then in this case, clearly state in the Detailed Specifications the interpretation of the requirements (i.e. the easier solution) that the performing organization intends to deliver.*

Watch out if you are caught (deliberately) hiding stuff that was definitely asked for, as this will antagonize the working relationship.

This is where a **project management "ethical dilemma"** occurs;

1) Whether to include any indication of the existence of the other possible solutions (i.e. the difficult ones), or hope that the customer's representatives will not realize the fact as the easier solution is implemented, or

2) Whether to highlight the alternatives?

Irrespective of the ethical decision made, **the features & functionality that the performing organization intends to deliver need to be clearly stated in the Detailed Specifications;** and thereby defining the project's **"scope boundary"**.

4.4.5.   Output: "Triple Constraints" Documents

Back in [Section 1.4] the notion of the triple constraints was described and it was stated how [Scope], [Time], and [Cost] are important primitives in the project management model that is often used to evaluate whether a project's is deemed a success or a failure.

Well, these triple constraints make a triumphant return via the project's planning documentation (and later on with reporting for the Monitoring & Control Phase). That is, [Scope] via the approved Detailed Specifications, the Work Breakdown Structure, and the Requirements Traceability Matrix, [Time] via the approved Project Schedule, and [Cost] via the approved Project Budget.

4.4.5.1.   [Time] ... Project Schedule

During this chapter on the Planning Phase, it was described how the project schedule would be created. However, by the time that the Executing Phase is underway and the Monitoring & Control Phase is in full swing, then multiple variations of the project schedule will be in use.

1) **Baseline Schedule** – is the "set-in-concrete" schedule (for this current release) which **established the project's deliverable milestone dates** and was presented to the project's primary-core stakeholders as proof that this project could be delivered on time, or to detail when exactly it is planned to be delivered.

    This Baseline Schedule will in the coming project phases be used as source data to establish the **Earned Value Performance Measures** against which the project's actual [Time] progress and [Cost] progress (expressed as a level of task effort) will be judged.

     For more information on Earned Value Performance Measures and the **Schedule Performance Index (SPI)**, refer to [Section 7.3.1] & [Section 12.3].

2) **Work-In-Progress Schedule** – is the **regularly updated** schedule used to record the **project's real-world progress** (and used as source data to calculate the current SPI). At the beginning of the current release, the Work-In-Progress (WIP) Schedule would have been copied from the Baseline Schedule. This WIP Schedule would be updated as the release cycle proceeds and thereby leaving the Baseline Schedule untouched.

3) **Experimentation Schedule** – is the schedule (a copy of either the Work-In-Progress Schedule or the Baseline Schedule) in which **different re-arrangements** of tasks, durations and resourcing levels are **tried out to understand the effect** on the project due to **changing circumstances** and to **consider alternate possible arrangements**.

4) **Rolled up Summation Schedule** – is the **high-level schedule that is presented to the project's primary-core stakeholders** for summarizing the project and its progress.

Potentially, one project schedule could be used for all of these purposes, but the easiest solution is to have different variant schedules with dedicated purposes. The use of each of these variant project schedules will be covered in more detail in [Chapter 12] on practical monitoring & control.

A PROJECT SCHEDULE IS ONLY A GUIDE TO THE PROJECT'S IMPLEMENTATION, AND SHOULD NOT BE TREATED AS GOSPEL...

To survive one must ...
reassess, revise and reapply.

4.4.5.2. [Scope] ... Work Breakdown Structure

In this Planning Phase chapter, the Work Breakdown Structure (WBS) was described as being analogous to a pile of labelled boxes that were stacked into columns.

These WBS boxes have characteristics of physical courier boxes in that each one is a container; holding a package (of work), being allocated resources (material and people) for its successful delivery, and having important information stamped on its side (i.e. a delivery date, delivery address, an id number for tracking & payment purposes), and finally each of these boxes has some associated monetary value.

It was shown earlier in this chapter, that there is a **tight binding relationship between the WBS and the Project Schedule, with one being used to help develop the other**.

Therefore, when the Project Schedule was baselined the corresponding layout of the WBS (and its boxes) should also have been frozen. This is analogous to; the contents of a courier package cannot be change once it has been collected for delivery.

Later on in the practical monitoring & control [Chapter 12], it is described how a copy of the WBS can be used as a presentation tool to highlight what components of the project are being worked on and which work packages have been completed so far.

LITTLE BOXES,
LITTLE BOXES,
LITTLE BOXES PILED UP
AGAINST THE WALL...
THE PROJECT MANAGER'S
JOB IS TO MAKE SURE
THAT THESE LITTLE BOXES
DON'T TUMBLE AND FALL.

4.4.5.3.   [Scope] ... Requirements Traceability Matrix

In this Planning Phase chapter, the **Requirements Traceability Matrix (RTM)** was briefly described; the major point being that the RTM **maps the coverage of the Customer Requirements with the features & functionality listed in the Detailed Specifications**.

Later on in the practical monitoring & control [Chapter 12], it is described how a copy of the RTM can be used as a tracking tool to highlight the project's progress and what features & functionality is being worked on and what has been completed so far.

4.4.5.4.   [Cost] ... Project Budget

There are three variations of the Project Budgets that could be used:

1) **Baseline Budget** – is **the "set-in-stone" budget** (for this current release) which **established the project's expected expenditure**. This was the budget presented to the project's primary-core stakeholders as proof of the financial viability of the project and the expected cost of delivery.

    This Baseline Budget could be used (in the coming project phases) as source data to establish the **Earned Value Performance Measures** against which the project's actual [Cost] progress will be compared. ... Else, this [Cost] progress could be determined via the level of task effort in the schedule, and subsequently what was in the budget.

    > For more information on Earned Value Performance Measures and the **Cost Performance Index (CPI)**, refer to [Section 7.3.1] & [Section 12.3].

2) **Work-In-Progress Budget** – is the costing sheet against which the project's **real-world expenditure is recorded**. This is also the budget (a copy of) in which different rearrangements of tasks, durations and resourcing levels are tried out to **understand the effect on the project's "cash flow"**. *... Try to even out the monthly spend.*

3) **Summation Budget** – is the **high-level budget as presented to the project's primary-core stakeholders** for depicting the project's progress.

I would recommend that ... costing documents **rigidly comply with the formatting required by the primary-core stakeholders**, because this budget will be used to compare against other projects and incorporated into grander-scheme programs, departmental, and business unit budgets.

4.4.6. Output: Risks & Issues Register

During the project's planning there were probably several risks & issues that were identified. These identified risks & issues should be noted down in the Risks & Issues Register; see [Section 10.1].

I would recommend that ... the **Risks & Issues Register be regularly updated** (at least once a week), **notable changes** should be **escalated** to the project's primary-core stakeholders, and the **decisions made recorded** for future reference.

4.4.7. Output: Stakeholder Management Plan

During the project's planning there were probably several more stakeholders identified. These identified stakeholders should be noted down in the **Stakeholder Register**; see [Section 10.3] for more information.

Though, don't be too surprised if the Stakeholder Register never comes into existence, let alone gets used. Because, for small to medium size projects / organizations, this Stakeholder Register can be seen as a nonessential nicety.

4.4.8. Output: Architectural Designs, Prototypes & Mock-ups

The project's implementation team cannot just run-off and throw together some software code, hardware components, infrastructure, and then honestly expect that the project will produce a quality deliverable that functions correctly and integrates well with other services-systems-applications. Most probably, some architectural design(s) and proof-of-concept prototype(s), and mock-ups will have to be built & evaluated. This will mean the involvement of Subject Matter Experts such as; system designer / business analysts, system implementers, system testers, and potential end users.

When coming up with the system / application's architecture **give significant thought to, how will different components & modules interface / interact and be tested**. Because, once a module's purpose, features, scope boundaries, and interfaces have been defined, then it is a relatively easy process to break-off this module so that it could be developed by whatever internal or external resources are available at the time; i.e. it can be handed onto others to design, develop, and test.

THE "THREE AMIGOS" OF DESIGN – DEVELOP – TEST MUST BE ADDRESSED IN ORDER TO ESTABLISH THE PROJECT'S REALISTIC PATHS TO SUCCESS.

4.5. Things to Watch Out for

When undertaking the Planning Phase there are a few things that have to be watched out for, these are:

✖ **Perfection procrastination ...**

Where the project is delayed due to the desire of some parties to have the Detailed Specifications (and other documents) **110% correct before moving on**.

Thereby, resulting in an extended game of "document tennis" where **documents are repeatedly passed back-n-forwards for review – rework – review**.

SOLUTION: documents should be **released as approved versions** that correspond to each release milestone. That is, each approved document version should only concentrate on the information that is necessary **for that particular release milestone**.

✖ **You know what I mean ...**

This situation arises **when one party** (either the customer or the performing organization) **knows the project's subject matter significantly better than the other party does.** Consequently, **"trivial details" are left out and not told because it is assumed that the other party knows these "simple facts".**

Subsequently, features & functionality can be misinterpreted, misunderstood, or even completely forgotten about. This situation is accentuated when it is the customer who possesses the superior subject matter knowledge (but doesn't explain it very well).

SOLUTION: a **Subject Matter Expert (SME)** from the more knowledgeable organization (*e.g. the customer*) should be made available and contactable at any reasonable time. This SME doesn't have to be a senior person but someone with at least a few years' experience and inside knowledge of the topic at hand, knows the project's purpose, has a grasp of the main features & functionality, and knows the standard operational processes & procedures used for this field of endeavour.

Thus, the SME is able to answer these "trivial" questions almost immediately or at least knows whom to ask in order to find the correct answers to these questions.

> *For example; SME "yes Development-Dave you are right, it is normally 10%, but under circumstance X we add an extra 5% surcharge."*

- **Not too little, not too much, just right ...**

Do not include features & functionality that has not been specifically listed in the Customer Requirements. This is because; to implement these would constitute additional work that is not being paid for by the customer; i.e. "working for free". This is especially true for a fixed-priced project, though for a time & materials project then this additional scope would need to firstly be approved by the customer's representative.

However, the product delivered needs to be "usable", so if there are **obvious oversights or misinterpretations then the customer should be notified that they need to consider having these additions included**.

> *For example; text input without the capability to copy-n-paste is a major usability oversight.* ... *A real face-palm moment.* ... *"Err, DERR".*

So, in addition to "Scope Creep" also look out for "Scope Shrinkage" as it can result in a disappointed & disgruntled customer when the product is delivered with evidently missing features & functionality. ... *That they should have been told about.*

- **Realistic estimates ...**

Realistic estimates for both [Time] & [Costs] are essential for the perceived success of the project; "not too short nor too long, not too low nor too high, but just right". Because no matter how good the [Quality] of the project's [Scope] outputs, if either [Time] or [Costs] are significantly different from what was told to the project's core-primary stakeholders then someone is not going to be too pleased with the outcome.

So ask those knowledgeable implementers for their opinions & estimations.

- **Stick to planning rather than playing ...**

  Sometimes there can be a tendency to **drift planning** activities over **into the actual implementation of the project**.

  > *For example; the architectural design is no longer an analysis of the system's potential design for estimation purposes but rather a complete design.*
  > *The user interface is no longer mocked-up but rather is discussed based on the visual appearance (even including the visual graphics).*

  As stated previously; **DO NOT use the Detailed Specifications to solidify the product / system's architectural & detailed design as this could lock the project into an non-optimal path during the Executing Phase**.

- **Play to your project team's strengths and not to management weaknesses ...**

  If the project team is full of agile experienced implementers then it is detrimental (and demoralizing) to plan the project out as a waterfall model just because management are most comfortable with that waterfall life cycle. Thus, **plan the project out based on how it will be implemented and not how it will be administered**.

  This may entail dispelling to management and the customer the virtues of the desired implementation method, or potentially translating the presentation from one implementation method to the representation of the other life cycle.

  Therefore, **plan to the Project Implementation Team's strengths, experiences, capabilities, and capacities**.

- **Plan the battle with the units at hand ...**

  Draw up the project's plans with consideration given to the [People] & [Resources] that have been allocated to the project, or are readily available to the project.

  DO NOT plan for resources that "might" be available; *e.g. the customer might be able to lend a couple of people to implement that part.* If these "maybe" [People] & [Resources] eventuate then that will be a bonus to the project, but if these don't appear then that is not a tragedy because these were not really expected anyway.

- **The customer is always right, until it is obvious to all that they're wrong ...**

  It must be remembered that the customer is the one who will eventually be paying for the project, hence if the customer is adamant of a specific feature or functionality then all that can be done is to present them with the counter-argument and counter-facts. However, if they are still insistent on that feature or functionality being a certain way then plan the project out their desired way. ... *Though, record the history of the discussion and the reasons for yours and their decisions.*

 **DO NOT plan contrary to the customer's decision** because if you were found to be correct then the results would be contract deviation compensation to the performing organization; whereas, if the customer proves to be correct then **the consequences could be severe.** ... *To your project management career.*

- **Quality ain't no afterthought ...**

  [Quality] cannot be added to the project by simply inspecting it in via Quality Control; rather [Quality] has to be built into the processes & procedures via Quality Assurance. Therefore, **plan for quality as a "continuously improving process", rather than as a tack-on to the end of the project**.

- **Buffering ain't no free lunch ...**

  Ensure that **reasonable buffers for [Time] and for [Cost]** have been **added to the end of each major release cycle**.

  Though, **DO NOT allow this buffering to be used up haphazardly**, but rather allocated to those requiring it for good reasons.

- **The only bad idea is the idea that is not shared ...**

  One cannot plan in isolation; other people must be actively involved (and encouraged to participate), their ideas heard & shared, and their opinions & concerns considered. *Especially seek the input & validation from those senior implementers.*

4.6.   Additional Thoughts

4.6.1.   Thought: Business Continuity and Disaster Recovery

More often than not, business continuity (i.e. continuing the business's operations during a crisis event) and disaster recovery (i.e. returning the business to normal operations) are completely overlooked when planning out the project. The assumption being that; this is not the project's problem but rather the domain of the IT Department.

However, what happens to the project when the development server dies and a replacement unit will not be available for a couple of weeks?

Oh, and it is going to take an additional week to re-establish the development environment. What then... will the project team just twirl their thumbs in the meantime?

Similarly, what happens if the project area / laboratory is inoperable and the project team members have nowhere suitable to work on their assigned tasks?

Is this really the project manager's problem?

The sad reality is that; when one of these types of issues derailed the project for a week, a fortnight, a month then it is not the IT Department who in the long-run is held accountable for the late delivery. As far as the customer's representative is concerned and even in the minds of the performing organization's senior management, it is the project's and thus the project manager's problem and not the IT Department.

Therefore, the project manager needs to give some thought to the project's business continuity and disaster recovery strategies.

4.6.2. Thought: Outsourcing

While outsourcing can be advantageous to the performing organization from a financial bottom line and operational efficiency perspectives there are a few things to consider:

1) If the decision is made to outsource the organization's core business then at least **maintain some in-house capabilities to perform that core business activity,** and also **retain the rights to perform that activity in-house when necessary**.

2) Must **maintain control over the activity being performed** by the outsource service provider on one's behalf. *Can't just let them do whatever they feel like doing.*

3) Must **maintain the standards expected for both Quality Assurance & Quality Control** of the outsourced activity. *Just whose name will be stamped on the deliverable?*

4) Must **protect the organization's intellectual property** even when it is the outsource service provider who created that intellectual property.

Outsourcing must be managed diligently and not taken for granted.

Example Case: the bad taste of outsourcing.

I will admit upfront that I am not a big fan of the performing organization outsourcing its core business (especially outsourcing overseas), possibly because I have often been engaged at the a$$-end of when outsourcing has been an unmitigated disaster.

The following are some example cases when outsourcing went wrong:

1) A legal disclaimer to be used on a new product's website was a copy-n-paste of an existing disclaimer for another product. However, very close to the go-live date it was found that buried away in the disclaimer was the name of a previous third party who was not involved with this particular product. To fix this problem shouldn't have been difficult as it was only a couple of words in error, and the legal department would not allow the product to be launched until this was rectified.

The contracted developer was contacted to get them to make this "very simple" change, but they responded that it would take 20 business days to make this correction. Subsequently, it came to light that this performing organization had outsourced that section of the development to an overseas business that had a mandatory 20 working day's Service-Level-Agreement (SLA) on any change requests. Thus, this simple correction was going to delay the product's release by at least a month given that the performing organization didn't have the contracted right (nor access) to make such changes. Consequently, this trivial fix evolved into an executive level issue due to the liquidated damages associated with the go-live date.

2) A performing organization outsourced its software development to an overseas company. The resultant product was delivered "exactly" as specified, but when the product needed to be modified to conform to the customer's actual "needs" it was then realized that the source code was hard-coded for the delivered functionality and hence the evolution of the product was going to be extremely difficult. Additionally, the commenting in the source code was very limited (not being the developers' primary language) making local maintenance a nightmare undertaking. Hence, this product could not be readily modified & adapted.

3) A business outsourced its product development to an overseas company.
A year after the completion of the project, (while meeting with potential customers in an evolving market) one of the performing organization's executives encountered a new competitor's "revolutionary" product that "felt-n-smelt" disconcertingly similar to their own product but at a greatly reduced price.

Think carefully about outsourcing non-core business (e.g. IT) that is a linchpin to the core business. While this outsourcing may be financially advantageous during normal operations, it can leave the business highly exposed during a crisis event due to slower response times when compared to that of having internal teams.

4.7.   Phase Completion Review

Now that the activities related to the Planning Phase have been completed, then some form of **Phase Completion Review** meeting would be held to **decide whether the project is in an acceptable state to advance onto the Executing Phase**.

Alternatively, the outcome of this **"Gate" review** could be to go back to the Initiating Phase and re-envision portions of the project again. ... Could the decision be that the project should be put on indefinite hold, or should the project be terminated for the betterment of all involved? That is, based on the resultant outputs of the Planning Phase [Section 4.4], and **when all things are taken into consideration** [Section 4.5] **will the decision be to "REVISE", "CONTINUE", "PAUSE", or "TERMINATE" the project?**

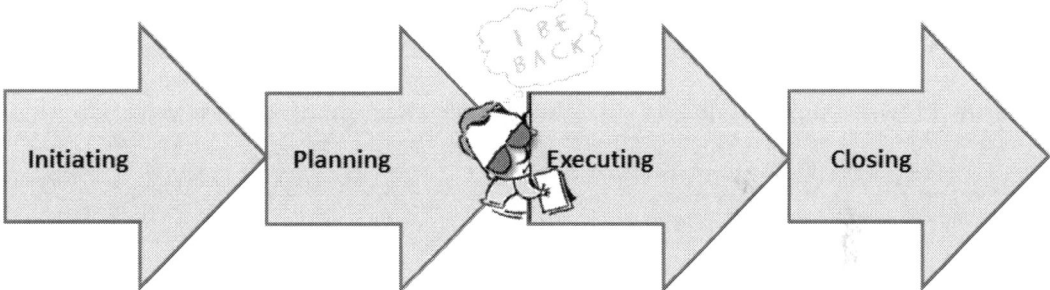

If the resultant decision from this Phase Completion Review meeting was **permission to proceed** (with the current release or milestone) then ensure that all of the authorization **signatures** are in place for the **Project Management Plan**, the approved **Detailed Specifications [Scope Baseline]**, the approved schedule **[Time Baseline]**, the approved budget **[Cost Baseline]**, and the approved **Acceptance Criteria [Quality Baseline]**.

NOTE   ❏ Due to the importance of the resulting decisions from this review, **it is crucial that this Phase Completion Review meeting involve representatives from the senior management of the performing organization, representatives of the customer organization, the project manager, and possibly selected subject-matter-experts from the project team.**

4.8.   Phase Completion Check List

1) Have you a **signed off Project Management Plan**?	✓
2) Have you an **approved Detailed Specifications** [Scope Baseline]?	✓
• Have you determined the Detailed Specifications **coverage of the Customer Requirements** via a Requirements Traceability Matrix?	✓
• Have you constructed a **Work Breakdown** Structure?	✓
3) Have you an **approved Acceptance Criteria** [Quality Baseline]?	✓
4) Have you an **approved Project Schedule** [Time Baseline]?	✓
• Did you include your own time (as project manager) in this schedule, did you also include holidays in the project's calendar?	✓
• Do you know the schedule's **critical path** and has appropriate buffering been added prior to major milestones?	✓
• Do you know the list of the project **milestones** per release?	✓
5) What type of **SDLC methodology** will be used by the Project Implementation Team, and does the schedule take this into consideration?	✓
6) Have the appropriate **[People]** & **[Resources]** been **allocated** to the project?	✓
7) Have you an **approved Budget** [Cost Baseline]?	✓
8) Have you planned for **contingencies** and have appropriate reserves been allocated for both the [Time Baseline] and the [Cost Baseline]?	✓
9) Have you sort **legal advice** relating to the engagement of any externally involved parties & activities; e.g. subcontracting, outsourcing?	✓
10) Have you obtained a Cost Authorization (Account) **Budget** for this project?	✓

11) Do you **know** who **the** project **team** [People] are? ✓

12) Have you assigned **roles & responsibilities** to the project team members? ✓

- Do they know their roles & responsibilities, and
  have they been appropriately trained for such activities? ✓

13) Have you **procured** the required **[Resources]** for the project team to be able
to start work on the project? ✓

- When will these resources be available / arrive? ✓

14) Have you considered how **communications** will occur internally and
externally with the project's stakeholders? ✓

15) Have you considered the project's **Quality Assurance** processes and
Quality Control procedures? ✓

16) Have you considered the **risks & issues** confronting the project? ✓

- Have you identified, analysed, and strategized about these risks &
  issues? ✓

- Have you created (and started filling in) the Risks & Issues Register? ✓

17) Has the project team investigated & reviewed the project's deliverables,
architecture, interfaces, and structure? ✓

18) Just checking, but do you have **signed off**; [Scope Baseline], [Time Baseline],
[Cost Baseline], and [Quality Baseline]; i.e. approved **Detailed Specifications,**
approved **Schedule,** approved **Budget,** and approved **Acceptance Criteria**? ✓

19) Do you think you have **planned adequately**, and
are you prepared if things change dramatically? ✓

20) Are you going to have a good night's sleep? ✓
Tomorrow you have the project's **Kick-Off Meeting**.

4/5. KICK-OFF Meeting

4/5.1. Overview

Well, you have planned for the big match (maybe this is your first as coach and not as the team captain), the playbook has been drawn up & distributed, the players have been training hard, and the field is set in readiness for the trials ahead.

But WAIT ... what about the pre-game huddle to go over the game plan one last time? There are also, the team's owners (i.e. the performing organization's primary-core stakeholders) and the team's sponsors (i.e. the customer's representatives) who want to meet the players and get one last visual confirmation that this is the team to do the job and you are the right person to guide that team to success.

That maybe an unusual analogy, however that is exactly what the Kick-Off Meeting is all about.

> THE KICK-OFF MEETING LETS EVERYONE
> INVOLVED KNOW WHERE HE OR SHE STANDS,
> AND WHERE EVERYONE FITS INTO THE PLAN.

4/5.2. The Agenda

As with any meeting, (especially ones involving the project's primary-core stakeholders and external parties) an agenda should be distributed beforehand so that the meeting's participants have reasonable opportunity to prepare.

In this agenda, list the topics to be covered during the Kick-Off Meeting, and include such things as:

1. **Welcome** to the kick-off meeting for project [INSERT NAME HERE].

2. **Outline** the project's objectives, major deliverables, and the key milestone dates.

3. **Summation-overview** of the **schedule** with respect to the milestones and the project's life span as per a calendar time scale.

4. **Summation-overview** of the **Work Breakdown Structure** (or architectural breakdown) with respect to the project team groups / sub-groups. This way, the meeting's participants have an idea of where they and others reside in the overall scheme of things.

5. **Introductions** to the meeting's participants / groups (though keep this brief).

6. **Project sponsor**'s gives their perspective of the project; i.e. the desired outcomes, perceived benefits, the importance of the project, and a short question & answer of the project sponsor by the meetings participants and vice versa.

The project sponsor and primary-core stakeholders may choose to leave at this point, though it is advisable that at least a representative or two remains to speak on their behalf.

7. **Project approach & organization** - a high-level outline of the project's schedule, an overview the proposed architectural design, a modular breakdown of the project's sections & stages, and detail the communications points & paths.
    This will consume the majority of the meeting's duration.

8. **Question & Answer** – based on all of the topics discussed up to this point.

9. **Concerns & Issues** – the meeting's participants should be allowed to voice their concerns and highlight any issues they envision with the project.

   Though keep this brief, but generate an actions list from the received inputs.

10. **Where to Next** – outline the upcoming project activities such as; soon to be approaching milestones, meetings, and deliverables.

11. **Clear Closure** – as with any meeting, close out by listing the meeting's action points, who has been assigned to these actions, and thank the meeting's participants for attending.

12. **Food & Drink** – don't underestimate the cost effectiveness of some nibbles, drinks, and even pizza for breaking down the barriers between the project's participants.

Beware of the potential negative image of having too grand a project kick-off celebration and especially with the production & distribution of project memorabilia such as team-shirts, caps, posters, and other types of "look at me" brandings.

If the project doesn't work out as well as was planned, then this "extravagance" can only serve to retrospectively advertise the project's mismanagement and wastefulness.

 I would recommend that ... the cost & effort involved with the kick-off meeting (and specifically with its associated social activities) should be miniscule when compared to the perceived pending cost & effort involved with the project's implementation.

A company paid-for lunchtime outing to a local (reasonably priced) restaurant or maybe an outdoors BBQ will suffice as a project kick-off marker.

4/5.3. Introductions All Round

This is possibly the first time that many of the project's participants have met; from the project's implementation team, to the customer organization's representatives, the performing organization's senior management representatives, outsourced third party contractors, and the vendors (if need be).

Thus, this is the perfect opportunity to:

1) **Outline** the participants' **roles & responsibilities** within this project.

2) **Build on** those **relationships** that will need to evolve during the life of the project.

I would recommend that ... because this kick-off meeting (and any other progress review meetings) involves your customer and your senior management, then before the meeting you should prepare your Project Implementation Team members with the potential questions that maybe asked and agree on the possible answers to these questions (i.e. those responses that reiterate the performing organization's proposed solutions & strategies).

You don't want them to offer counter proposals nor accidently expose any "confidential information" that is not in the performing organization's best interest to be known.

I would also recommend that ... some Project Implementation Team member's personal attire & behaviour might require some "adjustment" prior to the meeting, thereby giving them a professional appearance.

However, they still have to look, sound, and smell the part of capable implementers ... rather, "nicer" implementers.

No matter how technically brilliant the individual, if they have the appearance of a hobo, a slob, an anarchist, or a night crawler then this just doesn't come across as a competent individual and subsequently doesn't reflect well on the project team, their employer, and the project manager.

4/5.4. The PM's Time to Shine

NOTE ❏ This is probably not the project's first kick-off meeting; as a high level stakeholder meeting was probably conducted at the conclusion of the Initiating Phase (with the sign-off of the Project Charter), or part way through the Planning Phase (with the sign-off of the approved Detailed Specifications).

However, for the project manager this is probably the most important kick-off meeting because it will demonstrate your capabilities & capacities in both directions; to your primary-core stakeholders (i.e. senior management and the customer's representative), and to the Project Implementation Team.

FAILING   TO PLAN ADEQUATELY   IS
PLANNING TO FAIL DRAMATICALLY ...

INSUFFICIENT PLANNING
(OR RATHER THE PERCEPTION OF INADEQUATE
PLANNING) CAN EASILY DERAIL THE PROJECT'S
CHANCES OF BEING PERCEIVED AS A
POTENTIALLY SUCCESSFUL ENDEAVOUR.

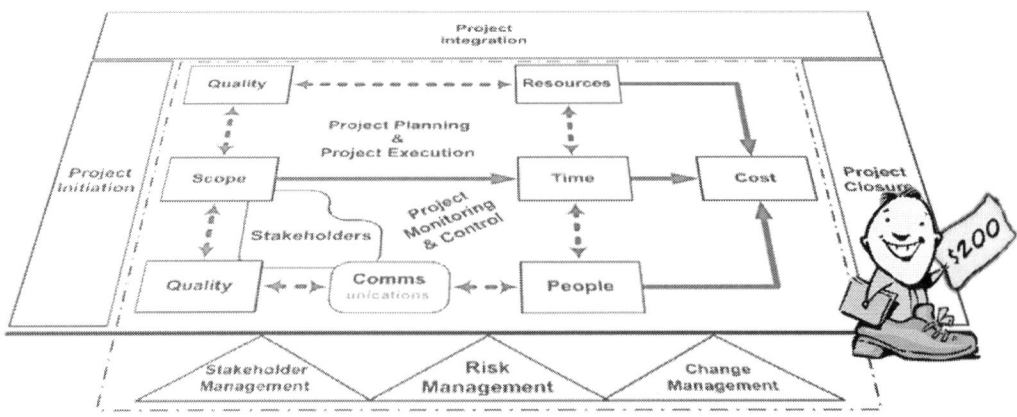

5. EXECUTING Phase

5.1. Introduction

In reality, **the Executing Phase will probably be the longest duration and financially the most expensive phase of the project's life cycle**, yet this chapter on executing the project's implementation is relatively short when compared to its actual duration.

The reasons for this chapter's brevity are simply because:

1) You, **as the project manager are no longer a "doer" (implementer) but rather a "thinker" (manager)**. Analogously, you are no longer the captain of the team but rather the coach who ~~issues orders~~ calls out ~~instructions~~ suggestions from the sidelines, thinks through the game's progress, and mulls over the game's strategies.

2) As the project manager, you have already mentally done the execution of the project's implementation when you were undertaking the Planning Phase.
    Hence, **from a project management perspective there is little involvement with the "Execution" other than to "Monitor & Control" the project's progress.**

PROJECT MANAGEMENT IS MORE ABOUT
INTERPERSONAL SKILLS THAN TECHNICAL.

IMPLEMENTER   MANAGER

Lessons Learnt: the difference of perspective on implementation and management.

The change in perspectives is big and sometimes uncomfortable for a new project manager who has moved across from being a technical implementer.

Take a moment to think this one through. ...

If you have the illusion that you can do both the project management and technical implementation roles at the same time for a prolonged period then just remember that. ... A project manager will spend about 80-90% of their time communicating in various forms; hence, how can you honestly expect to do much implementation work in the remaining 10-20% of the day?

I have witnessed project managers who believed that they could do both roles simultaneously, but this only resulted in them working extremely long hours, producing work substandard to their pre-promotional norms, damaging their work credibility due to their continual late or last minute delivery on commitments, and in some cases - wrecking their personal home life.

The simple truth is that; being a project manager is not like being a team leader.

No longer will you be "getting your hands dirty"; rather, you will be delegating others to "get dirty on your behalf", watching how they do their assigned work, and issuing corrective instructions when required. This is a **big change from the "hard technical" skills of an implementer to the "soft inter-personal" skills of a manager.**

The **1st Mistake of a new project manager** (ex-technical lead) **is to think that they can function as both implementer and manager.**

The **2nd Mistake of a new project manager is to think that their project team members are machines and thus trying to micro-manage everything they do.** ... *This will end badly for you.*

Example Case: Failure of trying to be both the Technical Lead and Project Manager

Once upon a time, there was a senior implementer – team leader who tried to simultaneously perform both the technical leadership role of doing tasks, and the project management functionality of orchestrating the completion of scheduled tasks.

Initially the project progressed along "nicely" with task after task marked as 100% done. The Project Implementation Team was transformed to operate in a hierarchical structure with the team leader at the apex of the pyramid, as the *"project leader"*.

However, as the project progressed into the core of the implementation work, tasks started to slip the scheduled completion dates. Yet, this project leader felt the need to report that those slipped-tasks had been done, even though these slipped-tasks were still being worked on as the next set of scheduled tasks were to begin. Eventually this covert *"backfilling"* of slipped-tasks built-up to the point where tasks that had not truly commenced were reported as partially completed and almost finished. ... Then CRASH; the facade of mistruths came tumbling down, and this *"nicely progressing"* project and this *"highly productive"* Project Implementation Team were found-out to be in real trouble (with only a relatively short time before the major delivery milestone).

This *"cooking of the books"* (misrepresentation of the truth) by the project leader was not initially a deliberate act of deception towards the project's senior management. Rather, this all started out as just trying to keep-up with acting as both the project's senior technical implementer and being the pseudo project manager responsible for progress. But as things slipped out of control, then an implementer's *"fire-fighting"* instinct kicked in, until the overworked team could no longer keep the issues contained.

In the worst case; the Project Implementation Team can self-destruct due to internal accusations of blame, and the project leader can become secretive & defensive when a dedicated Project Manager is assigned to determine what has gone wrong with this project. ... *Then the real dilemma is how to get this project back on track, yet not risk losing this once excellent technical leader and project team.*

5.2.   Overview

As illustrated below in [Figure 56], the Executing Phase is concerned with the darker shaded area of the Project Management Process model.

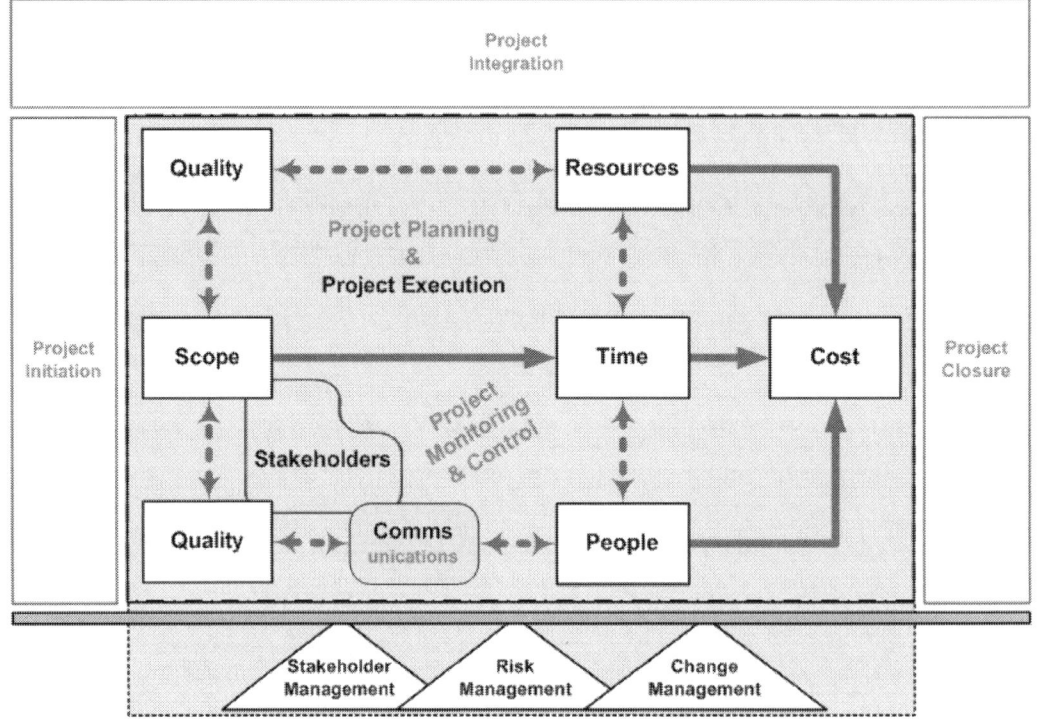

Figure 56: The Executing Phase of the Project Management Process model.

NOTICE how in [Figure 56] the Executing Phase is essentially identical to that for the Planning Phase in [Figure 33]. This similarity is by no means coincidental given the **symbiotic relationship** between these two phases, where changes to one phase will result in changes to the other phase.

TO DO OR NOT TO DONE ... NOW THAT IS
NO LONGER THE QUESTION FOR PM-YOU.

5.3.   Purpose

The purpose of the Executing Phase is to:

1) Perform those specified tasks to produce the project's agreed [Scope] of deliverables, within the agreed [Time], and keeping within approved [Cost] Budget.

2) Thereby achieving the project's objectives to the agreed [Quality] Acceptance Criteria and thereby satisfying the project's stakeholders.

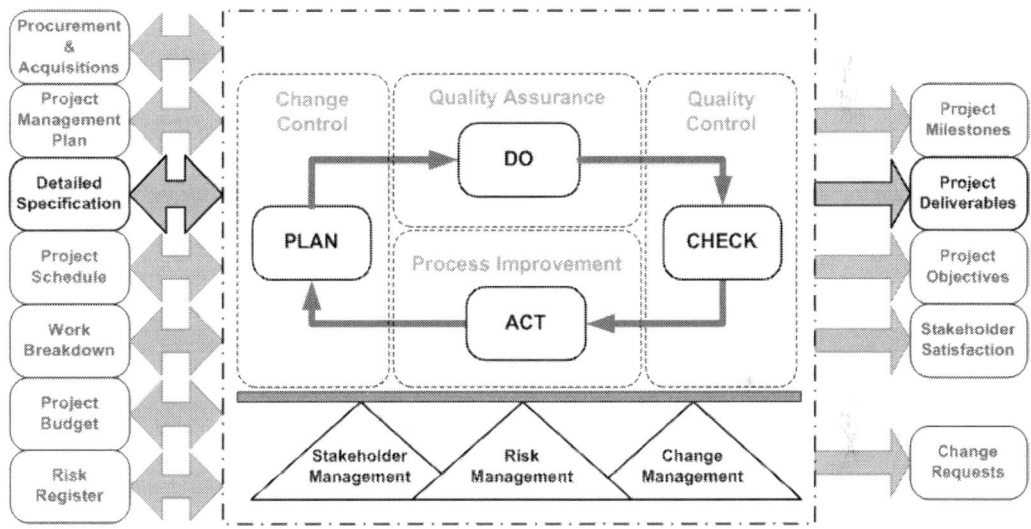

Figure 57: The Executing Phase as a "PLAN – DO – CHECK – ACT" process model with associated inputs and outputs.

NOTICE how in [Figure 57] the inner box contains a "PLAN – DO – CHECK – ACT" process model, rather than the "Scope – Time – Cost – Quality – People – Resources" process model as used in [Figure 56] and in the other previous phases' models. Also, notice the inclusion of Change Control, Quality Assurance, Quality Control, and Process Improvement, as well as how the Detailed Specifications has replaced the Project Management Plan as the primary input.

The reason for the change to a "PLAN – DO – CHECK – ACT" process model is that **the Executing Phase can be viewed from two distinctly different perspectives**:

1) **Performing the project's execution** – "<u>doing</u>" the implementation work (i.e. the implementer's role) will **refer primarily to the Detailed Specifications and hence is based on the "PLAN – DO – CHECK – ACT" process**.

2) **Managing the project's execution** – "<u>overseeing</u>" the implementation work (i.e. the manager's role) will **refer primarily to the Project Management Plan (specifically its associated subsidiary plans & documents)** and hence be **based on the "Scope – Time – Cost – Quality – People – Resources" process**.

If you originally come from an implementer's background then you are probably very proficient and knowledgeable on the "doing" aspect. Thus, it would be time inefficient and not beneficial for this book to cover your *"bread & butter*" implementer's knowledge base.

How unpleasant it was as an implementer to have your manager continually looking over your shoulder, as if doubting & questioning everything you did.

5.4. The Big Questions

The project Executing Phase answers the following questions:

1. How exactly **WILL** the **TEAM DO IT**?   ... *Plan*

2. The **TEAM** is **DOING IT**.   ... *Do*

3. The **TEAM** is **CHECKING IT**.   ... *Check*

4. Is the **TEAM DONE YET**?   ... *Act*

Umm, but these are not really questions but rather actions.

5.4.1. Question: "How exactly WILL the TEAM DO IT?"

For an implementer this question refers to such things as; architectural designs, hardware designs, software designs, database schemas, network designs, use cases, test plans ...etc. That is, those areas that this book will not be delving into; because, from a project manager's perspective, **this question was answered during the Planning Phase**.

5.4.2. Question: "The TEAM is DOING IT?"

For an implementer this question refers to such things as; software coding, hardware schematics design, hardware layout, prototyping, pre-production samples, debugging, module & component testing ...etc. That is, those areas that are out-of-scope of this book.

 Sometimes the Project Implementation Team will contain several highly skilled & experienced members, and hence declaring a team leader early in the project's life could result in bitterness within the team. A possible solution to this problem is for different people to be the "COP - Point Of Contact" for specific sections, i.e. Backend COP, Middleware COP, User Interface COP. Let the natural leader arise, and then appoint the most appropriate COP as the team leader.

5.4.3. Question: "The TEAM is CHECKING IT?"

For an implementer this question refers to such things as; integration testing, system testing, and verification testing. That is, those areas that are out-of-scope of this book.

5.4.4. Question: "Is the TEAM DONE YET?"

For an implementer this question refers to such things as; product verification testing, acceptance testing, and the **Post Implementation Review**. These areas will be covered in the Closing Phase in [Chapter 17] and more specifically in [Section 17.4.2].

Often in my role as a project manager, I have found myself involved with the daily run-around of dealing with the primary stakeholders' requests, while the project team gets on with the daily grind of implementing the project's plans.

5.5. Outputs

5.5.1. Output: Documents & Deliverables

The major outputs of the Executing Phase are the deliverables.

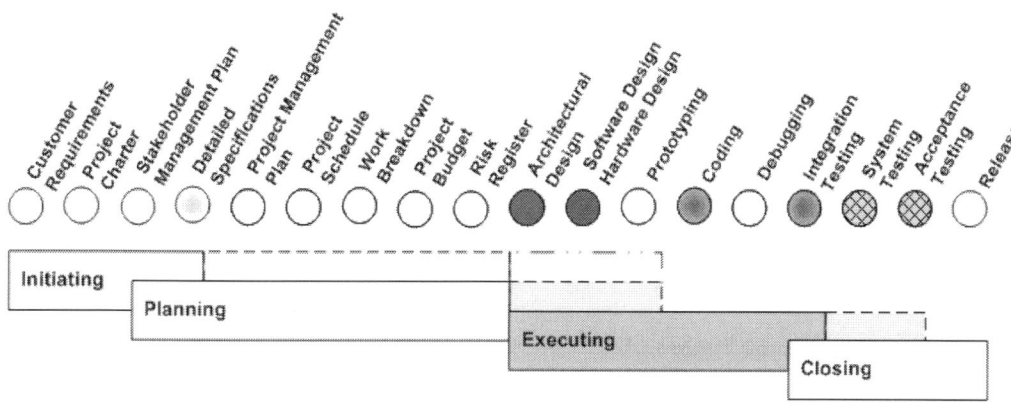

Figure 58: Documents & deliverables output of the Executing Phase.

⊠ **A project manager's role is NOT TO DO, but rather, to make sure IT GETS DONE.**

5.5.2. Output: PMBOK-SDLC Map

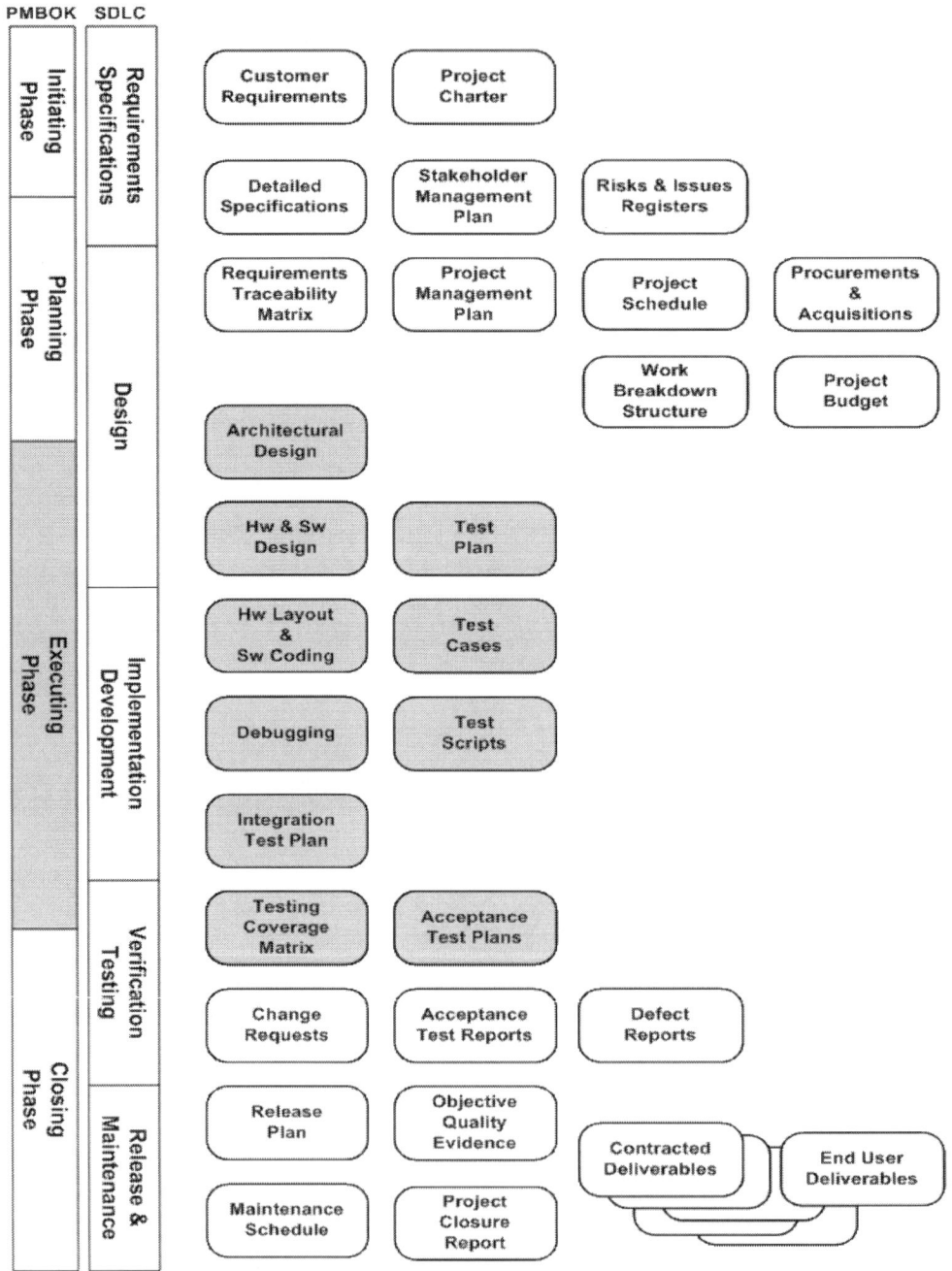

Figure 59: Relationship of PMBOK - SDLC and the documents & deliverables produced for the Executing Phase.

5.6. Things to Watch Out for

When undertaking the Executing Phase there are a few things that the project manager has to watch out for; as these will affect the project's performance, and hence the perceived worth of the project's manager:

✖ **A fixer-upper becomes a money pit ...**

During the processes of implementing the project deliverables, the project team may need to utilize other associated applications / systems for development and testing purposes.

For example; emulators, simulators, databases, test harnesses, automated test scripts, defect tracking systems, document management systems, version control tools, configuration management systems, networks, servers, virtual machines, etc.

It is likely that operational imperfections will be found with the particular associated utility application / system, and hence the project team member(s) may need to rectify or work-around any such flaws. While, the occasional "fixer-upper" is not necessarily detrimental to being able to successfully achieve the project's objectives (in fact this "fixer-upper" effort maybe essential to the project's success and even highly beneficial to the performing organization's other projects), such a "fixer-upper" can very easy snowball into a project-within-a-project. That is, a sizable amount of [Time], [Cost], [People], and [Resources] could be consumed with the creation, rectification, and upkeep of such an associated utility application / system. If the project team members are claiming this "fixer-upper" effort against the project's Cost Account Code, then this effort could substantially inflate the [Time] & [Cost] involved with the project.

Therefore, unless there has been a contingency included in the project's baselines, then this type of "fixer-upper" work should be limited as much as possible. Moreover, this work should be charged to a Cost Account Code independent of the project; *e.g. maintaining the in-house development environment.*

✖ **Building a limousine instead of a daily driver …**

There can be a tendency for some **implementer**s to **over-do their design, coding, and/or testing**. That is, the implementer does the analogical equivalent of building a limousine when what was called for was a cheap daily driver. This situation usually results when the implementer misinterprets what is required of them, or they personally consider the quality & grade of their produced deliverables as a reflection of their skills & prowess as an implementer.

While the pursuit of perfection is admirable, the problem with this tendency is that it; takes longer to complete that deliverable, and it often results in a deliverable that is overly complex which has greater potential for things to go wrong. Consequently, more testing [Time] will be required and more [Cost] will be involved.

✖ **Colouring outside the lines …**

There can be a tendency for some implementers to **take it upon themselves to implement additional functionality**. This situation usually results when the additional functionality is required for an upcoming release and since the implementer happens to be working on that particular area, they decide to implement that extra bit of functionality now rather than later. The problem with this tendency is that, this additional functionality was probably left out of the current release cycle for a very good reason.

For example; it may not have been decided exactly how that functionality should operate, and/or it has not been decided if that functionality will be required at all.

Hence, this additionally implemented functionality runs the risk of being redefined or being deleted in the future. Thus, that extra bit of implementation will have to be either reworked or removed, and this could introduce more defects and problems, let alone the additional [Time] consumed and the extra [Costs] involved with implementing & modifying that functionality each time round.

✗ Not looking at the big picture …

This situation is the opposite of "*colouring outside the lines*". In this case, the implementer **does not consider that the feature** they are working **on is going to be built upon** and therefore the implementer does not lay the foundation for expansion and reuse in the proceeding iterative release cycles.

> *For example; hard-coding application variables and subroutines so that the resultant functionality is specific only to the current release.*

Consequently, this bare minimalist interpretation of the functionality renders the work done on the current release cycle practically useless, and hence in the future this functionality will have to be re-implemented for an expanded purpose.

✗ While we're under the hood let's fix that too …

There sometimes is a tendency for software developers to **fix-up things around the code that they are working on**. These developers often refer to this as "**refactoring**", so as to improve the code's performance for the betterment of the deliverable. However, what this "refactoring" may translate to is restyling the code or completely rewriting / redesigning the code so that it conforms to their personal aesthetics; i.e. not for function but rather for form.

The problem with this tendency is that; it results in modifications that are not necessarily required, but these changes will introduce the potential for defects and new problems. Hence, this "refactored" code will have to be regression tested and thus require additional project [Time], which subsequently adds to the project's [Cost].

Thus the adage, "**if it ain't broke don't touch it**" needs to be applied.

✗ Scared of the monster in the forest …

This situation occurs when the implementer is so **pre-occupied with the complexities of some future feature** / functionality that they **lose focus on the current ones** that they are supposed to be working on.

✘ Dog with a bone …

This situation occurs when the implementer is so **pre-occupied with fixing one specific defect or implementing a certain feature** that the resultant tunnel vision is **to the detriment of all else**. Consequently, the implementation of their other assigned tasks is delayed indefinitely while they continually spend "a coupla days" resolving this issue.

What makes this scenario more intriguing is frequently this bone could be put-off to a later date. Nevertheless, many hours or days are lost in the struggle, yet it takes the project manager only a few minutes to conclude that it should simply be "dropped".

✘ Pulling rank … playing the, "I've got ## years of experience" card to dominate …

A project team member who uses their job title and/or say something like, "*listen here, I've got ## years of experience in this field*" to push their technical point, so as to win a discussion, is not a team player. This person is very detrimental to the team's morale & cohesive work-efforts because they stifle thinking and others willingness to contribute.

PROJECT MANAGEMENT … THE ACT OF BALANCING WHAT WAS PLANNED TO BE DONE VERSUS WHAT CAN REALISTICALLY BE DONE.

THE REALITIES OF IMPLEMENTING A SUCCESSFUL PROJECT

THE PLAN

5.7.   Additional Thoughts

During the Executing Phase as the project manager, you may want to remind the project team members about the following implementer's lessons learnt, as these will directly & indirectly affect the project's [Scope], [Time], [Cost], [Quality], [People], and [Resources]:

- **The early bird may catch the worm but it's the early builder that squishes the bugs ...**

  As soon as feasibly possible get demonstrable prototypes / proof-of-concept versions of the system or application working so that the customer can evaluate what they are going to receive. The project's other stakeholders will also need access to such early builds (and associated documentation) so that they can determine how their own parts are going to work and how best to interoperate with the system or application. These early builds are beneficial to:

  1) **minimizing the potential for misunderstandings & misinterpretations** between the project's various stakeholders; see [Section 3.3.1] and [Section 3.3.2],

  2) **highlighting any inconsistencies** between the Customer Requirements and the approved Detailed Specifications,

  3) **determining any omissions & oversights** with the project's planning and with the approved Detailed Specifications; see [Section 4.3.1],

  4) **allowing the other project stakeholders to get on with doing their own thing** such as; coding their parts, creating artwork, producing manuals & guides, accumulating source data that will be used with the application/system, etc.

 I would recommend that ... as soon as possible, frameworks and usable documentation for the system / application's external interfaces should be created and distributed to the relevant project stakeholders. For example; Application Programming Interfaces (APIs), Graphical User Interfaces (GUIs), libraries, communication protocols, and database schemas.

- **Deliver a bunch or deliver only once; it really depends on the SW/HW crunch ...**

  Given the relatively minimal cost & minimal effort involved with doing **software** compiles then it is highly beneficial **for the software development** portion of the project to use **an iterative or agile life cycle** to "**deliver often**"; see [Section 2.2]. That is, when some software part or a collection of software components are believed to be working correctly then these would be merged together into a testable build that could be easily handed over to the representative of the customer (or to others within the performing organization) to evaluate & provide timely feedback.

  While this **"deliver often" strategy is effective for software development**; for hardware development then this strategy is not always a viable option due to the potentially high cost & level of effort involved with producing each "one-off" prototype and pilot-run units. Consequently, **for hardware developments the classical waterfall life cycle** approach of designing everything up front **maybe a better option**.

- **Keep your feet planted firmly on the ground ...**

  Where practical, try to **keep the deliverable system / application** (and development environment) **as close as possible to its real-world usage state**.

  1) The system / application should be able to be **quickly transformed into a presentable deliverable**.

     For Example; the decision may come down from the project's primary-core stakeholders that the system / application needs to be shipped immediately to the customer, or demonstrated to a potential customer for evaluation, or to demonstrate project progress as a confidence builder (to both the primary stakeholders and even to the implementation team), or as the finished product.

  2) The system / application should be run often (and as soon as possible) on a **hardware configuration that closely approximates** (but not significantly superior to) **that used by the intended** customer or the **end users**.

For example; the deliverable will be perceived as a resource-hog if the customer's intended hardware platform is underpowered for the job.

3) The system / application should be **utilizing realistic data and handling realistic scenarios**; not solely limited to sample data and pristine "blue sky" conditions.

   For example; the size of each data package, the quantity of the data transferred per day, the number of transactions to be processed per hour, the timing & latency of the data exchanged, the number of users per session.

- **Don't poop in your own bed ...**

  DO NOT test the deliverable in the same environment that was used to develop or debug that system / application. The reason for ensuring the clean separation and independence of the development environment and the test environment is because:

  1) The tools and programs used to develop & debug the system / application may have **unknown installation components** hidden away on that platform.

  2) The resultant deliverable system / application may have **unknown compatibility issues** with other platform's configurations. *... E.g. network & system security.*

  Consequently, while the product has operated perfectly fine during testing, when it comes time for deployment to the destination platform then the deliverable system / application could fail dramatically due to these unknown dependencies.

- **Don't re-invent the wheel; just steal it off a colleague's car ...**

Plagiarism from sources within the performing organization should be encouraged because the reuse of designs, code, and hardware layouts will potentially save the project a considerable amount of implementation effort since that "borrowed" element has previously been examined, debugged, and tested.

"GOOD PROGRAMMERS KNOW WHAT TO WRITE, GREAT ONES KNOW WHAT TO REWRITE (AND REUSE)." ... ERIC S. RAYMOND (THE CATHEDRAL AND THE BAZAAR)

Be aware of where this "copied" material has been acquired from; because **"borrowing" from an inappropriate source** (especially an external one) **could result in costly legal problems.**

Also, examine the license Terms & Conditions of utilising Open Source as part of the project's source code.

While "plagiarism" can be highly beneficial to the project, **DO NOT go to the extreme of copy-n-pasting large chunks that "kind of does" what you are after** in the misguided believe that this thing can be "tweaked" and "rejigged" for the intended purpose. Because, what usually evolves from such "tweaking" is a Frankenstein application / system that is often unstable, is convoluted to code & debug, has superfluous parts that confuse the inner workings, and is very difficult to understand.

 I would recommend that ... instead of direct "copying" large portions of the material, the team member should take "inspiration" from such material.
That is, firstly understand what is going on, and then write their application using relevant extracts from the reference material.

How can you hope to debug that which you don't reasonably understand?

- **A comment a day keeps the defects away ...**

Comments in code (like with the project's other documentation) is not meant to be a chore, rather comments add value by contributing to its understandability. Therefore, **tailor comments specifically for those who will have to debug & maintain the code in the future**. Though, only write what has to be told, and don't reiterate that which is obvious because it will only consume valuable project implementation [Time].

However, **DO NOT assume that the code is self-evident enough to eliminate the need for comments** because sometime in the future when you and/or others are under pressure to resolve some critical defect, then will that obvious functionality still be so self-evident (at 2 o'clock in the morning, when this thing has to be out the door in a few short hours from now).

YOU NEED CHIANTI WITH THAT

FAVA BEANS

/* ALWAYS CODE AS IF THE GUY WHO
ENDS UP MAINTAINING YOUR CODE
WILL BE A VIOLENT PSYCHOPATH
WHO KNOWS WHERE YOU LIVE.   */

... MARTIN GOLDING

- **Keep it merged, meaningful, and maintained ...**

  Once each developer has written, commented, and tested their code portion, then this code should be merged into the code-base / repository that is being utilized by the other developers.

DO NOT allow any of the developers to procrastinate with the merging of their working code back into the code-base. Because the longer that a development branch is allowed to diverge from the code-base then the greater will be the difficulty, the effort required, and the [Time] & [Cost] involved with reunifying this source code.

Additionally, **non-merged code doesn't receive the same amount of inspection, evaluation, debugging, and testing as that used communally** by all of the developers and testers.

Therefore, each developer will have to set aside some time each day or week specifically for merging code between his or her development branch and the current code-base. Once the source code is contained in the code-base then it will have to be maintained so that it remains easy to understand, easy to debug, easy to modify and thereby extend its functionality & practicality of use.

NOTE   ❏ **The maintenance of the code & comments is** NOT the sole responsibility of the person who originally wrote that code; rather it is **the responsibility of the entire Project Implementation Team.**

DO NOT JUST MAINTAIN THE CODE ...
RATHER MAINTAIN THE CODE AND THE COMMENTS.

I would recommend that ... irrespective of whether the developer is tasked to maintain the source code or is just exploring it, if they happen to encounter a line or few of code that is not easily understood then they should "do their fellow developers a favour" and leave some comments explaining what they figured out about that section of code. Similarly, if they encountered remnant code & comments that are no longer appropriate then this should be marked for review prior to disabling / deletion.

However, don't rewrite (refactor) the code without prior consent from the senior implementers and the project manager.

 I would recommend that ... comments not be written after the coding is completed, but rather the comments should be written prior to and during the coding activity as a form of program outline and to clarify what the code is supposed to do (not does).

- **Implementation safety harness, because "do-overs" are expensive ...**

  Source control, version control, configuration management, document repositories, and (off-site) backups are essential to the good health of the project's implementation. Irrespective of the project's size & duration, there is never a valid excuse for not using these types of tools.

  Developer's should be able to easily rollback to an earlier version of the changes that they have made, merge code into & from other branches, and check-in & check-out components that they are working on.

  The development environment, source materials, and data should be restorable & redeployed within a few hours of a crisis event without greatly affecting the project's progress. If this is not the case or the project cannot be restored to at least where it was yesterday then the project's business continuity & disaster recovery strategies need to be seriously examined. Because **the further back in time that one has to go to retrieve the lost information then at least twice that duration will be required for the project to catch-up to where it should be today.**

- **Virtuality is better than reality ...**

When at all possible develop & test within virtual machine environments as this has several distinct advantages over working on physical machines.

1) **Virtual machines offer excellent business continuity strategies** because it is relatively easy to back-up the entire virtual machine, redeploy it to different physical hardware, and have the work environment up and running again within a few minutes when compared to the hassles of cloning physical computers.

   As an aside, there are significant advantages when using enterprise-grade virtualization products that offer real-time hot-swappable redundancy capabilities.

2) **Virtual machines enable the creation of quickly deployable standardized development & test platforms** without the need to reproduce the environment by reconstruction on each team member's physical machine. Thereby, saving on the project's ongoing [Time] & [Cost].

   That is, one or two people could spend hours to days getting the development / test platform "just right", then the resultant **Standardized Operating Environment (SOE) inside a virtual machine (VM)** would then be copied, issued, and deployed to the project's other participants so that their own identical environments can be up and running in a matter of minutes [Time].

   This SOE-VM would also enable remote locations to run the exact same development / test platform, by simply copying the required SOE-VM over the internet or by copying an image onto a portable hard-disk and then having this external-drive couriered to the remote personnel.

3) **Virtual machines offer the ability to "sandbox" different versions of operating systems and different editions of targeted applications** without the complexity of trying to have them cohabitate on the same physical machine.

4) **Multiple virtual machines can run concurrently on one hardware platform** thus saving [Costs] on the amount of physical hardware required for the project.

- **Continuing to do the same thing over and over again, yet expecting different results is insane ...** *However, many implementers & testers do exactly this.*

  It doesn't take an Einstein to realize that one.

  Hence, where possible try to **implement automated testing** (especially for smoke and **regression testing**) as this will add value to the project's quality control and subsequently free up time for the manual testing of new features & functionality, instead of using up valuable project [Time] going over the same "boring" tests again. *... AND AGAIN!!!*

- **Bandwidth & hard disk space is money ...**

  Occasionally a project will stumble because the hard-disk space on the server has run out (and compiles are failing), or the internet download limit has been exceeded (and implementers are not able to successfully remotely access the servers because their connections keep dropping out, and file downloads keep failing).

  Therefore, **DO NOT skimp on the quantity & quality of these types of IT infrastructure**, and keep a watchful eye on how much remains available.

- **A tradesman is only as good as their tools ...**

  Now, I am not espousing buying the entire project team the latest hardware, but at least spend a few dollars on ensuring that each team member's workstation has sufficient memory & hard disk space (to run those virtual machines), to do those compiles, and also consider having decent size monitors to work with.

  These are not massive capital outlays (and they can be reused on other future projects) but in a very-very short time the productivity benefits will exceed the financial outlay. Similarly, look at the software applications that are being used as newer versions (or other products) could provide significant productivity advantages.

  Who wants to use a decade old operating system & vintage applications!
  "A leading edge technologies company, my A$$ this is".

5.8.   Phase Completion Review

Now that the activities related to the Executing Phase have been completed, then some form of "~~Phase Completion Review~~" *Post Implementation Review (PIR)* meeting would be held to **decide whether the project is in an acceptable state to advance onto the Closing Phase**.

Alternatively, would the outcome of this **"Gate" review** be to go back to the Planning Phase and re-define portions of the project, would the decision be that the project should be put on indefinite hold, or should the project be terminated for the betterment of all involved. That is, based on the resultant outputs of the Executing Phase [Section 5.5], and **when all things are taken into consideration** [Section 5.6] **will the decision be to "REVISE", "CONTINUE", "PAUSE", or "TERMINATE" the project?**

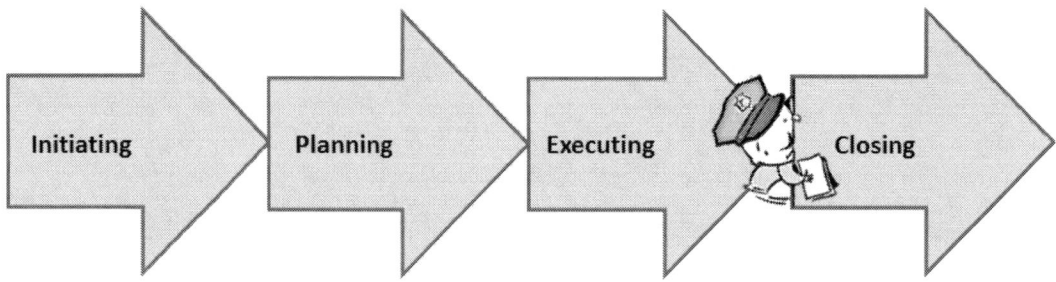

NOTE   ❐ Due to the importance of the resulting decisions from this review,

it is vital that this Post Implementation Review (PIR) meeting involve representatives from the senior management of the performing organization, representatives of the customer organization, the project manager, and selected members from the project team.

 For more info on **Post Implementation Review (PIR)** refer to [Section 17.4.2.1]

Post Completion Review (PCR) refer to [Section 17.4.2.2].

5.9. Phase Completion Check List

1) Are they implementing the work that has been assigned to this release; i.e. are they sticking to the plan?	**MC**
2) Are they only implementing the agreed [Scope] of this release?	**MC**
3) Are they following the agreed [Quality] processes & procedures?	**MC**
4) Are they producing the agreed deliverables by the expected delivery [Time]?	**MC**
5) Are they verifying that the deliverables are corresponding to the agreed [Quality] acceptance criteria?	**MC**
6) Are they providing relatively accurate [Time] reporting and [Cost] expense claims for the activities that they have been working on?	**MC**
7) Are you honestly prepared to no longer be a technical lead, but rather be the managerial lead?	**???**
MC... Monitoring & Control	

A BAD TRADESMAN BLAMES THEIR TOOLS;
A GOOD TRADESMAN IS RESTRAINED BY
THE TOOLS AT THEIR DISPOSAL. ... And by,

the limitations of the workspace made available to them.

KEEP ONE EYE ON THE ROAD,

KEEP THE OTHER EYE ON THE GAUGES,

AND BOTH HANDS ON THE STEERING WHEEL.

6. MONITORING & CONTROL Phase ... Theory Part 1

A Road Trip Analogy

6.1. Introduction

At this point one could continue with the sporting analogy; that the match is now underway, the team is playing to the game plan, the coach (i.e. project manager) is watching the game's progress from the sideline ready to issue instructions when the situation requires. ... But, let's start this Monitoring & Control Phase with a new analogy.

Example Case: a road-trip analogy

Consider the scenario that, you have to get home in time for a special anniversary celebration of going out to a "once-in-a-lifetime" performance that your partner purchased tickets for many months ago; i.e. the equivalent of an unmovable "drop-dead" End-Date project.

Miss this one and your new postal address will be Room 101 Dog House Lane.

The situation is that; after attending a very important meeting with a regional customer you have to drive 200 kilometres to get home, you've got 2 hours to get there, you've got a quarter tank of fuel, *"half a pack of cigarettes, it's dark, and we're wearing sunglasses"* The Blues Brothers, Universal Pictures, 1980.

you've left your wallet / purse on the kitchen bench, your mobile phone battery just flat-lined, and most importantly your relationship bliss is dependent on you making it to this concert on time and in a presentable form "OR ELSE !!!".

For this road-trip analogy, the project variables and project constraints are;

1) the distance to drive home is the equivalent of the project's [Scope],

2) the time remaining to get there is equivalent to the project's [Time],

3) the amount of fuel in the car is equivalent to the project's budget [Cost],

4) the misplaced money and dead mobile phone is equivalent to the project having no contingency reserve, and

5) the relationship aspect is the equivalent of the satisfaction of the project's primary-core stakeholder / customer.

The strategies available are to either;

• Speed home as fast as possible to make sure that there is time remaining, but risk running out of fuel somewhere along the way; i.e. the equivalent of running out of project budget.

• Drive conservatively to ensure that there is sufficient fuel left to arrive at the destination, but risk getting home too late to then be allowed into the concert; i.e. the equivalent of missing the delivery date and thus suffering the associated penalty clauses and liquidated damages.

Thus, drive too fast or too slow and you will lose. Hence, to succeed you need to keep an eye on; the distance left to travel [Scope], the time remaining [Time], the amount of fuel left in the tank [Cost], and not damaging yourself or the car [Quality].

Therefore, to make it safely to this important engagement you will need to judge the adjustments to the car's accelerator pedal to balance the speed verses fuel consumption versus the clock. This is the same for projects; do too much over [Time] work and assign too many [People] and consume too much [Resources] in an effort to complete as many of the deliverables as soon as possible in the given [Time] period, but run the risk of blowing the budget [Cost].

6.2. Overview

6.2.1. Model: Scope – Time – Cost – Quality – People – Resources

As illustrated in [Figure 60], the Monitoring & Control Phase is concerned with the darker shaded area of the Project Management Process model.

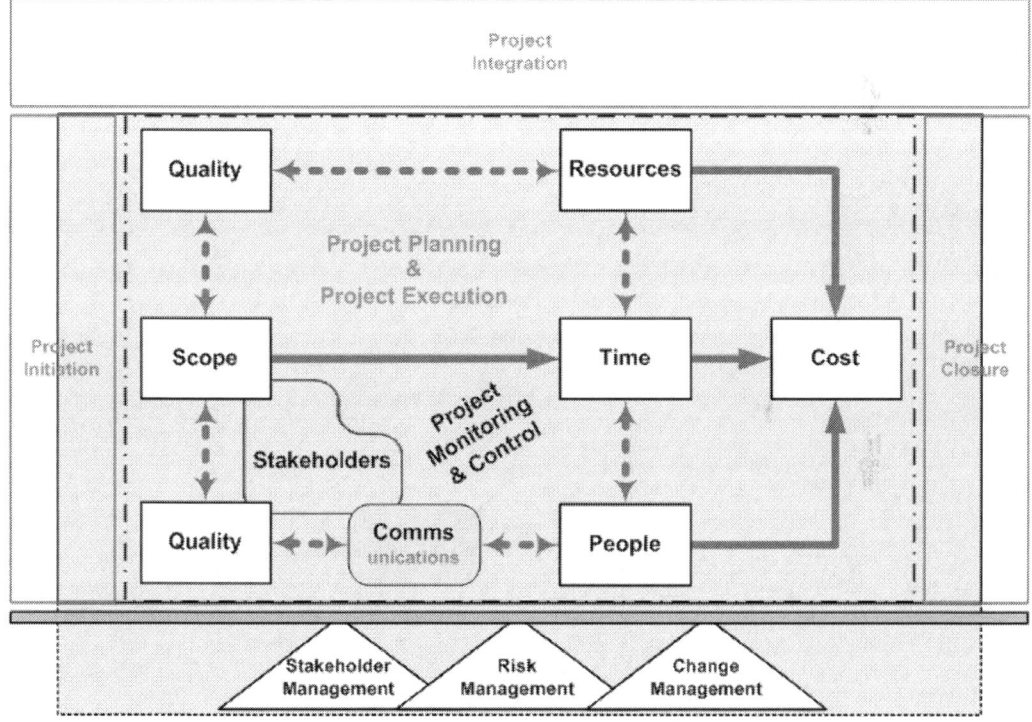

Figure 60: The Monitoring & Control Phase of the Project Management Process model.

NOTICE how in [Figure 60] the Monitoring & Control Phase covers the same territory as the Planning Phase and the Executing Phase, plus some of the Closing Phase and some of the Initiating Phase. This is no ordinary coincidence, as it is the **Monitoring & Control Phase** that **ensures that the project being executed is in accordance with that which was planned and agreed to be implemented.**

6.2.2. Model: Plan – Do – Check – Act

Thinking back to the road-trip analogy... a safe driver doesn't spend all of their time concentrating on the speedometer, tachometer, temperature, and fuel gauges; i.e. the "Scope – Time – Cost – Quality – People – Resources" process model in [Figure 60]. Rather, more often than not, the driver should be; planning where to position the vehicle on the road, *checking the mirrors,* moving the vehicle into a vacant space, checking the mirrors, and re-acting to the changing circumstances around the vehicle. That is, a "PLAN – DO – CHECK – ACT" process model as presented below in [Figure 61].

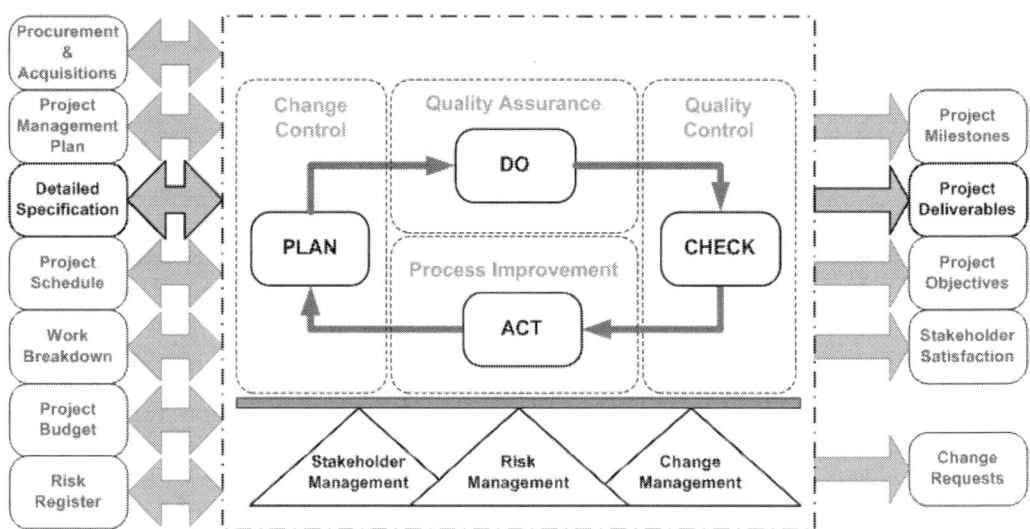

Figure 61: The Monitoring & Control Phase as a "PLAN – DO – CHECK – ACT" process model with associated inputs & outputs.

Subsequently, the **daily monitoring & control of the project** would be **based on the "PLAN – DO – CHECK – ACT" process** model [Figure 61].

Whereas, the **weekly/monthly project monitoring & control** would be **based on the "Scope – Time – Cost – Quality – People – Resources" process** model in [Figure 60]. Unfortunately, as the fuel (budget) starts to run out then many a driver (management) gives disproportionately more focus to what is happening with the gauges rather than concentrating on the events occurring on the road in front of them.

6.3.   Purpose

The purpose of the Monitoring & Control Phase is to:

1. **Continually track & review the project's progress & performance by quantifiable**
   Comparison of the **actual results against those planned baselines**.
   Traditionally this measurement process has been for [Scope], [Time], and [Cost] as
   illustrated on the next page in [Figure 62].

2. **Identification of any deviations from these baselines,**
   Evaluation of the causes for such deviations or the potential causes for
   future deviations (i.e. **variance analysis**), and if necessary **initiate actions in**
   Response to such deviations.

3. **Communicate the project's status** and its expected future situation to
   the relevant project's stakeholders.

Based on the above listed steps the following decisions can be made in
response to any deviations to the project's baselines:

1) Take **Corrective actions to bring** the project's future performance **back into line** with
   the expected performance.

2) Take **Preventive actions to reduce** the likelihood of **negative consequences** associated
   with these deviations.

3) Undertake **Defect Repairs to fix or replace** those components of the project that are
   not conforming to the expected performance.

4) **Update** the project's **Documentation and plans** to accurately represent the current
   and expected state of the project.

5) Seek authorized approval via a **Baseline Change Request** to **modify the planned**
   baselines to something more appropriate (given the project's current situation and
   the prevailing circumstances).

NOTE ❑ **Project monitoring** is concerned with **verifying** that the **project is progressing in accordance with the project baselines**. Whereas, **project control** is concerned with **ensuring** that the **project will conform to the project's baselines**.

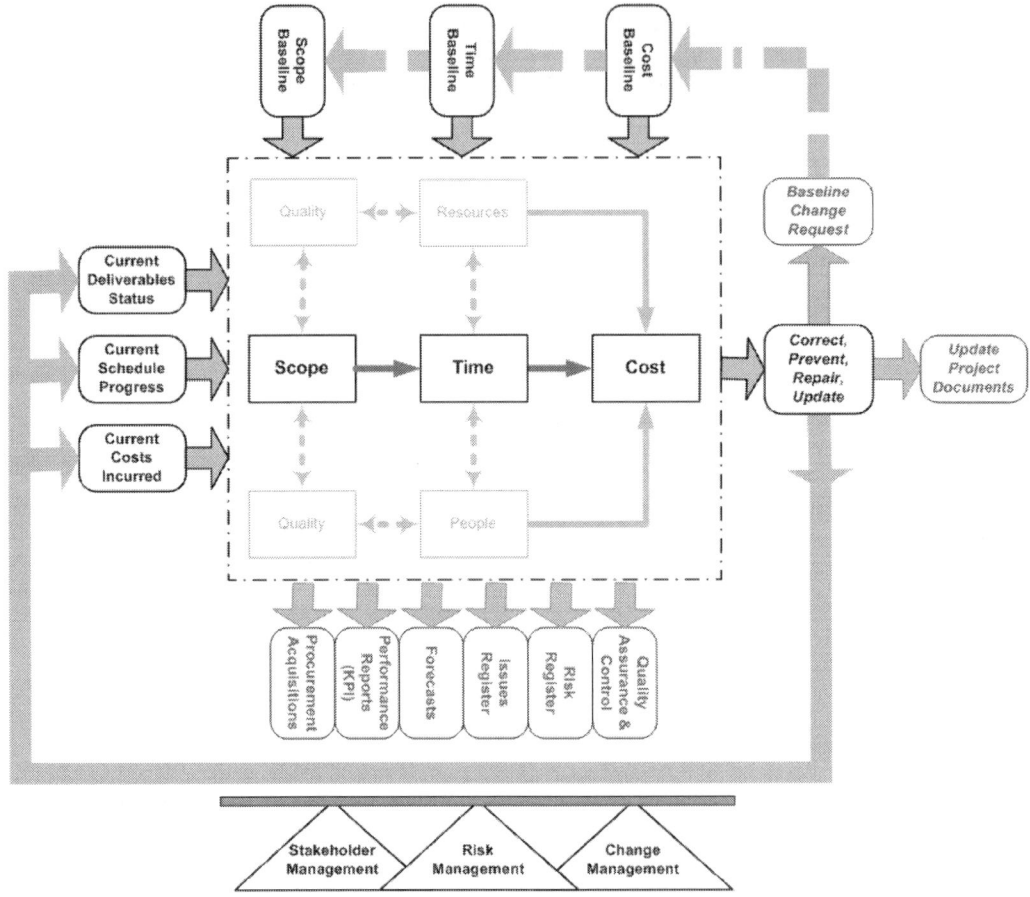

Figure 62: Process model for comparing the [Scope Baseline], the [Time Baseline], and the [Cost Baseline] with their actual performance values.

NOTICE in [Figure 62] the presence of the three risk triangles.

This is because, for the project manager, the **Monitoring & Control** Phase **involves a significant amount of** juggling; the stakeholder's needs (**Stakeholder Management**), the risks confronting the project (**Risk Management**), the necessitated changes to the project (**Change Management**), and all the **while coordinating the activities of the project's implementers**.

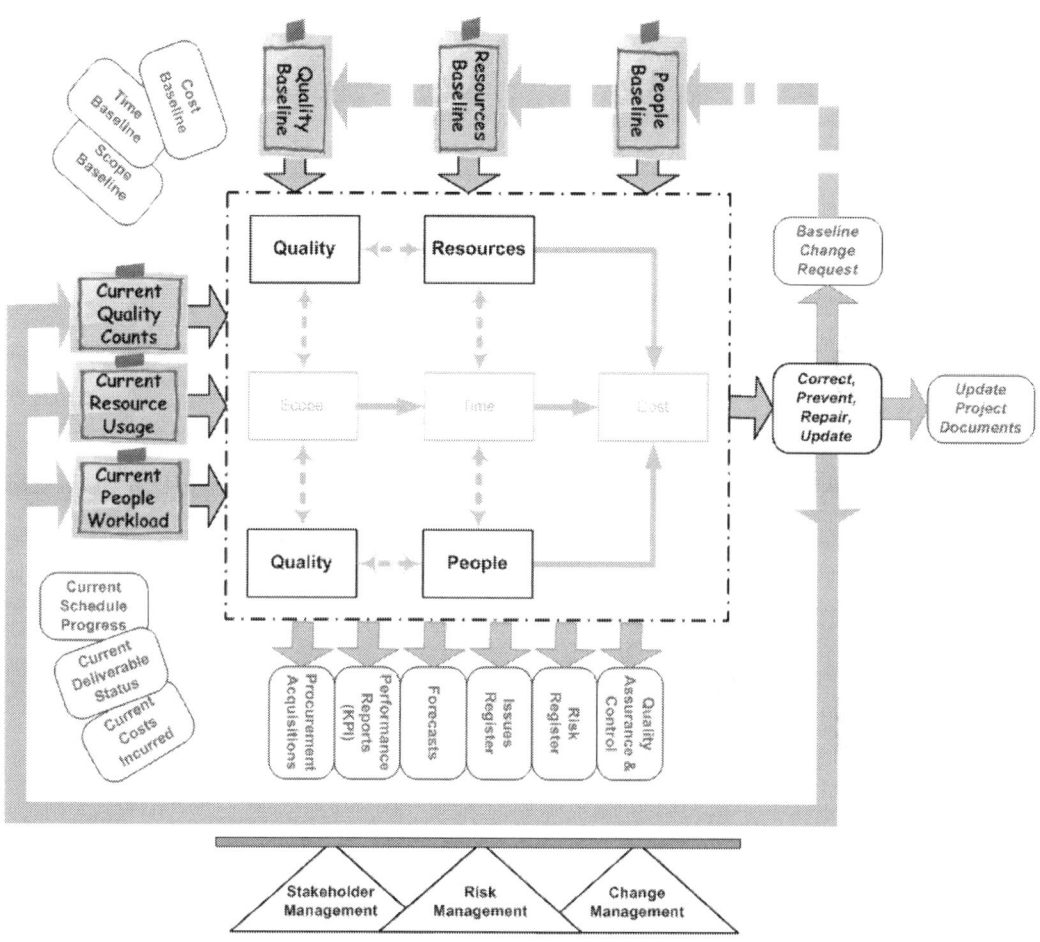

Figure 63: Process model for comparing the [People Baseline], the [Resource Baseline], and the [Quality Baseline] with their actual performance values.

The process model illustrated on the previous page in [Figure 62] only dealt with the three most basic project baselines / constraints of [Scope], [Time], and [Cost]; i.e. the three primary gauges that are used to judge whether a project is deemed a success or a failure. However the project's [Quality], [People], and [Resources] will simultaneously need to be monitored & controlled, as these do have direct flow-on effects on the [Scope], [Time], and [Cost].

An Iterative Arrangement

Now, suppose that these two process models in [Figure 62] and [Figure 63] were simplified into a common model for an individual project variable; as presented below in [Figure 64].

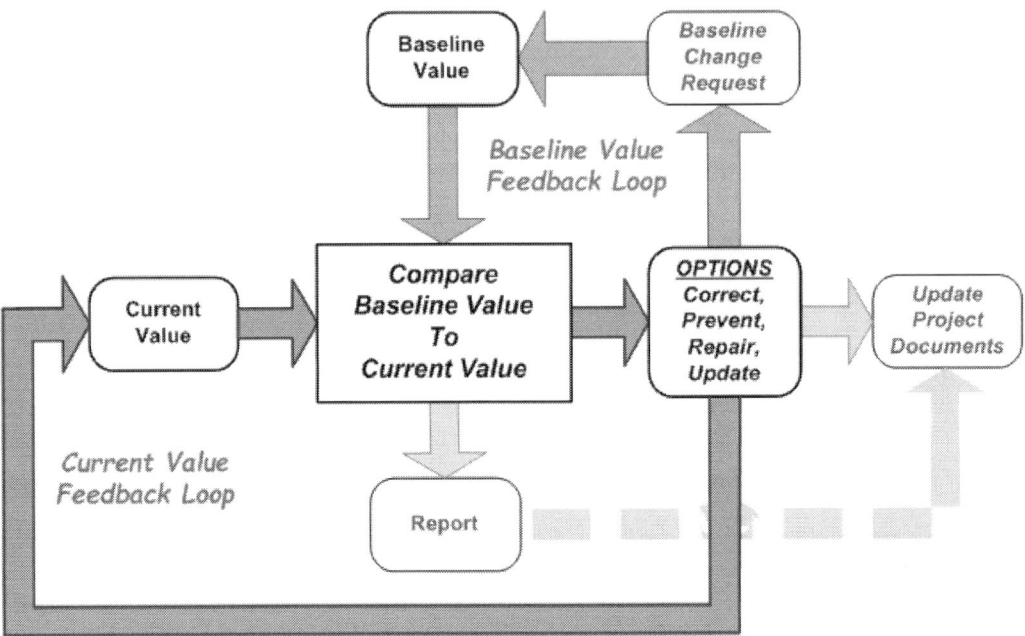

Figure 64: Monitoring & Control as a simple process model for an individual project variable.

NOTICE in [Figure 64] the presence of two feedback loops; one for the current value of that project variable and the other for the (planned) baseline value for that same project variable.

The "**current value feedback loop**" would be updated on a daily or weekly basis (during the Stand-up Meeting and/or Weekly Progress Report), whereas the "**baseline value feedback loop**" would be updated less frequently (possibly at the beginning & end of each agile-sprint, release cycle, or Monthly Project Update Report).

6.4. Perspectives

6.4.1. Perspective: Management Hierarchy

For monitoring & control, there are in essence **three (3) different levels of project management hierarchy**. Where, each of these management hierarchy levels has different purposes, and hence each level will be interacted with in a slightly different manner.

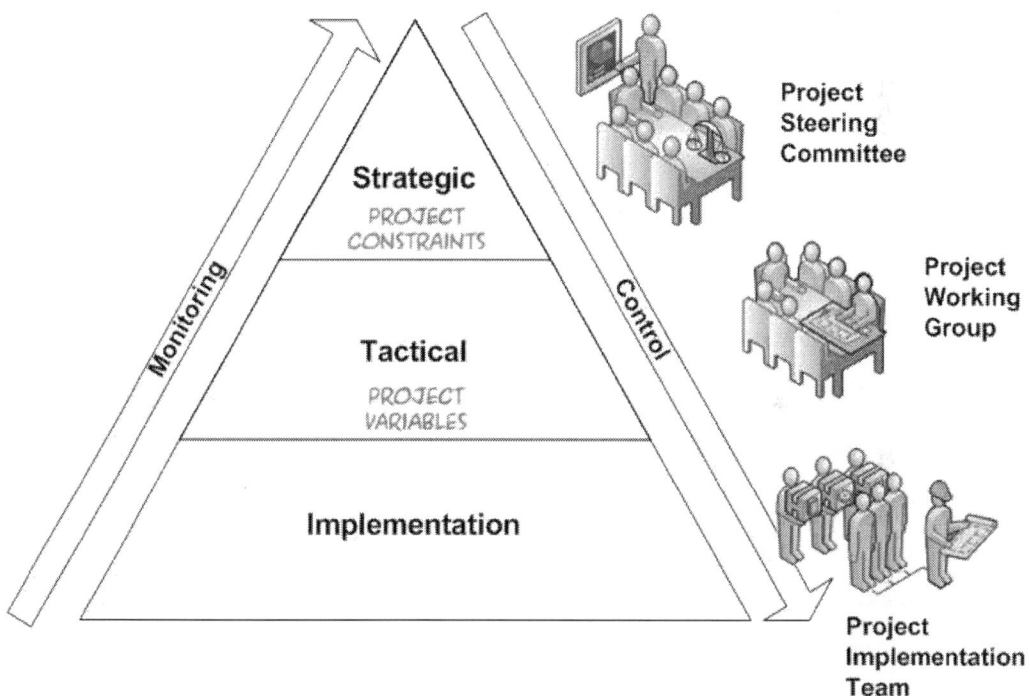

Figure 65: Three different perspectives on monitoring & control.

1) **Project Steering Committee** – are the project's primary-core stakeholders who are **generally drawn from the ranks of senior management** at the performing organization (and possibly from the customer organization). This committee is **concerned with the "big picture"** of how the project fits in with the business's strategies. Consequently, the Project Steering Committee **provides broad guidance on the general direction for the project**.

However, the steering committee does not usually get directly involved with the daily activities of the project or its daily management unless the project is about to or has careered off the proverbial cliff.

Though, the **most important purpose of this Project Steering Committee is to give approval for changes to the project's baselines** via authorization of the **Baseline Change Request (BCR)**.

Hence, the Project Steering Committee are *"**the masters of the project constraints**"*.

> *For example; permission to increase the project's agreed budget [Cost], extend the project's agreed delivery dates [Time], redefining the agreed [Scope], and redefining the [Quality] acceptance criteria. Such re-baselining usually involves negotiations between the performing organization and the customer about changes to the project's [Scope Baseline], [Time Baseline], [Cost Baseline], [Quality Baseline], and discussions about the associated compensation to the affected parties due to these changes.*

2) **Project Working Group** – are the project's primary & secondary stakeholders generally **drawn from the ranks of middle management** at the performing organization (and possibly from the customer organization). This working group **provides tactical directions & coordination of the implementation** of the "individual pictures" composing the project; i.e. they are *"**the masters of the project variables**"*.

Though, the **most important purpose of this Project Working Group is to coordinate the activities** of the project's implementation teams and any involved external parties.

3) **Project Implementation Team** – is composed of those persons (internal and external to the performing organization) that **do the "hands-on" execution of the project's activities**.

NOTE   ❒ Depending on the size of the performing organization and the scale of the project, **these three levels of project management hierarchy may not formally exist or be known by such names, though in essence these levels will be there**.

That is, **these three levels of project management hierarchy will always be present.** ... maybe the Project Steering Committee is solely the company's boss or department head, maybe the Project Working Group is some of the senior "doers" and "thinkers" in the organization, but there will always be some form of Project Implementation Team.

Therefore, irrespective of the size of the performing organization and the scale of the project, the overall roles & responsibilities of these three levels of project management hierarchy will be the same, and thus **monitoring & control will have to interact with these levels accordingly**.

6.4.2.   Perspective: Top-Down & Bottom-Up

For the project to be successful, **it is essential that the project's management hierarchy reflects the current managerial organizational structure** of the performing organization. Thereby, providing the most appropriate levels of authorization, governance, ownership, responsibility, and responsiveness. That is, **project control is exerted from the Top-Down,** see [Figure 65]. Whereas, **project monitoring (notifications)** of risks, issues, problems, and general reporting are **escalated from the Bottom-Up**.

1. The **Project Implementation Team** members (if a scrum stand-up meeting arrangement is being used) daily or every few days will (in a round-robin fashion) update their team-leader or the project manager about; their progress (and the tasks they are working on and have completed), the issues that they face, the blocking problems that they have encountered, and the risks that they perceive to the project being completed successfully.

2. The project manager and/or team-leader will compile this information from the Project Implementation Team and bring these before the **Project Working Group** who will resolve what they can and update the project's status accordingly.

3. The representative of the Project Working Group will then fortnightly or monthly present a summation of the project's progress, risks & issues, and problems faced to

the **Project Steering Committee** who (in accordance with the business's strategies) will resolve those things that they are empowered to do.

NOTE ❑ With each level of hierarchy traversed, the project's **information** becomes more **filtered and manicured specifically for the "needs & wants"** (and the actual requirements) **of the stakeholders** at that level.

6.4.3.   Perspective: Time Windows for Reporting

Because of these three levels of project management hierarchy there will be **three corresponding independent reporting cycles** and hence, there will be **three corresponding time windows for monitoring & control** of (1) Daily, (2) Weekly, (3) Fortnightly / Monthly.

That is, for each of the project management hierarchy levels:

- The **Project Implementation Team** will probably get together **daily** or every few days.

- The **Project Working Group** will probably get together on a **weekly** or fortnightly basis.

- The **Project Steering Committee** will probably get together (and be reported to) on a **fortnightly or monthly** basis.

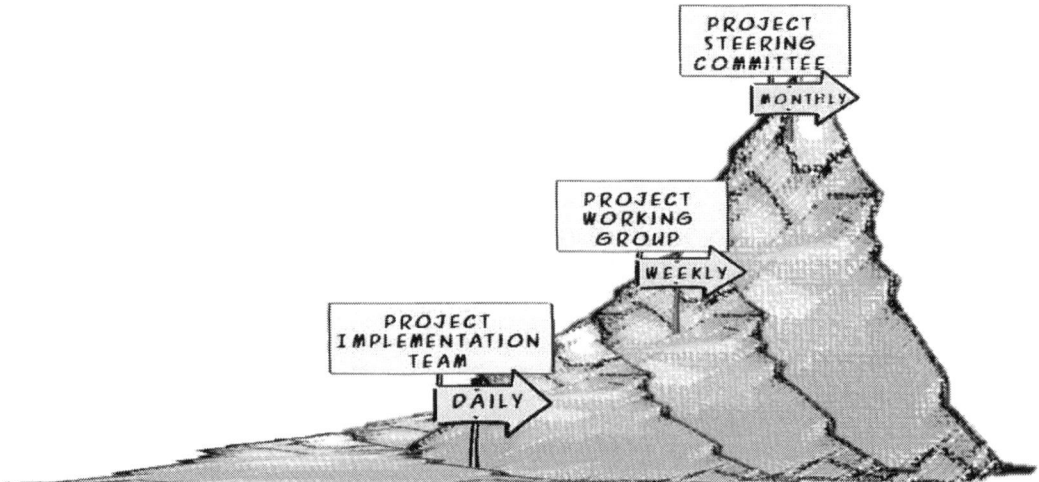

6.4.4. Perspective: Implementation Methodologies & Project Life Cycles

There are many different ways that the project's physical implementation can be undertaken, but for this book these were rationalized down to three overarching methodologies based on the two project life cycle models previously examined in [Section 2.2]. That is, the implementation methodologies of:

1) **Waterfall** – is where all of the project's features & functionality are undertaken as a logical sequence of tasks to be completed in accordance with the order laid out in the project's schedule baseline (possibly with only a single deliverable release cycle).

   Waterfall is built around an overwhelming desire for governance, accountability, and control. ... Possibly resulting in micro-management.

2) **Iterative** – is where a subset of the project's total features & functionality are implemented as a series of miniature waterfall release cycles of varying durations that culminate in each project delivery milestone.

3) **Agile** – is where a subset of the project's total features & functionality are pooled as use-cases (aka "stories") into short fixed duration "sprints". These pooled stories are then divvied up as tasks among the implementers (as they deem appropriate for completion within the timeframe of each particular sprint).

   Agile is built around implementation. One could even go as far as describing agile as being "grass-roots" when compared to the orchestration of the traditional techniques such as waterfall. Consequently, agile is highly adaptive to changing situations & circumstances that were not necessarily known when the project was planned out (let alone when the project was conceived).

NOTE   ❏ **The relationship between the project life cycle, the implementation methodology, and the management hierarchy will influence how the monitoring & control of the project will be undertaken.**

6.4.5. Perspective: Implementation Methodologies & Management Hierarchy

Each of the previously listed project life cycle methodologies better suits the needs of specific levels of the project management hierarchy.

The Project Steering Committee has a "big picture" view of how this project fits in with the performing organization's other projects / programs / business strategies and therefore the waterfall methodology (with its logically sequential nature of an approximately known duration) is better suited for laying out the roadmap of future endeavours and allocating budgets for such endeavours.

Thus, a **waterfall perspective is best for the Project Steering Committee**.

Let us "call a spade a spade"; many of the Project Steering Committee members are not usually concerned with the project implementation methodology that is being used. Rather, they are more interested in whether the project succeeds based on the traditional measures of [Scope], [Time], [Cost], and [Quality].

For the **Project Implementation Team** with their "hands-on" focus on executing the project's tasks, then **an agile methodology is the best fit**; see [Section 7.4].

This leaves the Project Working Group as the "middleman" between the Project Steering Committee and the Project Implementation Team, interfacing / translating between these two parties. For the **Project Working Group, the iterative methodology is the optimum fit** because it can be manipulated to provide the information necessary for the waterfall methodology and the control essential for the agile methodology not to become chaotic (or perceived as chaotic); see [Figure 66] and [Figure 67].

NOTICE in [Figure 67] that 'Project Charlie' is composed of three iterative releases, where Release-2 is composed of seven agile implementation sprints.

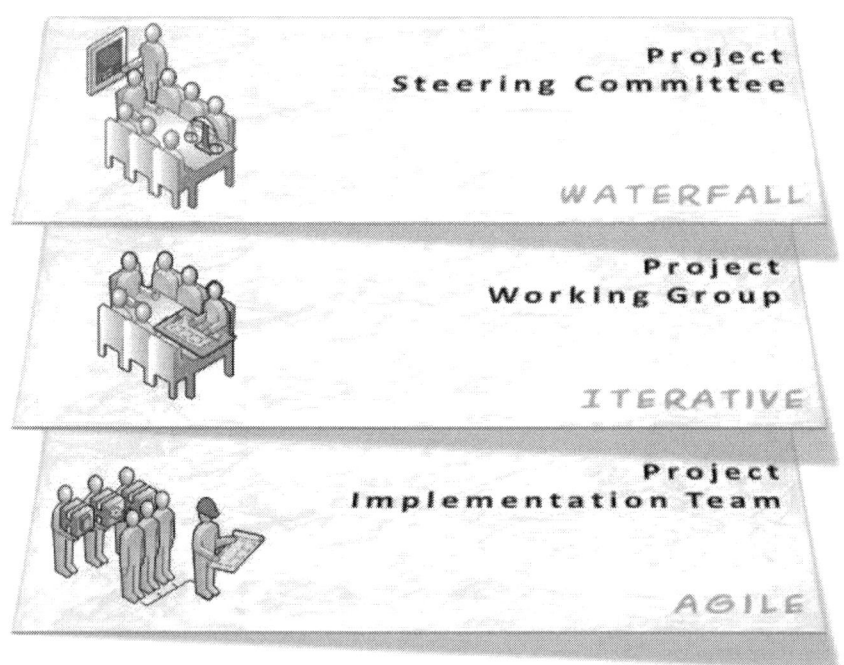

Figure 66: Relationship between the management hierarchy and the project life cycle methodologies.

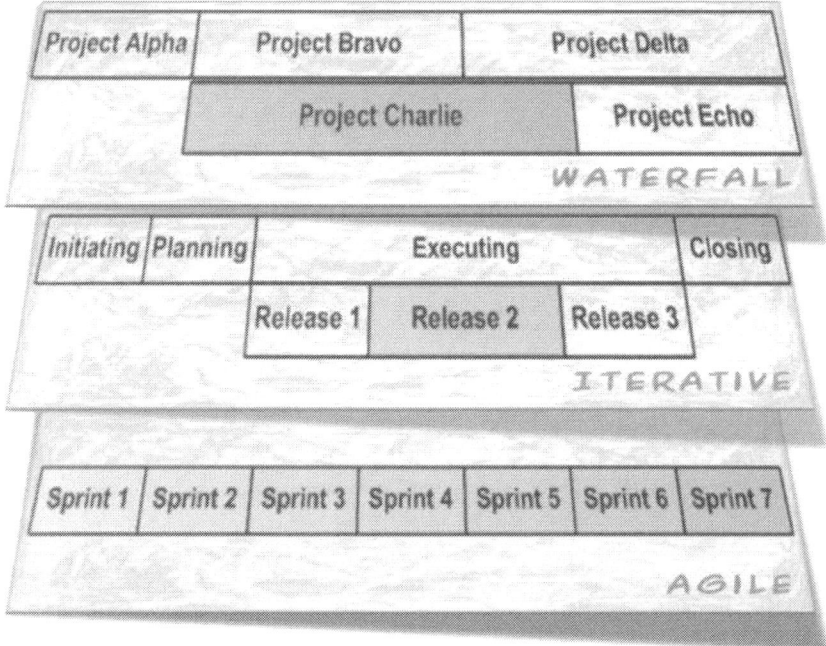

Figure 67: Relationship between the different project life cycle methodologies.

THE UNDERSTANDING OF THE PROJECT'S CURRENT SITUATION IS ONLY AS GOOD AS THE ACCURACY, RELIABILITY, AND RELEVANCE OF THE GAUGES USED TO REPRESENT WHAT IS HAPPENING NOW.

7. MONITORING & CONTROL Phase ... Theory Part 2

Watching the Gauges

As was explained back in [Chapter 6] and illustrated previously in [Figure 60], the project is composed of several variables & constraints that have to be managed throughout the entire life of the project. However, these project variables & constraints (in all probability) are eventually going to deviate away from their approved baseline values. Therefore, these **project variables & constraints will require regular and periodic monitoring & control**.

This current chapter will concentrate specifically on the equivalent of watching the gauges; i.e. being based on the traditional [Scope], [Time], and [Cost] measures of project success.

A.K.A. Keeping an eye on those Key Performance Indicators (K.P.I.).

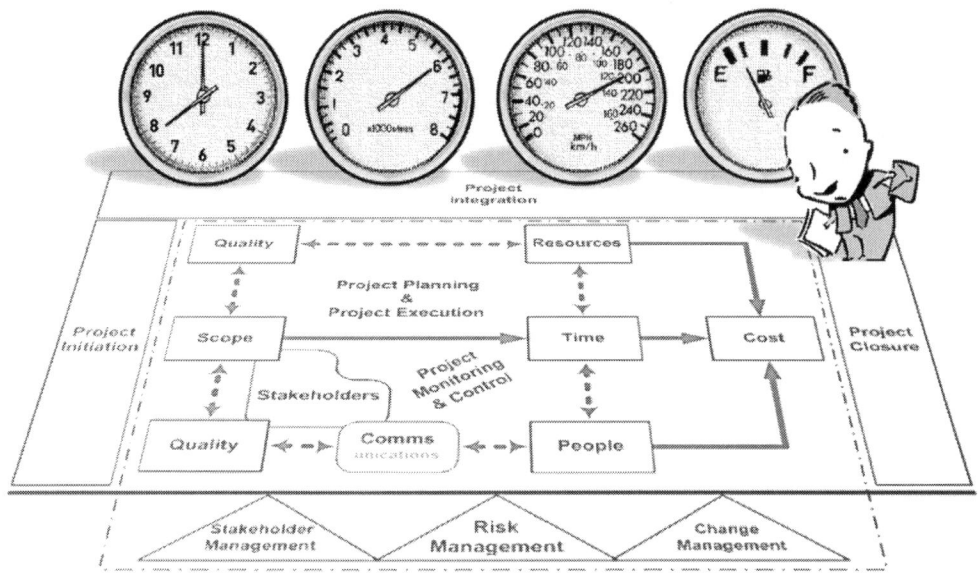

7.1.   [Scope] Monitoring & Control

The purposes of [Scope] Monitoring & Control are to:

1) **Confirm that only those agreed features & functionality [Scope] are being included in the appropriate deliverables** as specified in the [Scope Baseline]; i.e. keeping an eye on what is and is not going into each and every release.

   - This would also include, keeping track of what has and has not been included for each release in the past, present, and future. *...i.e. Configuration Management.*

2) **Ensure that only those approved changes are being incorporated into the relevant deliverables,** and to ensure that unapproved changes are not being snuck into releases (i.e. watching out for "scope creep" and "scope shrinkage").

 **If scope creep** was to occur virtually **unchecked then** the project will **slowly die from a [Time] & [Cost] haemorrhaging**.

Unfortunately this ain't an uncommon real-world event

---

3) **Manage the scope change process**, so as to review & approve requested changes, coordinate the incorporation of those approved changes into the [Scope Baseline] and into the relevant deliverables, and finally verifying that each approved change is present in the corresponding release(s).

4) **Determine and report on** how any **differences between the implemented [Scope] and the [Scope Baseline]** will affect the successful outcome of the project.

For the monitoring points (1) and (2) listed above, the Requirements Traceability Matrix (RTM) and the Work Breakdown Structure (WBS) become useful tools.

For the controlling point (1) and (3), a Change Management Process is required.

For point (4), this is where the project manager's analytical and presentation skills come into play.

Requirements Traceability Matrix (RTM)

The **Requirements Traceability Matrix (RTM)** will be used as the equivalent of a **"tick-off list"** to determine if any features & functionality have been overlooked, and to confirm the **"coverage" of the** Customer Requirements (as **agreed** to in the **Detailed Specifications**). That is, what was and was not agreed to be **within the project's Scope Boundary**.

Work Breakdown Structure (WBS)

The **Work Breakdown Structure (WBS)** will be used as **a visual indicator** of; **what features & functionality have been implemented**, is being implemented, or is yet to be included in the current release (as well as past, present, and future releases).

SCOPE CREEP ... THE ART OF MOVING SOMETHING FROM BEING "OUT OF SCOPE" TO BEING "IN SCOPE" WITHOUT HAVING TO COMPENSATE SOMEONE FOR THE PRIVILEGE OF DOING SO.

Before concluding this introduction to [Scope] Monitoring & Control there are
a few important points to consider:

- While [Scope] Monitoring & Control verifies that the agreed features & functionality
  are included in the appropriate deliverables / releases, **[Scope] Monitoring & Control**
  does not verify that the features & functionality delivered actually meet the purpose
  for which the customer intended. This type of "purpose-ability" verification is the
  concern of [Quality] Monitoring & Control.

> **Be wary of this demarcation between [Scope] and [Quality],**
> as **there has been many a disagreement** between the customer
> organization and the performing organization **due to a deliverable**
> **conforming to the specifications yet it did not conform to the**
> **purpose for which the customer had "intended".**

- **The scale of the difference between the [Scope Baseline] and the implemented**
  [Scope], plus the cause & effects of such variations **will influence how best to respond**
  to such deviations; i.e. **whether corrective, preventive, defect repair, or re-baselining**
  is required. Hence, if there is only a small difference then only a small response is
  required, whereas a large deviation may require drastic measures to be taken to
  get the project back on course.

- The decision made to react to the variations in [Scope], the reason for making that
  decision, and the plan for implementing that **decision have to be communicated to**
  the relevant stakeholders and documented accordingly for future reference.

- **Changes to the [Scope] will have a direct effect on the [Time]**, a ripple effect on the
  [People], [Resources], [Quality] and subsequently a cascading effect on the [Cost];
  see [Figure 68].   *... There ain't no such thing as an innocuous scope change.*

  For more information on the topic of practical [Scope] Monitoring & Control,
  please refer to [Section 12.1].

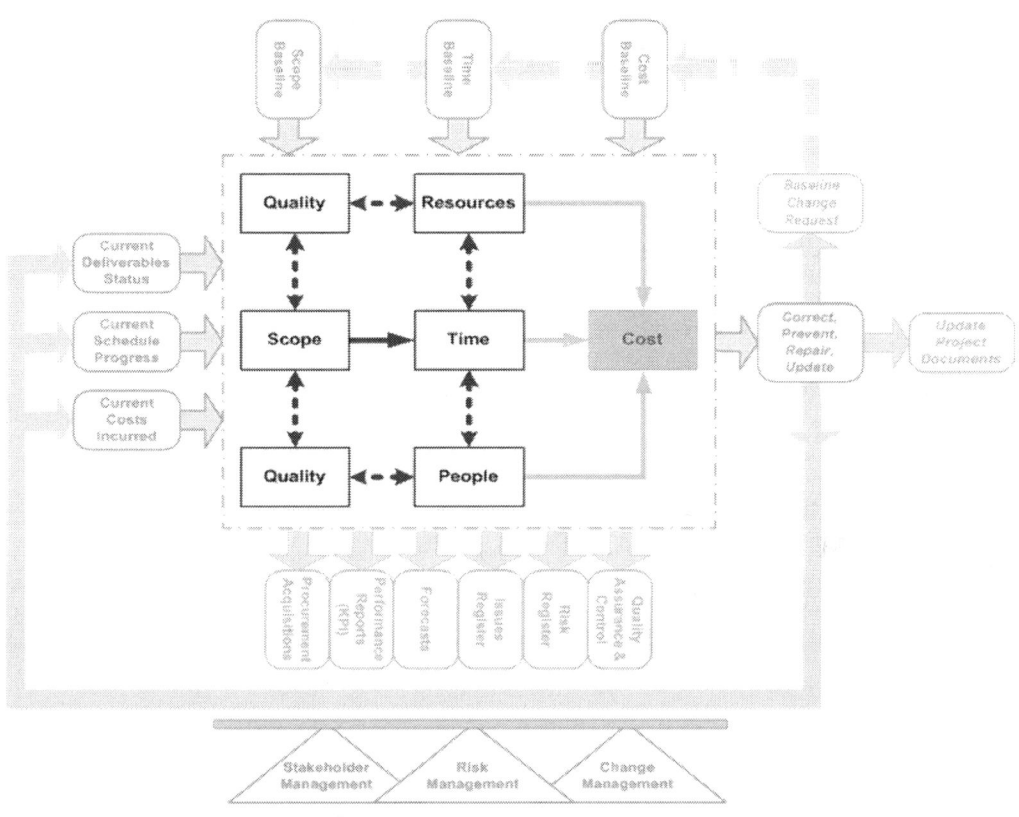

Figure 68: The direct effect of [Scope] on the monitoring & control of [Time] and the ripple effect on [People], [Resources], and [Quality] … and, eventually onto [Cost].

7.2.   [Time] Monitoring & Control

The purposes of [Time] Monitoring & Control are to:

1) **Keep an eye on the project's progress** by comparing over the same period (*e.g. weekly*) those completed and work-in-progress tasks against their counterparts in the schedule [Time Baseline].

2) **Manage changes to the [Time Baseline]** due to changes in the [Scope Baseline], as well as due to the changing situation & circumstances confronting the project.

3) **Perform variance analysis**, and thereby:

   - Determine the **cause** of any time and duration variations between the actual and baseline scheduled tasks.

   - Determine how any such variations will **affect** the project's successful outcome.

   IMHO, what is important is the timing and duration of those collective tasks composing the next milestone release, and not the individual tasks. Just keep delivering each release on Time, within Cost Budget, with Scope & Quality then the project's primary stakeholders ain't really gonna care about how precisely the project's Baseline schedule was followed for those individual tasks that made-up the greater whole.

4) When the situation requires, **perform corrective and preventive actions**:

   - **Decide** what actions are necessary to bring the project's [Time] duration and delivery dates back into line with the agreed [Time Baseline].

   - **Report** to the relevant stakeholders about what the chosen actions are and the reasons for taking such actions.

   - **Execute** those chosen actions.

   - **Confirm** that those chosen actions have been implemented.

Please refer to [Section 12.2], for practical [Time] Monitoring & Control.

7.2.1. Time: Schedule Duration Compression

At some point during a project's life, the schedule's duration may have to be reduced so that the resultant delivery dates are more acceptable to the project's stakeholders. There are a **few different techniques that can be used for reducing the delivery date**.

1) **"Crashing"** the schedule (**"C.P.R."**) – is compressing the schedule's duration **by shortening the time allocation for critical path** tasks via:

- **paying additional (C)ash** to the implementer to increase the priority for those tasks that have to be completed sooner, and allowing for additional overtime.

- **adding more (P)eople** to work on those tasks that have to be delivered immediately.

- **adding more (R)esources or extending the allocated usage** of those resources that are vital for the completion of those tasks.

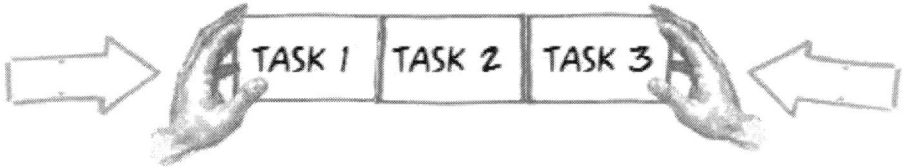

NOTE   ❑ The **"Crashing" technique only works for tasks on the critical path**; i.e. those tasks with zero "float" that if not completed by their due-date will directly result in the project / release being delivered late.

 The **"Crashing" technique is ineffective when applied to non-critical path tasks**. In fact, this technique if applied inappropriately will be detrimental to the project because it will unduly increase [Cost] and can have negative effects on morale.

---

2) **"Fast Tracking"** the schedule – is compressing the schedule's duration **by performing some tasks in parallel and glossing-over those tasks that are not evident in the deliverable**. However, this could potentially reduce [Quality].

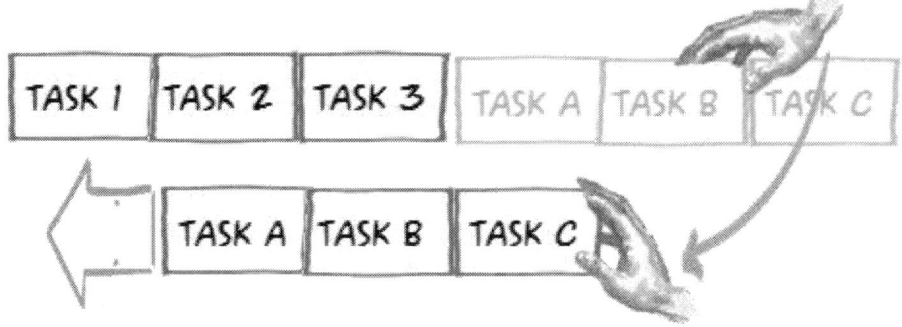

NOTE	❐ The "Fast Tracking" technique works best for tasks associated with the critical path, but it could be used en masse for all tasks in the release. This technique **requires that the shifted tasks are not sequentially dependent.**

❐ Those "non-evident" parts such as documentation, source code comments, and peer reviews are often the first victims of fast tracking. Followed by a reduction in unit testing and eventually a reduction in system testing. *... Eventually to a "who cares, just ship it" mentality.*

The **"Fast Tracking" technique can result in more problems than it was intended to solve**, because of the potential that rework will be required due to:

- the **consequential reduction in [Quality]**, and
- the possibility that the **deliverables may not have been built with future re-use in mind**, instead being tailored "hard coded" specifically for the targeted milestone.

Fast Tracking is inherently risky to the project's overall success.

---

3) **"Scope Minimization"** – is compressing the schedule's duration by coming to agreement with the customer's representative that **certain features & functionality will not be included in the time-sensitive release**; i.e. scope re-baselining. Instead these features & functionality shall either; be included in a later release, or be removed completely from the project's deliverables. ... *"De-scoped"*.

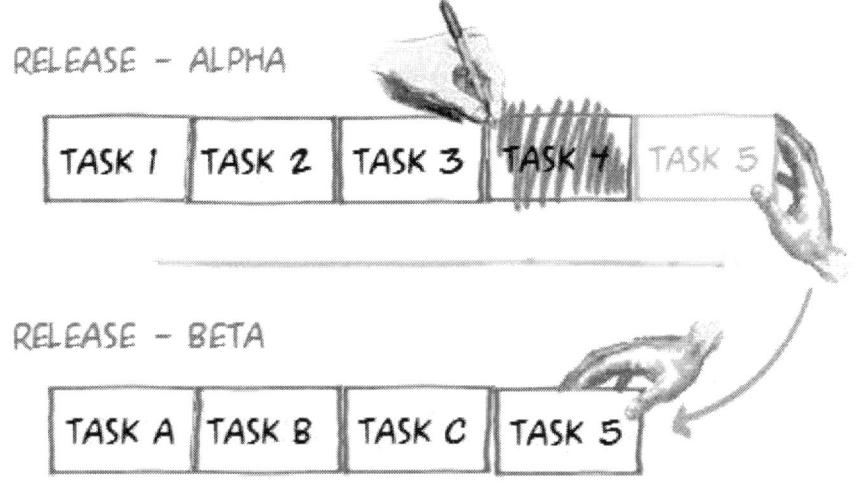

NOTE   ☐ The "Scope Minimization" technique is most effective with an iterative project life cycle.

 The **"Scope Minimization" technique could potentially result in the schedule duration problem being shifted towards the end** of the project as more and more tasks are reassigned to later releases. Alternatively, **compensation and penalty clauses** may be invoked for the minimalist deliverables with its missing features & functionality.

What's now in & what's now out, is possibly a future argument? So better, update the Detail Specifications for re-approval.

As with the project's [Scope Baseline], **change control is necessary when modifying the [Time Baseline]**. That is, the Baseline Schedule changes have to be requested, reviewed, approved, implemented, and then verified.

However, **change control is not necessary for the daily Work-In-Progress Schedule**, as to do so would prove to be burdensome and very time inefficient.

Changes to [Time] should never directly affect the [Scope].
Rather, if the project's duration has to be reduced then the performing organization and the customer's representative will need to **negotiate scope changes, else covert scope shrinkage may occur**.

While **changes to the [Time] will NOT have a direct effect on the project's [Scope],** **these changes will have a direct effect on the project's [Cost]** and a subsequent ripple effect on the [People], [Resources], and [Quality] aspects; see [Figure 69] below.

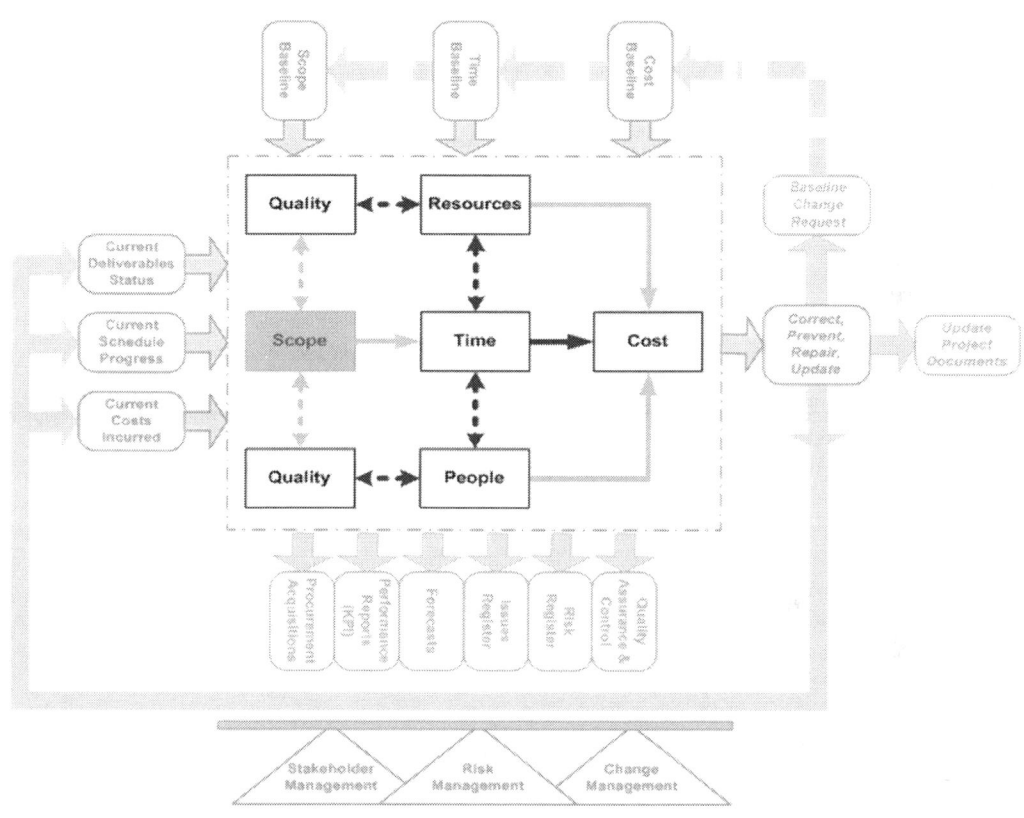

Figure 69: The direct effect of [Time] on the monitoring & control of [Cost] and the ripple effect on [People], [Resources], and [Quality].

SCOPE INFLUENCES TIME ... TIME EFFECTS COST ...
THOUGH NOT THE OTHER WAY AROUND,
ELSE CUSTOMER SATISFACTION
WILL NEVER BE FOUND.

7.3.   [Cost] Monitoring & Control

The purposes of [Cost] Monitoring & Control are to:

1) **Keep an eye on the project's [Costs]** by comparing over the same time-period (*e.g. weekly*) the actual costs against the authorized [Cost Baseline].

2) **Manage changes to the budget [Cost Baseline]** due to changes in the [Scope], [Time], [People], and/or [Resources], as well as changes to the financial circumstances confronting the project.

3) **Perform variance analysis**, and thereby:

   - Determine the **cause** of any cost variations between the actual spend and the budget.

   - Determine how any such variations will **affect** the project's successful outcome.

4) **Perform corrective and preventive actions**:

   - **Decide** on what actions are necessary to bring the project's [Costs] back into line with the approved [Cost Baseline].

   - **Report** to the relevant stakeholders what are the chosen actions and the reasons for taking such actions.

   - **Execute** those chosen actions.

   - **Confirm** that the chosen actions have been implemented.

Hey, the above list of **purposes for Cost Monitoring & Control** *reads* **very similar to** *that for* **Time Monitoring & Control***; was this just a cut-n-paste accident or is this similarity deliberate?*

The similarity between the purposes of monitoring & control for both [Time] and [Cost] is not coincidental. This is due to the **tight relationship between [Time] & [Cost].** However, **[Time] has a bi-directional relationship with both the [People] & [Resources] that also happen to flow directly into the project's [Costs].**

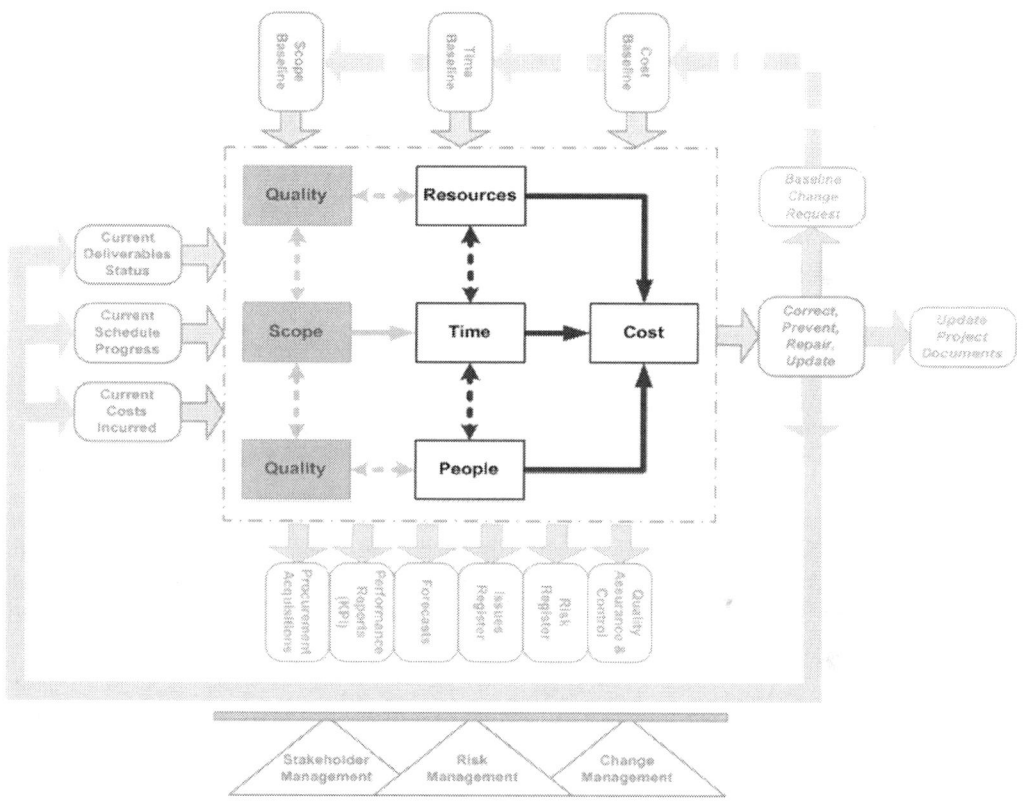

Figure 70: The direct effect of [Time], [People], and [Resources] on the monitoring & control of [Cost].

Given the relationships illustrated above in [Figure 70], it should not be surprising that if there is any change to [Time], [People], or [Resources] then this will have a direct effect on the project's [Cost].

 For more information on the topic of practical [Cost] Monitoring & Control, please refer to [Section 12.3]; especially for Earned Value Performance.

7.3.1. Cost: Earned Value Performance Measures

Earned Value Management (EVM)

- **Planned Value (PV)** ... is the authorized / baselined budgeted value in terms of monetary units and/or labour hours that were planned to be incurred up to the current status date (when the measurements were taken) for the performance of a planned number of scheduled tasks / WBS work-packages. That is, **PV is what was the planned spend to date, to get the project to where it was planned to be by now**.

  (PV) is also known as, **Budgeted Cost of Work Scheduled (BCWS)**.

  > *For example; in 6-days of this 10-day project we planned to have spent $7000 and be 60% of the way through the project. In this case, PV = $7000.*

- **Actual Cost (AC)** ... is the actual "real" cost in terms of monetary units and/or labour hours incurred up to the current status date (when the measurements were taken) for the work that was actually "really" performed on scheduled tasks / WBS work-packages (including both completed and work-in-progress). That is, **AC is what has actually been spent to date, to get the project to where it really is now**.

  (AC) is also known as, **Actual Cost of Work Performed (ACWP)**.

  > *For example; so far in 6-days of a 10-day project we've spent $8742 to get 43% of the way through the project. In this case, AC = $8742.*

- **Earned Value (EV)** ... is the budgeted "Expected Value" in terms of monetary units and/or labour hours that were anticipated / planned to have been incurred up to the status date for the work actually performed on scheduled tasks / WBS work-packages (including completed and work-in-progress). That is, **EV is what was planned to have been spent to date, to get the project to where it really is now**.

  (EV) is also known as, **Budgeted Cost of Work Performed (BCWP)**.

For example; so far, in 6-days of a 10-day project, we have only got 43% of the way through the project, and for this 43% we planned to have only spent $5500. In this case, EV = $5500.

- **Cost Variance (CV = EV - AC)** ... measures the difference between the "Expected Value" (i.e. Earned Value) and the "Actual Cost" of the work performed up to the current status date. That is, **CV is how far above or below the [Cost Baseline] Budget to get the project to the point where it really is now**.

  For example; in 6-days of this 10-day project we have actually spent $8742 for those number of tasks that were actually performed (i.e. to get 43% of the way through the project), while we planned to have only spent $5500 on those tasks.

  CV = EV - AC = $5500 - $8742 = - $3242   *NEGATIVE CV IS BAD.*

- **Schedule Variance (SV = EV - PV)** ... measures the difference between the "Expected Value" (i.e. Earned Value) and what was the "Planned Value" for the same point in the project's duration. That is, **SV is how far ahead or behind [Time Baseline] Schedule of where we really are now, against where we planned to be by now**.

  For example; in 6-days of this 10-day project, we planned to have completed $7000 worth of tasks, but we have only completed $5500 worth of tasks.

  SV = EV - PV = $5500 - $7000 = - $1500   *NEGATIVE SV IS BAD.*

- **S-Curves** ... as depicted on the following page in [Figure 71], the S-Curve is a graphical representation of; Actual Costs (AC), Earned Value (EV), and Planned Value (PV).

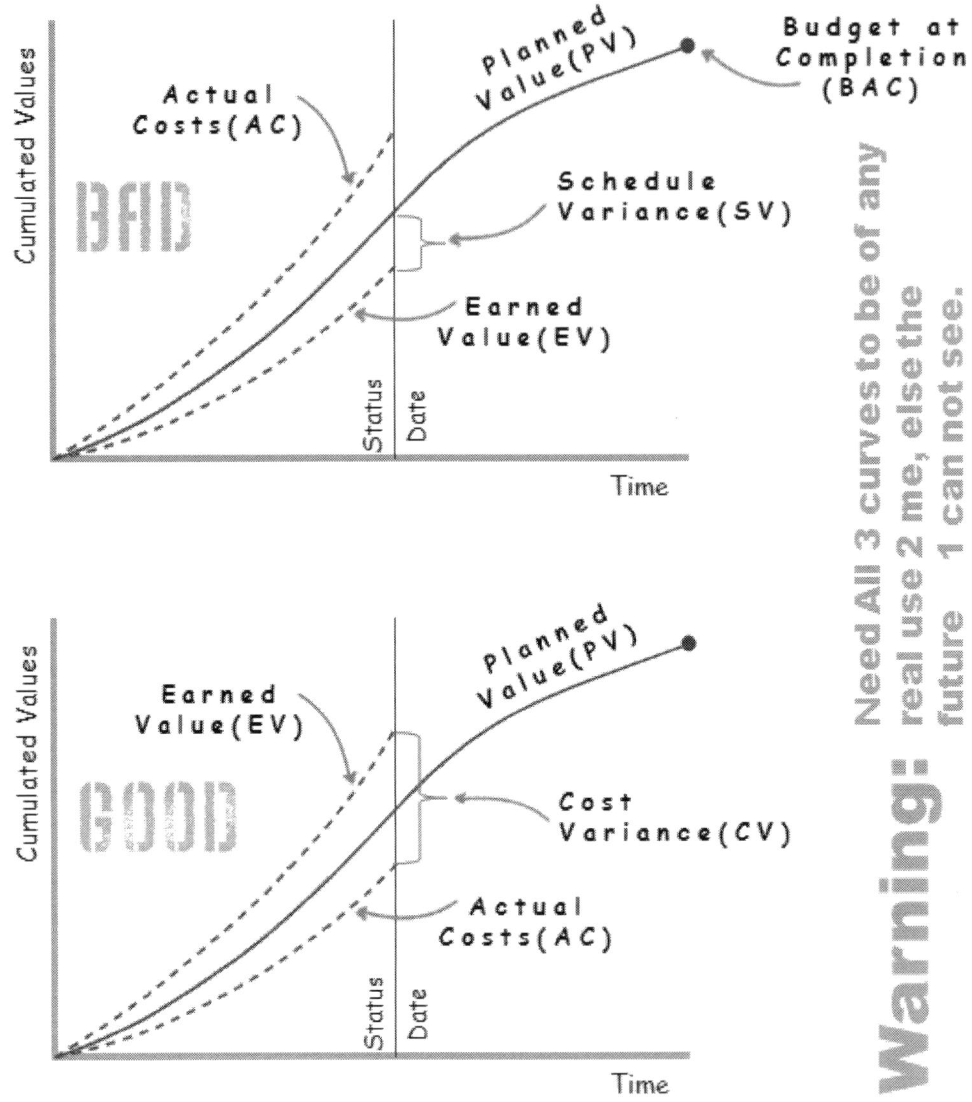

Figure 71: Actual Costs (AC), Earned Value (EV), and Planned Value (PV) versus time mapped as Earned Value Management S-Curves[5].

 **Need all 3 curves (PV, AC, and EV) to be plotted at the same time** to be any real use as an indicator of the project's performance progress.

[5] Based on PMI 2008, *"A Guide to the Project Management Body of Knowledge (PMBOK Guide), 4th Edition"*.

Why is the cumulated cost curve shaped like an 'S', rather than as a straight line?

This 'S' shape is because at the beginning of the project there are only a few people involved (i.e. charging time to the project's Cost Account Code), whereas

by the Executing Phase the entire Project Implementation Team is on board.

However, as the project begins to conclude the team will begin to be dispersed to work on other projects so that there will be only a few people remaining.

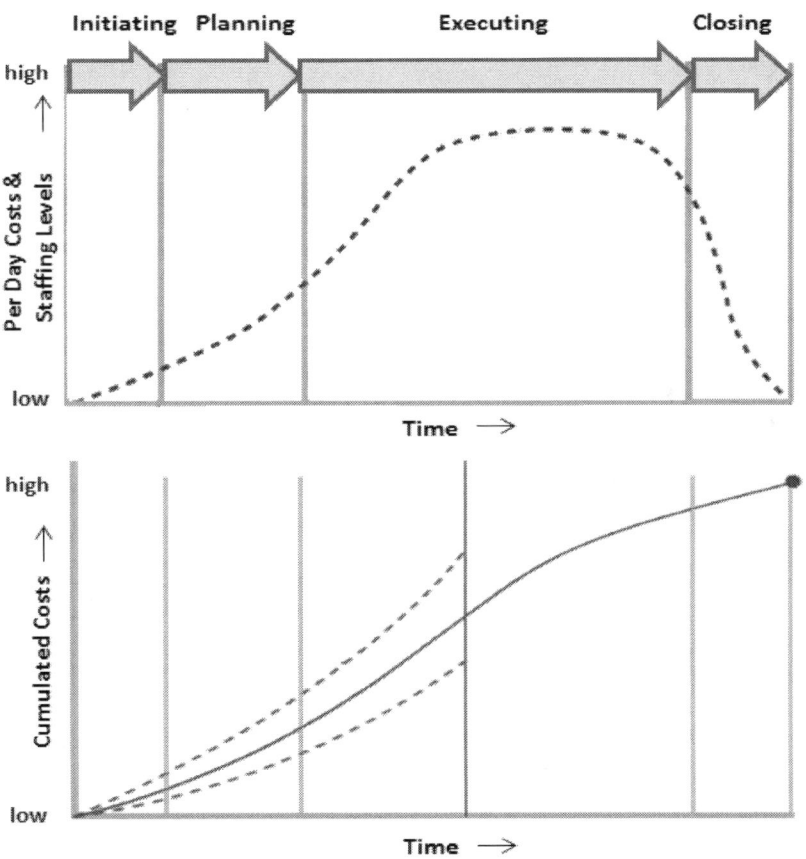

Figure 72: Staffing levels compared to the accumulated cost S-Curve.

NOTE ❑ For a project with a fixed / obligated work effort *(e.g. 40 hours per week)* then the S-Curve would in fact be a rising straight line.

- **Cost Performance Index (CPI = EV / AC)** ... determines up to the status date whether the project is currently exceeding its allocated budget by comparing the ratio of the "Expected Value" (i.e. Earned Value) against the "ACtual value" (i.e. Actual Cost). That is, **CPI denotes how good or bad is the project's performance when compared to the [Cost Baseline] budget** (in terms of monetary units / labour hours), **to get the project to where it really is now?**

  > For example; in 6-days of this 10-day project we've spent $8742 when
  >
  > we expected to have only spent $5500 for the number of tasks performed.
  >
  > CPI = $5500 / $8742 = 0.63   *CPI LESS THAN ONE BAD.*

  CPI of less than one (< 1.0) indicates a **cost overrun** for the work performed so far.

  CPI of greater than one (> 1.0) indicates a **cost underrun** for the work performed.

  CPI GREATER THAN ONE IS GOOD ... or it could mean that we are not working at full capacity and possibly, the budget burn-rate is so awesome because we are not doing any work on the project at present.

- **Schedule Performance Index (SPI = EV / PV)** ... determines how well the project is currently progressing through the scheduled tasks / WBS work-packages by measuring the ratio of the Earned Value against the Planned Value, for the same point in the project's duration. That is, **SPI denotes how good or bad is the project's performance when compared to the [Time Baseline] schedule** (in terms of monetary units / labour hours), **where the project really is now, against where it was planned to be by now?**

  > For example; in 6-days of this 10-day project we planned to have spent $7000,
  >
  > but based on the number of scheduled tasks actually performed we have
  >
  > earned / achieved only $5500 worth of work.
  >
  > SPI = $5500 / $7000 = 0.79   *SPI LESS THAN ONE BAD.*

  SPI of LESS than ONE (< 1.0) ... less progress has been made than was planned for.

  SPI of GREATER than ONE (> 1.0) ... more progress has been made than was planned.

  SPI GREATER THAN ONE IS GOOD ... or it could mean that we are exceeding our working capacity and possibly, the budget burn-rate is frighteningly scary because we are working so far in advance of the planned schedule.

Need Actual Costs (AC) and Earned Value (EV) both GOOD

When the **Actual Costs (AC) is greater than** the **Planned Value (PV)** then this is **potentially BAD** because the project is "**over budget**".

However, when the **Actual Costs (AC) is less than** the **Planned Value (PV)** then this is **potentially GOOD** because the project is "**under budget**".

When the **Earned Value (EV) is less than** the **Planned Value (PV)** then this is **potentially BAD** because the project is "**behind schedule**".

However, when the **Earned Value (EV) is greater than** the **Planned Value (PV)** then this is **potentially GOOD** because the project is "**ahead of schedule**".

Why is this (maybe) "potentially" good or bad?

Both Actual Costs (AC) and Earned Value (EV) have to be "GOOD" at the same time else, the project is "BAD".

Actual Costs (AC) ... (ACWP)	Earned Value (EV) ... (BCWP)	Project Status	Road Trip Analogy
GOOD	BAD	Under worked	Driving too slowly; won't get home in time even though there will be fuel remaining in the tank.
BAD	GOOD	Over worked	Driving too aggressively; won't have enough fuel to make it home even though there will be time to spare.
GOOD	GOOD	Optimal Performance	Correct speed and fuel economy to arrive home in time to make it to the milestone event.

However, these performance measures of GOOD & BAD are only as useful as the accuracy of the quantified data used to determine these. For example; incorrect percentage complete on scheduled tasks and over/understated hours worked during the period will result in a misrepresentation of the project's current true condition.

Earned Value Forecasting Techniques

- **Budget At Completion (BAC)**

  What is the expected total value (in terms of monetary units and/or labour hours) when all of the project's planned tasks / WBS work-packages are completed.

  What is, the total Planned Value when the project is finished?

- **Estimate To Complete (ETC)**

  From where the project has got up to (by the current status date), what do we calculate is required (in terms of monetary units and/or labour hours) to finish off the project's remaining planned tasks / WBS work-packages.

  What do we have to spend from here on in, to get this project across the finish line?

- **Estimate At Completion (EAC = BAC - CV)** ... *if 100% efficient performance.*

  $EAC = BAC / CPI$ ... *if continuing at the current rate of performance.*

  Based on the project's actual progress up to the current status date, what is estimated to be the total value (in terms of monetary units and/or labour hours) when all of the project's planned tasks / WBS work-packages are completed.

  What do we now calculate as going to be the total cost when this project finishes?

- **To Complete Performance Index (TCPI = (BAC - EV) / (BAC - AC))**

  Compares the value of the remaining work to the value of the remaining budget.

  What cost performance is required from now on to deliver the remaining work within the planned budget?

- **Estimated Duration At Completion (EDAC = PTPT + Slippage)** ... *100% efficiency,*

  else $EDAC = PTPT / SPI_{Time}$ ... *where* $SPI_{Time} = (time\ planned\ to\ spend) / (actual\ time\ spent)$

  The project's **Estimated Duration At Completion (EDAC)** equals the **Planned Total Project Time (PTPT)** (*i.e. the baseline schedule's total duration*) **plus slippages** (*i.e. actual duration of worked performed so far, less planned duration of that same work*).

* *Management Reserve* = $EAC - BAC$

Example Case: Earned Value Performance Measures, Not Always So Welcome.

Once upon a time, there was a performing organization that would continually get nasty surprises when it was discovered that some of its projects were grossly late and over budget when compared to the actual progress towards completion. It was as though, one part of the organization (the development department) was focused solely on measuring project progress based on [Time] & [Scope], while another part of the organization (the finance department) disassociatedly focused on measuring progress based on the organization's operational [Costs]. That is, there was a disconnection between the projects' realities and the reported factualities. Hence, it was a hit & miss affair for this performing organization's senior management to determine whether it was going to achieve its projected quarterly / yearly revenue targets, and subsequently whether the performing organization would achieve its profitability aspirations.

A "professional" program manager was brought on-board, and after analysing the project management situation and past performances, this program manager rolled-out a weekly regime of Earned Value Performance Measures (EVPM) and Key Performance Indicators (KPIs) for all projects within the performing organization. A template EVPM / KPI spreadsheet was issued to all of the project managers & senior management, and training provided on the utilization of these tools so that the same measurement rules & methodology would be utilized consistently across all of the organization's projects. Alas, within a few weeks, the true state of each project's progress was numerically apparent, and all of the currently active projects' performance trends clearly highlighted in a single report with "Traffic Light" indicators; see [Section 12.3.3] & [Section 12.3.4].

However, the organization wide utilization of EVPM did have some negative pushback:

- Some members of staff felt that the "*imposed*" EVPM were burdensome to their project management workload. ... "*The curse of my Monday mornings*".

- Some members of staff, who were used to ad-hoc (reactionary) management of their projects, felt that the "*imposed*" EVPM impinged on their management style.

- Some (senior) members of staff did not like that the EVPM exposed the truth about their projects' pending outcomes, and thus be interpreted as revealing their own project management insufficiencies / incompetencies (which for years had been hidden away behind the complexities of disassociated departmental reporting).

Simplified Cost Formula

For now, just remember this simplified cost formula.

$$COST_{\text{TO DATE}} = (PEOPLE_{\text{RATES}} \ X \ TIME_{\text{USED}})$$
$$+$$
$$(RESOURCE_{\text{UNITS}} \ X \ AMOUNT_{\text{USED}})$$
$$+$$
$$(FIXED \ COSTS_{\text{TO DATE}})$$

As a novice project manager, you will most probably only be responsible for the project's [Scope], [Time], [People], [Resources] and not directly the [Cost].

The reasoning behind this arrangement is;

> *If the project's agreed [Scope] was delivered on [Time] using only the [People] and [Resources] that were allocated then the project would be delivered within the allocated budget [Cost].*

This is a fair assumption; being based on the principle that during the Planning Phase in [Chapter 4] the project's [Cost] and hence the allocated budget was appropriately calculated using the project's agreed [Scope], the allocated [People], and the acquired [Resources] that were necessary to successfully complete each of the scheduled tasks within the agreed [Time] frame.

The budget allocation is only as good as the estimates that were provided for [Time], [People], and [Resources] used to calculate the project's [Cost].

BAD Estimates = BAD Budget = FAILED Project   = *BAD Cash Flow*

7.3.2. Cost: No Direct Relationship To Quality

Notice that up to this point, none of the cost metrics have dealt with the quality of the project's deliverables. As illustrated below in [Figure 73], this is simply because there is **NO DIRECT RELATIONSHIP BETWEEN [COST] AND [QUALITY]**.

Rather, **[Quality] is achieved via the intermediaries of [Scope], [People], and [Resources]**.

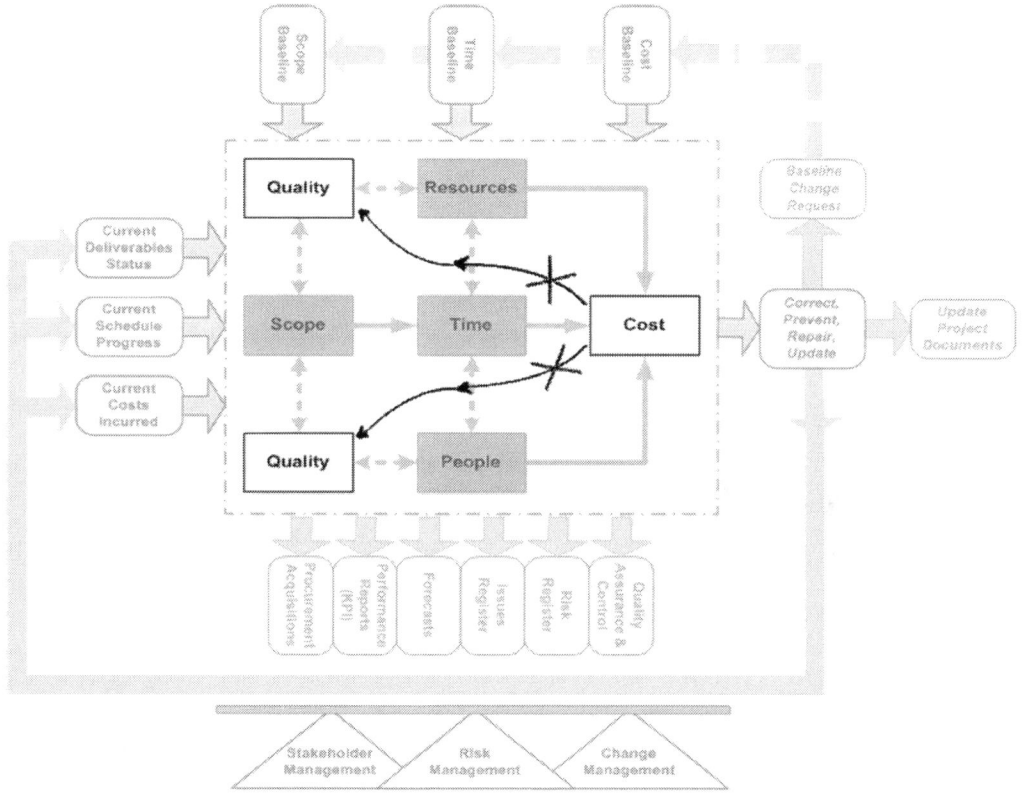

Figure 73: No direct relationship of [Costs] to the [Quality] of the deliverables.

COST DOES NOT REFLECT QUALITY.

High costs do not ensure high quality deliverables.

Conversely, low costs do not necessarily mean low quality.

However, as will be described later on in [Section 8.1], cutting [Cost] will most probably reduce [Quality] because the active reduction in [Cost] would be either via a reduction in [People], [Resources], and/or [Scope] which will all directly affect [Quality].

BAD ESTIMATES   =   BAD BUDGET

BAD BUDGET   =   FAILED PROJECT

FAILED PROJECT   =   BAD CASH FLOW

BAD CASH FLOW   =   A DYING COMPANY

HOW DOES A PROJECT GET TO BE MILLIONS OVER BUDGET AND YEARS LATE?

...ONE DAY AT A TIME,
...ONE PERSON AT A TIME,
...ONE RESOURCE AT A TIME,
...ONE SCOPE CREEP AT A TIME.

7.4. Agile Implementation with Monitoring & Control

7.4.1. Agile: Introduction

There is sometimes a perception that because a project is using an agile methodology then the previous three sections on [Scope], [Time], and [Cost] Monitoring & Control no longer apply in the same way that these did for the traditional methodologies of iterative and waterfall.

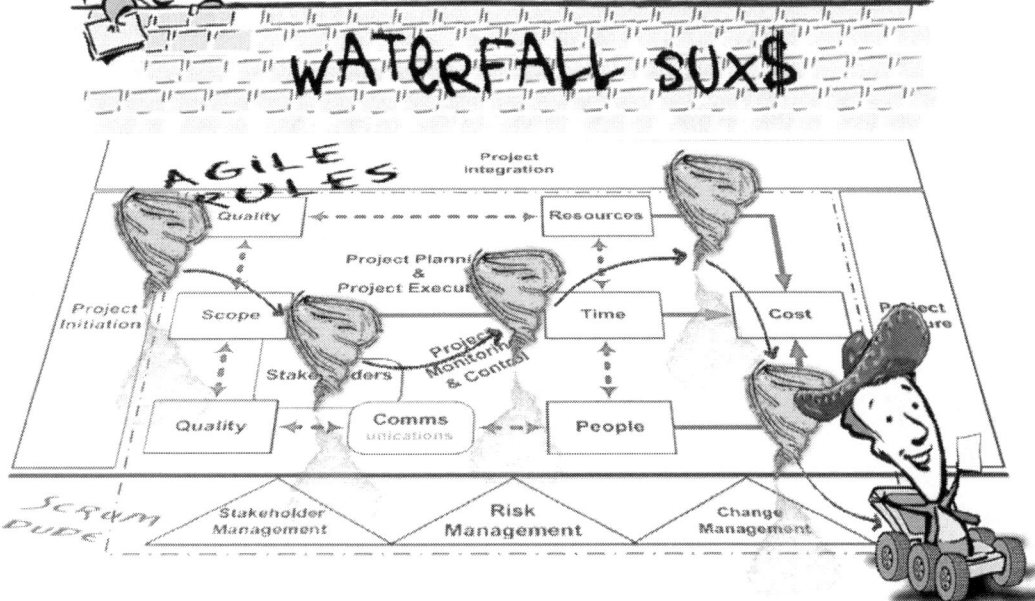

 **[Scope], [Time], and [Cost] Monitoring & Control are just as important for agile based projects** as they are for the traditional iterative and waterfall projects.

7.4.2.   Agile: Overview

Scrum-ish *… is not Scrum per se, but Agile adapted to your needs.*

Scrum-ish is a lightweight (minimalist) agile process that **utilizes consistent duration mini-iterative release cycles called "sprints"**. Though, an upcoming sprint does not necessarily have to comply with this consistent duration as it could be shortened or extended so as to align the sprint's end date with an external event (*e.g. the Christmas holiday break*), or aligned with a milestone for another project, or aligned with timings for groups within the performing organization or the customer organization.

7.4.3.   Agile: Process

Scrum-ish agile method has some differences from traditional iterative methodologies. For example; **scrum is "time boxed"** (i.e. has a fixed timeframe) and hence each sprint has a **defined end date** that it must be concluded by **irrespective of whether the release's deliverables are ready or not**.

A sprint is the equivalent of a drop-dead End-Date mini-project.

AUTHOR'S NOTE

As stated previously, this book will NOT be delving deeply into agile. However, this book will be "adaptively" using ideas from these agile methodologies (more specifically from Scrum & Kanban) and tailoring the perspective on these towards project management.

The **number of sprints in the project should NOT be predefined,** as this would impose a fixed limitation that could result in the perceived necessity to cram features into the final few sprints. That is, the **number of sprints required** by the project should be **continually reassessed.** This reassessment should be **based on past experiences with this & other projects, and on the "velocity" of earlier sprints**; i.e. through the wisdom of hindsight. Due to this reassessment, future sprints could be added or deleted.

Though, if additional sprints are required then this could be an indication that there are underlining problems with the project; *e.g. under estimations of the amount of work involved, signs of scope creep, or evidence that insufficient [People] & [Resources] have been allocated to the project.*

What is "Velocity"?

The **velocity is how many tasks or "story points" (use cases) are being completed per sprint**; i.e. how much work is being done per sprint. This velocity value should be used to calculate exactly how many tasks / stories can be delivered practically & feasibly in the next sprint.

If you know the velocity of how much work is being done per sprint and given that the sprints are of a constant duration, then one should be able to extrapolate exactly how long it is going to take to get to the next milestone, and subsequently how long it will take to get to the finish of the project.

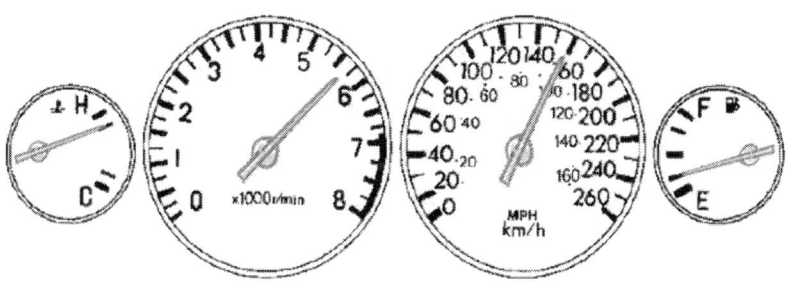

How is the Scope divided up among the sprints?

The content of each sprint is not methodically planned out in advance by the project manager. Rather, at the start of each sprint a **"Planning Meeting" is held between the "Product Owner"** (i.e. the representative of the customer or the primary-core stakeholders) **and the Project Implementation Team.**

At this Planning Meeting, the **Product Owner** presents a prioritized set of high-level requirements that they **"would like" to be delivered** based on their perceived value to the customer (i.e. Return On Investment). This **prioritized list** is called a **"Product Backlog".**

Based on past experiences (i.e. how effective were the last few sprints) the Project Implementation Team (optimally of about 7-8 cross-functional persons) then select a reasonable number of features that they are prepared to commit to completing by the end date of the upcoming sprint. Often these chosen items are a collection of **functional units / "stories"** that the Project Implementation Team intend to demonstrate at the conclusion of that sprint.

Where each story is a scope entry for this mini End-Date project.

Once the **limited number of features are selected,** then these **go into the "Sprint Backlog". NO ONE is allowed to change the content of the Sprint Backlog until the next sprint**. Thereby, preventing scope creep and in essence enforcing a form of *"wait until the next sprint"* change control.

This is the same as freezing the approved Detailed Specifications per milestone release as used in the traditional iterative and waterfall methodologies.

NOTE   ❏ **It is NOT the project manager**, but rather the members of the Project Implementation Team that decide the amount of effort that will be required to complete each selected story. Hence, **it is the team members that commit to delivering each story.**

Each story may be broken up into multiple tasks of approximately half a day (minimum) to a couple of day's duration (maximum).

These **tasks are NOT assigned** specifically to anyone, **but rather the individual members** of the Project Implementation Team will **pick from the Sprint Backlog the tasks that they intend to implement**.

Where a story is the equivalent of a work package in the Work Breakdown Structure.

Sounds like the inmates are running the asylum.

The **Sprint Backlog may be physically represented by notes attached to a "Task Board"** with lined-off sections for; 'To Do' (i.e. the Sprint Backlog), 'Work In Progress' (WIP), and 'Done'. These last two columns could be further subdivided into 'Development' and 'Test'; see [Figure 74].

It is sensible to add extra columns as deemed necessary to best represent the implementation processes & procedures that are being used by the performing organization; *e.g. an additional column representing quality control.*

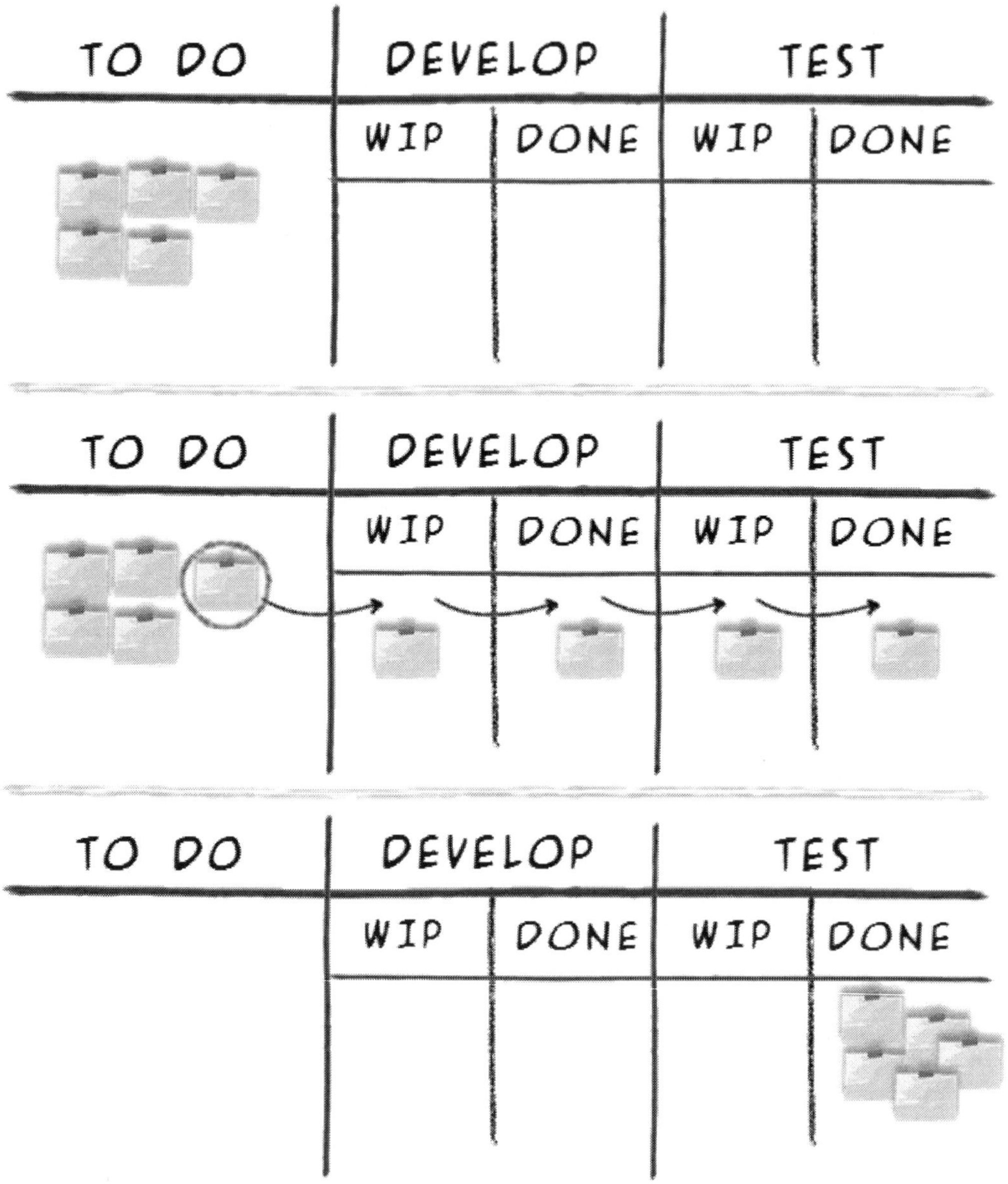

Figure 74: An agile Task Board and the traverse of a task across the board.

A note / card representation of an **individual task / story passes from left-to-right across the board** as it is worked on (WIP – Work In Progress) **until it has been verified** by testing that it has been **successfully "Done"**. These **cards could be different colours** to represent specific types of tasks, defects, enhancements, or criticality.

Each team member should update the Task Board with their tasks' progress at least once a day, though ideally they should do this a few times a day.

 I would recommend that ... the Task Board be presented in a physical form and "publically viewable" by all within the performing organization. In a physical form, the Task Board (on the wall) will be acting as a visual representation of the current sprint's execution, and as an indicator of the project's progress, thus far.

IMHO, the use of a software application to represent the Task Board is NOT such a good idea, because it hides this project information away from "public" scrutiny and accessibility.

Though, with a physical Task Board you will want to ensure that the individual tasks are securely attached and not just stuck-on, because you may or (may not) discover that a detached task or two had fallen to the floor and drifted away to only be realized when the associated feature / function is enquired about by the customer's representative.

If an individual task / story is found to not function as specified (i.e. "bad"), then its corresponding card (with the required changes or defect noted down on the card) would be **placed back into the** 'To Do' section of the **Sprint Backlog**; see [Figure 75].

Therefore, the motion across the Task Board is similar to that of waterfall except on an individual task / story basis.

This is akin to lean development where one feature (in its simplest form) is specified, designed, developed, and then tested.

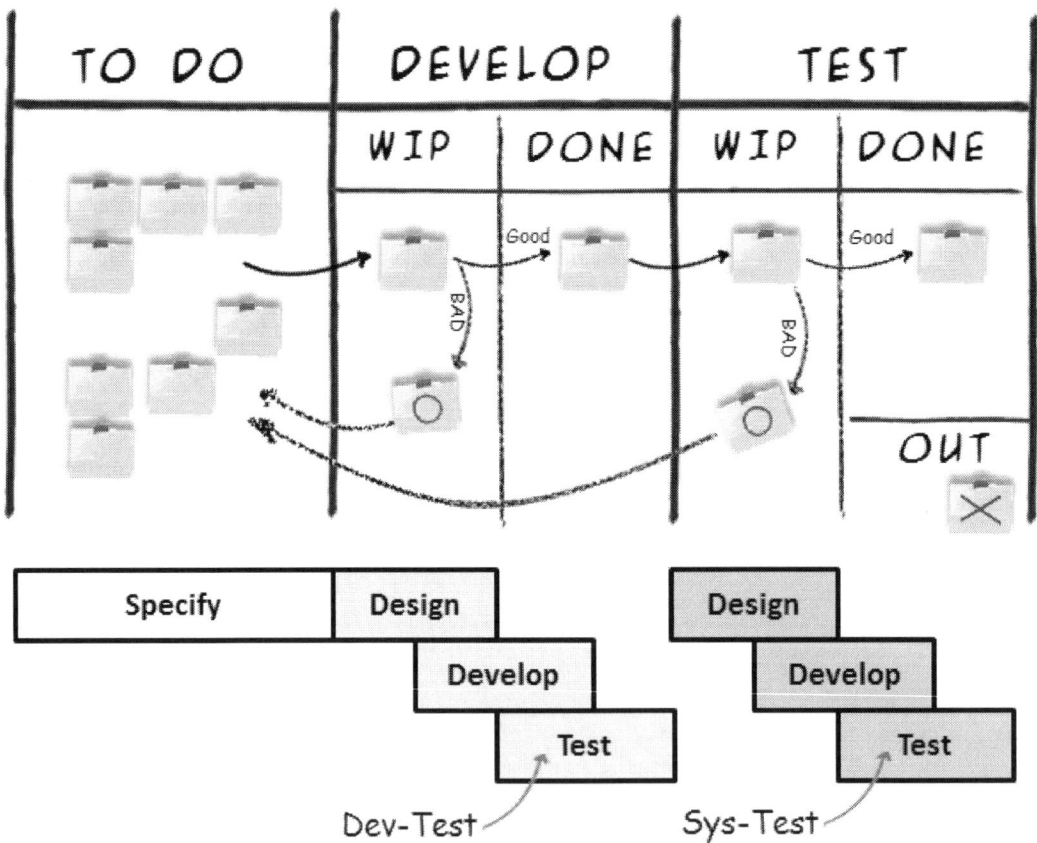

Figure 75: Agile Task Board motion with its own little waterfall sequences for both development and testing.

The 'Develop' and 'Test' sections of the motion across the Task Board in a practical implementation would be as illustrated previously in [Figure 75], where the operations on an individual task / story would be further subdivided.

1. **'Develop'** implementation of the individual task / story would have its own phases of design, develop, and tested by the implementer. Where, the testing would entail unit & integration testing of the implemented functionality in 'Dev-Test'.

2. **'Test'** implementation of the individual task / story would have its own phases of design, develop, and tested by the system testers. Where, the design and development would be of test harnesses, test data, and test scripts to verify the correct operation of the feature from both an individual unit perspective as well as being part of a use-case.

3. If during the sprint, it was decided that it would be best not to implement a task/story then that task/ story would be moved to the 'OUT' of sprint area. *... be put in Prison*

Theoretically, with Scrum there would not be a distinct demarcation between those who develop and those who test as the team is supposed to be cross-functional. However, such separation may be sufficient to satisfy the concerns of waterfall methodology traditionalists.

If for any reasons, some of the sprint's **functionality** is **not completed** (i.e. **not ready to be demonstrated** or to be deployed) **by the end date of the sprint then** this functionality would be **removed from the sprint's release**.

This incomplete functionality would be:

1) quickly tied-off (i.e. concluded or stubbed-out) so that it will not have any visible effect on other functionality in the current sprint,

2) excluded from being available and accessible in the release of the current sprint,

3) returned to the Product Backlog (i.e. the list of "yet-to-be delivered" functionality),

4) to be completed in a later sprint.

At the conclusion of the sprint, the Task Board and the Sprint Backlog would be cleared and an empty one used for the next sprint.

 I would recommend that ... any features dropped from a sprint or any features found to have defects during the demonstration to the project stakeholders, then these features should be included in the next available sprint or sprints. Else, the project team will be perceived as lacking control and not taking the project seriously.

Sorry, I am still not convinced ... Now this definitely sounds like the inmates are running the asylum, and **deciding for themselves what is in, what is out, and when things are done.**

With the members of the Project Implementation Team deciding on and choosing the tasks for themselves, deciding on how to design-develop-test each task then this will **build team ownership** of the project and **promote self-organization** within the team instead of the perceived drudgery of being assigned to work on "someone else's project".

This type of **"team buy-in"** will potentially give the project manager more time to concentrate on; the stakeholder's needs, the risks confronting the project, and the necessitated changes to the project. ... *And, allowing the project manager to be assigned to simultaneously manage other projects.*

7.4.4.   Agile: Management Hierarchy

At this point, it should be evident that **agile techniques are very much oriented towards the Project Implementation Team**; i.e. are "grass roots" based.

This is a change in perspective to that for the traditional iterative and waterfall methodologies, which are more focused on exercising managerial control. This difference between agile versus iterative – waterfall even extends to the makeup of the agile project team, where there are (theoretically) three roles:

1) **The Team** who implement the execution of the project.

2) The **Product Owner** who represents the interests of the project's primary-core stakeholders. Where the Product Owner acts as the single voice for the project sponsor and the customer.

3) The **Scrum Master** who ensures that agile procedures & good practices are abided by.

The Scrum Master functions as both a team leader and a mentor.

So with agile there is no project manager role?

This is not necessarily correct, as there is still a role for the project manager.

However, instead of directly managing the project team members' daily activities **the agile project manager functions as an intermediary between the Project Implementation Team, the Project Working Group, the Project Steering Committee, and any other stakeholders**.

And as always, the **project manager removes those obstacles that are preventing the Project Implementation Team from delivering**. This could be anything from; acquiring specific people or resources, dealing with customer relationships and internal political issues ... and concentrating on the project management activities outlined in [Chapter 10].

7.4.5. Agile: Monitoring & Control

How will you know if they are really working on what they are supposed to be, if they are self-managed?

As mentioned earlier, the Task Board provides a visual form of monitoring & control. However, if one is still not confident that this self-managed Project Implementation Team is really doing their job (i.e. not knowing until they do or don't deliver on their commitments) then there is also a **"Burn Down Chart"** which can be used to "publically" display (internally to the performing organization) the team's **progress of working through the Sprint Backlog**.

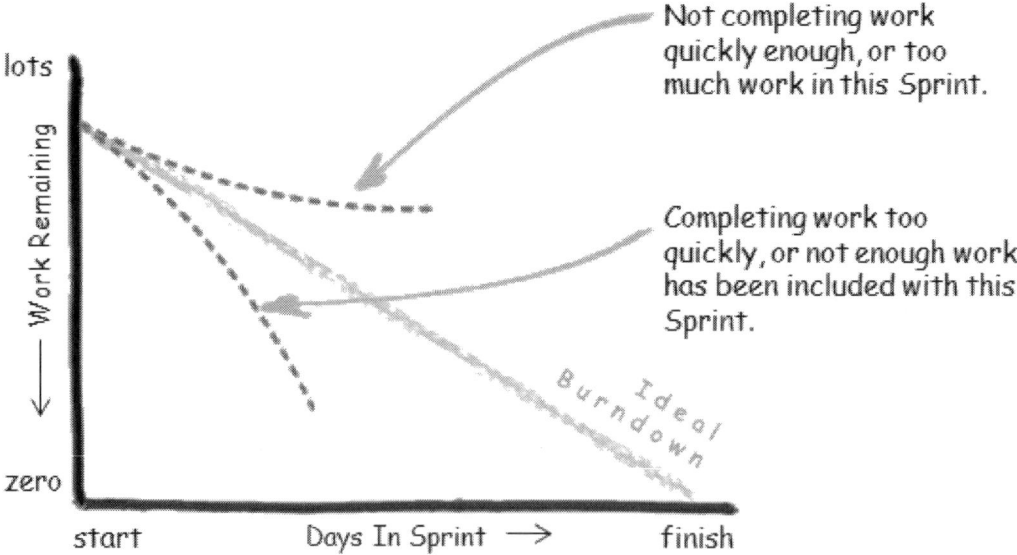

Figure 76: Scrum Burn Down Chart.

This Burn Down Chart has characteristics of the [Cost] Monitoring & Control's Earned Value Management S-Curve previously illustrated in [Figure 71].

However, instead of the **S-Curve** comparing progress via **[Cost] versus [Time]**, the **Burn Down Chart** compares the progress via **[Scope] versus [Time]**.

Comment: The uncanny similarity between the Scrum Burn Down Chart and the Earned Value S-Curves.

While **the Burn Down Chart is NOT an Earned Value S-Curve,** it does have similarities

if the Burn Down Chart's vertical axis was inverted; see [Figure 77].

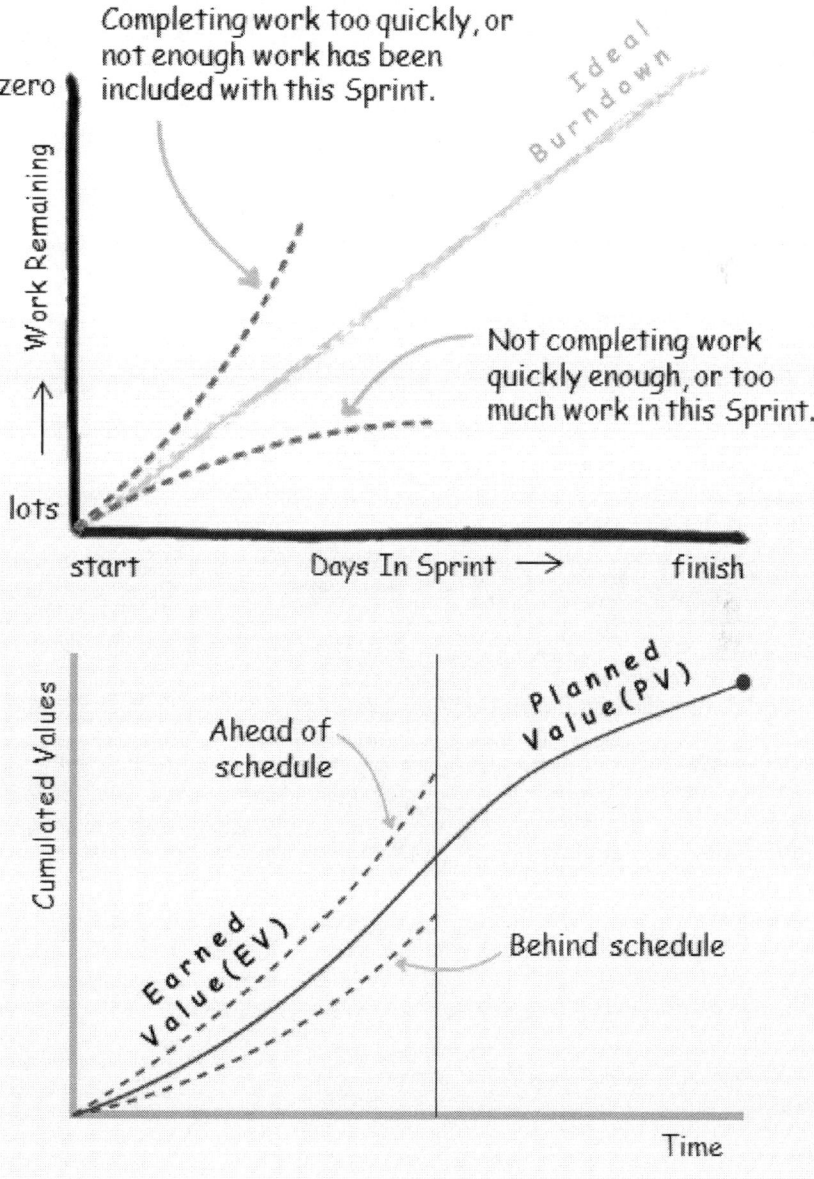

Figure 77: Comparison of an inverted Burn Down Chart to Earned Value S-Curve.

Both the Earned Value S-Curve and the Burn Down Chart visually represent the progress through the sprint / release; primarily indicating whether tasks are being completed on time or not.

However, the difference being that the **Earned Value S-Curve is concerned with cumulated [Costs]**, whereas the **Burn Down Chart is concerned with the daily updated count of tasks [Scope] or work [Time] remaining**.

That is, the **S-Curve** is concerned with **the money** that the performing organization has **expended** in the pursuit of delivering that revenue producing product.

Whereas, the **Burn Down Chart** is concerned with **the work** that has been **done** towards delivering usable product portions to the customer that could right now earn revenue for the performing organization.

Hence, both the agile method and the traditional methods of iterative & waterfall have quantifiable ways of indicating the project's progress.

What about progress reporting?

Each workday at the same time and in the same place (preferably where the Task Board is located) all of the members of the Project Implementation Team should come together to hold a "**Stand-up Meeting**" for a **time-limited maximum of 15 minutes**. To ensure that this meeting is short and sharp; all meeting participants are require to physically stand-up for the entire meeting, and being late is NOT acceptable.

During this stand-up meeting in a circular fashion, each team member summarizes:

1. **What** they had **achieved** during the last 12 working hours?   ... *Round 1.*

2. **What** is **"blocking" them** from doing their work in the next 12 working hours, and have they identified any risks to the sprint's timely completion?   ... *Round 2.*

3. **What** they **plan to work on** in the next 12 working hours (a work day and a bit)?

             ... *Round 3.*

This **Stand-up Meeting should function as a "centralized information exchange"** between all of the Project Implementation Team members and as **"a forum by which to ask for help"**. Also, participation in the Stand-up Meeting forms an individual's commitment to "The Team".

 For the Project Implementation Team, I would highly recommend that ... the daily Stand-up Meeting be used for both agile and non-agile projects.

Yes, that is correct; the daily Stand-up Meeting should also be used for, traditional iterative and waterfall projects.

 You will probably find holding the Stand-up Meeting in the morning will be more beneficial than holding the meeting at midday or in the afternoon. This is because, with the morning stand-up meeting, most people will concentrate on what they are going to do for the rest of the day instead of what have they done. Thus, a morning stand-up meeting is forward looking rather than backwards focused.

One of the problems that I have seen with agile projects is that incomplete functionality is just pushed from one sprint to the next; similarly, rework of requirements and defect fixes are pushed into proceeding sprints, hence **come the final sprints there can be a significant amount of work crammed into such a short duration.** *Consequently, I have witnessed this type of final sprints* cramming **result in a very dissatisfied customer** because **not all of the functionality they required was present.**

True, though this is also true for both the traditional iterative and waterfall methodologies. However, this does highlight a perceived problem with agile, especially from those who are from these "old school" methodologies. That is, the project's end date seems to be constantly being pushed out as unforeseen problems & issues are encountered.

As advocated way back in [Section 2.2.2] for iterative life cycles;

> *"Do the project's make-or-break features (i.e. use cases) as early as possible in the order of iterative cycles".*

If the Project Implementation Team **concentrated firstly on doing those high priority features,** then those unforeseen problems & issues encountered later in the project's life would obviously be associated with lower priority features (because the high priority ones would have been delivered earlier).

Though, this does raise a couple of interesting questions:

1) Do these end-of-project issues really have to be dealt with, given that these relate to lower priority features?

2) Do these end-of-project issues result in an acceptable Return On Investment for the customer?

Reviewer's Comment ... Suggested reading.

I can tell that you have based the agile part on how we do it. ...
I would suggest that your readers have a look at the following books:

"Scrum and XP from the Trenches" ²⁰⁰⁷ by Henrik Kniberg.

"Kanban and Scrum, making the most of both" ²⁰¹⁰ by Henrik Kniberg and Mattias Skarin.

 **Agile methods only work when there is a single voice representative of the customer.** That is, not a committee.

Additionally, this single voice needs to be either available at any time or embedded with the Project Implementation Team.

7.4.6. Agile: Advantages

The advantages with agile techniques as described in this section are:

1) **The high priority features are delivered in the early stages** of the project's life and therefore these features can be utilized via a usable product before the project is even completed. Hence, this **provides a significant Return On Investment for both the performing and customer organizations**.

2) With the **low priority features, being delivered towards the end** of the project's life then there is the **possibility to drop these low priority features** as they comparatively provide **minimal Return On Investment**.

3) The **regular periodic delivery of usable functionality** (with relatively short durations between deliveries) means that the **customer can be more involved with the evaluation** of the deliverables and therefore **request changes & corrections sooner,** rather than later as can be the case with the iterative and waterfall methodologies.

4) With **usable deliverables being provided sooner and more often,**
the customer could decide at say 80% of the way through the project that,
"hmmm, this product does exactly what I want it to do so let's finish off here".

AGILE = HIGH PRIORITY FEATURES
DELIVERED A.S.A.P.

 **Customers who are used to the traditional** iterative and waterfall **methodologies often have a real problem with the concept of dropping these low priority features**, as though on-pain-of-death every feature must be there irrespective of what it is practically worth to the end-user.

IT IS ONLY BLIND LUCK THAT CAN HOPE
TO AVOID THAT WHICH IS PRESENT
BUT GOES UNNOTICED.

8. MONITORING & CONTROL Phase ... Theory Part 3

Watching the Road Ahead

Back at the beginning of the Monitoring & Control Phase in [Chapter 6], there was an analogy of the project as a cross-country road trip.

The previous chapter concentrated specifically on the equivalent of watching the gauges; i.e. being based on the "**Scope – Time – Cost – Quality – People – Resources**" process model; as illustrated below and reproduced from [Figure 60].

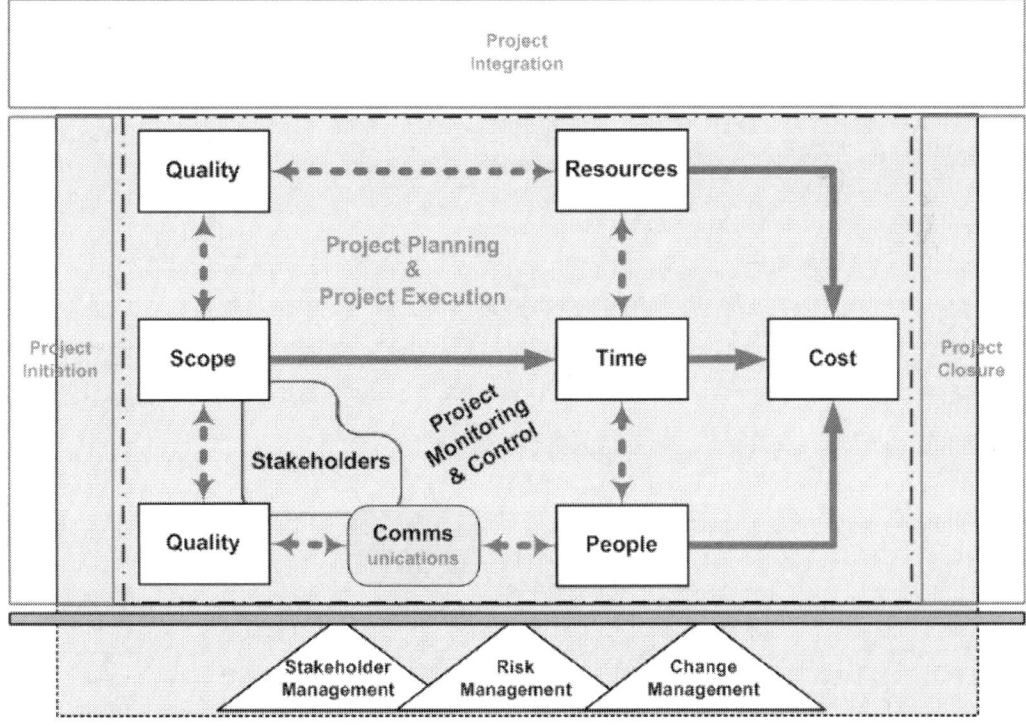

Figure 60: The Monitoring & Control Phase of the Project Management Process model.

This current chapter on the Monitoring & Control Phase will concentrate specifically on the "**PLAN – DO – CHECK – ACT**" aspect, as presented below in [Figure 78].

That is, this chapter will examine the at-hand planning of what to do next, doing that next thing, checking that it was done properly, re-acting accordingly if what was done does not correspond to what was planned to be done, and then doing this cycle again, and again.

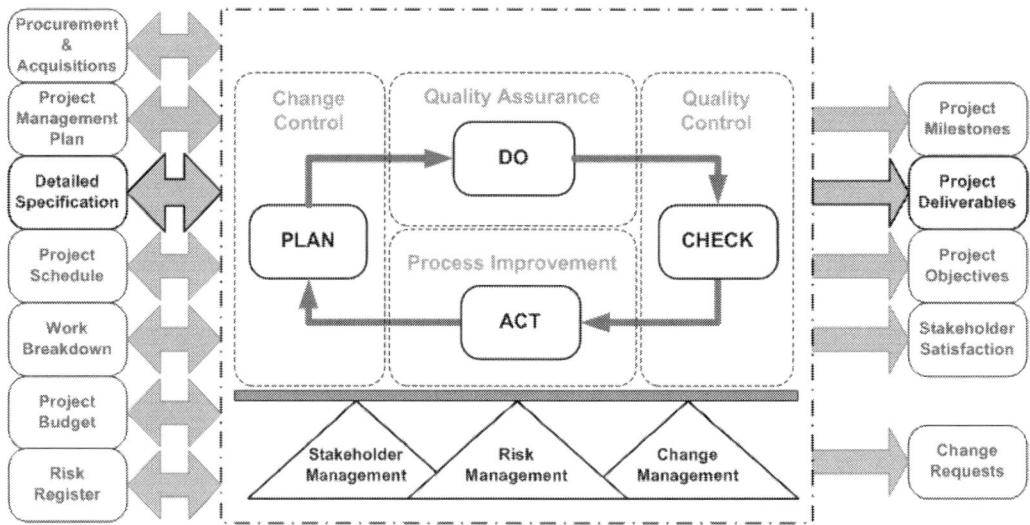

Figure 78: The Monitoring & Control Phase as a "PLAN – DO – CHECK – ACT" process model with associated inputs & outputs.

NOTICE in [Figure 78] the presence of the boxes for Quality Assurance, Quality Control, Process Improvement, and Change Control.

8.1. [Quality] Monitoring & Control

Way back in [Section 4.3.4.2] of the Planning Phase, the concept of PLAN – DO – CHECK – ACT was briefly mentioned.

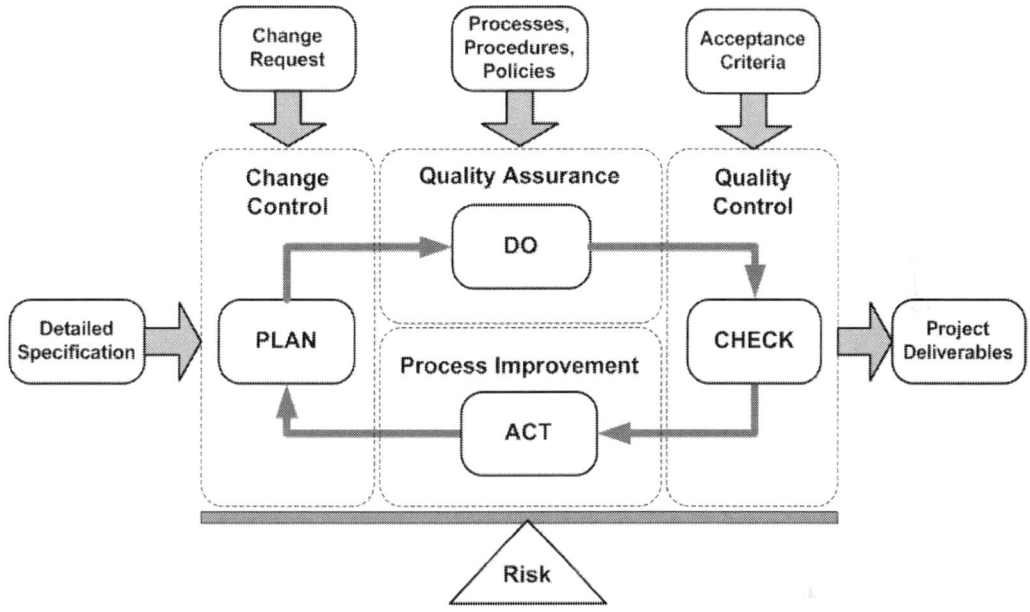

Figure 79: Relationship of Quality Assurance, Quality Control, Process Improvement, and Change Control as a "PLAN – DO – CHECK – ACT" process model.

Diagrams [Figure 79] & [Figure 80] depict the **iterative nature of [Quality] Monitoring & Control** and highlight the following **four aspects** that compose it:

1) **Quality Assurance – is concerned with the processes, procedures, and policies used to produce the project's deliverables**, so that the resultant deliverables **will conform to the agreed definitions & descriptions as documented in the approved Detailed Specifications**.

2) **Quality Control – is concerned with evaluating & verifying** that the project's deliverables **do conform to the Acceptance Criteria that were agreed** to between the representatives of the customer and performing organizations.

3) **Process Improvement / Continuous Improvement** – is concerned with;

- **Modifying the Quality Assurance** processes, procedures, and policies as a result of non-conformances,

- **reducing waste** in the processes & procedures,

- **eliminating those activities** in the processes & procedures that **don't add value** to the production of the project's deliverables,

- **continuously implementing small improvements** to the quality processes, procedures, and policies instead of via big disruptive changes,

- and where & when feasible, applying applicable industrial **"best practices"**.

4) **Change Control – is concerned with the review & approval of Change Requests and proposed Process Improvements**, then ensuring that such changes and improvements are **implemented in a timely fashion**.

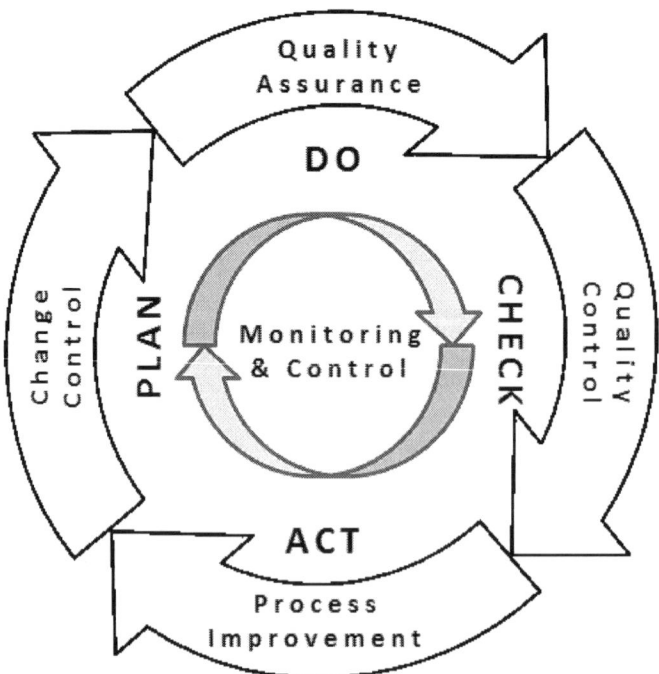

Figure 80: [Quality] Monitoring & Control as a "PLAN-DO-CHECK-ACT" model.

Reviewer's Comment ... Quality is no academic treatise.

Having reviewed the draft of this section, I have to say ... it is some dry material that feels like an academic treatise rather than a guide to practical project management.

Okay then, I will start by putting it another way. ... If the work was implemented poorly due to inadequate Quality Assurance, then when the project deliverables are tested in Quality Control or (worse) by the customer during User Acceptance Testing then there will be a significant number of Defect Reports generated. This will most probably mean that; you as the project manager (for a small to medium size organization) will be spending a significant portion of your days & evenings trawling through all of these Defect Reports. ... Reading each Defect Report, trying to understand what happened, what was expected to happen (which means cross checking it with the functionality described in the approved Detailed Specifications), and trying to determine whether this is a new defect or a repeat of an existing one. ... Then you will have to communicate back & forth with both the defect raisers and the members of the Project Implementation Team that were involved with its implementation. Subsequently, a considerable amount of your time and maybe even that of your customer representative's time will be consumed.

At this point, you as the project manager will come to the conclusion that; *"yes these are defects and these will have to be fixed, which is NOT GOOD!!!"* ... You also realize that; the schedule will have to be reworked to find [Time] to fit in these defect fixes, and [People] & [Resources] will have to be assigned / reassigned to implement these fixes. This will mean that, other tasks in the schedule will have to be put on hold, which will mean delays to the project's progress. Consequently, future delivery milestone date(s) could be missed.

In turn, the customer will not be happy; subsequently, your senior management will not be amused. In response, members of the Project Implementation Team will have to do overtime work if the project is to have any chance of being delivered on time.

And, you can bet that **they will not be too pleased ... and all thanks to POOR QUALITY**.

Example Case: a bakery analogy of [Quality] Monitoring & Control.

Let us step away from the road trip analogy for a moment.

Back in [Section 4.3.4], quality was described as being;

> *"the taste of the cake and not the processes that go into baking the cake"*.

While this is true from the customer's perspective, this is not the case for the cook; i.e. not true for the performing organization.

For the bakery to remain a profitable business venture then the ingredients used and the processes involved with baking the cake are essential. There is no economic value in spending the early hours of the morning cooking numerous cakes, pastries, and breads only to have to refund the customers' money because the resultant food "tasted awful". Alternatively, giving the customers a nasty bout of food poisoning and possibly the resulting financial penalties, statutory fines, legal litigation, and reputational damage.

For this bakery; a quality control procedure of sampling the final products does little to reduce the impact on the business of bad cooking processes & procedures, as any tainted food (and any other foods produced in the same batch or on the same day) will have to be safely disposed of as waste, while having no acceptable products to sell in the meantime.

Therefore, the quality assurance processes & procedures involved with cooking the food are essential to the continued profitability of the business. This is the same for any project; **the cost of prevention via Quality Assurance is significantly less than the cost of detection & correction via Quality Control.**

⊠ "Prevention" during the act is more cost effective than "Inspection" after the fact.

Sounds like a slogan for a family planning or health advisory campaign.

With this cooking example; the bakery business cannot afford to use only the best ingredients (which are potentially more expensive), and using costly cooking processes (which possibly take longer to produce) because if the customers are not prepared to pay that extra price "for quality" then the bakery risks that these customers will go to some other provider whose products are cheaper.

Quality is NOT the pursuit of Perfection.

That is, **quality is NOT about producing project deliverables that precisely meet the customer's "needs & wants".**

Rather, **quality is delivering exactly what was agreed to between the performing organization and the customer's representative(s) as recorded in the signed-off / approved Detailed Specifications.**

Hence, quality deliverables are bound to conformance with the approved Detailed Specifications and the agreed Acceptance Criteria.

This is not the statement about quality that I was expecting to read!

What about, the deliverables fitness for use, and meeting the customer's needs?

WARNING: *Meeting the customer's "needs" exactly as they "want" and not as was "agreed to" (as per the Detailed Specifications) can result in "chasing moving goal posts", and hence is a sure way to blowout the project's budget & duration.*

Now don't take this the wrong way; because **delivering what the customer perceives as a quality product is essential for the project to be considered a success**, and to ignore the customer's satisfaction is done at one's own peril (as there could be serious consequences for the performing organization), however ...

However, one must remember that **the customer is not the only primary-core stakeholder of the project. ... There is also the performing organization's senior management (and the business's shareholders) who will be concerned with the profitability of the endeavour**.

As stated back in [Section 1.3], the determination of whether a project is deemed a success or a failure is based entirely on the expectations, perspectives, and opinions of the project's primary stakeholders.

Therefore, for the project to be successful from the primary-core stakeholders' perspective then the project has to:

1) produce **deliverables that meet** the customer's requirements **as agreed to in the approved Detailed Specifications**,

2) produce **deliverables that are fit-for-use in accordance with the agreed Acceptance Criteria**, and

3) is a **profitable venture** for the performing organization and the customer organization.

THE SKILL WITH RUNNING A PROFITABLE ENDEAVOUR IS BALANCING THE DELIVERY OF A QUALITY PRODUCT ᵛˢ· PRODUCTION COST

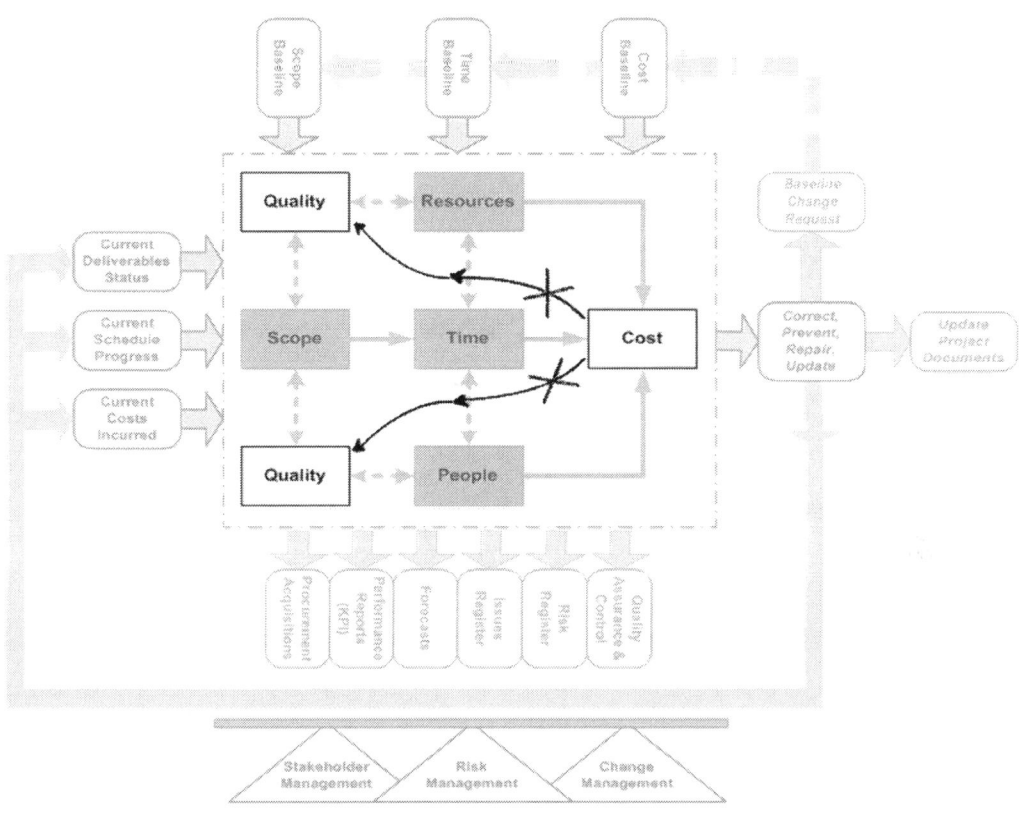

Figure 73: No direct relationship of [Costs] to the [Quality] of the deliverables.

As illustrated previously in [Figure 73] and replicated above, there is no direct relationship from [Cost] to [Quality]. That is, **increasing [Costs] doesn't necessarily increase [Quality], though increasing [Quality] will probably increase [Cost]** via the flow-on effect of any necessitated [Scope] changes, involvement of additional [People] & [Resources], and the [Time] required to implement said [Quality]. See the illustration in [Figure 81].

So, what this diagram illustrates is that

there is no direct path from Cost to Quality? *But, there must be some relationship between Quality and Cost, or how else could quality affect the customer's satisfaction yet the project still be a profitable venture.*

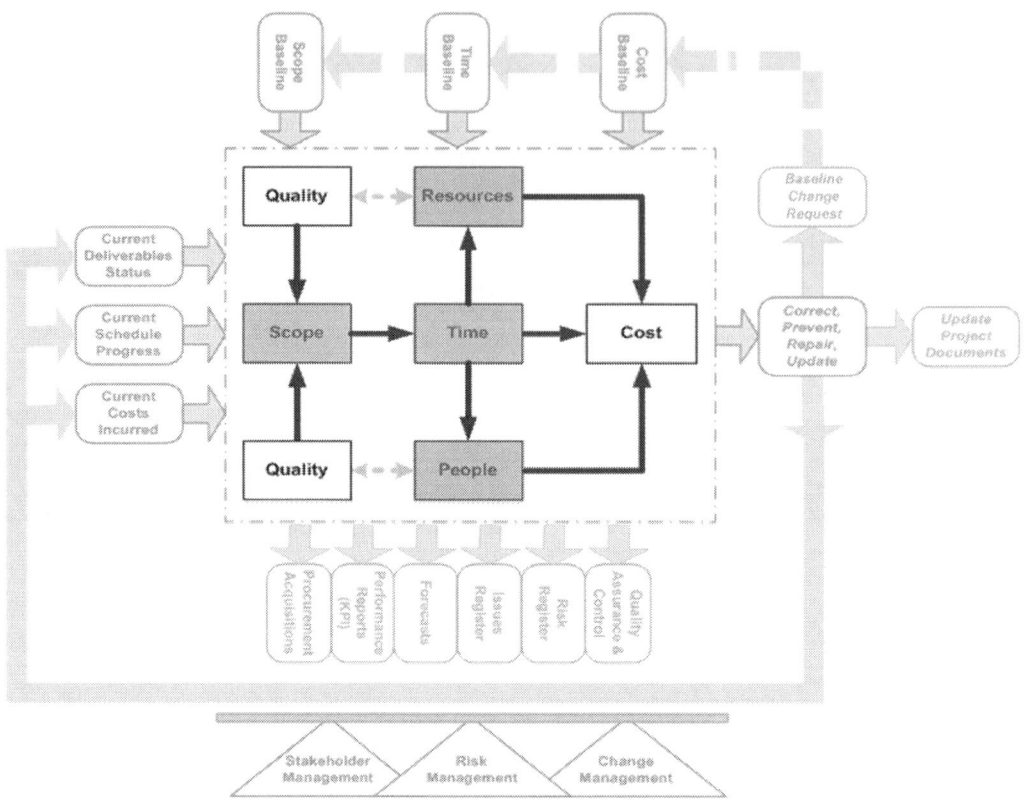

Figure 81: A semi-direct relationship of [Quality] to the [Costs] of delivery.

This relationship between [Quality] and [Cost] is known as the **"Cost of Quality"**, and it **is the total cost of all efforts pertaining to achieving quality over the entire life of the project**. That is, the **"Cost Of Doing It Right"** plus the **"Cost Of Getting It Right"**.

COST OF DOING IT RIGHT

COST OF = (COST OF CONFORMANCE)
QUALITY   +
          (COST OF NON-CONFORMANCE)

COST OF GETTING IT RIGHT

Cost of Conformance

Cost of Conformance **is the costs incurred during the project's life to avoid failures**.
There are two parts to Cost of Conformance:

1) **Costs of Prevention (Quality Assurance)** – is related to ensuring that the project
   deliverables are going to conform to the approved Detailed Specifications and
   the agreed Acceptance Criteria; i.e. **keeping failures from reaching the deliverables**.

   > *For example; training of the Project Implementation Team members so that*
   > *they have the appropriate skillsets and use appropriate methodologies to*
   > *produce quality deliverables, providing them with appropriate equipment to*
   > *produce quality deliverables, providing them with the time required to do their*
   > *job correctly, unit testing each implemented feature, and time to document*
   > *what is necessary to pass knowledge onto those who will be involved with the*
   > *project or supporting the deliverable in the future.*

2) **Costs of Appraisal (Quality Control)** – is related to verifying that the project
   deliverables do conform to the approved Detailed Specifications and the agreed
   Acceptance Criteria; i.e. **keeping failures from reaching the customer**.

   > *For example; module & product testing each deliverable against*
   > *the test specifications, and inspection of the final deliverables.*

   NOTE ❏ Unit testing is ensuring that individual components of source code or
   hardware are functioning correctly. ... *All implementers should unit test*

I would recommend that ... unit testing be done as part of the
project's Quality Assurance processes & procedures and
NOT something that is done during Quality Control, and definitely
NOT left to the end of the sprint/release/milestone delivery.

Refer to [Chapter 13.1] for practical [Quality] Monitoring & Control.

Cost of Non-Conformance

Cost of Non-Conformance – **is the costs incurred during & after the project's life due to failures that did occur**. This would involve verifying & appraising the failure (as well as determining the causes) and then fixing up those non-conformances when compared to the approved Detailed Specifications and the agreed Acceptance Criteria.

There are two parts to Cost of Non-Conformance:

1) **Costs of Internal Failures** – relates to failures, defects, issues, and problems **found by the performing organization**; i.e. internally "behind closed doors".

> *Includes such costs as; rework (modifying it to function correctly), scrapping (throwing it away & starting again), and refactoring (modifying it to improve its internal readability, reusability, or structure but not changing its evident functionality).*

2) **Costs of External Failures** – relates to failures, defects, issues, and problems **found by the Customer**; i.e. externally in public view and scrutiny.

> *Includes such costs as; warranty repairs, product recalls, refunds, hot fixes, update patches, service packs ... compensation payments, litigation, class action suits, the negative impact on the performing organization's reputation, and reduced goodwill.*

 **DO NOT underestimate the cost of poor quality and the effect that it can have on** the performing organization. ... Especially on the **financial bottom-line.**

Yep, think about the negative impact on the organization of a product recall. Not only is there the cost of undertaking the recall and the associated cost of rectifications, but there is also the legal liability and the damage to public image. ... Which can easily destroy a lot of hard work that has been built up over those previous years of business operations.

Lessons Learnt: Dealing with quality now, rather than having to pay for it later.

It is better to resolve discovered non-conforming features & functionality (and differences of opinion over whether such non-conformances constitutes either a defect or a change request) before the deliverables have been handed over to the customer, rather than trying to resolve non-conformances after the customer has "gone live" with the deliverables; i.e. during the warranty period.

Recall back in [Section 2.5], where it was stated that the further that the project is into its life cycle then the greater would be the cost of undertaking changes.

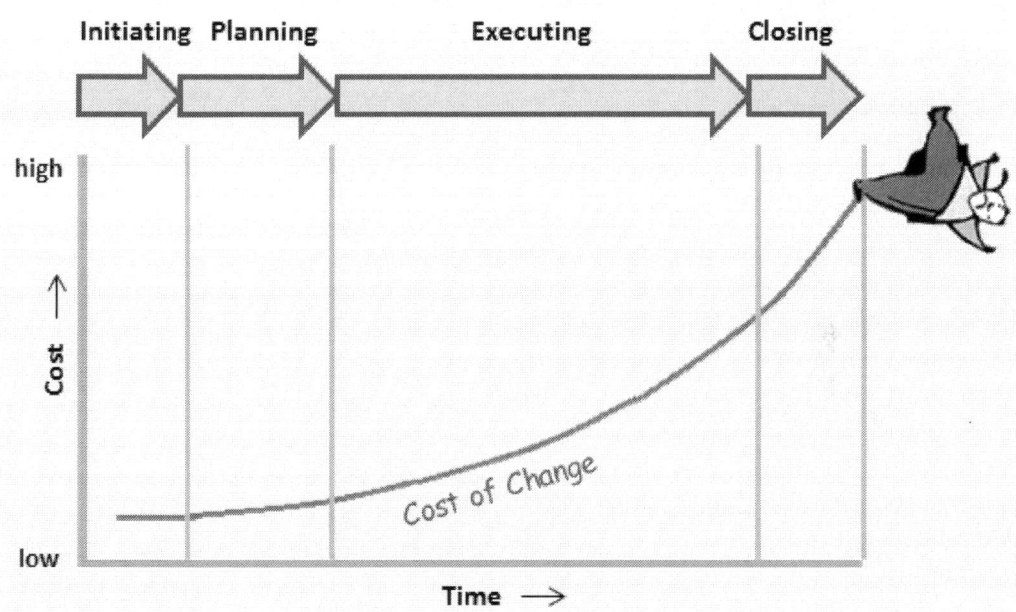

Additionally, by the time that an "in-the-field" problem arises then most probably the Project Implementation Team could have already been disbanded or departed. Hence, the resolving of this problem could be a lot more difficult to coordinate and cost penalizing than it would have been to correct / prevent / repair / mitigate this problem during the project's Executing Phase or the Closing Phase.

8.1.1.   Quality: Assurance

The PMBOK® Guide's section on quality assurance describes the inclusion of (independent) quality auditing to ensure that the appropriate quality processes, procedures, and policies are being followed during the implementation of the project's deliverables; see [Section 17.4.4.1].

Let's be honest ... unless otherwise mandated by the project's contract or stipulated by the customer organization's industrial standards & guidelines ... for a small size project and more specifically for a small sized performing organization, an internal audit (let alone an external audit) of quality processes & procedures *"ain't gonna happen"*. Given that, there probably is not even a quality authority to drive such an audit. Consequently, quality audits & quality assurance methods will have to take the form of; peer reviews, code walk-throughs, unit testing, use of tools & applications to statically analyse the source code & compiled code, plus the use of tools & applications to dynamically analyse the product's operations.

IMHO, these "self-assessment" quality audits should occur irrespective of the organization's size or the project's size. It is just good practice to; Reassess, Revise, and Reapply.

DO NOT underestimate the impact that the Project Implementation Team member's individual past experiences, training, and skillsets will have on the quality of the produced project deliverables.

 I would recommend that ... where possible try to encourage collaborative work arrangements between team members such as pair programming and master-apprentice mentoring.

⊠ **Quality Assurance is only as good as the "proficient" team of people executing it.**

8.1.2.   Quality: Control

In a medium to large size performing organization there probably is a dedicated **"independent" quality control team** (i.e. **VV&T – Verification, Validation, and Testing**) **who would confirm that the project's deliverables do conform to the agreed Acceptance Criteria.** These project's deliverables would be acceptance tested as part of a continuous stream of work, as the VV&T Team would scrutinize each project's deliverables prior to signing off on their part as authorization to handover the deliverables to the customer.

For a small project or with a small performing organization there may not be access to such (semi) independent quality control resources, in which case **the Project Implementation Team will have to perform their own product verification**.

For more information on the various types of project testing, refer to [Section 13.1.1].

NOTE   ❑ When the Project Implementation Team has to become its own product verification team then extensive testing of every individual requirement may not be possible due to limitations on the availability of [Time], [People], and [Resources]. Therefore, if the implementers have all been unit testing their own development components, then **scenario / use-case testing would be appropriate to verify the conformance of the project's deliverables**. Where each test scenario would **cover multiple test cases** via the execution of targeted hypothetical stories.

> *For example; open editor program, create new file, type text into editor, copy & paste other text into editor, save the file, close the file, open the file again, confirm that the entered text has been saved and is displayed in the open file, modify the text, save the file, close the file, close the editor, open again and confirm the new contents is correct.*

⊠ Quality Control is only as good as the "independent" team of people confirming it.

8.1.3. Quality: Process Improvement & Change Control

❖ <u>Process Improvement</u> / Continuous Improvement – is **determining the cause & effect of non-conformance and inefficient & ineffective quality processes or procedures**. Followed by the recommendations that will; correct, prevent, repair, or mitigate against the reoccurrence of these issues related to the quality of the deliverables.

❖ <u>Change Control</u> – is **reviewing** those proposed process improvements and then **deciding** on which ones should and should not be implemented, **organizing** the timely **implement**ation of those chosen improvements, **verifying** that those **chosen process improvements** have been implemented, and **auditing** that this process improvement **continues to be used** on a regular basis.

For more information on Change Management, please refer to [Section 10.2].

⊠ The desired result of process improvement and change control is to reduce the cost of quality.

Let us face facts … for small to medium size projects and for small to medium sized performing organizations, the orchestration of the process improvement and the change control will probably be driven by the project's management and not by some dedicated department or group.

 I would recommend that … the project manager should listen to the concerns & opinions of the Project Implementation Team (and the Project Working Group), because they are very close to the frontlines of the project's implementation, and hence they will on a daily basis encounter the problems & issues with the project's execution & delivery.

This is where a technique such as the agile-scrum stand-up meeting becomes very useful; see [Section 7.4].

Lessons Learnt: Quality and my days as an implementer.

Take a moment to think about this [Cost] – [Quality] relationship ...

When you were a member of the Project Implementation Team, you probably never really thought about how the quality of your work and that of your colleagues directly affected your employer's financial position, and consequentially how it affected your work situation & circumstances.

Think about just how much money is wasted annually due to the [Time], [People], and [Resources] required for rework, defect fixes, verification of defect resolutions, and ongoing maintenance? This [Time], [People], and [Resources] could have been better utilized on other projects and endeavours.

TWO ROADS DIVERGED IN A YELLOW WOOD,

AND SORRY I COULD NOT TRAVEL BOTH

AND BE ONE TRAVELER, LONG I STOOD

AND LOOKED DOWN ONE AS FAR AS I COULD

TO WHERE IT BENT IN THE UNDERGROWTH.

...

YET KNOWING HOW WAY LEADS ON TO WAY,

I DOUBTED IF I SHOULD EVER COME BACK.

EXTRACT FROM, "THE ROAD NOT TAKEN", ROBERT FROST, 1916

9. MONITORING & CONTROL Phase ... Theory Part 4

The Road Not Taken

9.1. [People] & [Resources] Monitoring & Control

Back in [Section 7.3] for [Cost] Monitoring & Control, it was described how there is a bi-directional relationship between [Time], [People], [Resources], and that this relationship has a direct effect on the project's [Cost]. See [Figure 70] reproduced below.

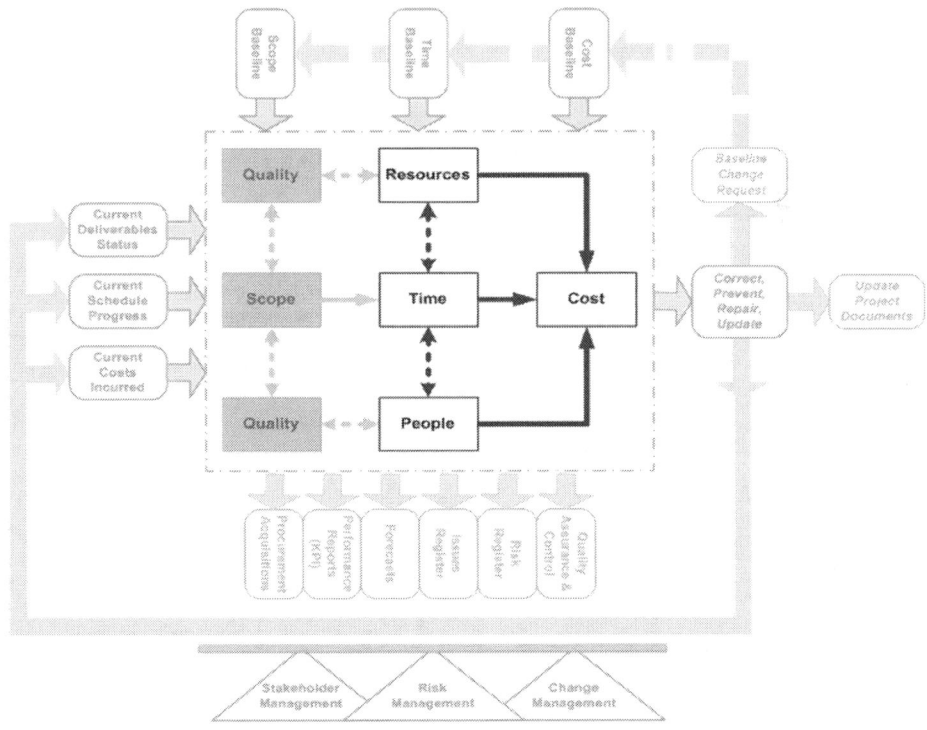

Figure 70: The direct effect of [Time], [People], and [Resources] on the monitoring & control of [Cost].

These [People] & [Resources] will over [Time] transform the project's [Scope] into physical deliverables. However, these [People] & [Resources] *"don't come for free"* as these have to be paid for [Cost], plus these will also require supporting infrastructure (such as a location to operate, electricity, lighting, communications, etc) which also add to the [Cost].

9.1.1.   Potential Cost Of Lost Opportunities

NOTE   ❑ All of these [Costs] will have to be covered by the revenue raised by the projects that are / have been undertaken by the performing organization.

Example Case: The "Potential Cost" of late projects on Profitability and Survivability

Once upon a time, there was a small software development company whose operations were based around the **continuous flow of agreed-price bespoke projects** where the revenue from the project just completed went towards paying for the work on the project that was currently being implemented, and also contributed to covering the upfront costs for the initiation of the next project that had just entered the queue.

If the completed 1st project did not generate sufficient revenue (**inwards cash flow**) and the remaining **cash reserves** of the business, **did not exceed** the **outwards cash flow** [Cost] of the [People] & [Resources] and Operating Costs of undertaking the 2nd and 3rd projects then the **company would run out of money**, and thus **go out of business**.

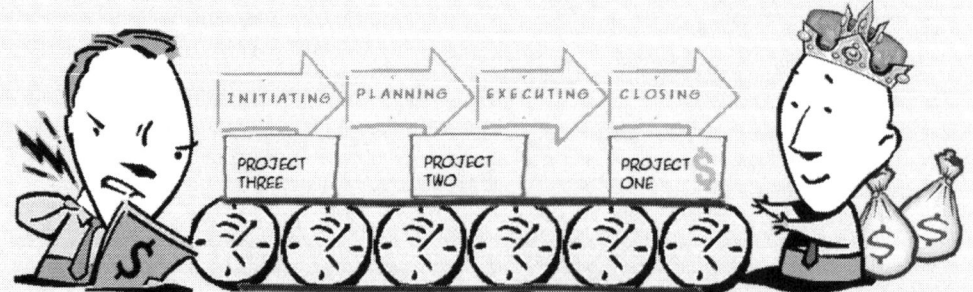

This small company had won a fixed-price contract to deliver a software application by a specific end date. The implementation plans & schedule had been accepted, the budget had been allocated, and the necessary personnel and resources had been acquired.

On average, the staff were being paid $50 per hour per person based on a 7½ hour work day, and the associated operating cost (of premise rental, equipment leasing, electricity, insurance... etc.) was calculated as a fixed cost of $375 per day.

Therefore, the Real Costs (i.e. explicit costs) of the workers' "bums-on-seats" was:

Real Costs = Staff Costs + Operating Costs + *Materials Used & Equip Hire*

= (N x $50/hour x 7½ hours) + ($375/day) + *Resources Used*

The performing organization's senior management considered this Real Cost as acceptable given that all of their fixed-price contracts were based on a charge-out rate of $150 per hour per person (i.e. "revenue") which should result in a profit per project.

Profit = Revenue − Expense   = *Contract Price − Real Costs*

However, if for some reason the performing organization was unable to work on the scheduled project because it was tied-up finishing off a late predecessor project, then the expected revenue would not be earned due to that delayed project and hence, this **missed revenue** could be considered as a "**Lost Opportunity Cost**".

Lost Opportunity Cost = (N x $150/hour x 7½ hours) Per Days Delay

Therefore, when the next revenue-raising project was delayed due to the current project not being delivered on time, then there was a "Potential Cost" to the performing organization of:

Potential Cost = Real Cost + Lost Opportunity Cost

So, for this small company with its 10 staff, when the current project was overdue (but continued to be worked on), and as a result the next project was prevented from commencing as was expected / planned then the business was potentially losing:

Potential Cost Per Day   = ($3,750 + $375) + ($11,250) = $15,375

Potential Cost Per Week   = $76,875

Potential Cost Per Month = $307,500

That is, a monthly Real Cost of $82,500 and a Lost Opportunity Cost of $225,000, which equates to a Potential Cost of $307,500 per month due to the late project.

NOTE: ❏ While this Potential Cost does not necessarily correspond to exact dollar amounts coming out of the performing organization's bank account, it does translate to how profitable a financial year the performing organization has.

Cost Type	Number of Staff	Number of days that the project is overdue			
		1 day (7 ½ hours)	1 week (5 days)	2 week (10 days)	1 month (20 days)
Staff Cost ($50 / hr)	2x people	$ 750	$ 3,750	$ 7,500	$ 15,000
	4x people	$ 1,500	$ 7,500	$ 15,000	$ 30,000
	6x people	$ 2,250	$ 11,250	$ 22,500	$ 45,000
	8x people	$ 3,000	$ 15,000	$ 30,000	$ 60,000
	10x people	*$ 3,750*	$ 18,750	$ 37,500	$ 75,000
Operating Cost	$375 / day	*$375*	$1,875	$3,750	$7,500
Lost Opportunity Cost ($150 / hr)	2x people	$ 2,250	$ 11,250	$ 22,500	$ 45,000
	4x people	$ 4,500	$ 22,500	$ 45,000	$ 90,000
	6x people	$ 6,750	$ 33,750	$ 67,500	$ 135,000
	8x people	$9,000	$ 45,000	$ 90,000	$ 180,000
	10x people	*$ 11,250*	$56,250	$ 112,500	$225,000

You may be thinking that this Potential Cost would be recovered when the next project rolls off the production line, but ... if your employer was literally operating from "hand-to-mouth" as the money was paid into their bank account, then **any delays in turning out revenue-earning projects could have significant consequences to the business's cash flow, and subsequently to theirs (and to your personal) financial survival.**

9.1.2. Effects of [People] & [Resources] Monitoring & Control

A **significant proportion of the Real Costs and the Potential Costs** of the project are composed of **[People] work hours and their associated rates of pay**, plus the spend on the **amount of [Resources] used** to get the project to where it currently is at.

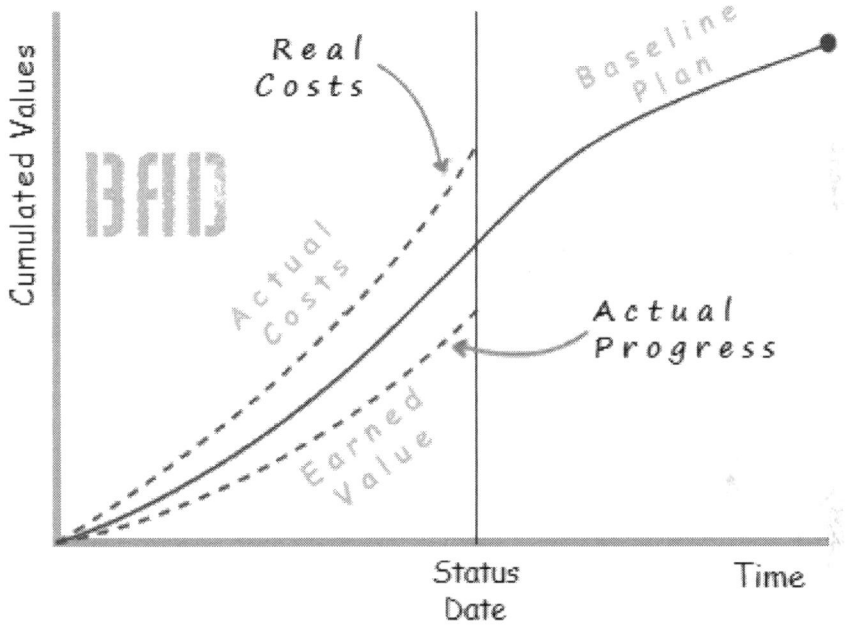

Figure 82: Relationship of the Earned Value S-Curve to the project's Real Costs and Actual Progress.

IF at the current point in time, there were **too many hours worked and/or too many material resources consumed for the actual progress made** (when compared against the planned baselines) ... THEN this could **result in a greater financial outlay for the performing organization** than they had accounted for and (more importantly) **were prepared to cope with**; as illustrated over the page in [Figure 83]. This **excessive financial outlay could mean** that the performing organization experiences a **negative cash flow crisis,** that **could consequently affect the business's solvency**, and subsequently affect the personnel via delayed wages, retrenchments, or even non-payment for services rendered.

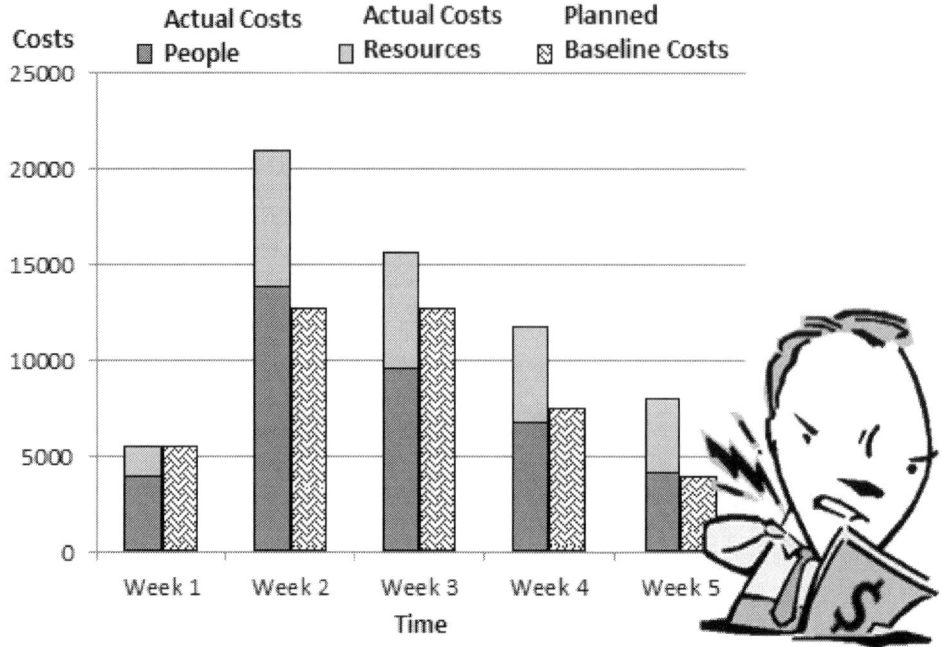

Figure 83: Relationship of Actual Costs (People & Resources) to the planned baseline weekly costs.

Thus, the project manager has to be aware of the amount of [Scope] progress that is being made in the [Time] passed, when compared to the amount of work being performed by the [People], the amount of [Resources] being consumed, and the associated [Cost] charged. ... And all the while, striving to meet the delivery milestones.

This is especially important when these milestones are keyed to trigger the project's partial payments, as a missed milestone can cause major problems.

Hence, **the sooner that excessive [People] & [Resources] utilization is determined and/or any deviations away from the planned baselines are identified then the sooner** that management at the performing organization will **be able to take the steps necessary to limit or counter the potential damage that could result**.

 For more information on the topic of practical [People] & [Resources] Monitoring & Control, please refer to [Section 14.1] and [Section 14.2].

War Story ... Sustainable Cash Flow or pending company death.

I was once employed by this software development company. We had the most excellent development team; but alas, due to circumstances beyond the developers' control, the business missed a major delivery milestone, and consequently did not receive the associated partial payment. Subsequently, the business ran into a "Cash Flow Crisis", and as a result, all of the staff (including myself) were not paid for that past month's work.

When the business owner did eventually pay us our outstanding wages, it was weeks too late for some of us. Hence, irrespective of how much the owner reassured us that, "this situation will never occur again", this was the "writing on the wall" as far as many of us were concerned. Thus over the preceding weeks, one-by-one we left for more financially secure jobs.

R.I.S.C. MANAGEMENT IS DEALING WITH WHAT HAPPENS WHEN EVERYONE WAS PLANNING ON SOMETHING ELSE OCCURRING.

10. MONITORING & CONTROL Phase ... Theory Part 5

Changing Lanes, To Avoid A Crash

Back at the beginning of these monitoring & control chapters there was the analogy of the project as a cross-country road trip. Based on this analogy, the previous chapters concentrated on the equivalent of watching the gauges (i.e. the "Scope – Time – Cost – Quality – People – Resources" process model), and watching the road (i.e. the "PLAN – DO – CHECK – ACT" process model).

This current chapter on **monitoring & control** will concentrate on the analogical equivalent of **taking corrective actions** to swerve to miss something that has suddenly moved out in front of the vehicle. That is:

(1) **Risk Management**, ... *Risk & Issues Management*

(2) **Stakeholder Management**, and

(3) **Change Management**.

NOTE ☐ Procurement Management could also be thought of as a 4th line to the list above. However, for this book, procurement management is being considered as part of the [People], [Resources], and [Quality] project variables & constraints, as well as being incorporated into Risk Management, Stakeholder Management, and Change Management.

R.I.S.C. – Risk, Issues, Stakeholders, Changes

As illustrated below in [Figure 84], the purpose of these three management areas is to prevent & mitigate against the "Scope – Time – Cost – Quality – People – Resources" process from coming unbalanced.

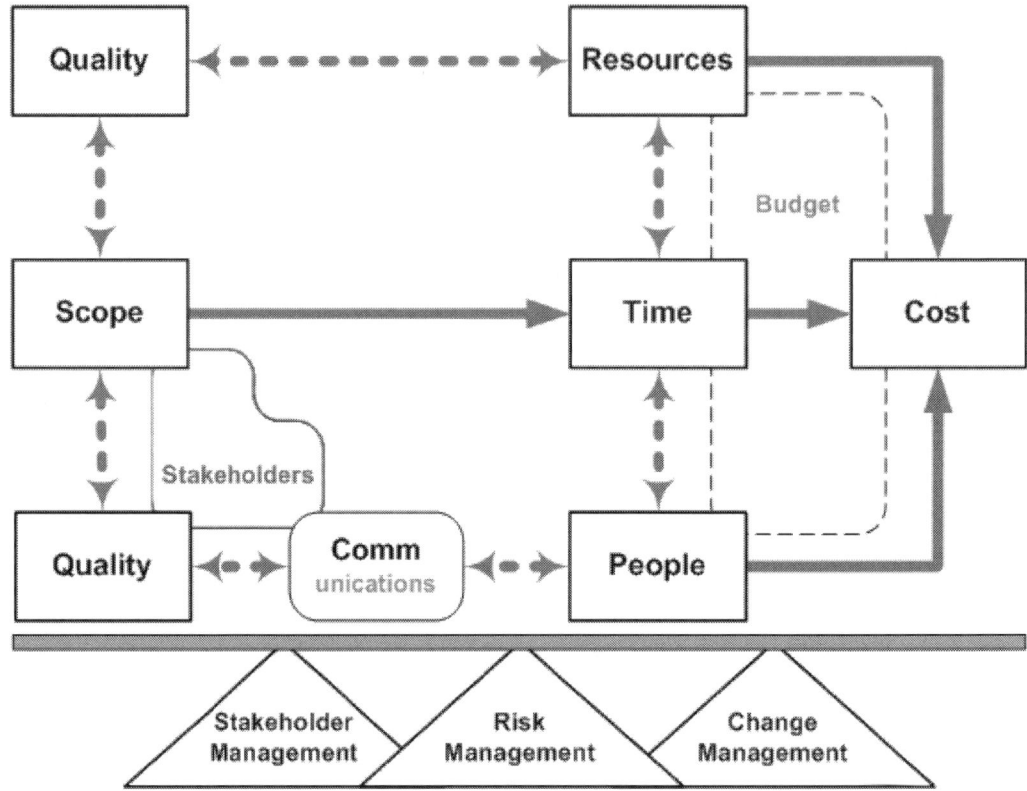

Figure 84: Relationship of Risk Management, Stakeholder Management, and Change Management to the Project Management Process model.

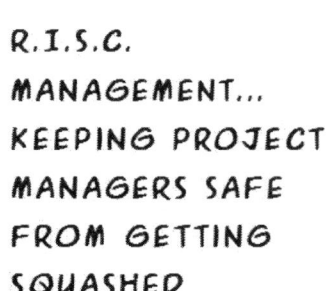

R.I.S.C.
MANAGEMENT...
KEEPING PROJECT
MANAGERS SAFE
FROM GETTING
SQUASHED

10.1.   Risks & Issues Management

Risk Management involves the following steps, illustrated below in [Figure 85]:

1. **Identify** the risks associated with the project.

2. **Qualitative analysis** of each identified risk; i.e. "**subjectively**".

3. **Quantitative analysis** of each identified risk; i.e. "**numerically**".

4. **Plan the response** to each of these identified risks.

5. **Monitor** each of these identified risks **and then control the responses** to each of these identified risks.

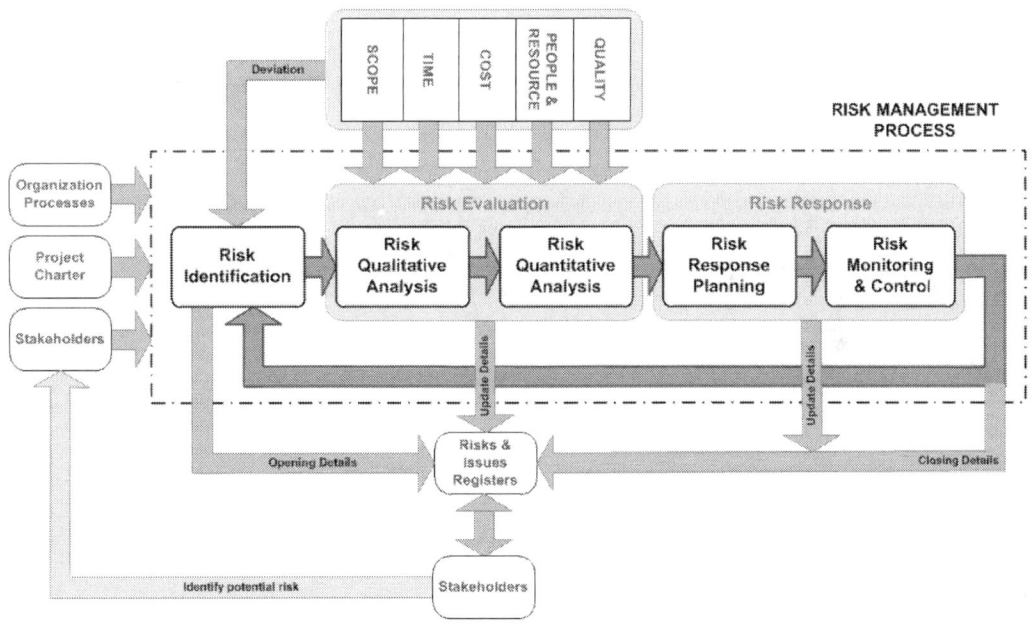

Figure 85: Risk Management Process.

As the project progresses through its life cycle, the risks confronting the project will change; some risks will become more important, others will simply fade-away, and new ones will come into being. Therefore, **Risk Management has to be an ongoing iterative activity performed regularly throughout the entire life cycle of the project,** and not just at the beginning of the project during the Initiating and Planning Phases.

10.1.1.   Risk: Identification

Before starting on the identification of the project's risks there is an organization specific aspect that has to be taken into consideration.

❖   **Risk Tolerance / Risk Appetite** – how risks are perceived and handled by the performing organization are very much dependent on their risk tolerance / risk appetite; i.e. **how much risk they are willing to accept**.

Risk tolerance is a reflection of the performing organization's primary-core stakeholders' perception of "Risk versus Reward".

For example; are they prepared to fast-track a project so as to have the first product to a new market even though this rapid product development could mean that the product will have to be significantly reworked for sustainable volume manufacture?

The level of risk tolerance will greatly influence how that performing organization will respond to risk, and more importantly how they will not respond to some risks.

Thus, an organization's **risk tolerance will influence how much of a "risk-taker" they are compared to being a "risk-avoider"**.

PLAY IT SAFE FOR A SECURE RETURN
OR RISK IT ALL
FOR A GREAT REWARD (OR FALL).

Types of Risks

- **Execution Risks vs. Operational Risks** – as stated back in [Section 1.2], the undertaking of a project is not operational work, but rather a limited duration unique endeavour that produces a one-off set of deliverables.

  However, what if because of the **execution of the project** (or as a result of the project's deliverables) this **affected the organization's operational activities**.

  For example; a small change (to access a database or an adjustment to the network architecture) to suit the project was to inadvertently result in operational applications being broken. Resources required for the project were drawn away from their regular operational duties. Alternatively, operational shortages could require that some project personnel and/or resources be diverted to revenue raising Business As Usual (BAU) operations.

- **Private Risks vs. Public Risks** – what happens if because of the undertaking of the project (or because of the project's deliverables) there is an **impact outside of the performing organization and/or outside of the customer organization**; i.e. to the general public?

  For example; environmental (ecological, pollution), political (creates mass local unemployment), religious (clashes with the local beliefs, faiths, and values), infrastructure (causes traffic stoppages and delays).

  These public risks could result in very expensive / negative consequences for the customer organization and for the performing organization; *e.g. legal, criminal, financial, public image, reputation, goodwill.*

NOTE ❑ While these above listed risk types relate to the performing organization as a whole these also relate to the project which is a microcosm of the greater performing organization. Consequently, **risks of one type will affect / induce risks of another type.**

Preparation To Identify Risks – The Tools At Your Disposal

Before starting to identify the project's risks, a few things need to be prepared:

1) **Risks Register** – a container of the **details** of **the risks identified**. This Risks Register would typically consist of a spreadsheet that includes such information as;

   - A **unique identifying number** for each risk; *e.g. Risk Item / Risk Number.*

   - A semi-detailed **description** (or **short identifying summation**) of the risk.

   - The **date** when the risk was **identified** and (optionally) **who identified** the risk.

   - An identifier of **who has been assigned responsibility** for the risk.

   - The **current status** of the risk; *e.g. OPEN, CLOSED, ISSUE.*

   - The **date** when the risk last changed status.

   - The perceived **impact / severity** of the risk; negligible, minor / minimal, moderate, major / critical, severe / catastrophic.

     o Possible **[Cost] impact** due to this risk becoming an issue.

     o Possible **[Time] impact** due to this risk becoming an issue.

   - The perceived **probability / likelihood** of the risk becoming an issue; rare / not likely, unlikely but plausible, may happen / moderately likely, likely, almost certain.

   - The **response** to be implemented; accept, avoid, mitigate, transfer. … *Share.*

   - Any relevant **comments** about this risk from any of the project stakeholders.

   NOTE   ❐ **The Risks Register has to be continually updated with the latest information and the decisions that have been made relating to each risk.**

   It is more practical to have the Risks Register and the Issues Log (that follows) in the same spreadsheet, just on different worksheet pages; i.e. a Risks & Issues Register.

2) **Issues Log** – risks and issues could be separated into two discrete registers; i.e. a Risks Register and an Issues Log. **When a risk becomes an issue, its details would be copied into the Issues Log**. In the Risks Register, leave each issue's risk entry as it was at the time of the move to the Issues Log. ... Though, clearly indicate that the risk has become an issue. As a visual indicator of its transfer, you may wish to cross-out (strikethrough) that risk's entry in the Risks Register.

Due to project auditing [Section 17.4.4.1] and Post Implementation Reviews of the project [Section 17.4.2.1], **never delete a risk or an issue entry from the register / log irrespective of whether that risk / issue has been resolved or is no longer appropriate, instead just mark its current status.**

3) **Risk Matrix** – draw up a 5x5 grid with the **impact / severity** of the consequence along the X-Axis and the **probability / likelihood** of a risk along the Y-Axis; see [Figure 86].

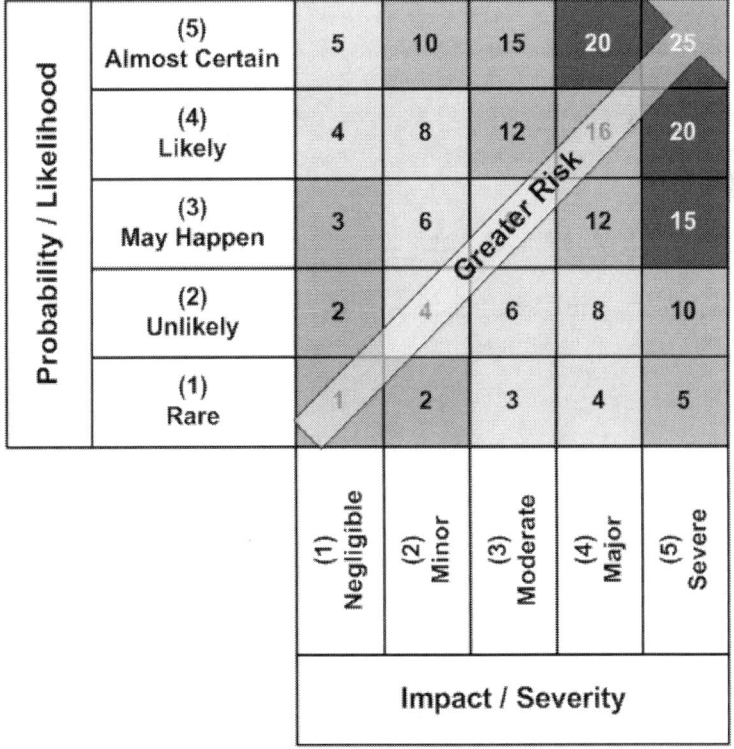

Figure 86: Example of a Risk Matrix.

Those risks with "greater urgency" would be closest to the top right corner of the grid, whereas those risks with "lower immediacy" would be near the bottom left corner. However, **the position of a risk in the Risk Matrix is not static and throughout the life of the project, a risk could move to a different position on the grid** (i.e. become more or less urgent to be dealt with). Alternately, a risk could **disappear entirely from the grid** (i.e. be resolved or eliminated), and a **new risk could appear**.

4) **Risk Breakdown Structure (RBS)** – Using a similar concept to the Work Breakdown Structure (WBS), though this time with each identified risks as the little boxes. Once these boxes of risk have been piled in the corner of the imaginary sorting room then stack these boxes based on some meaningful categorization; *e.g. internal : external, hardware : software, technical : management, physical subsystems, componentry, applications, etc.* This **grouping of risk by categories could potentially unearth "clustering" of risks that have common cause(s). Subsequently, the fixing of the root cause(s) could potentially eliminate a concentration of risks.**

You could also colour each risk box based on the colour of where it resides on the Risk Matrix. Hence, are there clusters of high priority "RED" boxes in certain category stacks?

Risk Identification Techniques

- **Brainstorming** – is one of the most basic risk identification techniques.
  With brainstorming **a group of people** with relevant knowledge of the project (and its associated subject matter) would **get together to "workshop"** (discuss) what could potentially affect the project's successful execution. This group session can be highly effective at stimulating the identification of other risks and providing more in depth information about each of these identified risks.

- **Interviews** – When getting a brainstorming group together is not achievable (*e.g. no common available time in the participants' calendars, or extremes of differences in knowledge, perspectives, bias, interests, and background*) then **one-on-one interviews** maybe a productive option.

- **Delphi Technique** – is the **distribution of a questionnaire** to the project's stakeholders (and associated subject matter experts) to **obtain written feedback as to their thoughts & opinions on the project's potential risks**. Once they have submitted their completed questionnaires, the information contained within their responses would be accumulated & distilled into another round of questionnaires and response reviews. Hence, after a few review cycles a relatively detailed collection of risks would evolve.

  This Delphi technique has some advantages when; it is next to impossible to get all of the participants together (*e.g. different time zones and great distances make brainstorming or interviews impractical*), there would be a clash of egos or bias opinions on the part of certain participants which would stifle the effectiveness of brainstorming, or anonymity of the participants is necessary. The **disadvantage of the Delphi technique is the inherent lack of dynamic stimulus of interactive thought**.

- **Checklist** – is the use of a **"tick-a-box" checklist of common risks** associated with the performing organization's industry and/or collection of previous projects.
  If the current project is similar to previous projects then based on past experiences it would be reasonable to assume that the risks encountered previously could also occur for the current project.

    > *For example; I.T. infrastructure failures, personnel skill set failures, outsourcing failures, supplier failures, policies & procedure failures, regulatory & statutory non-compliance, environmental failures.*

- **Planning Phase Processes** – were any risks identified during the conduct of the Planning Phase activities? See [Section 4.3.1], [Section 4.3.2], and [Section 4.3.3].

    o **[SCOPE]** – Detailed Specifications, Work Breakdown Structure, Scope Boundary.

    o **[TIME]** – Project Schedule and Critical Path Analysis.

    o **[People]** & **[Resources]** – procurement and availability.

    o **[Cost]** – funding, contract type, contingency reserves.

- **SWOT / SLOT Analysis** – is identifying the project's **Strengths, Weaknesses / Limitations, Opportunities, & Threats**. That is, what factors internal & external to the project will affect the project being able to achieve success.

- **Re-evaluation of Assumptions**, **Hypotheses**, and **Scenarios** – when the project was originally conceived (and the Project Charter was written) there were some assumptions made, and scenarios envisioned. Since the Initiating Phase and the Planning Phase some of these considerations may no longer be appropriate or may have escalated in importance. Hence, it is now time to re-examine these considerations to determine whether they still apply as they previously did.

- **Graphical Techniques**

  - **Flow charts** (i.e. work flows, process flows, system flows) depict the sequence flow from one component-process-system to the next. Therefore, consider what could "block" the flow.

  - **Cause & Effect Diagrams** (i.e. fishbone diagrams, Ishikawa diagrams) depict those factors that could affect particular parts of the project, and hence result in undesirable outcomes.

  - **Timeline diagrams** depict the relationship of this project and it's milestones to other projects' milestones, work-stream bookings, and calendar events.

⊠ Late risk identification is a project killer.

10.1.2. Risk: Analysis

Once the risks associated with the project have been identified (and their details have been entered into the Risks & Issues Register), then the next step is to "qualitatively" and "quantitatively" analyse these risks.

What is the difference between "Qualitative" and "Quantitative" risk analysis?

❖ **Qualitative risk analysis** "describes" the impact and likelihood of the identified risk.

❖ **Quantitative risk analysis** "numerically defines" the identified risk as having a **measurable impact** on the project being able to achieve its objectives;
e.g. an impact of X-financial amount and Y-days additions to the project's duration.

How do I go about doing risk analysis?

1. **Probability (Likelihood)** – how likely is it for each identified risk to occur?

2. **Impact (Severity)** – how much is it going to hurt the [Scope], [Time], [Cost], [People], [Resources], and [Quality] if this identified risk does occur?

3. **Risk Priority / Risk Ranking / Risk Index** – how does this identified risk compare to the other identified risks associated with this project?

$$\text{Risk Priority} = \text{Impact X Probability} = Severity\ X\ Likelihood$$

4. What is the **Time Criticality** of responding to this identified risk?

   - Are there **restrictions imposed by the timing of events**, such as a pending delivery milestone?

   - **The sooner that the risk is likely to become an issue then the more immediate will be the "urgency"** of finding a response to that risk.

5. What is the **"Perceived Urgency"** of establishing a response to this identified risk?

NOTE ❐ While there are methods to analyse and prioritize risks, there are also **the project's stakeholders whose own risk tolerances will influence their concerns & biases towards the "perceived urgency" of dealing with certain types of risks**; i.e. areas of preferential concern.

❐ Unfortunately, **qualitative risk analysis is prone to bias** (potentially being based on opinions & hearsay), **rather than the realities of facts.**

A down-to-earth guide to risk analysis.

1. **What risks** are perceived as associated with this project?
   - Having checked that this risk has not been covered before, give the identified risk an identifying name and take the next number in the Risks Register.

   - In the Risks Register describe each identified risk, what would cause this risk to happen, and provide additional comments & opinions about this risk.

   - Who knows most about this risk, and who will be responsible for resolving this risk? Record this information in the Risks Register.

2. **What is the "perceived impact"** to the project if this risk was to occur; i.e. **how severe** would be the consequences on the project?
   - Score this risk's impact on a scale of 1-5 where <u>1 is least impact</u> and <u>5 is most impact</u>; i.e. 1 = negligible, 2 = minor, 3 = moderate, 4 = major, 5 = severe.

   - What is the possible **monetary impact** due to this risk becoming an issue?

   - What is the possible **duration impact** due to this risk becoming an issue?

3. **What is the "perceived probability"** of the risk occurring; i.e. **how likely** is it for the risk to become an issue?
   - Score this risk's likelihood on a scale of 1-5 where <u>1 is least likely</u> and <u>5 is most likely</u>; i.e. 1 = rare (very unlikely), 2 = unlikely, 3 = may happen, 4 = likely to happen, 5 = almost certain (imminent).

4. Map each identified risk into the **Risk Matrix**; see [Figure 86] back a few pages.

Or alternatively calculate the Risk Priority / Risk Ranking / Risk Index where;

$$\text{Risk Priority} = \text{Impact X Probability}$$

Risk Priority = Severity X Likelihood

IDENTIFIED RISK	IMPACT (RATE 1-5)	X	PROBABILITY (RATE 1-5)	=	RISK PRIORITY (1-25)
[Risk 1]	[1-5]	X	[1-5]	=	[1-25]
[Risk 2]	[1-5]	X	[1-5]	=	[1-25]
[Risk 3]	[1-5]	X	[1-5]	=	[1-25]

5. **Record** the calculated **Risk Priority** in the **Risks Register**.

6. Add a box for each identified risk to the **Risk Breakdown Structure (RBS)** and
   position it in the appropriate pile / category.
   Possibly colour code each risk-box based on its risk priority; i.e. the colours being
   the same as that used in the Risk Matrix.

7. Is there a **concentration of risks** starting to form in the Risk Breakdown Structure;
   i.e. a disproportionate growth in the column for a certain category of risks, or
   a disproportionate number of high priority colour in certain categories of risks?

8. With the risk having been prioritized & categorized, determine **how "urgent" it is to
   establish response strategies & plans** for each of these identified risks.

   *The perceptions of those affected stakeholders and that of the
   primary stakeholders will also influence the urgency of dealing with
   each identified risk; i.e. "political priority" versus calculated Risk
   Priority.*

9. During this analysis, did any risk response strategies become apparent?

Due to the **stakeholders' Risk Tolerances, "political priority", and "perceived urgency" of dealing with the identified risks**, there is a need for this risk analysis to be based on reliable and substantiated numerical data. Thus, **some performing organizations will have predefined numerical descriptors for grading both the impact and the probability**.

These quantitative risk analysis numerical values will **clearly differentiate between each level of impact & probability when compared to the approved baselines.** Thereby, the risk rating / grading is based on hard-n-fast number ranges rather than on opinions.

> *For example; an impact of a budget overrun when compared to the [Cost Baseline]*
> *of 0-5 % is 1 = negligible, 5-10 % is 2 = minor, 10-20 % is 3 = moderate,*
> *20-30 % is 4 = major, and 30 %-greater is 5 = severe. Similarly, for late running*
> *schedules [Time Baseline], the number of defects reported [Quality] ... etc.*

> *For example; a likelihood of 1-in-3 chance is 5 = almost certain,*
> *1-in-10 chance is 4 = likely, 1-in-20 chance is 3 = may happen,*
> *1-in-50 chance is 2 = unlikely, and 1-in-100 chance is 1 = rare.*

The thing to realize about these values used to distinguish between the different levels of impact & probability is that **these chosen numbers are a reflection of the organization's risk tolerance.** Thus I would suggest that as the project manager; you should find out what are your organization's risk rating rules & grading definitions, and thereby be consistent with how the organization's other projects (past, present, and future) are evaluated.

NOTE ❑ **Risk analysis is a suitable time to seek expert advice about the risk**;
i.e. engage Subject Matter Experts and those closest to the pending issue for their insight.

❑ **Risk analysis is an iterative process** where each time through the cycle a clearer & more accurate understanding of the identified risks will be obtained, new risks will be identified, and some old risks will recede in significance. See [Figure 85] a few pages back.

10.1.3. Risk: Response Planning

Now that the project's identified risks have been analysed (and these details have been updated in the Risks & Issues Register) the next step is to establish strategies & plans for dealing with these identified risks, and assign who will be responsible for each risk's strategies. The following are the generic ways to **respond to risks**:

1) <u>Accept</u> the risk – **acknowledge that the risk could occur**.

   - A **passive approach** would be to do nothing, and thus worry about the risk "*if and when*" it does become an issue.

   - A **proactive approach** would be to compensate for the risk's possible impact; *e.g. by adding management reserve to the project's budget and adding buffering to the project's delivery milestone dates.*

2) <u>Avoid</u> the risk – taking actions to **prevent the risk from evolving into an issue** by removing or reducing the cause of that specific risk.

   For example; bringing in specialists/subject matter experts to concentrate on the area of concern, clarifying the project's requirements to all those involved, ... and by reducing the project's [Scope], increasing the project's duration [Time], increasing the project's budget [Cost], increasing the number of [People] working on the project, increasing the number of [Resources] available to the project.

3) <u>Mitigate</u> the risk – take steps to **lessen the impact of the risk**, to take steps to **reduce the likelihood of the risk occurring**, and/or **having contingency plans in place**.

   For example; improved implementation processes (Quality Assurance), more exhaustive system testing, in depth Quality Control testing, using better tools, peer reviews, software source code checkers, the use of proof of concepts & prototypes.

4) <u>Transfer</u> the risk – **to a third party who is more capable of dealing with it**.

   For example; outsourcing, contracting the work to a specialist, insurance.

NOTE ☐ **While transferring the risk does transfer the day-to-day responsibility for managing that risk, it does not transfer the ultimate responsibility for the risk** nor does it necessarily eliminate the risk, nor prevent the risk becoming an issue.

5) <u>Share</u> the risk – where the burden of the risk is distributed amongst the project stakeholders. E.g. each participant chips in a bit to the project's contingency reserve, or a combined emergency response team.

Each risk response can be thought of as its own "PLAN – DO – CHECK – ACT" mini-project, as mentioned in [Section 2.6.1]. Where, each identified risk's response plan will require:

- Someone to be **assigned overall responsibility for resolving the risk**,

- **[Scope] boundaries defined** for the implementation & verification,

- **[Time] & [Cost] budget allocated** for the implementation & verification,

- **[People] & [Resources] assigned** for the implementation & verification,

- **How will it be verified** that "The Plan"' was **implemented correctly** [Quality],

- The risk response plan will have to be **authorized & approved**.

10.1.4. Risk: Monitoring & Control

Now that the response strategies & plans have been created for the project's identified risks (and these details have been updated in the Risks & Issues Register) then:

- **Regularly monitor** these risks, *e.g. weekly during the Project Working Group meeting.*

- **Activating** the appropriate risk **response plan** when the situation requires,

- **Evaluating the effectiveness** of the engaged risk response plan,

- Making appropriate **adjustments to** the risk **response plans** where necessary,

- **Identifying** when a risk has changed importance / is no longer relevant / a new risk has appeared.

- Also, the changes to risks & issues will have to be **periodically reported** to the relevant project stakeholders, *e.g. during the monthly Project Steering Committee meeting.*

 I would recommend that ... the risks & issues reporting methodology & format used should be consistent over the entire life of the project, so that different time periods can be easily compared.

 I would also recommend that ... where possible try to be consistent with the risk management methodology & format used for all of the projects being undertaken by the performing organization, so that any underlying problems with how the performing organization is conducting its projects maybe detected and identified.

Risk Monitoring & Control is not just filling in and updating the Risks & Issues Register on the morning of (or evening prior to) the project's status review meeting. Yet, Risk Monitoring & Control is often treated in this way, as an afterthought chore.

 For more information on practical <Risk> Monitoring & Control, please refer to [Section 15.1.1] and [Section 15.2].

10.2.   Change Management

There is a saying that; "*the one thing in life that is constant is change*", and this adage is so true for projects.

Change is inevitable during the project's life ... to expect otherwise is foolhardy. ... Thus why expecting the project's implementation to precisely follow a micromanaged waterfall plan is just down-right stupid.

Nevertheless, the project team cannot just run-off willy-nilly and make the change without some form of control, as this would result in scope creep which would cause the project to hemorrhage to death financially & duration wise, see [Section 4.3.1]. Instead the proposed change has to progress through an **authorized Change Management Process** (aka "**Perform Integrated Change Control**" in PMBOK terms), as illustrated in [Figure 87].

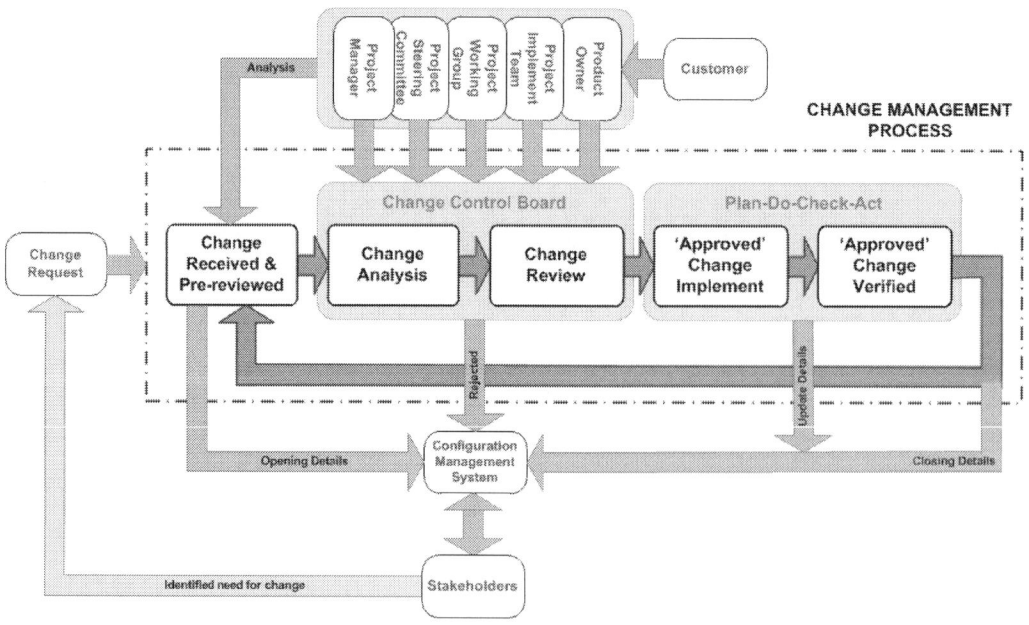

Figure 87: Change Management Process.

NOTE   ❐ Just like Risk Management, **Change Management has to be undertaken over the entire life of the project** and not just during the Executing Phase.

Hey, that diagram for the **Change Management process** *looks* **very similar to that for Risk Management***; why is this?*

The similarity between the process models for Change Management and Risk Management is not coincidental. This is because; **any change can inadvertently introduce** undesirable side effects, and hence introduce some form of **risk to the project being successfully completed**.

In an ideal-world; the project would flow sedately through the process of Define – Design – Develop – Test, but in the real-world this is not the case because we cannot accurately predict the future. That is, something will happen that was not expected, or in hindsight something could have been done differently / better, or what the customer thought they "wanted" was not actually what they really "needed".

This is where Change Management comes into play by:

1) Ensuring that the change will be **feasible & beneficial** to the project.

2) Ensuring that the change has been thoroughly **thought through**, so that it **does not impact negatively** on something else in the project.

3) Ensuring that the change is **implemented at an opportune moment**.

4) Ensuring that the change has been **implemented correctly**.

There are two related aspects to Change Management, these are:

- **[Baseline] Change Requests** – in [Section 10.2.1].

- **Defect Reports** – in [Section 10.2.2].

ANY CHANGE CAN INADVERTENTLY INTRODUCE RISK TO THE PROJECT'S CHANCES OF BEING A SUCCESS.

10.2.1. Change Management of: Requests

The [Baseline] Change Request process described here is illustrated in [**Figure 88**]:

1. A [Baseline] Change Request can be submitted by any project stakeholder who (given the current situation and the prevailing circumstances) finds or believes that; **some feature or functionality [Scope] defined by the approved Detailed Specifications will not result in the desired outcome, the agreed [Quality] Acceptance Criteria is inappropriate, the scheduled [Time] needs new milestone dates, or the budget [Cost] needs adjusting.** ... *Yep, Change Requests aren't just for changing [Scope].*

2. The [Baseline] **Change Request is received by the performing organization** from any of the project stakeholders. This [Baseline] Change Request should be in a **written or text-based form and not received solely by verbalized statements** or hearsay.

    The performing organization may have existing paper forms, email templates, or an online data entry system established specifically for accepting Change Requests. Whichever is used; the desired result is to have **consistently formatted Change Requests**, so that these can be **easily compared & interpreted**.

    Each [Baseline] **Change Request** would then be **retained in some form of Change Management System**; i.e. **Configuration Management System (CMS)**.

DO NOT accept or act on Change Requests expressed verbally, because if these details are not sufficiently recorded then at a later date it will be difficult to agree on whether this change was due to contract deviations or clarification of existing requirements.

---

3. **A preliminary analysis & review of the received** [Baseline] **Change Request** would be conducted (possibly by the project manager) to **determine if**; any **additional information is required** to clarify the Change Request, to determine whether this Change Request is a **duplicate** or very similar to an existing one, or whether this **Change Request should in fact be a Defect Report.**

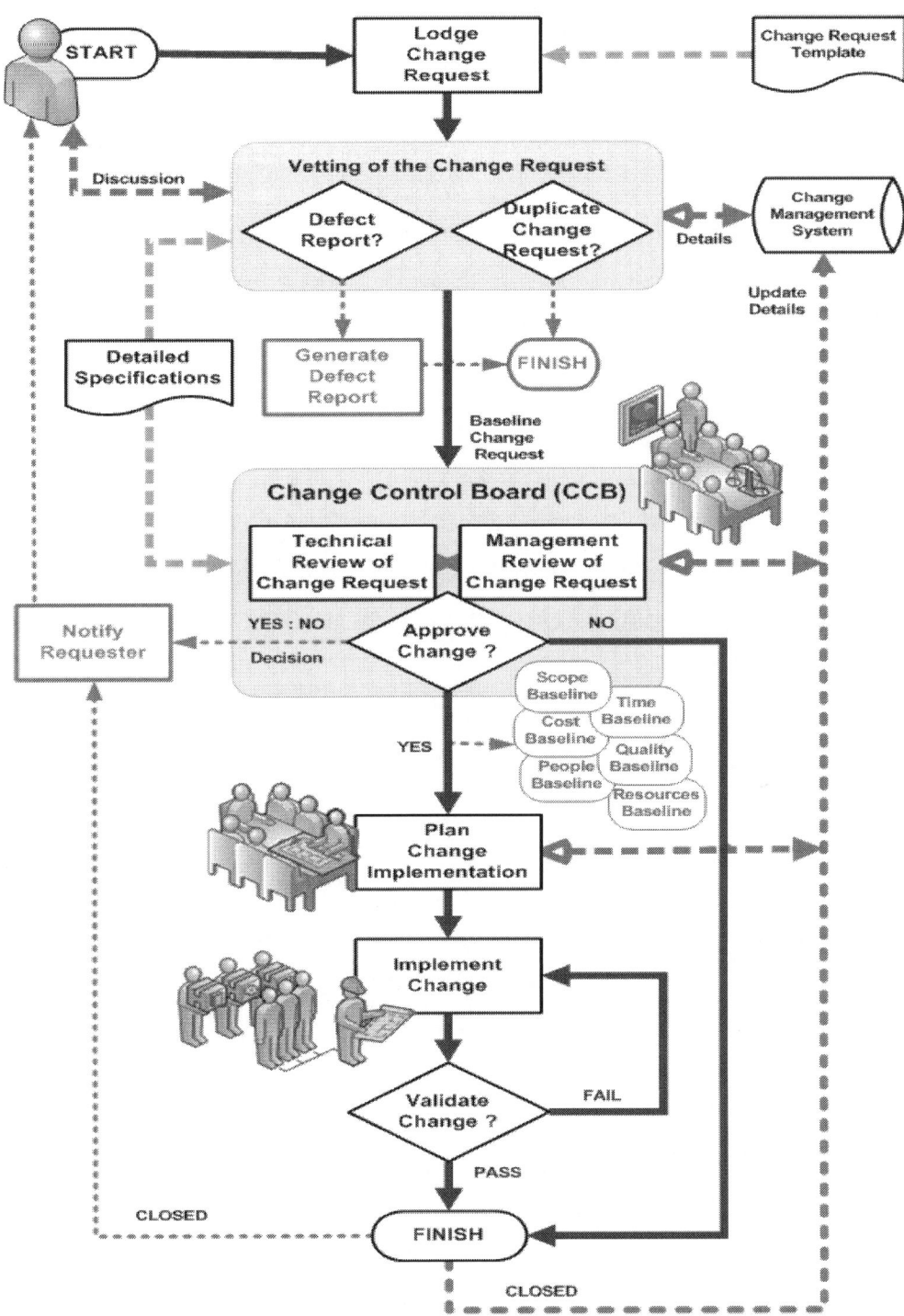

Figure 88: A simplified [Baseline] Change Request workflow.

4. The **vetted** [Baseline] **Change Request** would be **reviewed by the Change Control Board (CCB)** to determine whether it should be **accepted or rejected**.

    Based on an analysis of the "Return On Investment" for the change, a decision would be made as to whether its implementation should be **authorized** or put on hold.

    - **"Return On Investment"** means that; **if this change was included then what benefit or opportunity would this change give when compared to its impact on the project's [Scope], [Time], [Cost], [Quality], [People], & [Resources]**. ... *And, what are the potential risks that could result due to (or due not to) engaging in this change?*

      The decision on the "Return On Investment" may require that the Change Request be pre-examined prior to the Change Control Board meeting.

    - The **Change Control Board** would ideally be **composed of representatives** from the Project Steering Committee, relevant members of the Project Working Group, appropriate members of the Project Implementation Team, and representative(s) of the customer organization. The Change Control Board could meet on an as needed basis, at predefined periodic intervals, or at the commencement / completion of the release milestone to go over the outstanding Change Requests.

 I would recommend that ... the Change Control Board has members with different perspectives & expert judgment, so that the benefits & disadvantages (pros & cons) of the change can be openly discussed & intelligently analyzed.

The Change Control Board should have a representative of the customer (or the product owner) to ensure that the customer's perspective is being taken into consideration, and that decisions are not being "rubber stamped".

5. **If the Change Request was rejected then the person / group who proposed the change may be notified of its rejection and the reasons why** it was rejected. Based on this feedback the requester may make modifications to their Change Request and resubmit it for further consideration. The rejected Change Request's details should be updated in the Change Management System including the reasons for rejection.

6. **If the Change Request was accepted then it would be translated into adjustments to:**

   - the **[Scope Baseline]** in the project's (re)approved Detailed Specifications,

   - the **[Time Baseline]** in the project's (re)approved schedule,

   - the **[Cost Baseline]** in the project's (re)approved budget,

   - the **[People Baseline] & [Resources Baseline]** in the project's (re)approved resourcing calendar & plans,

   - the **[Quality Baseline]** in the project's (re)approved Acceptance Criteria / test plans, plus any other affected project documentation.

This [Baseline] Change Request would have consequently resulted in the project being **"re-baselined"**. That is, modifying the approved plans to accommodate the change and then "re-freezing" these baselines in accordance with the authorized change.

Depending on the implementation methodology being used, these changes could entail:

- For **agile**; changes would be translated into new – revised – removed **story points** and these story points would be placed into the Product Backlog, for the Project Implementation Team to choose to implement later on when they deem it most appropriate.

- For **iterative**; changes would be planned into **future releases**, and hence the project variables & constraints adjusted accordingly for those future releases.

- For **waterfall**; changes would be integrated into appropriate points in the **re-baselined** project plans.

 **DO NOT implement the change without some form of (written) agreement on the recompense that is to be afforded to the affected stakeholders**; *e.g. new delivery milestone dates and compensation payments / refund to the performing organization or the customer.*

---

7. **Only those [Baseline] Change Requests authorized** by the Change Control Board *(including obtaining the signature of the customer representative)* would then **be implemented** by the Project Implementation Team as either a **corrective action, a preventive action, and/or as a repair.**

> I would recommend that ... if the customer is not prepared to compensate the performing organization for the change, then negotiations should be entered into for the purpose of dropping an appropriate lower priority feature from the [Scope].
>
> Else covert scope shrinkage may occur where some random feature(s) is not implemented due to [Time] and/or [Cost] constraints which befall the performing organization due to having to undertake the change.

8. **Verify that only those authorized changes have been incorporated in accordance with the agreed Acceptance Criteria**, and that these changes have been incorporated into all relevant areas of the project; i.e. perform Quality Control activities.

9. **Document the completed change(s)**.

10. (Optionally) inform the change requester that the change has been incorporated.

NOTE   ❑ **It is essential that all involved parties (especially the implementers) are communicated to about the change** and its progress; i.e. the Change Request should be continually updated in the Change Management System. Where necessary, individual parties may have to be personally updated.

     ❑ **If the change does not affect the [Scope Baseline], the [Time Baseline], the [Cost Baseline], and/or the [Quality Baseline] then a formalized Change Management Process may not necessarily be required.**

When not to undertake Change Management?

The following would clarify some of the situations when Change Management would be ~~highly beneficial~~ *mandatory*, and should include the customer's involvement:

1) Any project documents that must be **agreed to (signed off)** between the customer and the performing organization; *e.g. the Detailed Specifications, the Project Management Plans, the Acceptance Criteria, and the Project Charter* (*though the Project Charter should not really have to be changed*).

2) Any project deliverables that form **contractual obligations**.

3) Any corrective, preventive, or repair actions that will require **changes to the** approved **baselines**; i.e. changes to the project constraints.

NOTE   ❑ In the absence of a formalized Change Management Process, then a couple of persons (probably the project manager and the representative of the customer organization) have to be authorized to accept & reject Change Requests.

Do defect reports need to go through a Change Control process?

All Defect Reports (especially those generated externally to the performing organization) **will need to go through some form of Change Management Process** for the following reasons:

1) Sometimes **[Scope] changes are inadvertently snuck through as Defect Reports**. This often happens when the customer finds functionality that doesn't comply with their "needs", "wants", or "expectations". That is, it is not as they "intended it to be", or it is not as they now "think it should be". ... *Contrary to what was agreed to.*

    However, the implemented functionality is not based on the Customer Requirements, but rather on the agreed & signed-off version of the approved Detailed Specifications. Hence, there may very well have been a misunderstanding & misinterpretation between the customer organization representatives and the performing organization.

    Though, recall back in [Section 4.4.4], it was recommended that the **approved Detailed Specifications should be the arbitrator on the validity of Change Requests & Defect Reports**, and NOT any other document such as the Customer Requirements.

Watch out for [Scope] changes snuck through as defects.

Check the Defect Report against the approved Detailed Specifications and NOT against the Customer Requirements.

2) There are **several categories of defects with differing priorities**.

Without a Change Control Process, these would be implemented (or not implemented) in a haphazard manner that would affect the [Time] & [Cost] of each deliverable. Additionally, this defect fix could potentially introduce instabilities at an inopportune moment in the delivery cycle.

10.2.2. Change Management of: Defects

The Defect Reporting & resolution process described here is illustrated in [**Figure 89**]:

1. A Defect Report can be submitted by any project stakeholder who happens to find or believes that **some feature or functionality of the project's deliverables does not conform to the agreed Acceptance Criteria / approved Detailed Specifications**.

    That is, the implemented [Scope] will need to be changed in order that the project deliverables will conform to the approved [Scope Baseline].

2. The **Defect Report is received by the performing organization** from any of the project stakeholders. This Defect Report should be in a **written or text-based form and not received solely by verbalized statements** or hearsay.

    The performing organization may have existing paper forms, email templates, or a centralized repository (**Defect Tracking System**) established specifically for accepting, handling, and administrating Defect Reports. Preferably, there will be a clear separation between those defects reported internally to the performing organization from those defects found externally by the customer organization.

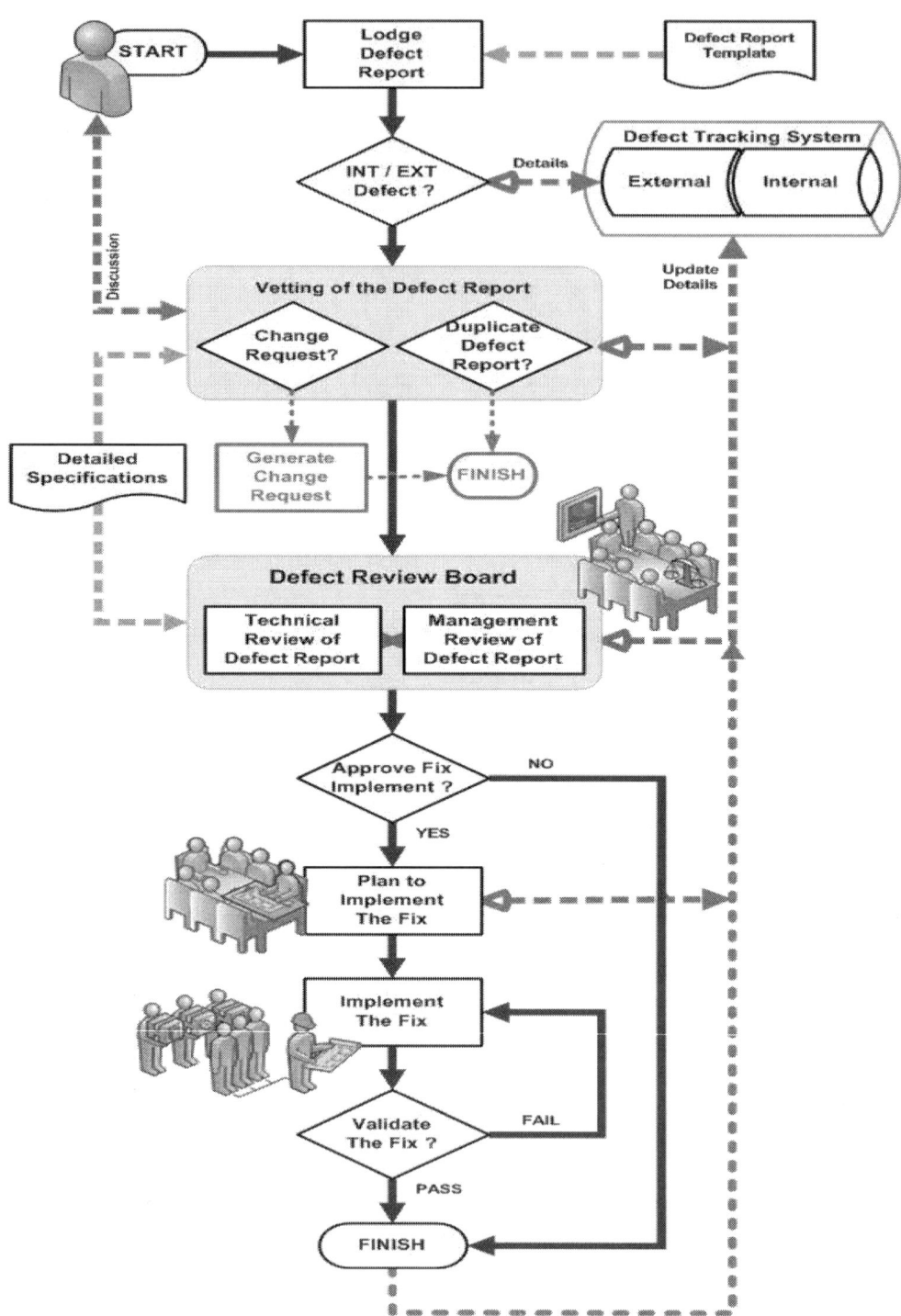

Figure 89: A simplified defect processing workflow.

 **DO NOT accept / act upon Defect Reports expressed verbally**, because without a recorded description of the defect then it may be difficult for the implementer to know exactly what defect is to be fixed, and under what circumstances did the defect occur.

3. **A preliminary analysis & review of the received Defect Report** would be conducted (possibly by the project manager) to **determine;** if any **additional information** is **required** to clarify the Defect Report, whether this Defect Report is a **duplicate** or very similar to an existing one, and whether this **Defect Report should in fact be classified as a Change Request**.

Ooh, I'm experiencing Déjà vu with the change request process.

4. The **vetted Defect Report** would be **reviewed by the Defect Review Board** to determine; whether they agree with the preliminary analysis that this Defect Report is a **"unique" defect** that should be accepted / rejected, is a **duplication** of an existing defect, or a **Change Request**.

- The review of the Defect Report will often involve comparing the report's **description of the noncompliant functionality against that described in the approved Detailed Specifications**.

- The Defect Review Board would ideally be composed of relevant members of the Project Working Group, appropriate members of the Project Implementation Team, and possibly representative(s) of the customer organization.
  The **Defect Review Board would** generally meet at a predefined periodic interval (*e.g. weekly, fort-nightly, or monthly*) to **discuss those outstanding defects to be implemented in the current pending release or in future releases**.

Based on an analysis of the defect, a decision would be made as to whether it is **appropriate to authorize the repair of this defect**.

5. The **Defect Report** may be **reject**ed because:

- It was essentially a **duplication of an existing Defect Report**.

  It would be marked as such and a reference to the "original" Defect Report for this issue included in its documentation. This "new" duplicate Defect Report would then be 'CLOSED'. The "original" Defect Report would be updated with any new information that was contained in the "new" duplicate Defect Report.

- It constituted a **Change Request** and thus is not a defect.

  Hence, this Change Request would be handled via the Change Management Process that was described back in [Section 10.2.1]. This particular Defect Report would be documented as a Change Request and then 'CLOSED'.

  NOTE   ☐ Depending on the **Defect Tracking System** being used (and the processes & procedures in place), then the person who reported the defect may or may not be notified of the defect's new status.

6. If the **Defect Report** was **accepted and authorized for repair** then the implementation of such repairs would be **planned for an appropriate time & place in the implementation cycle** (potentially specifying who will be responsible for the repair) so as to ensure that only those "authorized" defect repairs would then be implemented by the Project Implementation Team.

7. Once the **defect repair** has been **implemented, it would be confirmed** (by the "independent" verification & validation test team) that indeed the affected functionality does now **comply with the agreed Acceptance Criteria** and with its definition in the **approved Detailed Specifications**.

8. The **Defect Report's documentation in the Defect Tracking System** would be **updated with information about the repairs** and its status would be set to 'CLOSED'.

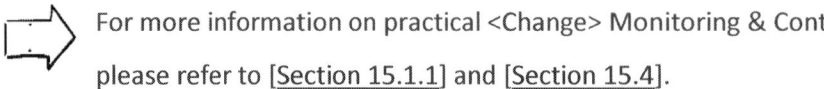

For more information on practical <Change> Monitoring & Control, please refer to [Section 15.1.1] and [Section 15.4].

10.2.3. Change Management of: Quality

At this point, it must be noted that **Change Management** is not a stand-alone activity.
**Change Management has a symbiotic relationship with both Risk Management &
Stakeholder Management, and also ties into the [Quality] processes & procedures** being
utilized by the performing organization. See [Section 8.1] and [Figure 79] in that section.

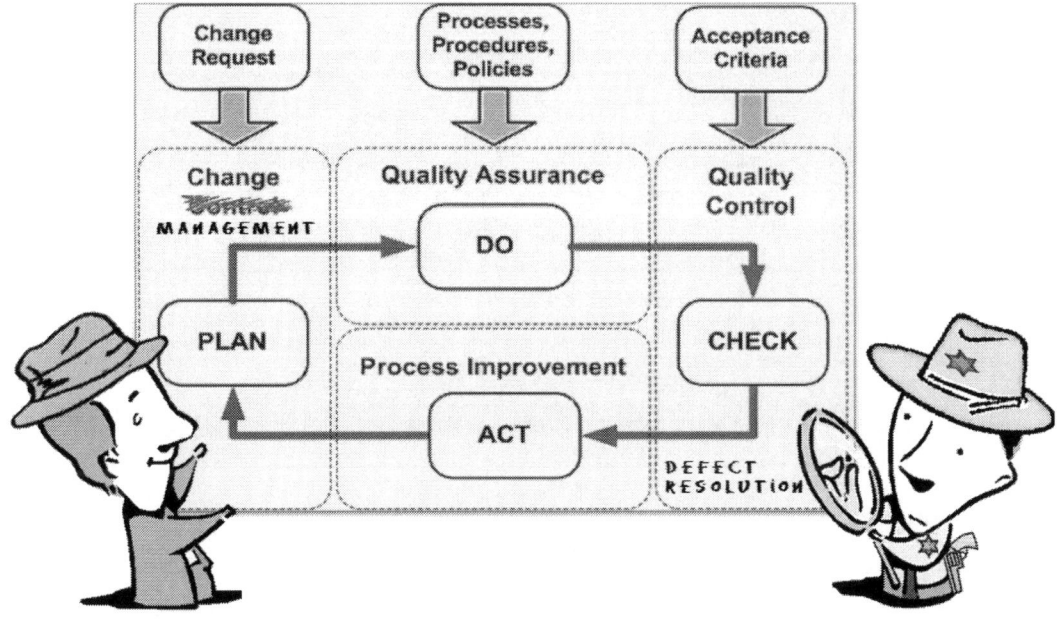

CHANGE MANAGEMENT IS
THE ORCHESTRATION OF CHANGE CONTROL.

DEFECT RESOLUTION IS
A CATALYST FOR CHANGE MANAGEMENT.

10.3. Stakeholder Management

Finally, those tiresome passengers in the backseat of the car, those annoying drivers in the other vehicles, those suicidal pedestrians on the sidewalk, that power-crazed highway patrol officer, and the service station attendant who would rather be somewhere else. Yes, this long arduous road trip analogy for monitoring & control is nearing the end of its journey.

As stated way back in [Section 1.3]; **for a project to be deemed a success or a failure is very much dependent on the expectations & perspectives of those stakeholders who have an interest in the project**. Where, "interest" means that these people are **either positively or negatively <u>affected by</u> the project, or** these people **have a positive or negative <u>effect on</u> the project's outcomes**. In addition, back in [Section 2.4] it was stated that during the early phases of the project life cycle (i.e. the Initiating Phase and the Planning Phase) these project stakeholders have the greatest influence on the outcome of the project. Therefore, as those who decide whether the project is deemed a success or a failure, these **stakeholders should never be taken for granted. Rather, they should be managed throughout the entire life of the project**, especially during the early phases.

Hence, the project manager needs to:

1) know what these **stakeholders expect**,

2) how to keep these **stakeholders informed**,

3) how to provide these stakeholders with **what they need to know** (so that they can make their own informed decisions),

4) **when** these stakeholders **need to know it**, and

5) to make sure things are done by when these stakeholders **need them to be done**.

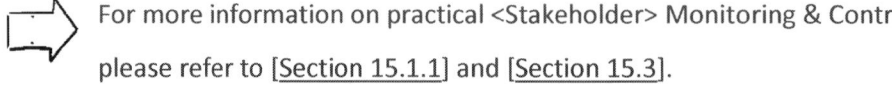 For more information on practical <Stakeholder> Monitoring & Control, please refer to [Section 15.1.1] and [Section 15.3].

How do I go about managing stakeholders?

Stakeholder Management involves the following steps shown in [Figure 90]:

1. **Identify** the project's stakeholders and determine their roles & responsibilities with respect to the project.

2. **Assess** these stakeholders:

   - **Expectations, perspectives, concerns, "needs & wants".**

   - **Categorize** these stakeholders into generic groupings.

3. **Analyse** these stakeholders by:

   - **Mapping** the **relationship** between the project and the stakeholders based on their **power, interest, position, uncertainty, influence, and impact**.

   - **Examine** the stakeholder **characteristics** that will influence how they should be managed.

4. **Strategize** how best **to engage** & manage these stakeholders so that they will **collaboratively participate** in the project.

5. **Monitor** each of these identified stakeholders and
   control the responses to and interactions with each of these identified stakeholders.

NOTE ❒ These steps would be **continually undertaken in a cyclic fashion during the entire life of the project**.

STAKEHOLDERS SHOULD NEVER BE TAKEN
FOR GRANTED. ... RATHER, THEY SHOULD BE
MANAGED THROUGHOUT THE ENTIRE LIFE
OF THE PROJECT, FROM START TO FINISH.

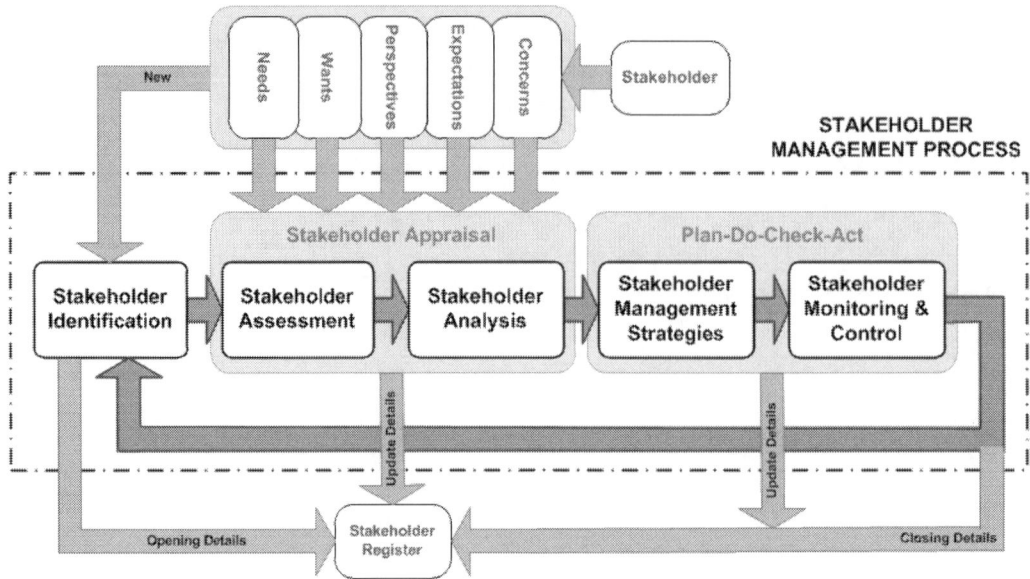

Figure 90: Stakeholder Management Process.

NOTICE how in [Figure 90] that the Stakeholder Management Process has the same process flow structure as [Figure 85] for the Risk Management Process and [Figure 87] for the Change Management Process.

These identical flow structures are not coincidental, given that:

1) These all **relate to specific aspects of risk** to the project; i.e. confrontational risk & issues, risks due to change, and risk due to the project stakeholders' perceptions & expectations.

2) These all have **identical phase ordering** of identification, analysis, plan / strategize, implement, and monitoring & control.

3) These all have **some form of tracking & information feedback**; *e.g. Risks & Issues Register, Change Management System & Defect Tracking System, and Stakeholder Register.*

4) These all have a **cyclic flow** that has to be undertaken over the entire life of the project.

10.3.1. Stakeholder: Identification

Stakeholder identification is determining those persons that are <u>involved with</u> the project, being <u>affected by</u> the project, and/or have an <u>effect on</u> the project.

Back in [Section 2.4.2], some of these project stakeholders were identified simply by considering how they obviously related to the development life cycle being used for the project; see [Figure 19] and [Figure 20] which have both been reproduced in this section.

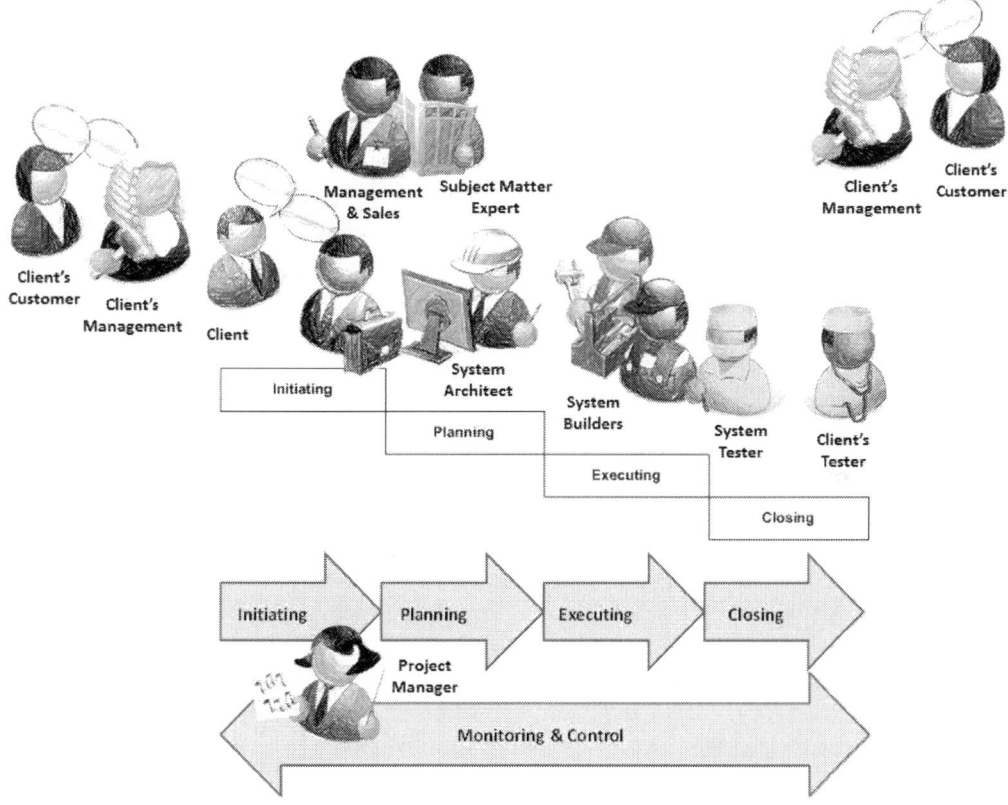

Figure 19: Stakeholders common to a waterfall project life cycle.

Alternative methods for the initial stakeholder identification would include; groupings within the Work Breakdown Structure, the system / applications architecture, the process flow of the system, and use case analysis.

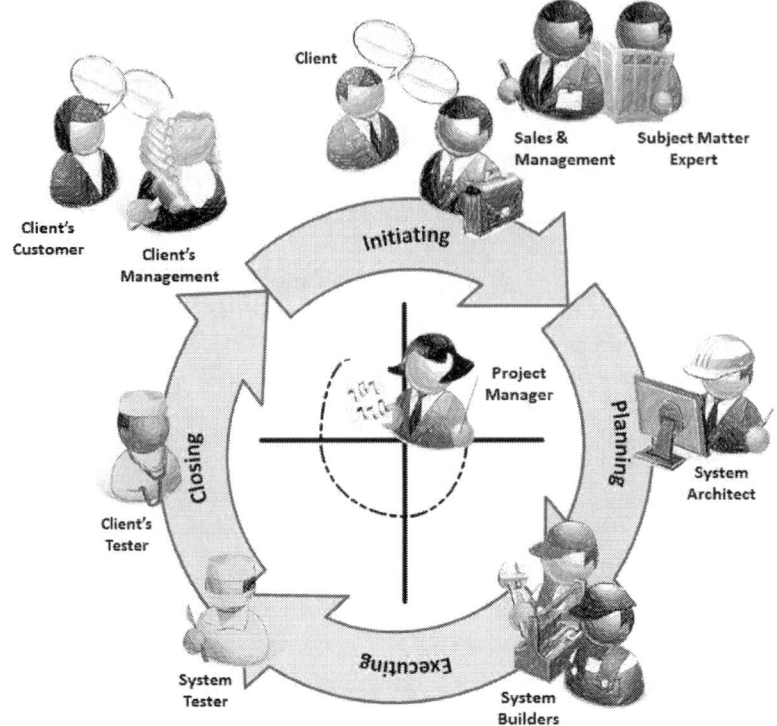

Figure 20: Stakeholders common to an iterative project life cycle.

This simple technique has identified some common SDLC project stakeholders:

- the (client) customer's representative,

- the (client) customer's management,

- the customer's customer (end user),

- the (client) customer's tester / product verification team,

- the performing organization's customer interface (account manager) sales and senior management,

- the performing organization's subject matter expert / business analyst,

- the performing organization's system architect / design authority,

- the performing organization's system builders / implementation team,

- the performing organization's system testers / product verification, and

- the performing organization's project manager.

NOTE   ❐ There is **one inherent group within the performing organization that is often overlooked**. ... Until, an issue arises that directly involves them.

This group is the **IT Department** who provide the underlying infrastructure that the project is built upon; i.e. analogous to the roads that the project drives on.

How about including in this stakeholder list:
- ‡ *consultants, contractors, and subcontractors,*
- ‡ *vendors and suppliers,*
- ‡ *manufacturing,*
- ‡ *publishing and printing,*
- ‡ *storage and distribution,*
- ‡ *web hosting, data centres, and help-lines,*
- ‡ *accounts and finance,*
- ‡ *legal and compliance,*
- ‡ *marketing and public relations,*
- ‡ *the general public,*
- ‡ *the families of the project team members,*
- ‡ *government and statutory authorities,*
- ‡ *quality auditors,*
- ‡ *the environment and ecosystems,*
- ‡ *the performing organization's owner / shareholders,*
- ‡ *financial markets,*
- ‡ *the governing body of the stock exchange.*

Based on this list, it should be apparent that there are numerous possible project stakeholders, both internal & external to the performing organization. Some of these stakeholders will be inherent to every project undertaken by the performing organization, whereas others will be unique to the individual project and its current situation and the prevailing circumstances.

Is there a process for identifying stakeholders?

Try answering the following questions and see what names pop-up:

1) **Who is directly affected by the project NOT having been already implemented**; i.e. prior to the project's existence?

2) **Who will be directly participating with implementing the project**?

3) **Who will be directly participating with concluding the project**?

4) **Who will be directly affected during** the implementation of the project?

5) **Who will be directly affected by** the project's **outcomes & outputs**?

Well, that should give a list of the stakeholders **"directly" involved** with the project.

However, there are also those persons who will be **"indirectly" affected** by the project.

Therefore, the above questions should be asked again for "indirectly".

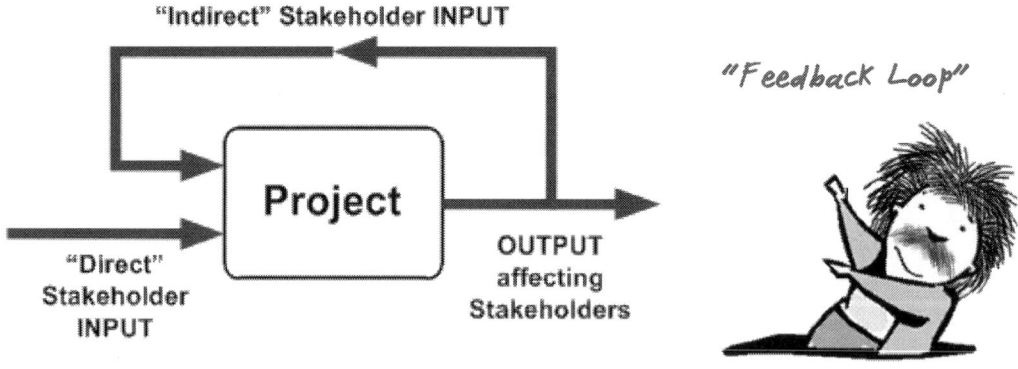

"Feedback Loop"

Figure 91: Direct & Indirect stakeholder inputs to the project.

Every action has a resultant reaction.

In [Figure 91], **"direct"** refers to an **immediate input** into the project, whereas **"indirect"** refers to an **input into the project that was introduced via the feedback because of the effects of the project's outputs & outcomes.**

IMHO, these indirect feedback loops cause many a surprise issue for the project. ... Aka "fire-fighting" situations.

Once these stakeholders have been identified, it would be beneficial to list them in the **Stakeholder Register**. Also in this Stakeholder Register should be included such things as each stakeholder's; contact details, their interests, their involvement with the project, their potential influence on the project, whether they are positively / negatively impacted by the project, their roles, their level of responsibility & authority, and eventually a list of strategies on how they should be managed.

Some of the **information collected about the stakeholders will be highly sensitive and even controversial**.

The **leaking of this information** (i.e. the Stakeholder Register) could result in an adverse reaction from specific stakeholders and **cause damage** to both the project and the performing organization.

THE KEY TO SUCCESS, IS KNOWING THOSE WHO MUST BE ADDRESSED.

WARNING: DO NOT overlook the importance of the Project Implementation Team (i.e. the characters wearing the sombreros) because these persons positive participation is essential to the project's chances of ever being successfully delivered.

10.3.2. Stakeholder: Assessment

First thing to realize about stakeholder assessment is that **each stakeholder has differing**:

- **Needs** – a **minimum** of what they **require to be satisfied**.

- **Wants** – a **maximum** of what they **desire to be satisfied**.

- **Concerns** – an **interest in** the effect on them or on others.

- **Expectation** – a **belief** about what is **to occur** and what is **to eventuate**.

- **Perspective** – the way they look at the situation and how they **interpret & understand** the project.

NOTE ❑ **How can you possibly understand their expectations, perspectives, concerns, "needs & wants" unless you first let them articulate these?**

The only way to obtain this kind of insight is to interact with the individual stakeholders by being both a good communicator & listener, see [Section 11.3].

Generally speaking, those **stakeholders with a positive expectation & perspective** of the project **will either try to help** the project along, **or at least not hinder** its progress.

Whereas, those **stakeholders with a negative expectation & perspective** of the project **will try to either actively impede** the project's progress, or they will **passively resist**.

NOTE ❑ Negative proponents of the project can surprisingly occur within the customer's own organization, even though that organization will be obtaining benefit from the project and its outcomes.

For example; some members of their staff may not want the new system deployed, but rather prefer to continue using the existing systems due to their own short-term interests (of what they already know & are proficient at), and not what is best in the long-term for their employer's scalability, sustainability, supportability, profitability.

Example Case: The product owner double agent's covert sabotage.

Once upon a time, I worked on this project developing a new piece of telecommunications equipment for a national Telco. The new equipment was going to have many times the connected line capacity when compared to the incumbent equipment and at significantly reduced per line cost and unit price.

The customer organization's senior management wanted the new equipment to be rolled out as soon as possible, but their (product owner) manager in charge of the department that would use this new equipment appeared to be continually delaying its deployment by what felt like nit-picking stalling tactics.

It turned out that; while the new equipment may have proved great benefit to the customer organization, the product owner's department and his performance was not being judged based on these considerations but rather he / they were being evaluated purely on service availability and meeting their Service Level Agreements.

Thus, from the product owner's perspective, it was "better the devil you know" incumbent equipment rather than "the devil you don't know" new equipment.

While some stakeholders will have similar expectations, perspectives, concerns, "needs & wants", others will be markedly different. Additionally, as stated back in [Section 10.1], **every stakeholder has their own level of risk tolerance that will also affect these stakeholders' characteristics**.

How can the project possibly succeed if the stakeholder's expectations & perspectives cannot be reconciled?

This is when the project manager's skills as a communicator, listener, negotiator, and conflict resolver will come to the fore; see [Section 11.3].

Categorization

There are many and varied ways that these identified stakeholders could be subdivided into groups. However, the important thing is to subdivide these identified stakeholders to **enable a quick & effective response to those stakeholders who are considered to have a higher priority when compared to those who have been deemed a lower priority.**

As the project manager, your day is going to be busy enough with managing the project without the additional burden of dealing with every single stakeholder's concerns, needs and wants. You just won't have the time for much handholding. Therefore, these stakeholders will have to be categorized & prioritized based on their potential positive & negative effects on the project being deemed a success.

Way back in [Section 2.4.1], the categorization & groupings of stakeholders were depicted as; primary-core, secondary-strategic, and environmental stakeholders as illustrated previously in [Figure 17] and [Figure 18] reproduced below.

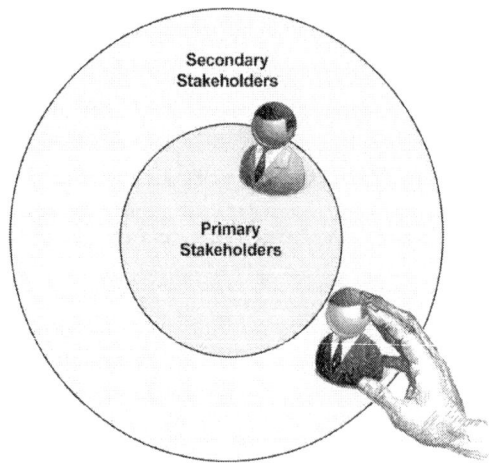
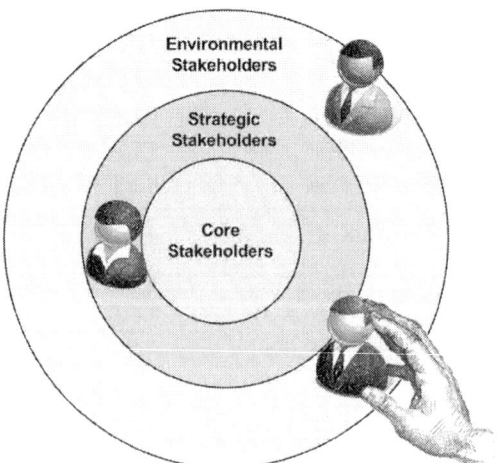

| **Figure 17: Stakeholder primary-secondary categories model.** | **Figure 18: Stakeholder core-strategic-environmental groupings model.** |

NOTE   ❐ **During the life of the project, a stakeholder could change categories** between core, strategic, & environmental or even appear & disappear.

10.3.3. Stakeholder: Analysis

Relationship Mapping

Irrespective of how it is decided to categorize these stakeholders, there are a few things that are relevant to all of these stakeholders:

1) Their "**power**" over the project's outcomes; i.e. **their capability to get things done to either help or hinder the project's progress**. This power usually comes about because of their level of authority and their role in the management hierarchy.

2) Their "**interest**" in the project and its outcomes; i.e. they are greatly **concerned for the project** or (at present) are not the least bit interested.

3) Their "**position**" about the project; i.e. they **support** (champion) the project or they **oppose** (strive for its demise).

4) Their "**uncertainty**" / predictability; i.e. how **likely** are they **to react as expected**.

5) Their "**influence**" over the project's outcomes; i.e. **their ability to effect change** and their capability to get others to do what they desire with respect to the project and its outcomes. *... They may be referred to as a "change agent".*

   For example; one individual environmental stakeholder has minuscule influence over the project, but a few thousand of these environmental stakeholders acting together as one consolidated group do wheeled a considerable amount of influence.

6) The "**impact**" on them by the project and its outcomes; i.e. **how will they be affected** ... this is not their impact on the project (see "influence").

A method used to depict these relationships is via **Stakeholder Matrices**, where each stakeholder is mapped onto a specific grid. However, it must be remembered that an **individual stakeholder's location on each grid can dynamically change over the life of the project**.

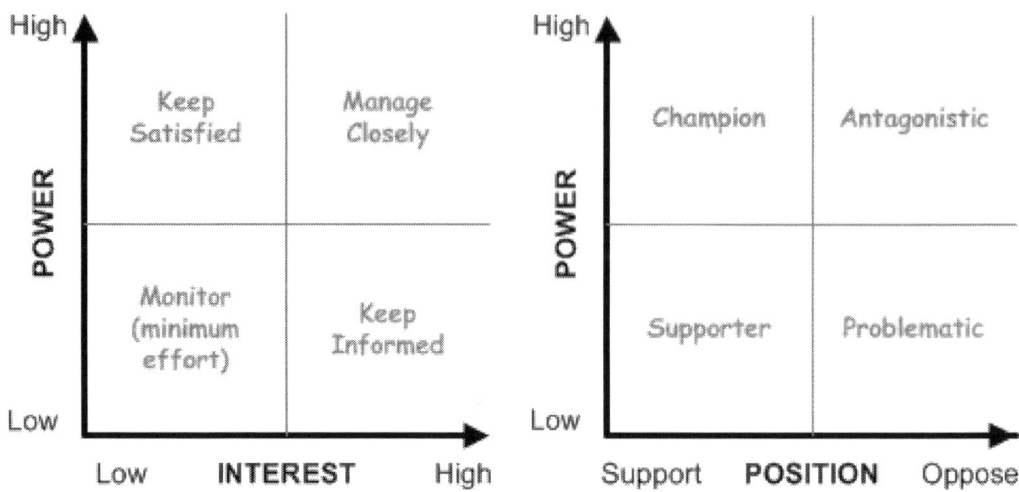

Figure 92: Power versus Interest.   Figure 93: Power versus Position.

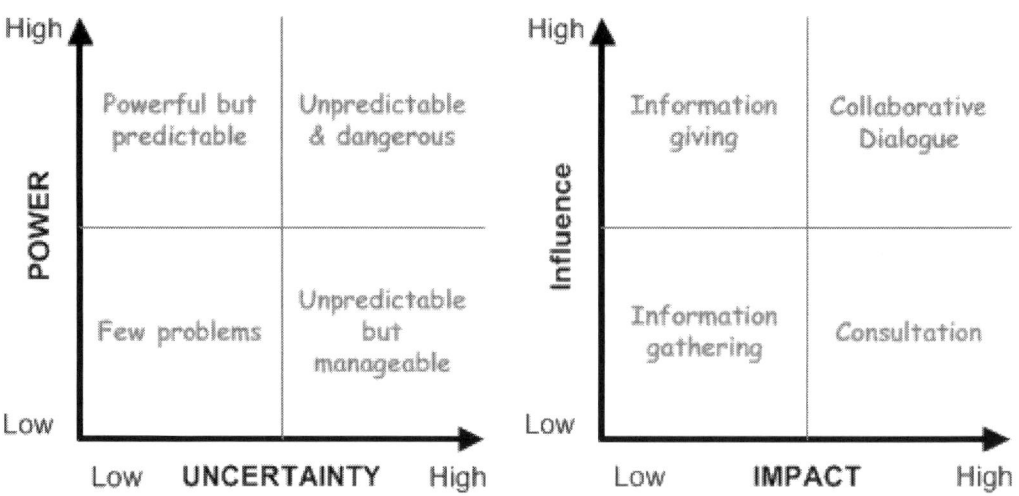

Figure 94: Power versus Uncertainty.   Figure 95: Influence versus Impact.

NOTICE that on each of the above grids there are grey comments about what should be done with the stakeholders in that quadrant, or comments about the characteristics of stakeholders in that quadrant. Later on, these quadrant comments will be used to decide on the relevant strategies to be used to manage each stakeholder.

NOTE ❑ The arrangements of these Stakeholder Matrices presented in this book are NOT the definitive collection. There are various alternative arrangements of the x-y axis, as well as entirely different grids; *e.g. "uncertainty" replaced with "predictability", a grid for "Importance versus Influence", a grid for "Importance versus Position" ... etc.*

I don't like to include "importance" as a parameter because it could subconsciously blind me to a future problem stakeholder ... "Oh don't worry about them; they're not important enough".

The arrangement of Stakeholder Matrices used in this book were chosen so as to align Stakeholder Management with the concepts for Risk Management, and hence to align the grids with the characteristics of the Risk Matrix; see [Figure 96] and [Figure 86].

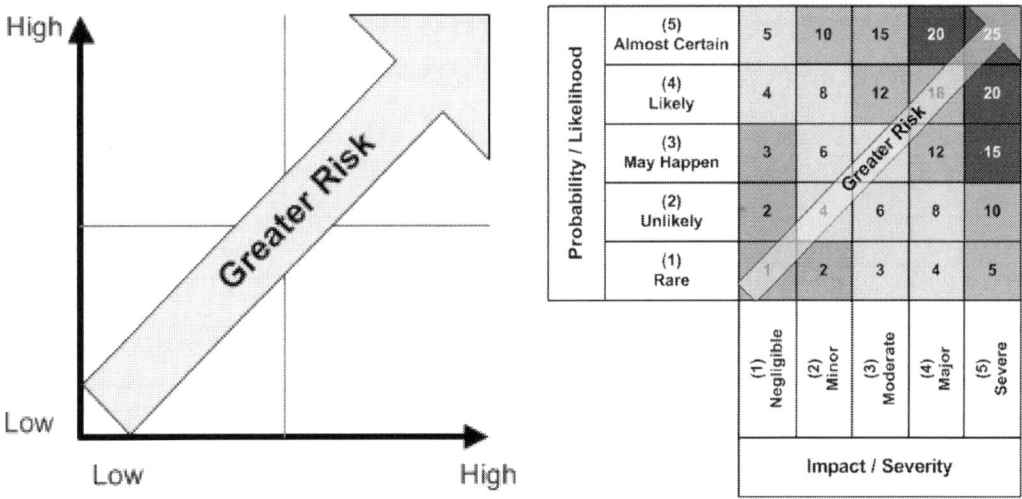

Figure 96: Stakeholder Matrix outline.   Figure 86: Example of a Risk Matrix.

Instead of a 2x2 grid, later on a 5x5 grid will be used for the Stakeholder Matrices.

For this book's variants of the Stakeholder Matrices, those stakeholders with higher ~~priority~~ risk would be closest to the top right corner of the grid. Whereas, those stakeholders with lower ~~priority~~ risk would be near the bottom left corner of the grid.

How do I map stakeholder relationships?

There are four key steps involved with the mapping part of stakeholder analysis:

1. **Identify** who are the project's stakeholders, determine their role in the project, and include them in the Stakeholder Register.

2. **Quantify** each identified stakeholder's **power, interest, position, uncertainty, influence, impact** ... etc. That is, assign numerical values for each of these aspects by interviewing & interacting with each of these stakeholders. Also, discuss each of these stakeholders with representatives from the Project Steering Committee, The Project Working Group, and even with members of the Project Implementation Team.

3. **Tabulate** the results to obtain an overall understanding for each of these identified project stakeholders.

4. **Map** the identified stakeholders onto each chosen variant of the Stakeholder Matrices.

THE KEY TO SUCCESS, IS KNOWING THE PRIORITY OF THOSE TO ADDRESS.

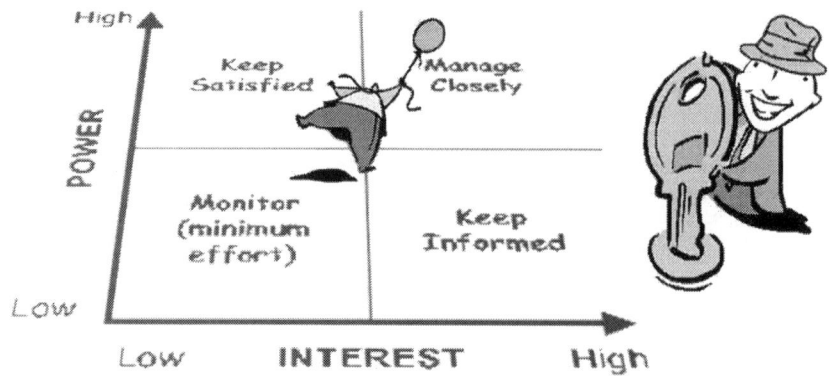

A down-to-earth guide for mapping stakeholder relationships.

1. **Who are the project stakeholders**, both directly and indirectly?

   - Include these stakeholders in the Stakeholder Register, along with their; contact details, roles, level of responsibility & authority.

2. **What is this stakeholder's "power"** to advance or hinder the project's outcomes?

   - Score this stakeholder's power on a scale of 1-5, where 1 is least powerful and 5 is most powerful.

3. **What is their "interest" / concern in** the project and its outcomes?

   - Score this stakeholder's interest on a scale of 1-5, where 1 is least interested and 5 is greatly concerned.

4. **What is their "position" for or against** the project?

   - Score this stakeholder's position on a scale of 1-5, where 1 is project champion and 5 is project nemesis; i.e. 1 = project sponsor, 2 = supporter, 3 = neutral, 4 = opponent, 5 = strives for its demise.

5. **What is their level of "uncertainty"** or what is your level of "**predictability**" about how they will react to the project, its outcomes, to change, and to risks?

   - Score this stakeholder's uncertainty on a scale of 1-5, where 1 is very certain and 5 are highly unpredictable.

6. **What is their "influence" / effect on** the project's successful outcome?

   - Score this stakeholder's influence on a scale of 1-5, where 1 is least influence and 5 is most influence.

7. **What is the level of "impact" on them** by the project and its outcomes?

   - Score this stakeholder's impact on a scale of 1-5, where 1 is little good / bad and 5 is major good / bad for them.

8. **Tabulate** these values per stakeholder.

Stakeholder	Power	Interest	Position	Uncertainty	Influence	Impact
Prj Sponsor	5	4	1	1	3	3
Worker	4	5	2	2	5	2
Supplier	2	1	3	3	4	1
End user	1	2	4	3	2	5

For this example case; while senior management (i.e. the project sponsor) is very enthusiastic about the project's new CRM system that will replace multiple proprietary databases & applications, there is the occasional grumbling from some end-user staff about the pending changes, but generally, these users don't seem too interested, also the supplier says the host-servers should turn up on time.

9. **Draw up the relevant 5x5 grids**, preferably with the worst characteristic (and hence, the greatest risk) in the top right corner and the best characteristics (least risk) in the bottom left corner. Thereby, having similar characteristics to the Risk Matrix.

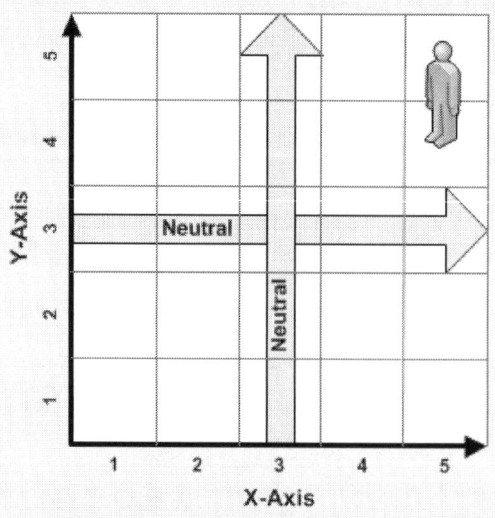

Figure 97: Outline of stakeholder grid.

Risk Matrix

10. Based on the numerical data that has been collected, **plot each stakeholder onto the Stakeholder Matrices.**

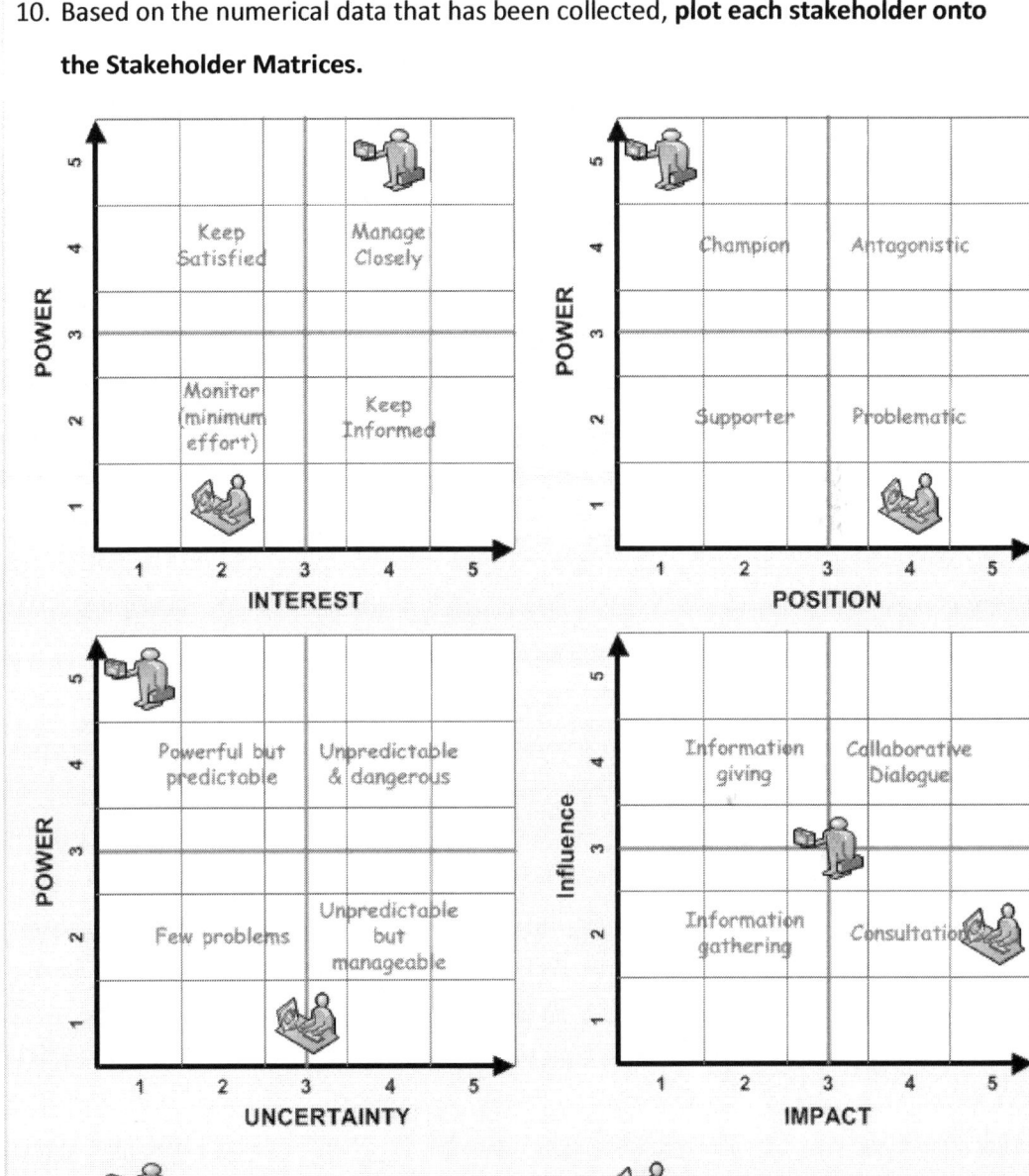

Figure 98: Example Stakeholder Matrices in use.

Include the collected information into the **Stakeholder Register**; i.e. interests, involvement, potential impact, potential influence, and eventually a list of strategies on how each identified stakeholder should be managed.

Characteristics Examination

Given that each of the identified stakeholders have now been mapped onto the Stakeholder Matrices, some conclusions can be drawn about each of these stakeholder's characteristics and hence what management strategies should be used on them.

This is where the grey comments on the template Stakeholder Matrices come into play.

Stakeholder	Power Vs. Interest				Power Vs. Position			
	Monitor (minimum effort)	Keep Informed	Keep Satisfied	Manage Closely	Supporter	Problematic	Champion	Antagonistic
Project Sponsor				X			X	
Worker				X			X	
Supplier	X				X	X		
End User	X					X		

Stakeholder	Power Vs. Uncertainty				Influence Vs. Impact			
	Few Problems	Unpredictable But Manageable	Powerful But Predictable	Unpredictable & Dangerous	Information Gathering	Consultation	Information Giving	Collaborative Dialog
Project Sponsor			X		X	X	X	X
Worker			X				X	
Supplier	X	X					X	
End User	X	X				X		

Matrix	Quadrant	Description of the stakeholder
Power Vs. Interest	Monitor (minimum effort)	• They don't care much about the project (until such a time, as they perceive that it is affecting them). • They have limited means to affect the project.
	Keep Informed	• They care about the project, but they have limited means to affect the project. • They can help with the finer details of the project as they most probably understand the inner-workings of those bits that matter to them.
	Keep Satisfied	• They don't care much about the project, but they do have the means to affect the project if they perceive that it is adversely affecting them.
	Manage Closely	• They care about the project & its outcomes, and they have the means to affect the project if & when they choose. • Their views & opinions must be given the highest level of consideration because their opposition could make or break the project.

Matrix	Quadrant	Description of the stakeholder
Power Vs. Position	**Supporter**	• They endorse the project, but they are not really in a position of Power to be able to do much that will be advantageous to the project's success. • Unless these stakeholders can be "mobilized" to act as one then their support will not be of significant use to the project.
	Problematic	• They oppose the project, but they are not in a position of Power to do much about it. • Individually they are not a significant risk to the project's success, but if they became mobilized then they could influence the attitudes of more important stakeholders, and hence they could become a "political force" to be reckoned with.
	Champion	• They support the project, and they are in a position of Power to affect the project's successful outcome. • They are a significant ally to neutralize those potential antagonistic opponents of the project.
	Antagonistic	• They oppose the project, and they are in a position of Power to be able to do something about the project if they so desire. • They present a significant risk to the project's success, especially if they have the groundswell of support from the problematic stakeholders.

Matrix	Quadrant	Description of the stakeholder
Power Vs. Uncertainty	**Few Problems**	<ul><li>They are relatively predictable, but they have limited means to affect the project.</li><li>They most probably will not change their current views of the project, so they should be managed based on their other matrices' characteristics.</li></ul>
	Unpredictable But Manageable	<ul><li>They are relatively unpredictable, but they have limited means to affect the project.</li><li>They will probably change their current views of the project if they are encouraged to do so.</li></ul>
	Powerful But Predictable	<ul><li>They are relatively predictable, and they have the means to affect the project.</li><li>They most probably will not change their current view of the project, so their views & opinions must be given a high level of consideration because once they are "won-over" they will probably continue to support the project.</li></ul>
	Unpredictable & Dangerous	<ul><li>They are relatively unpredictable and they do have the means to affect the project.</li><li>They probably will change their current view of the project, so try to encourage them to support the project rather than letting them drift into opposing it.</li></ul>

Matrix	Quadrant	Description of the stakeholder
Influence Vs. Impact	Information Gathering	• They are relatively unaffected by the project and its outcome, though they have limited means of influencing the project's outcomes. • They are a source of information on what probably will and will not affect them.
	Consultation	• They are affected by the project and its outcome, but they have limited means of influencing the project's outcomes. • They most probably understand the inner-workings of those issues that affect them, and hence they should be consulted before they become problematic or antagonistic.
	Information Giving	• They are relatively unaffected by the project and its outcome, but they do have the means to influence the project if they choose to do so. • They are a source of information as to the possible objectives of the project.
	Collaborative Dialog	• They are affected by the project and its outcome, and they do have the means to influence the project. • Their views and opinions must be given the highest level of consideration because their opposition can make or break the project.

NOTE ❑ **The stakeholder's position on these matrices is based on the perceptions of the person(s) undertaking the analysis and this may not represent the true relationships with the stakeholder.** Hence, this analysis will have to be iteratively revised & reviewed over the duration of the project's life, and based on continual interaction with these stakeholders.

By using an arrangement of Stakeholder Matrices that ties into the Risk Matrix model (as utilized in this book), then any stakeholders residing in the top right of the grids are potentially high sources of risk for the project; see [Figure 99].

Using each Stakeholder Matrix in [Figure 98], and based on the numerical values assigned to each quadrant's squares in [Figure 99] a weighting of the "riskiness" priority for each stakeholder can be accumulated, as demonstrated below.

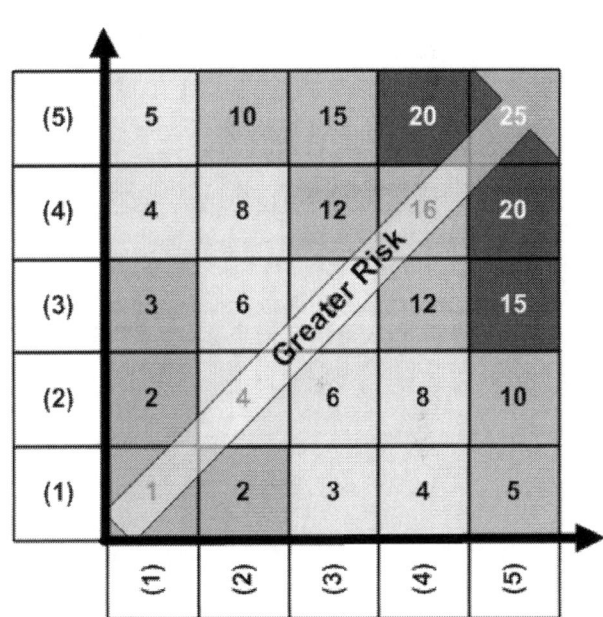

Figure 99: Stakeholder Matrix as a risk grid.

Stakeholder	Power Vs. Interest	Power Vs. Position	Power Vs. Uncertainty	Influence Vs. Impact	(Priority) Total
Project Sponsor	20	5	5	9	39
End User	2	4	3	10	19

10.3.4. Stakeholder: Management Strategies

As previously demonstrated with the stakeholder analysis example, a set of priorities will be produced based on the associated "riskiness" of each stakeholder. This prioritized list should form the basis for determining the ordering of the strategies to be used to manage these stakeholders.

There are three **facets to Stakeholder Management strategies**:

1) **Continuous bi-directional communications** with the stakeholders.

2) **Continuous collaborative effort to reconcile** the stakeholders' expectations, perspectives, concerns, "needs & wants" with the project's objectives.

3) **Continuous implementation of strategies** that are based on the principle of **maximizing the positive influences** on the project while, **minimizing those negative impacts** on the project's stakeholders.

By actively managing these stakeholders, the project manager is better able to **balance the providing of benefit to the stakeholders while still achieving the project's objectives**, and hence prevent or mitigate those stakeholder related risks & issues confronting the project.

10.3.4.1. Strategy: Bi-Directional Communications

The lack of frequent, "solid & well grounded" communications can be a catalyst for misunderstandings, misinterpretation, and mistrust amongst the project stakeholders. In addition, **a lack of communications can be a source of much angst and unhappiness.**

Ever noticed on the evening news when irate stranded passengers are interviewed about a major problem with some airline or the failure of the public transport system. ... More often than not; the passengers don't complain as much about the breakdown, but rather their anger is focused more at the lack of regular updates about what is going on. For them, even to be told that, "there is no new information at this time" is markedly better than no information at all. Hence, keep your stakeholders regularly informed, even when there is nothing new to tell them.

DO NOT under estimate the part that effective & efficient bi-directional communications has to play in managing the stakeholders' expectations & perspectives.

COMMUNICATE AS EARLY AS POSSIBLE AND AS FREQUENTLY AS POSSIBLE

How do I communicate effectively & efficiently with stakeholders?

Providing effective & efficient communications is built upon the topics covered in [Section 11.3] for the project manager roles as a communicator & listener.

Here is a short list of some of the communications actions that a competent project manager should do:

1) **Communicate clearly and openly** about the project; what are its objectives, its outcomes, its benefits (how will these exceed its short-term detriments), who will most probably be affected by it, and how will they be affected.

2) **Listen to the stakeholders'** expectations, perspectives, concerns, "needs & wants"... but especially to **their concerns**, as these concerns will **highlight** what these stakeholders **perceive as the risks / threats to them** because of the project.

3) **Acknowledge that these stakeholders have been heard**, that their concerns have been noted, and not simply pay "lip-service" to them.

4) **Communications** should be **tailored specifically for the targeted stakeholders** so that it **contains predominately the information that is relevant to them**, in a form that they can understand and utilize. ... *Not alien gibberish, that only frustrates.*

5) **Establish relevant communications channels** such that stakeholders are able to; voice their concerns, provide feedback, ask questions, seek additional information, and access the currently available information when they require.

NOTE ❏ How often and how exactly each of these stakeholders would be communicated with, would be recorded in the Stakeholder Register.

Though in reality, for small to medium size projects and organizations, I haven't seen the Stakeholder Register being used; rather, it is done by gut-feel and the use of established processes & procedures.

The Dangers of Over Providing Information

As part of the communications process, progress updates would be provided to the various project stakeholders; *e.g. WPR - Weekly Progress Report, PSR - Project Status Report.*

For these progress updates to be of any real benefit and to satisfy these stakeholder's expectations then it is essential that these updates be:

1) **provided by when** the information **needs to be known**, … *weekly, fortnightly.*

2) **contains** the relevant details of **what needs to be known**, … *known risks & issues, schedule progress (SPI), cost progress (CPI), expected milestones.*

3) **presented in a form** that can be **readily understood & utilized** by the targeted audience, … *suggest using a Key Performance Indicator (KPI) spread-sheet that is standardized across all of the organization's projects.*

4) **presented consistently** from one reporting period to the next.

However, there is an important caveat of exactly how much information should be provided to these stakeholders.

DO NOT provide these **stakeholders** (especially the customer's representative) **with more information than**;

1) **was agreed to be provided** (i.e. planned for),

2) **within the boundaries** of what they **need to know**.

Because the over-providing of information will most probably result in that particular stakeholder manifesting the expectation that such detailed information will be supplied henceforth.

Consequently, the ongoing production of this additional information will **unnecessarily over-burden the project manager and the project team**.

The over-providing of information can also encourage the customer's representative to try to micromanage the project, and more specifically the activities of the project team. Subsequently, this type of micromanagement would greatly limit the flexibility of the Project Team and the project manager to adjust the project's plans & execution to respond to the project's ever-changing situation and the prevailing circumstances.

> *For example; the customer responds with "But, I thought the plan was for the team to be working on activities ABC and not XYZ?", "Is there something going on that I should be aware of?"*

Thus, the **over-providing of information can be just as detrimental** to the project (and to the project manager) **as** providing **too little information**.

The following recommendation is not entirely ethical ... however, due to the "political necessity" of the current situation and the prevailing circumstances, it maybe necessary to tell a "little white lie" to one or more of the project's stakeholders.

I would recommend that ... **whenever a "non-truth" needs to be told then it should be wrapped in as much of the truth as possible.**

This way the non-truth will be easier to swallow by the recipient, and if the recipient happens to question this non-truth then it can always be neutralized by you "realizing" that it was a misunderstanding or an "innocent mistake" with the information that was presented.

"A LITTLE INACCURACY SOMETIMES SAVES A TON OF EXPLANATION" ... SAKI (H. H, MUNRO)

AND CAN GREASE THE WHEELS OF CONVENIENCE.

Though, be very wary of outright lying to the project stakeholders (especially to the customer's representatives) as this is a sure-fire way to poison the relationship with that particular stakeholder and their immediate bosses.

The Dangers of Email Communications

Some projects may have a dedicated Communications Management Plan, but often with small to medium size projects / performing organizations, this is forgone. Instead, the resources at hand form the foundation for the communications processes & procedures.

> *For example; email suites such as Microsoft Outlook (and its shared calendars) are used to schedule meetings based on when the relevant stakeholders would be available. Regular meetings with the relevant groups (e.g. the Project Steering Committee, the Project Working Group, and the Project Implementation Team) can be prescheduled well in advance of the meeting's occurrence.*

 I would recommend that ... in your Email application you:

1) Create a mailbox folder specifically for each project that you are involved with.
2) Within each project folder create subfolders for all email exchanges directly to stakeholders external to the performing organization; i.e. an "Incoming" folder (copy in relevant emails from the inbox) and an "Outgoing" folder (copy in the relevant messages from the sent folder).
3) Create an "Internal" folder for all email exchanges that are exclusively internal to the performing organization.
   That is, those emails not involving the customer organization.

 I would recommend that ... each project email message start with a subject line that clearly identifies the project and provides some indication of what the email is about. At the beginning of the body of the email quickly reference what project this email is related to and the topic to be discussed.

 Remember, **work emails (and memos)** can be used as legal evidence ... hence, **treat them as legal documents**. Therefore, **before you click 'SEND' rethink the contents and** to whom the email maybe read by, and especially **who exactly is being included in the 'CC' list.**

Work emails are not like personal emails. ... *Lest we forget.*

The mindset for constructing a work email is different to that for a personal email. In the majority of cases, with a personal email the communications is just between the sender and the receiver. Whereas, with a work email the audience is not just limited to the addressed recipient, but also to anyone else they may happen to forward the email; i.e. are they going to flick it to their boss, another department, or another colleague who needs to utilize the information contained within that email.

 I would recommend that ... the body of the work email should contain enough pertinent information so that a "forward" recipient without an indepth knowledge of the communications exchange can obtain an understanding of the essence of what is going on. This could even be taken a step further where the contents of the email is crafted specifically to preempt the other parties possible response.

RETHINK THAT NEXT CLICK ON 'SEND'
BECAUSE WHO ELSE IS GOING TO
READ THAT EMAIL'S CONTENTS?

10.3.4.2. Strategy: Collaborative Effort To Reconcile

As this section title reads, a collaborative effort to reconcile the project's objectives with the stakeholders' expectations, perspectives, concerns, "needs & wants".

To do this; **the strategies employed will have to "build the relationships" with the project stakeholders by involving their participation in the project, and** (where appropriate & practical) by also **involving these stakeholders in the decision-making process** (or at least considering their interests). Thereby resulting in resolutions that will be mutually beneficial to both the project stakeholder and the performing organization ... and potentially forming reciprocal arrangements.

NOTE   ❐ With any project, there will be a **few critical stakeholders who if their support for the project can be won then they will serve as a catalyst for winning over other stakeholders**. However, once these stakeholders' support has been acquired then the next battle is maintaining their support for the duration of the project / release milestone.

How do I reconcile stakeholders with the project?

Reconciling these stakeholders' expectations, perspectives, concerns, "needs & wants" with the project's objectives is built upon those topics covered in [Section 11.3] for the project manager roles as a negotiator, an influencer, and a conflict resolver.

...

IF you **understand** the project's **stakeholders THEN** you will **have an appreciation of how they will potentially react to changes** related to the project and to the project's outcomes.

Thereby, enabling advanced planning on how the performing organization will act & react to these stakeholders' expectations, perspectives, concerns, "needs & wants".

The following table highlights some of the strategies that should be considered for each quadrant of the Stakeholder Matrices depicted in [Figure 98].

Matrix	Quadrant	Stakeholder Strategy
Power Vs. Interest	**Monitor (minimum effort)**	• Don't invest too much effort into this group of stakeholders, because generally they are not really interested in the project or what you have to say. • Provide generalized "educational information" outlining the benefits & outcomes of the project so that they are aware that something is going on. • Monitor these stakeholders, and only react to them if they start drifting towards opposing the project or when the project needs defending.
	Keep Informed	• Keep this group adequately informed about the issues that matter to them. • Balance their interests against those of high-powered interested groups ... while making them feel as though they are contributing to the project. • Monitor these stakeholders to be aware that no major issues are arising with them.
	Keep Satisfied	• Provide fully to these stakeholders' informational needs, but don't provide excessive amounts of communications that could bore them or worry them, because they may react burdensomely. • Respond to these stakeholders' requests in a timely fashion, but don't be confrontational as you want to maintain their neutral stance. • Involve these stakeholders according to the issues that matter to them.
	Manage Closely	• These are the key stakeholders of concern, so deal with them both proactively and reactively. • Provide fully to these stakeholders' informational needs ... making all efforts to satisfy their needs so as to allay their concerns. • If these stakeholders so desire, then get these stakeholders fully engaged with the project.

Matrix	Quadrant	Stakeholder Strategy
Power Vs. Position	**Supporter**	• Provide educational information that encourages their continued support and involvement. • Monitor these stakeholders to be aware that no major issues are arising, and react to them if they start drifting away from supporting the project.
	Problematic	• Provide educational information that encourages these stakeholders' positive involvement ... potentially moving them towards supporting the project. • Monitor these stakeholders, and react if they start mobilizing their opposition to the project or if they start drifting towards supporting the project. • Be prepared to "negotiate & modify" the project's plans / objectives if these stakeholders start to organize opposition to the project.
	Champion	• Provide educational information that reinforces their support, and encourage their involvement. • Be prepared to put work into these stakeholders' relationships so as to keep them aligned with the project's objectives, as they're the project's allies. • Ask these stakeholders to highlight the benefits and importance of the project to those problematic & antagonistic opponents of the project.
	Antagonistic	• Need to "identify & understand" the nature of and the source of their opposition to the project. • Need to construct good working relationships with these particular stakeholders so as to mitigate against an "army of hostility" to the project. • Need to carefully monitor, and then counter any opposition they raise against the project. • Need to "negotiate & compromise" the project's plans / objectives where appropriate so as to "appease" these stakeholders.

Matrix	Quadrant	Stakeholder Strategy
Power Vs. Uncertainty	Few Problems	• Don't invest too much effort into this group of stakeholders, because generally they are not going to change from their current stance. • Provide educational information as to the benefits of the project's outcomes, and encourage their involvement when it is practical to do so.
	Unpredictable But Manageable	• Provide educational information as to the benefits of the project's outcomes, and encourage their involvement when it is practical to do so. • Monitor these stakeholders, and react if they start to either mobilize their opposition or support for the project.
	Powerful But Predictable	• Need to "identify & understand" these stakeholders' desires for the project's outcomes. • Need to "win over" these stakeholders early on in the project's life. • Need to give a high level of consideration to their "views & opinions".
	Unpredictable & Dangerous	• Need to "identify & understand" these stakeholders' desires for the project's outcomes. • Need to construct good working relationships with these stakeholders so as to ensure a "coalition of support" for the project instead of inducing an "army of hostility" against the project. • Need to "negotiate & modify" the project plans / objectives where appropriate so as to "appease" these stakeholders. • Need to carefully monitor, and then counter any opposition that these stakeholders may raise against the project. *"Keep your friends close, but keep your enemies even closer".*

Matrix	Quadrant	Stakeholder Strategy
Influence Vs. Impact	Information Gathering	• Provide educational information as to the benefits of the project's outcomes. • Monitor these stakeholders, and react if they start feeling they're being impacted on by the project.
	Consultation	• Keep these stakeholders adequately informed about the issues that affect them, and encourage their involvement so as to be aware that no major opposition is mobilizing. • Try to limit the negative impact on them, but not at the expense of high priority objectives.
	Information Giving	• Note these stakeholders' concerns, and respond in a reasonably timely fashion. • Look for these stakeholders' input into the issues that are of concern to them.
	Collaborative Dialog	• Need to "identify & understand" these stakeholders' concerns. • Need to give a high level of consideration to these stakeholders' views, opinions, and objections.

NOTE   ❏ **The more powerful the stakeholders that can be engaged to support the project** (especially during the early phases of the project's life) **then the more likely it is that the project will be successful.** This is because; these stakeholders will often have access to the [People] & [Resources] that are needed for the project to be able to succeed.

❏ Irrespective of whichever stakeholder strategies are chosen to be implemented, try to **fairly distribute who shall receive the benefits and who shall bear the burden.** As an unfair distribution (i.e. favouritism) could result in those disadvantaged stakeholders being transformed into unpredictable problematic & antagonistic sources of risks to the project being a success.

10.3.4.3. Strategy: Implementation

The **execution of Stakeholder Management strategies has to be prioritized based on the potential "riskiness" that is associated with each particular stakeholder.**

If opposition starts to arise with the more powerful & influential stakeholders then one must be prepared to negotiate, resolve conflicts, modify the Stakeholder Management strategies, and if necessary adjust the project's plans & objectives accordingly.

On occasions, the resolution of the stakeholder's expectations, perspectives, concerns, "needs & wants" will result in [Baseline] Change Requests. These [Baseline] Change Requests will have to progress through the formalized Change Management Process.

As these changes progress (i.e. as these [Baseline] Change Requests are being implemented) then it is **important to keep the relevant stakeholders informed of what is and is not going on.** Thereby, **keeping them "in-the-loop".**

If while implementing the project and/or the Stakeholder Management strategies; a "conflict of interest" arises or it is apparent that there will be negative impacts (legal, ecological, or sociological) on other parties outside the performing organization and/or outside the customer organization then ...

 I would recommend that ... as soon as practically possible, the project manager should make sure that senior management are aware of these concerns.

Also, it is in your own best interest to provide written notification of these concerns to your senior management.

 **Be very wary of** stakeholders (especially inside your own organization) who place **short-term profitability** and/or "political gains" **ahead of the potential long-term viability** of either the performing organization and/or the customer organization. Refer to [Section 3.5] for things to watch out for, as these could wreck the project's chances of success.

10.3.5. Stakeholder: Monitoring & Control

Now that the stakeholder strategies have been created for the project's identified stakeholders (and their details have been updated in the Stakeholder Register).

Yeah right, from my personal experience that never happens.

The Stakeholder monitoring & control steps are:

1. **Periodically monitor** these stakeholders.

2. **Evaluate** the **effectiveness** of their associated stakeholder strategies.

3. When / where necessary make appropriate **adjustments to** their associated stakeholder **strategies**.

4. **Identify** when new stakeholders have appeared and when stakeholders have changed their "riskiness" priority.

If these stakeholders were not managed correctly (or not managed in a timely fashion) then their expectations, perspectives, concerns, "needs & wants" could evolve into risk & issues for the project. ... *So, you have been warned.*

NOTE ☐ **When faced with pending failure, people will (openly or covertly) revert to what they know works best for them**. Therefore, try to figure out if and when they believe a failure is happening, and if possible try to determine what their "revert" response would entail.

IT IS THE COMPLETE PROJECT TEAM, AND THE INDIVIDUAL MEMBERS OF THAT PROJECT TEAM WHICH MAKE OR BREAK THE PROJECT, AND CONSEQUENTIALLY ... WHAT MAKES OR BRAKES YOUR CAREER AS A PROJECT MANAGER.

SO, DON'T TAKE THE "LITTLE PEOPLE" FOR GRANTED.

11. MONITORING & CONTROL Phase ... Theory Part 6

Watching the Drivers

While all of this monitoring & control of the project variables and project constraints is essential to the successful outcome of the project, there is another aspect of monitoring & control that must be continually examined by the project manager. This aspect is the (internal and external) human relationships involved with the Project Implementation Team and the project's participants.

However, this chapter will not be re-examining the [People] project variable [Section 9.1], nor will it be considering Stakeholder Management [Section 10.3]. Rather, this chapter will focus on seeing "The Team" as composed of "human persons" and not simply as an assignment of "human resources".

Potentially this humanitarian aspect may be a bit of a change of perspective for you; especially if you have come from a purely technical background where you possibly never had to really deal with the management of "interpersonal relationships".

The important things to remember about this humanitarian aspect is that;

1) It is the **people who transform the project from mental concepts into physical deliverables**, and

2) These **people are NOT machines**, and **they should never be treated merely as interchangeable parts**.

   IMHO, to try to treat them as numbered resources and to attempt to micromanage them will limit their true potential and subsequently restrain your managerial capabilities and the organization's capacities.

11.1. Project Manager: A Humanities Perspective

As stated back in [Chapter 5] for the Executing Phase, *"being a project manager is not like being a team leader"* ... the big difference is the ... *"change from the hard technical skills of an implementer to the soft inter-personal skills of a manager"*.

 The tragic truth is that; **good technical skills and implementation prowess does not necessarily translate into project management competency**.

Actually, what made you an outstanding implementer may instead make you a dismal project manager.

There is more to being a project manager than making sure that the [People] have completed their assigned tasks on [Time], within the budgeted [Cost], have produced [Quality] deliverables, and driving them to work harder when they have fallen behind.

 **DO NOT consider the project hierarchy as a pyramid where** you (as **the project manager**) **sit at the top** issuing decrees to your minions down below.

Rather, consider the project manager to project team arrangement as a **flat hierarchy** (possibly involving some form of mentoring) where the project manager takes care of the coordination / management aspects of the project's life cycle, while **"The Team"** takes **care of the technical aspects**.

It is beneficial that the project manager be the project team member who handles all of the non-technical matters such as responding to the stakeholder needs & wants, and deals with the inter-stakeholder politics.

Here is another way of thinking about this. ... Imagine a planet with its moon and satellites. You as the project manager are not the planet and the team is not the satellites, but rather the team is on planet "Project" and the project manager is in the star-ship patrolling the space around this planet.

Your mission as the ~~project manager~~ "Planetary Protector" is to monitor & control the interactions with those on the planet; filtering the communications to & from the project's primary-core stakeholders (*e.g. the behind the scene "politics", alternatives being considered, and future changes to the project*), shooting down those risks & issues which could potentially endanger Planet-Project or inhibit the inhabitants from successfully doing their tasks. Thereby the project manager's role is to shield the project team; only allowing through to the planet those things that they need to know, and ensuring that they do receive those things that they require to successfully complete the project.

HOUSTON...
WE'VE GOT NO
PROBLEM HERE.

11.2. Project Manager: has to Earn Respect

Have you ever witnessed how dysfunctional & ineffective an organization can become when promotions are based more on "who you know and not what you know", or "he who has been there the longest is next in line for promotion". Just look at a business where the owner's sons & daughters are all of the department heads, or the director / vice-president of this & that. Often these businesses (by the 2nd or 3rd generation) become self-indulgent & non-adaptive because all of those people with potential (but not having the right connections nor the advantageous background) eventually leave to pursue other opportunities in the hope that their skills & knowledge will be better appreciated, and subsequently to be rewarded for their actual contributions to their employer.

 **DO NOT expect to receive the respect of the project team just because the title of Project Manager has been embossed on your business card**. To obtain the respect of the project team will require you as **the project manager to be worthy of their respect**.

How do I earn their respect?

There are many and varied ways of obtaining the respect of the project team, though here are just a few:

1) **Know Your Stuff ...**

   That is, know what you are talking about; to this end:

   - **Know the basic principles & concepts behind project management, and be familiar with the implementation methodology being used** by the project team. This does not necessarily mean that you have to be certified in this or that, but it does mean that you understand how & why the project is to be undertaken in specific ways, and the "relevant" project management practices to be used.

     Sometimes this can be very difficult to accomplish if the performing organization (and/or the senior management) have no formulated project management methodology per se; rather, operating via the philosophy of "make it up as we go", or management by "telepathy". Introducing some forms of recognizable project management methodology (especially EVPM and KPIs) into such an organization could turn out to be an uphill battle with some conflicting opinions and even dogged pushback on the validity of such approaches.

   - **Know the basic details of the project** that you are supposed to be managing. This means knowing; the overview of the project, the assumptions made about the project, the scope boundaries of the project, the constraints imposed on the project, the priorities associated with the project, the generalized requirements of the project (or at least where to quickly find these and extract the relevant information), the project's deliverables, and the relevant dates for the project's milestones.

     That is, **become very familiar with the Project Charter, Detailed Specifications, and Customer Requirements**; see [Section 3.3].

- **Know the basic details of the architecture of the project's product / system / application that is being implemented.**

  This does not mean that you have to become a subject matter expert, but it does mean that as the project manager, you are able to draw a relatively simple block-diagram / sketch explaining how it all fits together and more specifically you are able to point to what sections certain project team members (or sub-teams) are working on and are responsible for.

- **Know the basic details of the technologies & processes behind the project's product / system / application that is being delivered.**

  Again, this does not mean that you have to become a subject matter expert, but you will need to at least have some understanding of what the project team members (and the project stakeholders) are talking about. ... Instead of, you as the project manager, looking and sounding like a "complete moron" who knows nothing about what is going on and what the project is all about.

But, I've been told that **a project manager doesn't have to know anything about the technological aspects, nor know anything about the technical processes involved.**

War Story *... You gotta know what you're talking about.*

I know of a customer whose senior management requested that the project manager be reassigned off of "their project" because the more that the project manager spoke then the more apparent it was that the project manager was "talking out of their ass" and did not know nor understand the basics of the technology that underpinned the project. Hence, they strongly insisted on a replacement who was "a lot more technically savvy".

Honestly, how can the project manager be sure of the task estimates that are being provided by the implementers (more specifically by external contractors), if the project manager doesn't reasonably understand the technologies & terminologies, and the processes involved with that particular project?

Similarly, a project manager who continually gives the implementers too short durations to complete tasks, when someone with a bit of knowledge about the subject matter would know that these durations were ridiculously too small. If the project manager (due to their lack of technical understanding) continually underestimates or grossly over estimates task durations then this will result in the implementers considering the project manager as a "Jerk" (or more colourful language meaning devoid of intelligence).

A manager who (due to their own technical knowledge inadequacies) responds to the slightest technical inquiry with a signature brush-off line such as, "So what's your plan", will downright infuriate the implementers, and thereby only gain their absolute contempt.

YOU CANNOT HOPE TO MANAGE THAT WHICH YOU DO NOT KNOW OR REASONABLY UNDERSTAND.

2) **Know what is going on ...**

That is, **being aware of what is happening in the project team, around the project, and within the performing organization**. The team will not be too appreciative if any nasty surprises befall the project or them. Therefore, keep a watchful eye on the risk & issues confronting the project. This will mean that you as the project's manager will have to **learn to "network"** with other parties within the organization.

3) **Passionate ...**

The team needs a project manager who is **enthusiastic about the project, about the project team**, and about the project's **chances of a successful outcome**.

4) **Be a good first date ...**

That is, **show some interest in the team's personnel**, especially when they or you are new to the project (or to the organization).

- With your new project team together; **introduce** yourself, give a brief background to your work history and what relevance it has to "their project" that you will be managing. That is, try to **build up their confidence** that you know what you are doing, and demonstrate that you will quickly get up to speed on what you don't know.

- In small groups or in one-on-one individual sessions find out; what are each of the team **member's names** (remember this and write it down on a mud map of where they usually sit), what is **their role** (within the project and within the organization), where do they reside in the overall structure of the project, what part of the product / system / application architecture are **they working on or are responsible for**. Partake in "**small talk**" and get to know about them (their background, their interests, and pick up on their home life situation ... "*do they have kids*"). Hence, **show some interest in them as "individual persons"**.

5) **Give praise when praise is due ...**

That is, when a team member does a good job, is working hard, is working competently on work outside their skills / knowledge base, then a **"pat on the back"** simple **acknowledgement of their efforts** will go a long way. Though, ensure that this positive feedback is provided **as soon as possible after the performance occurred.** However, **DO NOT give praise out so readily, or for mundane / expected achievements as this will diminish the impact of that praise**.

I once had a manager who the only time he personally communicated with subordinates (other than by email decree) was to bark out orders or angrily chastise someone for something he perceived that they had done wrong. He never showed any sign of appreciation. ... What a "Despotic Jerk".

6) **Give credit where credit is due ...** *And, DO NOT steal other's recognition.*

That is, when the project team does well (especially when this performance is noticed by senior management and/or by the customer's representative), then **ensure that the recognition goes to the team or the relevant project team members**. Also, let senior management know of the good performers, and let the good performers know that their efforts have been noticed by senior management and/or by the customer.

Never ever, steal the credit from your team members.

Because, if they find out about your pilfering then in retaliation they could slacken off in their efforts, complain about you as the manager, or quit the team (and potentially the organization).

I and these other guys once worked our butts off on this project and at the company sponsored thank-you lunch, this sales manager we had never seen before stood-up and made this speech like he was the one who single-handedly drove the delivery of this project. We were not impressed to say the least. ... What a "Credit Stealing Jerk".

7) **The buck stops here ...** *And, NOT throwing them under the bus.*

That is, while the project manager should pass the credit down to the team, the project manager should **never pass the blame for the project's failures onto the team, nor single out individual team members for blame**. Remember that; it is the project manager who is responsible for the project's execution, and hence it is the project manager who is ultimately accountable for the project's successes & failures.

Never ever, blame your team members for the project's failure.

Because, they were only implementing "your plan", and if they were not implementing "The Plan" then what were you as the project manager doing, cause evidently it was not managing them.

---

8) **Bearer of bad news and defender of the truth ...**

A project manager has to be **prepared to give bad news about the project as soon as reasonably possible**, so that something can be done to resolve the issue, and **defend the team when they are being unjustifiably persecuted for such bad news**.

EXPERIENCE IS SOMETHING YOU REALLY GAIN
ONCE YOU HAVE FAILED ... HOWEVER
IF YOU ARE TOO AFRAID OF
THE PUNISHMENT FOR FAILURE
THEN YOU WILL NOT TRY THAT WHICH
COULD POTENTIALLY RESULT IN FAILURE.

9) **Be a project manager not a project accountant ...**

There is more to being a project manager than dealing with the "hard skills" of; ticking off each line in the Requirement Traceability Matrix and the Detailed Specifications, marking 100% complete for each task in the project's schedule, or getting every digit correct in the earned value calculation.

⊠ **The project manager is a manager of relationships.**

The project manager needs to master those "soft skills" of dealing with human beings. Such skills as interpersonal communications, showing empathy & concern, leading & motivating, negotiating & influencing, problem & conflict resolving.

 **DO NOT be a "hard skilled" project accountant; instead become ... a "soft skilled" project manager of relationships.**

A quart of hard skills and a gallon of soft skills.

---

10) **Don't be a pushover, rather stand-up for the team ...**

A continual "yes-man" project manager does not serve the best interests of the project team. The project manager who is too afraid to question or disagree with what is being said & proposed by senior management (or by the customer representative) even when there are justifiable reasons to the contrary is a detriment to the project's success. This inability to occasionally be a "why-man" and even a "no-man" **will result in the project team continually being committed to losing situations.**

No one wants to be on a perpetually losing team.

11) **Be confident, not arrogant ...**

No one wants to work for a "Wimpy Loser". Hence, the project manager needs to **present a demeanour of confidence in one's own abilities and in the team's abilities**. However, **NOT** to the point of being perceived as **arrogantly overconfident**, because this type of personality can lead to making decisions & committing to things that should never have been agreed to. Additionally, the arrogant manager is more likely to ignore the input, opinions, advice, and recommendations from the "little people" (i.e. those considered subordinate to their own titled-position).

12) **Steadfast but Flexible ...**

The project team will quickly lose respect for a project manager who continually changes the project's direction at the "slightest breeze"; *e.g. distractions, disruptions, possible scope changes, unrelated business politics.*

The project team needs the project manager to be **flexible in adjusting** the project's **direction when the current situation and the prevailing circumstances require, but stubborn enough to continue on course** to achieving the project's objectives.

We once had a "headless chicken" of a manager who bounced us around from one task to the next ... "Work on this, no-no-no work on that". We ended up totally ignoring the "twit" and worked on what we thought was appropriate to get the project finished.

13) Be consistent and non-preferential with how people are treated ...

The project manager has to be **uniform with the ground rules for acceptable behaviour** that is applied to and expected of all project team members, and not one rule for this person & another rule for everyone else.

That is, **no apparent favourites among the project team members**.

14) Be a good listener, not a know-it-all ...

No one wants to work for someone who thinks they know everything and who doesn't listen to anyone (aka *"I'm the smartest person in this room"*). Because, after a while of having their input and past experiences continually ignored, many a project team member will start to question why they should even care about the project let alone bother communicating with the project manager. Once this communications breakdown has occurred then problems, risks & issues confronting the project can go unmentioned which could result in some "arrogant know-it-all" project manager being splattered by the proverbial bus; see [Section 4.3.4].

11.3.   Project Manager: has Many Faces

For the project manager to be able to successfully monitor & control the project's human relationships, **the project manager must firstly "know thy self"**.

Secondly, the **project manager will need to learn to be versatile & capable of** wearing many different ~~faces~~ hats when the situation requires. This not only entails **performing different roles but also being able to see the varying perspectives of all parties who interact with the project** (both internally & externally to the performing organization).

To be successful the **project manager must be; a communicator, a listener, a negotiator, an influencer, a conflict resolver, a problem solver, a planner, an organizer, a delegator, a motivator, a leader, and a team builder**. ... But, no longer an implementer.

11.3.1. Project Manager: is a Communicator

As previously stated back in [Section 2.6.2], "*a project manager will spend 80-90% of their time communicating in both written and oral forms*".

Therefore, the project manager has to be able to communicate effectively & efficiently with all of the project's stakeholders and especially with the project team.

COMMUNICATIONS IS A TWO-WAY STREET

 For more information on communications, please refer to [Section 4.3.4].

 I would recommend the following as some simple techniques for being a "good communicator":

1) **Make your message relevant to the receiver**, present it, and speak it in a language that is **appropriate to the receiver's level of understanding**. If they are confused by its complexity or bored by its simplicity then they will just stop listening as their thoughts wonder off to better things. … *Hmm, what's for lunch?*

2) **Layer your message**; start with a **simple summation of the outcome or major points**, then (if required / requested) **fill in the details**, and **conclude with the recommendations or actions**.

3) **Repeat and summarize the key points** as you go. You know the topic at hand so summarize the key points at the end of each major section so that they note what is important (IMHO).

4) **Request feedback and questions**. This way, you can cover in more detail areas of concern to the message receiver.

5) If they did not understand your message, it is not their fault. You may have to **repackage how you present the information**.

6) If you are presenting information to senior management, remember that their available time is in demand, therefore your **timing and message duration is essential**. So, keep your **message efficient & relevant to their "wants & needs"**.

11.3.2. Project Manager: is a Listener

Effective communications is bi-directional, where **someone is** telling (**transmitting**) **while the other party is** listening (**receiving**), then vice versa, and versa vice.

While the speaker's role is to effectively & efficiently transfer the information, the listener's role is to **actively take in the information, and then provide some form of feedback to indicate that the information was received & understood**. Subsequently, the project manager needs to demonstrate that they are an "active listener" who takes in what was said by the other party and **then (most importantly) evidently acts upon this information**. *... And, reinforcing this listening via active feedback.*

 I would recommend the following as a few simple techniques for being a "good-listener":

1) Take a notepad and **jot down bullet point notes** as they speak. This has a few advantages; they can see that you are taking notes therefore they perceive that you must be listening to them, by taking notes you can adjust how fast they will talk because if they are an observant speaker they will try to let you catch up with what they are saying, by taking notes you are actively doing something so that you are not likely to glaze over, and by taking notes you don't have to constantly look them in the eye (if your personality is uncomfortable with continual eye contact).

2) **Do not interrupt the speaker** during their message as this can disrupt their chain of thought and this may confuse their message. However, if it is a long monologue then at the obvious end of a topic, interrupt them with the equivalent of "*excuse me, so what you are saying is ...*" followed by a quick summation of your notes. This firstly breaks up the monologue, and secondly this forces you to pay attention to what they are saying because you already know that you will be mini-testing yourself along the way.

3) **Ask them questions** about that which you don't understand.

4) Whenever possible **act on what they have said**. This way, they will perceive that they were really being listened to.

11.3.3. Project Manager: is a Negotiator

The reality is that; not every situation can be "win-win", and therefore the project manager has to be able to **work towards a "compromise" solution that is the best possible outcome for the involved parties** (and not just "winner-take-all" for a limited few).

To be a good negotiator involves:

1) Trying to **understand the situation that is causing the problem.**

    This means trying to see the situation from all parties' perspectives.

    "You will never understand a person until you have walked a mile in their shoes", or "carried some of their burden upon your own back".

2) **Determine what each** of the involved **parties** *(really)* **"wants",** and more importantly, **what each involved party** *(really)* **"needs".**

    NOTE ☐ **"Wants" is the high point of what the party is after.**

    ☐ **"Needs" is the low point of what that party will accept to be satisfied.**

    If all involved parties "needs" are at least being met (in some way) then a "compromise" solution is potentially achievable.

    ☐ *"Greed" ... to want more than one could possibly need.*

    DO NOT discount good old fashion "Greed" to hinder the negotiation towards an amicable resolution.

3) "Play the ball, not the player" ... that is, **concentrate on the issues & concerns** of the involved parties and **DO NOT focus on the emotional position that they are taking.** This is because, by focusing on their issues & concerns, one can obtain an understanding of their "needs" rather than be diverted by their "wants".

    ⊠ **Satisfying stakeholder "needs" is a key to success.**

4) Realize that people's **personal characteristics, ethnicity, and cultural background will influence** how they will **approach any negotiations**, and more importantly **how far they will be prepared to compromise**; *e.g. "a need to save face".*

5) **Ensure that all involved parties are receiving a "fair-deal",** because if one or more parties feel that they have been taken unfair advantage of, then the negotiated agreement / arrangement will not last very long. Subsequently; everyone will be dragged back into negotiations, however this time round, everyone will probably be in a lot less malleable mood because their hands (i.e. advantages & disadvantages, strengths & weaknesses) will have already been revealed during the previous rounds of negotiations.

6) "Haggle" ... all negotiations involves some form of bargaining.
   DO NOT make a concession without at least appearing to be giving up something of value (in the other party's opinion), because this will help form an unwritten obligation for the other party to also give up something of value in return.

   "Come on man, your killing me at these prices".

7) "Don't be a pushover" ... DO **NOT be the party who is continually surrendering their "needs & wants" in order to try to "reach a compromise"** because the other party is likely to continue taking until there is nothing worthwhile left to give.

   "Give an inch, then another inch, and another ... and the next thing you know is that you'll have given away a mile".

8) "Don't overplay your hand" ... **DO NOT over sell your position,** because if you **are caught out** then you may be **perceived as being in a weaker position** than you actually are. As a result of this perception, the other party may think that they themselves are in a stronger position and hence they may be a lot less willing to compromise because they are confident that you will be forced to give in first.

9) Be a **good Listener & Communicator.**

11.3.4. Project Manager: is an Influencer

A good project manager is able to affect people and events without necessarily being directly involved, as though by using some unknown cognitive force.

Okay, we all can't be Jedi-Masters of Influence, but here are a few useful techniques:

1) **Know what you want to accomplish ...**

   You cannot redirect them towards your objectives unless you understand what you are actually trying to achieve:

   - **See your perspective** – that is, getting them to understand & acknowledge your view of reality, that they are going to give their best while working on the project (in accordance with the Detailed Specifications).

   - **Modifying their behaviour** – that is, getting them to change their behaviour, **based on your perception of its desirability or undesirability**.
     This is done by using positive enforcement to encourage the desirable behaviour and ~~negative reinforcement to~~ discourage the undesirable behaviour.

     IMHO, if you have to resort to negative reinforcement then you have lost the capability to sustain long-term influence over them.

2) **Know what they "need" and what they "want" ...**

This is a similar idea to negotiation; you cannot redirect their self-interests towards helping you to achieve your objectives, unless you understand **what is in it for them**.

- **Rewards**; i.e. self-esteem, self-confidence, sense of achievement, praise & gratitude from others, recognition & respect.

- **Gains**; i.e. financial, material, experience, knowledge.

- **Safety**; i.e. job security, stability of employment, and belonging to a group.

This sounds like **Maslow's Hierarchy of Needs.**

Having determined what their "needs & wants" are, then your **message can be better tailored towards matching these**. Thereby, being able to emphasize "what is in it for them", rather than "what is in it for you" (the project manager and the performing organization). In addition, by helping them accomplish their "needs & wants" they are more likely to reciprocate and help you achieve yours.

3) **Look, listen, learn, and ask ...**

How else can you know their "needs & wants" unless you pay attention to what they are saying, and are not saying? You will also have to observe what work and social life constraints would prevent them from being able to do what you require them to do. Therefore, when the time & circumstances are right; state your proposal, actively listen to their responses, ask them questions, and then actively listen to their answers.

Hence, to be a good influencer requires the project manager to, firstly be **a good observer, a good communicator, and a good listener**.

4) **Listen to and be open to counter proposals ...**

Sometimes to get what you "want", you will have to **re-evaluate and re-adjust your proposal based on their responses and counter-offers**. Noting that, it is easier to get someone to do what you "want" when they are involved in the decision making.

5) **Keep focus ...**

That is, **keep their attention on what you "want" them to do**, and get them to **commit to a date** by when they will do that which they have agreed to do.

However, this is a two-way-street; so if you have made a commitment to do something for them then it is advantageous for you to do exactly that (by when you said it would be done), and thereby create a reciprocal expectation on both parties.

6) **Communicate openly, clearly, calmly, and confidently ...**

- No one will bend to your will if they do not clearly understand what you are wanting of them, so **precisely** inform them of exactly **what you "want" (in terms that they will understand)**.

- **Without ambiguity and multiple interpretations** ... *"was that delivery on the 1st of January, or was that delivery on the first working day in January, given the Christmas and New Year's holiday break"*.

- **Calmly & confidently**, state what you "want", without doubt or hesitation.

  "This is not the droid you are looking for".

7) **Do unto others, as you want them to do unto you ...**

That is, before you can get someone to do what you "want", they must at least "tolerate" you and may even "like" you.

If you treat them badly or, are unfriendly towards them then there is not much chance you will get them to voluntarily do what you asked, let alone to the standard that you expect.

Yep, if they "hate your guts" then they will keep putting it off for as long as they can; and, when they do get around to doing it, they probably will only put in a half-hearted effort. ... Well, wouldn't you?

"You catch more flies with honey than with vinegar".

So remember that an influential project manager is:

✓   polite & diplomatic, courteous & tactful,

✓   sociable & friendly, generous & kind,

✓   concerned with the plight of others,

✓   shows respect & gratitude,

✓   gives others their undivided attention,

✓   shows appreciation for others efforts,

✓   strives to resolve the problems at hand,

✓   shares one's victories and successes with others,

✓   compliments people and praises them when they do a good job,

✓   when possible takes a minute to give them a helping hand,

✓   When necessary goes out of the way to help them.

An influential project manager is NOT:

✘   anti-social, snobbish, arrogant, know-it-all,

✘   mean, nasty, unkind, insensitive to other people's predicaments,

✘   puts people down, belittling, degrading, intimidating, bullying, threatening,

✘   back-stabbing, bitchy, rumour monger,

✘   steals the credit from others,

✘   passes the blame onto others,

✘   Is selfish, show-off, look-at-me,

✘   Ignores others problems.

✘   Ignores the fact that a problem exists

And, does not manage by avoidance.

8) **Mummy, Daddy, I want ...**

That is, children are brilliant practitioners of getting their own way by; directly stating (strongly) what they "want", then persistently asking for what they "want".

Many a parent has given in rather than to continue resisting such repeated onslaught.

Admittedly, this is a childish tactic but obviously, it does work. Ever noticed how the little darlings try to choose a time and place where you are either unprepared or will quickly answer with the response they are after (all too just shut them up).

For an adult the lessons learnt are to; **be persistent, state clearly what you "need", what you "want", and choose an appropriate time to ask.**

 **DO NOT manipulate others by devious means**, because when a person feels that they are being wangled into doing something that they would not normally want to do then they will eventually resent doing such a thing and other subsequent things.

DO NOT dictate and reframe from micro-management, because a bit of freedom will help in satisfying their own "needs & wants".

---

9) **Time your criticism ...**

Occasionally the person you have influenced may not have done exactly what you have requested, or they may have done this poorly. In such cases, negative feedback will be necessary. However, **how this negative feedback is administered will greatly affect how much influence you will be able to exert on them in the future.**

A misdirected or inappropriately timed negative feedback (given their situation and their prevailing circumstances) **is a sure-fire way to lose** their respect and reduce their willingness to give the criticizer that extra bit of effort in the future.

Also, criticism is a two-way street ... so you unfairly criticize them, what's to say that they ain't now criticizing you to their workmates.

10) Criticism that can be invested ...

That is, if you must criticize someone then do it in a **constructive manner**, so that the recipient will **learn from the experience** and not be resentful or defensive.

- **Always make sure that you are "absolutely correct" before unleashing your criticism on someone.** Make sure that you have all of the facts and that you **understand the situation and the circumstances** before you give your criticism.

THE BLACKENED POT DOES NOT GET THE RESPECT OF THE ABUSED KETTLE.

- **Never criticize something that cannot perceivably be changed** (i.e. you cannot criticize a cat for not being a dog), or because they didn't perform as well as a certain other person would have / did.

  How much does a child hate being negatively compared to their sibling? ... "Oh why, can't you be more like ... ".

- **DO NOT criticize them for** dealing with an issue / problem the way that they did if that issue / problem resulted from **your own failings** to do something.

  If they made multiple requests of you (as their manager) to perform some action and/or to provide them with the appropriate resources, but **you failed to deliver in a timely manner, then it is not appropriate** for you **to reprimand them for taking the initiative to do what needed to be done** to complete their assigned tasks.   *... Especially when you, told them to "sort it out yourself".*

- **Never "badmouth" theirs or your predecessor's performance** (i.e. the person who previously did theirs or your role) as this will raise suspicions about what you are currently saying "behind their back" about theirs and others performance.

  Let alone what negative things you will be saying about them in the future. ... Bad mouthing the departed is a sure-fire way to lose the respect of those who remain, and marks you out as a poor manager.

- **Never turn the criticism into a personal attack** on their character.

  Stay with the facts and be precise with the criticism.

  DO NOT deviate off-course into an attack on their personality (or worse their **culture, ethnicity, religion, believes, sexuality, or gender**).

- **Criticise as soon as possible after the event that initiated the criticism.**

  Though, limit the amount of criticism that is dispensed at any one time or is dispensed over a relatively short period of time, else you will be perceived as someone who cannot be pleased (or worse, you will be perceived as a bully).

- **Get straight to the point** … be very clear, precise, and honest about the criticism.

  DO NOT waste time "buttering them up" and

  DO NOT skirt around the crux of the issue, afraid of saying the truth.

- **Be calm & clear,** because being frustrated, angry, or out-right hostile will only make the criticism's recipient defensive to whatever has to be said.

- **Be serious the entire time** that the criticism is being given.

  Criticism is NOT a joke; to treat it as such will be interpreted as not showing real concern.

  DO NOT be sarcastic about the issue, as this will make you appear as an "Insensitive Jerk".

- **Let them respond to the criticism**, so that they can "tell their side of the story", as there may be key information that is missing, or the situation & circumstances may justify why they did what they did.

- Having dispensed the criticism, now **make it clear what is expected of them**.

- Finally, if **punishment** must be meted out then ensure that this punishment is **proportional to the crime** that they have committed.

  Not too lenient, not too harsh, but just right. And, remember that, criticism is a two-way street … so, who is also criticizing you.

11.3.5. Project Manager: is a Conflict Resolver

Whenever two or more people are involved with something, there will always be some form of conflict. Generally this conflict will stem from their individual "needs & wants" (for resources / territory / priority), for acceptance of their opinions, over principles or believes, due to poor communications, misinterpretations, or misunderstandings.

Surprisingly, often the simplest solution to conflict resolution is to **"listen" to the parties involved, "communicate" back** to them your interpretation of what they have said, and then **ask "questions"** about the situation and the circumstances of the issues. Sometimes, just the **"airing of the grievance" is enough to initiate the Conflict Resolution Process**.

Let each involved party state their case without being interrupted, unless it is to politely tap them back on course because they are going over the same point again & again; e.g. *"I understand what you are saying, is there anything else you wish to add?"*

Also, ensure that the "discussion" concentrates specifically on the **current issues** at hand and **does not degenerate into personal attacks, nor encompasses ancient history, nor misdeeds from the past**.

Show some indication to all of the involved parties that you understand what they have said (even if you happen to disagree with their points of view); e.g., *the occasional head nod, and "hmm, yes, I see".* Then ask them how this issue can be resolved.

The solution for how to satisfactorily resolve the issue may now be apparent, given that all parties have had the opportunity to express their viewpoints, and air their differences.

Thus, **conflict resolution has to be undertaken in an open constructive manner**, using those techniques that were previously laid out for influencing & negotiating, as well as listening & communicating.

Nah, all conflict resolution can't be that simple?

No, not all conflicts are that easy to resolve; however, there are other techniques that can be used for conflict resolution [Ref: Thomas & Kilmann], these are:

1) <u>**Withdrawing / Avoiding**</u> **– is where one of the conflicting parties retreats from a real or potential conflict / disagreement**.

   This technique deploys tactics such as; not bringing up the issue, sweeping the issue aside by changing the subject, or saying it will be dealt with later (though fully intending not to do so).

   "Let's discuss that tomorrow", but we never do.

   These tactics can thus be used to; postpone the conflict until one is better prepared to handle the situation, cool-off the parties involved (especially the aggressor), or simply to not have to deal with such a trivial issue.

   "He who runs away, gets to fight another day". ... However, "left to themselves, things tend to go from bad to worse".

   If used inappropriately or continually this Withdrawing / Avoiding technique can result in bottled up tensions, and conflicts simmering away beneath the surface of the relationships. It could very well boilover into a larger conflict than those that were being avoided in the first place.

2) **Smoothing / Accommodating – is where one of the conflicting parties forgoes their own "needs & wants" so as to oblige the other conflicting parties.**

This technique deploys the tactics of modifying or completely changing ones position or opinion to coincide with the other parties; possibly aligning with the consensus, the subject matter expert, the superior, or the dominant personality.

If used inappropriately or continually this Smoothing / Accommodating technique can result in the obliging party being seen as weak and a pushover. In addition, this technique may conceal the underlying cause of the conflict between the involved parties. This technique could also be used as an extension of the Withdrawing / Avoiding technique.

3) **Collaborating – is where the conflicting parties work together to find a "win - win" resolution that is mutually beneficial to the involved parties.**

This technique involves; having open discussions about the conflict and its impacts on their respective objectives, with all involved parties contributing to the discussion and the proposal of alternative solutions, and finally common agreement on the chosen solution.

If used inappropriately or continually this Collaborating technique can result in the involved parties being bogged down in "design by committee" which may not be able to achieve much in the available time.

"When a committee designs a horse, the outcome is a mule".

4) **Forcing / Dominating / Competition – is where one or more of the conflicting parties adopt an assertive posture where their "needs & wants" are placed before the other parties**; potentially to the other parties' detriment due to a "win - loss" resolution.

This technique is based on capitalizing one's; authoritarian position within the management hierarchy, influence within the organization, dominance in the field/market, superior subject-matter knowledge, or simply by aggressive behaviour.

If used inappropriately or continually this Forcing / Dominating / Competition technique can result in animosity between the conflicting parties, resentment by the "losing" parties, a feeling of superiority & dominance by the "winning" parties, a feeling of disempowerment by the "losing" parties, reduction in morale & motivation, and an antagonistic atmosphere amongst the involved parties.

This technique should only be used to get things moving along when a quick decision is essential, such as during a crisis event.

"All tyrants eventually get strung up by their subjugated peasants."

5) **Compromising** – **is where the conflicting parties work together to find a "no win – no lose" resolution that is mutually satisfactory to all concerned.**

This technique involves; discussing the problem (possibly with the involvement of a neutral third party as mediator), each party gives up some portion of their "needs & wants", bargaining & negotiating occurs between the parties, and finally the parties come to an agreement / arrangement / understanding.

"A compromise is where everyone gets something that they are not truly satisfied with."

This compromising technique should be used when the conflict is escalating but there is still time to come to a resolution.

As the conflict resolver **DO NOT** (appear to) **take sides**.

Rather, **be a "resolution facilitator"** and act as an **intermediary for communications** between all of the involved parties.

How will I know which technique I should use?

The choice of conflict resolution technique is dependent on:

- **How intense is the conflict?**

  Is it, light-hearted jousting or is someone about to get a gun?

- **How relatively important is the conflict?**

  Is it, a battle over the last pizza slice or is it a fight over the last life-vest on a sinking ship?

- **Is the conflict drawing in unrelated parties?**

  Could you make some money taking side-bets from the onlookers or is it about to turn into a mob riot?

- **Will resolving this conflict achieve anything beneficial in the long term or short term?**

  Is it, just posturing & positioning in the group?

  Will this competition & rivalry be good for innovation and advancement towards achieving the business's objectives?

I would recommend that ... the project manager should **attempt to resolve conflicts within the project team as soon as possible, preferably in private and confidentially.**

Also, while the project team and the project manager may know that the conflict is nothing more than friendly mucking around, others outside the team may interpret it as the outbreak of uncontrolled hostilities. It is not a good image for the project team nor for the project manager, as this demonstrates a lack of discipline & control.

If the conflict persists then DO NOT be afraid to escalate the conflict resolution to a formalized resolution process possibly involving senior management.

The last thing you want is threats of physical violence or actual violence in the workplace.

11.3.6. Project Manager: is a Problem Solver

The project manager should live to solve problems, either by themselves or via the coordination of others. If you have come from a technical implementer's background then you are probably already very adept at problem solving, and hence do not have much to learn here. ... But wait, there is a difference; whereas previously you solved inanimate technical problems using **"hard skills"** (i.e. analytical, mathematical, mechanical, logical), now you will be solving issues related to human beings and their inter-relationships using **"soft skills"** (i.e. communicator, listener, negotiator, influencer, conflict resolver, motivator).

As with the solving of technical problems, **an ad-hoc solution for people problems will often generate other unforeseen problems.** One way around this is to employ a **"PLAN – DO – CHECK – ACT" problem solving** methodology:

(1) determine the real cause of the problem and not just the associated side effects,

(2) break the problem up into its constituent parts and analyse the alternative solutions,

(3) choose a solution (or sequence of solutions) to be implemented, and

(4) determine whether the enacted solution has produced the desired results.

However, often the **best problem solving result involves a collaborative approach with all of the involved parties.**

"HE THAT IS GOOD WITH A HAMMER TENDS TO THINK EVERYTHING IS A NAIL."

ABRAHAM MASLOW

11.3.7. Project Manager: is a Planner

Given the size of [Chapter 4] on the Planning Phase, is it really such a surprise that being a good planner is a prerequisite for a good project manager.

11.3.8. Project Manager: is an Organizer

The project manager has to be able to coordinate with the project's stakeholders for the necessary [People] & [Resources] to be available and ready to perform the agreed tasks [Scope] in accordance with the agreed [Time] schedule, and to ensure that the approved [Cost] budget is available so that there is sufficient cash flow to pay for the implementation. That is, to **orchestrate others to transform "The Plan" into reality**.

If the project manager is a poor organizer, then no matter how wonderful the planning, the project will still come to a grinding halt. This is because, someone or something will not be available when required; consequently, everyone and/or everything else will have to wait. Hence, there will be people standing around wondering what they should do now, there will be equipment gathering dust and material resources going to waste, there could be unpaid invoices & bills piling up. ... And, all the while the project's [Time] clock keeps ticking down, and [Cost] expenditure keeps tallying up. Subsequently, the project will operationally be a complete debacle, irrespective of what "The Plan" dictated.

To succeed, **the project manager will have to be able to "compartmentalize" sections of the project into work packages that when necessary can be delegated (and effectively handed-over) to other [People] to complete.**

11.3.9. Project Manager: is a Delegator

The project manager has to be able (and willing to) designate activities to other project stakeholders to undertake. This delegation could be downwards to the individual project team members, side-wards to other project managers (or equivalent), and when the situation requires higher authorization then upwards towards the project's primary-core stakeholders (i.e. to senior management).

However, **delegation does not mean handing over to the delegation-receiver the final responsibility for the activity as this remains with the delegator; i.e. the responsibility remains with the project manager.**

 **Be very wary of delegating upwards towards the project's primary-core stakeholders,** because doing this too often could start them wondering what is the point of the project manager.

Therefore, **only delegate upwards those essential decisions** related to sensitive risks & issues, and notable changes to the project's [Scope Baseline], [Time Baseline], [Cost Baseline], & [Quality Baseline].

In addition to reducing the workload on the project manager, delegation has another useful benefit. By **delegating** some of the work **to subordinates,** the project manager is **training their potential replacements**.

But, why would I want to make myself replaceable?

Well, one way to make yourself un-promotable is to make yourself irreplaceable in your current role. By not delegating and keeping all of the work to yourself, then you, the project manager is burdening himself or herself with a potentially excessive workload, and leaving the project with no contingency if said project manager is unavailable for some period of time. In addition, not only is the project manager limiting their own promotional opportunities but also they are curtailing that of their subordinates, which could negatively affect the motivation of the project team.

DON'T BE IRREPLACEABLE ... IF YOU CAN'T BE REPLACED, THEN YOU CAN'T BE PROMOTED ... SO GET USE TO DELEGATING AND MENTORING YOUR SUCCESSORS.

Hence, every member of the project team should be mentoring his or her own replacement, including you, as the project manager.

11.3.10. Project Manager: is a Motivator

The project manager has to be able to drive the project team members to be committed to delivering the project within the agreed [Time], within the approved [Cost] budget, and conforming to the agreed Acceptance Criteria for [Scope] and [Quality]. To achieve this, the project manager needs to "motivate" the project team members' combined efforts.

How do I motivate the project team?

It is, the exact same things that motivate you each workday.

> To wake up before the dawn to that incessant alarm clock. ... To drag yourself out of that nice warm comfy bed. ... To have a shower & groom while your eyes are still welded shut. ... To apply that sociable face. ... To put on that not so comfortable outfit & shoes. ... To jealously kiss your partner on the forehead while they get a few minutes more sleep. ... To say goodbye but not be heard as the youngest one plonks their butt down in front of that big-ass TV screen that you never seem to get to watch, while the older one giggles away at something indecipherable that their virtual friend just typed.

> To walk to the bus stop or train station in dark miserable weather, only to stand like emperor penguins for an hour or more with people squished into your personal space. Alternatively, to listen to advertisement-interrupted senseless breakfast radio as you stare at the stream of glowing red taillights snaking off into the distance as you crawl in traffic away from your little kingdom in the suburbs.

> To enter that open planned cubical farm of half-height padded white walls where everyone can see every little thing you do and hear everything you say. ... To sit down in that less than ergonomic chair surrounded by an anaglyphy of post-it notes and faded photos of when your kids thought you had a relevant opinion on life.

> To finally, sip on that re-boiled brown pick-me-up in your brown stained "world's best" coffee cup, while having breakfast at your assigned desk, as you are confronted by a barrage of unanswered emails.

> *Thanks, I feel so depressed now. ... Think I'll go get another coffee.*

What really does motivate you (and your project team members) to come to work each day, and give your all for the project's success? ... Then repeat it all again tomorrow and the days after that, for weeks – months – years to come?

Financial & material gains,
now that is a real strong motivator.

Are you sure, that it is the hourly thought of coins falling into your piggy bank that motivates you to do your best at work?

Sorry, but you are gravely wrong because money, bonuses, and even share options are NOT a motivator, rather these are initiators to action (i.e. a work place "hygiene factor"). Though, if you are currently unemployed or in an underpaying job then money will be a good motivator to find that new job. However, when you have a relatively secure job that pays you at an acceptable level then money is no longer a motivator. ... Rather, payment is an expectation, an inane right (given that daily drudgery of coming to work).

Secondly, you will probably be paid weekly / fortnightly / monthly and those bonuses are only handed out at the end of the year / half-year. Hence, these material gains are not directly associated with your daily work efforts, rather these incentives are determined by the annual / biannual review of your performance (if and when these do happen). Therefore, material gains do not drive you (or your project team) to give all for the project.

Oh, and if your project team consists of contractors then they probably will not get any performance bonus let alone a performance review. Instead, they are often just shown the door when it is done.

Thirdly, the project manager probably won't have a direct say over what the project team members are individually paid, as these people are probably assigned to the project.

Hence, **financial remunerations & material gains are NOT a motivating tool in the project manager's arsenal.**

Oh by the way, there will be no pay rises this year (for the majority if not all of the staff) ... not even an inflation matching CPI increase. Moreover, the work place will be demotivated for some time to come. For sure, it will create a few disgruntled employees, some of whom will contemplate heading for the exit door (with how many others in tow).

Rewards & recognition, now that is a great motivator.

These are NOT relevant motivators because (like material gains) these are based on past performance and do not spur the project team on to perform today. Some companies try to use rewards & recognition such as "employee of the month" or "most valuable employee" to stimulate performance. While these maybe acceptable motivators in a competitive environment such as sales; in a SDLC project environment these **"one winner takes all" motivators** would in fact be **detrimental to the project team's cohesion and morale, because these motivators encourage individual self-interest over teamwork**.

Secondly, as with the material gains for the rewards & recognition, the point of application of these supposed motivators is greatly removed from the moment when the performance occurred. The reality is, for such motivators to be beneficial then these have to be applied throughout the entire project's life, and not just at the end of major milestones.

Thirdly, as the project manager, you probably can only make recommendations about which project team members should be considered for such rewards & recognition.

Oh, and company rules probably dictate that contractors are not eligible for such rewards & recognition. So, there goes that as a motivating tool for a portion of the project team.

Hence, **rewards & recognitions are NOT a motivating tool in the project manager's arsenal.** Rather, **rewards & recognitions are "anti-demotivating" tools, for the performing organization and not for the project manager to deploy**.

How do I motivate the project team then?

To motivate the project team, the project manager needs to create a work environment that is beneficial to each project team member's intrinsic & extrinsic motivations, while minimizing those demotivational factors; [ref: Herzberg].

❖ **Intrinsic motivation – is generated within the individual project team member due to the type of work that they are performing;** i.e. a sense of achievement & worthwhileness, plus a feeling of personal growth.

❖ **Extrinsic motivation – is generated externally to the individual team member** and includes such motivators as; financial & material gains, rewards & recognition, competition, as well as threats & punishment.

❖ **Hygiene factors** – are not motivators as such, but if these factors were not present then the project's work environment would be less "healthy"; i.e. a less enjoyable place to work. Also referred to as "**demotivating factors**". ... *"I hate this job"*.

As described previously, these extrinsic motivations are not tools that the project manager can effectively utilize, because these are administered after the project team member's performance has occurred. Hence, the project manager needs to **concentrate on those intrinsic motivators that are closer to the time when the work activities were performed or** (even better), **prior to the activity being performed** so that it is beneficial to the project's quality assurance processes.

While the **project manager deals with the intrinsic motivators**, the **project's primary-core stakeholders should deal with the extrinsic motivators**.

Whereas **everyone**; including the project manager, the primary-core stakeholders, and the project team **need to deal with the hygiene factors**.

What intrinsic motivators are in the project manager's arsenal?

To intrinsically motivate the project team, the project manager needs to create a work environment that encourages and is a catalyst for; job satisfaction & self-esteem, a sense of accomplishment & achievement, and a sense of belonging & fellowship.

Maslow's Hierarchy of Needs

This all points to motivators based on **Maslow's Hierarchy of Needs**, illustrated below in [Figure 100].

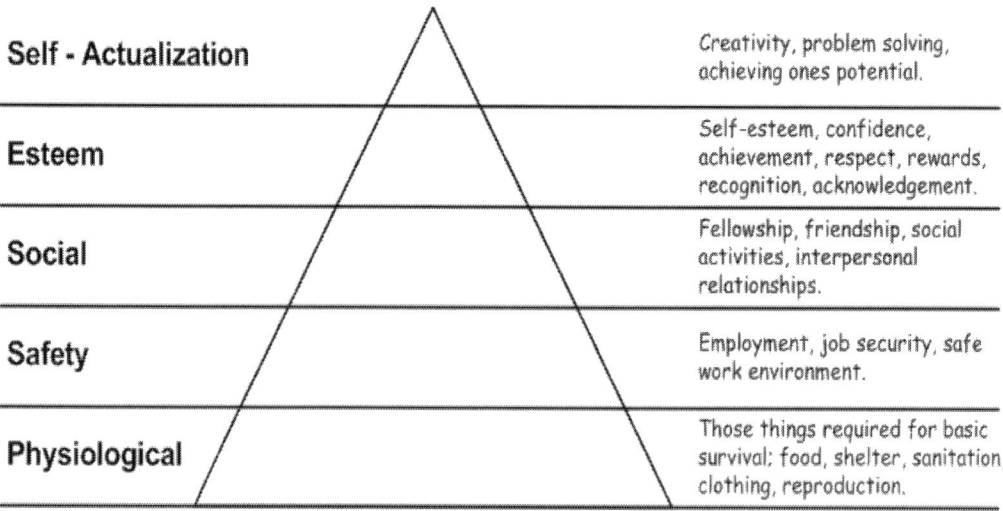

Figure 100: Maslow's Hierarchy Of Needs.

The techniques that the project manager could use would include, but not limited to:

1) **Be aware of** (cognizant to) each team member's **efforts & contributions** towards the project's objectives; i.e. to know what they are doing and what they have done.

2) **Give praise** to the relevant team members as soon as possible after they have done a good job. ... *The proverbial "pat on the back for a job well done" can do wonders for a person's self-esteem and the sense of being appreciated.*

3) Provide **timely feedback** (and if needed **constructive criticism**) on their performance.

4) Provide **challenging tasks** and not just repetitive (boring) activities.

5) Provide **new or different roles & responsibilities**.

6) Provide **opportunities for individual growth & career advancement**.

7) Provide **acceptance of their proposals & ideas**.

...

8) Ensure that they have a **balance between their work life and their home life**.

EVERYONE HAS TWO PARTS TO THEIR LIFE ...
WORK AND HOME ... AS LONG AS ONE OF THESE IS
OKAY THEN EVENTUALLY THE OTHER WILL COME GOOD,
AND YOU WILL END UP WITH A HAPPY WORKER.
HOWEVER, IF BOTH OF THESE ARE NOT OKAY, THEN
YOU WILL HAVE A CONSTANTLY UNHAPPY WORKER ...
WHO WILL CONTAGIOUSLY MAKE THOSE AROUND THEM
UNHAPPY TOO.

11.3.11. Project Manager: is a Leader

One thing for sure ... leadership is NOT a popularity contest, nor a beauty contest, nor who is the most charismatic, nor who is the smartest in the room, nor who has the most degrees framed up on their office wall.

How can the project be a success, if the project manager cannot lead / **focus** the project team's **efforts towards the common goal** of producing the project's deliverables within the agreed [Time] and approved [Cost] budget while still conforming to the agreed Acceptance Criteria?

As stated back in [Section 5.5.1]

> *"A project manager's role is NOT TO DO,*
>
> *but rather, to make sure IT GETS DONE".*

Therefore, the project manager has to be able to **communicate a common vision** to the project team, while **inspiring them to perform effectively & efficiently** to achieve the project's objectives.

Management Styles

The project manager needs to determine **what style of manager** they are predominately likely to be; [ref: McGregor]:

❖ **Theory X manager – will intrinsically need to micromanage every little thing that the team members do.** This style of manager will be driven by some innate mistrust of the team members that they will not do their jobs, because they are lazy unmotivated slackers who will (whenever possible) go out of their way to avoid work and they will cheat & lie about the work that they have & have not done.

Hence, the **Theory X manager will be highly restrictive with the delegation of work and the allocation of responsibility** to the project team members, even going as far as to **silo team members in discretely demarcated job roles**. This will all be based on the manager's belief that the only way to get these "workers" to do their jobs (other than by the pay-cheque) is by the **use of rigidly defined hierarchical work structures and draconian work Processes & Procedures** (including prescribed punishments).

In a performing organization where there is a desire to be people oriented instead of purely procedural (e.g. agile instead of production line), the **Theory X manager can have a devastating effect on the project team's morale.** This is because, the Theory X manager often believes that whenever something goes wrong, it is the "workers" fault "because they must not have abided by the processes & procedures", instead of considering the possibility that the failure was due to the prevailing circumstances and the current situation. ... *And not possibly due to the manager's own failings.*

For a performing organization that is using an agile implementation methodology then the **Theory X manager is completely at odds with the underlining principles of agile**, that it is the team and not the manager that runs the Executing Phase of the project.

The **Theory X manager is more suited for the traditional waterfall** (and some iterative) implementation projects. ... *And when the project is very similar to those projects that came before, i.e. "just paint it a different colour".*

❖ **Theory Y manager – will trust the project team members to do their jobs, and feels confident that they will not let down the team or the manager.** This style of manager will be driven by the belief that the team members will be self-motivated to do their job as best that they can, given the prevailing circumstances and the current situation, and based on the training that they individually and collectively have.

Hence, the **Theory Y manager will be prepared to delegate work and allocate responsibility to the team members**, even going as far as **encouraging cross-functional team ~~structure~~ arrangements**. This will be based on the manager's belief that to motivate the team members to perform at their best then they have to be presented with challenges, problems to solve, opportunities for growth & advancement, diversity in roles & responsibilities. ... *As well as timely rewards and recognition of their efforts and achievements.*

In a people orientated performing organization, the Theory Y manager will likely have a beneficial effect on the project team's morale. Hence, the **Theory Y manager** is what **is required for a project being implemented using an "adaptive" agile methodology.**

In a procedural (waterfall) performing organization, the Theory Y manager could be driven mad by the bureaucracy and inflexibility. ... And, downright frustrated by the inability of the organization's senior management to (dynamically) adapt to the changing situation and the prevailing circumstances that now confront the project.

However, a project manager is not very likely to be all Theory X or all Theory Y, but rather composed of a few drops of one and a sprinkling of the other.

That is, **all managers are a mixture of both styles, rather than as a sliding scale with Theory X at one end and Theory Y at the other.**

IN THE RED CORNER, WE HAVE ANGRY-MAN THEORY X; IN THE BLUE CORNER, WE HAVE SMILEY-MAN THEORY Y. WHO WILL WIN THE BOUT FOR PROJECT MASTERY?

As the project nears the finish date of a major delivery milestone, **be wary of the Theory X type of senior staff member** (possibly from outside of the project's immediate management structure; *e.g. sales & marketing or customer relations*) **who comes "sweeping in" and tries to** directly or indirectly **micro-manage the project to completion.**

Based on their intervening actions, it is apparent that this Theory X Senior Person does not have faith in the Project Implementation Team / project manager being able to successfully deliver to the next major milestone date. *Even when this doubt is contrary to the reported facts that; the project has thus far been progressing nicely (according to the Weekly Progress Reports and the Key Performance Indicator numbers), the Project Implementation Team have been working effectively – adaptively dealing with all of the risks & issues that have confronted the project, and the project team's morale is high.*

The truth is that; it is not the project, the project team, nor the project manager that is the cause of concern. Rather, the issue is with this **"seagull" manager**'s own personal insecurities of having their direct / indirect fate being in other's hands. [ref: Blanchard]

This issue is exacerbated when the project is operating under an "adaptive" agile methodology, which is at odds with that Theory X Person's preference for conformity to a waterfall structure.

I would recommend that ... the project manager should use the **management chain-of-command** (i.e. the program director, head of programs) to "shush-off" this "seagull manager" as soon as they start intervening with how the project is being executed. Because; once this Theory X type starts having **reactionary changes made to the project's tasking, then the project is more likely to stumble & fall,** and the team's cohesion & confidence will decline rapidly.

At which point, this Theory X type is likely to place the blame for any subsequent project failures on the project team and the project manager. ... Not cognisant to the fact that they themselves completely turned the project upside down.

Leadership Styles

The project manager needs to realize **what leadership styles** are compatible with their own personality traits; [ref: Lewin, Lippitt, White]:

1) <u>**Authoritarian / autocratic**</u> **– is where the leader is central to all decision making**; i.e. an absolute ruler come dictator.

   While this style of leadership does result in quick decision making it does not necessarily result in the correct or optimal decision, because this type of leader often does not listen to or accept input & opinions from those whose station is subordinate (or they consider as subordinate) to their own hierarchical position.

   Team motivation & participation, as well as respect are often induced via threats of punishment and negative reinforcement. However, some form of authoritarian leadership style maybe beneficial during the project's Initiating Phase and especially during the project's Closing Phase when "perfection procrastination" (i.e. being 110% correct before moving on) can occur.

2) <u>**Democratic / participative**</u> **– is where the leader consults with the group before making the decision, or the decision is based on the consensus of the group.**

   Team motivation & participation is via self-interest to fulfil the vision and objectives. This leadership style is most beneficial during the Planning Phase and the Executing Phase. Hence, why it is the basis for agile implementation techniques such as Scrum.

3) <u>**Laissez-faire / free rein**</u> **– is where the leader removes oneself from the decision making process instead leaving that to the group or some subordinate member(s) of the group.**

   Team motivation & participation is again via self-interest to fulfil the vision and objectives. This leadership style can leave the project team wondering what the purpose of the project manager is. However, this style maybe beneficial if the project manager is outwards facing dealing with the primary-core stakeholders (and external issues) while a senior project team member (such as the team leader) is inwards facing dealing with the technical issues.

Based on the project's current situation and the prevailing circumstances, the project manager will need to apply differing portions of each of these leadership styles.

- In a "crisis" where a **quick decision is paramount**, then an **authoritarian style** is highly beneficial.

- A **democratic style** would be beneficial during the project's Planning Phase and the Executing Phase **where the input of different minds could produce better alternatives and realistic solutions** (yet to be perceived by an individual project manager).

- A **laissez-faire style** maybe optimal for the **day-to-day functioning of the project team**, while the project manager deals with those external issues confronting the project.

When **selecting the appropriate style of leadership to be applied to the current situation and the prevailing circumstances**, the project manager will **also need to take into consideration** what is more important at that moment in time; **completing the objective (i.e. being task oriented), or maintaining & sustaining a working relationship with the project's team members (i.e. being relationship oriented)**.

This means that sometimes the project manager will have to **"make hard decisions"** that put achieving the project's objectives before the individual interests of members of the project team.

LEADERSHIP IS NOT THE WIELDING OF POWER... RATHER LEADERSHIP IS ORCHESTRATING + INFLUENCING OTHERS TOWARDS ACHIEVING A DEFINED + TANGIBLE OBJECTIVE.

What makes a good leader?

Before listing what does and does not make a good leader, it must be realized that the **current situation and the prevailing circumstances is often a determining factor as to whether a person is** (in hindsight) **a good or bad leader**.

There has been many a political leader (who according to the opinion polls) were consider by many of their constituents to be incompetent to say the least, yet when war or natural disasters befall the nation / state then they stood-up and proved to be the right person at the time to do the job. Similarly, there has been many a great crisis-leader who has fallen from grace during civil times.

IMHO, national heroes do not make good political leaders, because they may not be prepared to make those hard decisions that could make them unpopular with their constituents. Then again, most present day politicians are more concerned with their polling numbers (and their financial backers' wants) rather than they are with making good societal decisions.

THE DIFFERENCE BETWEEN A LEADER AND A FOOL IS NOT THE WILLINGNESS TO CHARGE INTO ADVERSITY, BUT RATHER THE SENSE TO KNOW WHEN + WHEN NOT TO CHARGE.

What are the common traits of good leaders?

✓ Self-confidence, high self-esteem, and a belief in themselves.

  Or, at least they present a persona of having such characteristics.

✓ They are a good communicator of their vision, and they are focused on precisely communicating those objectives that are achievable in the short to medium term.

✓ They are able to persuade others to follow them (either by appealing to emotions, beliefs, intellect), as they drive towards achievable short to medium term objectives.

✓ They earn the respect & trust of their followers, and they respect & trust their followers to do what is required / requested of them.

✓ They provide their followers with a feeling of purpose, well-being, relative safety, security, and stability.

✓ They adapt to the realities of changing situations and the prevailing circumstances.

✓ (When necessary for the greater good of the whole) they are prepared to make the hard decisions irrespective of the consequence to themselves or to their followers.

✓ They take personal responsibility for the outcomes, the good and the bad.

What good leaders are NOT is:

✘ Narcissistic – arrogant, self-absorbed, egotists, with an unassailable appetite for power.

✘ Toxic – leaving their followers (i.e. the project team and the performing organization) in a worst state then when they first started to lead them.

11.3.12. Project Manager: is a Team Builder

It is essential that the project manager strive to encourage collaborative teamwork within the project team, as **it is not the project manager, but rather the team that transforms "The Plan" into reality**. That is, the project manager can only provide guidance to their hands. *... Not cut every plank of wood and hammer in the nails, no matter how good it feels to get one's hands dirty again.*

Hence, the success of the project is very much dependent on the teamwork that is generated within the team, and therefore, how well the project manager encourages & fosters an environment that emphasizes team-building & team-work over individual self-interest & self-gratification.

IT'S NOT WHO HAS THE MOST TALENTED INDIVIDUALS,
BUT RATHER WHO IS ABLE TO GET THESE INDIVIDUALS
TO FUNCTION AS A COHESIVE UNIT ... AS A TEAM.

11.4.   The Project Team

11.4.1. Team: Acquisition

We have all heard the following line so many times before; but have you ever stopped to really think this through?

"A CHAMPION TEAM WILL BEAT A TEAM OF CHAMPIONS".

For a project to be successful, this adage is so true. The project can have the perfect plan, a superb project manager, and the best available team members. ... Nevertheless, without the "right blend" of people to execute that plan, then the project is destine for failure (or more specifically not to perform as optimally as desired).

Though, a team of under skilled players will always lose the game.

At this point, I could go into "team inventory" [ref: Meredith Belbin] and describe the nine team roles; i.e. plant, resource investigator, coordinator, shaper, monitor evaluator, team worker, implementer, completer / finisher, and specialist.

However, for the majority of cases, you as the project manager will most probably have very little (if any) input into the composition of your project team, as the team members will have been assigned by the performing organization's senior management (possibly based on the fact of who was available at the time).

Yep, sometimes you just have to go with who you have been given.

Additionally, if you were involved with the recruitment of new people (permanent, part-time, or contractors) to work on the project, then you most probably will have not encountered their true nature, but rather their "spit-n-polish" persona that they presented at the job interview. Therefore, let's ditch this team inventory theory, and instead concentrate on the apparent personality of the team members that are at hand.

I would recommend that ... **the project team be built with people that have different personalities**, because while a team of highly intellectual clones (or malleable worker drones) may be successful in the short term this grouping will lack the spark to continually generate new ideas & initiatives (nor continue to be productive).

Additionally, after a while this work environment could feel very stale, monotonous, and void of recognizable human life.

The successful project team needs a blend of both technical minded and socially minded individuals. You should not build a team with all technical clones, nor a team with just social communicating "chinwaggers". A good mix of both types will result in a project team that has good morale, spirit, work ethic, mateship, and commitment to each other.

 Do not underestimate the beneficial impact that meaningless social chitchat and horseplay can have on incubating a cohesive teamwork environment.   ... *A team that plays together stays together.*

11.4.2. Team: Development

A team is composed of individuals who come together to achieve some common objective. As with any group of individuals that comes together, the resultant team will naturally go through five distinct stages of relationship development; [ref: Tuckman & Jensen].

These stages of team development are:

1. **Forming – is when independent persons come together** as assigned team members and learn about the project and its objectives, learn about their individual roles & responsibilities within the team, and learn about each other. As with any relationship, **this is the stage when people start to reveal themselves**; i.e. their values, beliefs, priorities, abilities, and their quirkiness (as found in many an SDLC implementer).

   If a pre-established project team has a heap of new people added to it (in a very short amount of time), or there is a marked change in the workplace location, then this Forming stage may very well happen again.

 **DO NOT expect a great deal of productivity in producing project deliverables during the Forming Stage,** because the team will be relatively dis-functional as they act as individuals trying to be accepted, and determining their position within the group; i.e. who is who within the team, and the shaping of a "pecking order".

2. **Storming – is when the team members start to "feel" their way around each other and around the project in general.**

Differences in team roles & responsibility will become clearer, and these will be formally & informally agreed upon by the team members.

Differences in opinions, perspectives, and ideas will be raised, considered, debated (argued), and the "winning" ones will be acted upon. This is also when the ground rules for expected & acceptable behaviour will formally & informally be established. These differences could potentially result in highly detrimental conflict between team members, and (in a worst case) conflict with the project manager and the project's other stakeholders. To resolve any such conflicts will require the guiding hand of the project manager, and the maturity of the project team members to communicate openly with each other.

 **Unless the team members' start to collaborate, compromise, and their self-interests converge, then the Storming Stage can be a very long dark miserable road for all persons involved with the project.**

This Storming Stage is when the project manager's skills need to come to the fore as; a communicator, a listener, a negotiator, an influencer, a conflict resolver, a problem solver, a planner, an organizer, a delegator, a motivator, and a leader. This is also the stage when the project manager's leadership style (as authoritarian, participator, and laissez-faire) will greatly influence the long-term resolution of any such conflicts.

 **DO NOT expect a great deal of productivity in** producing project deliverables at **this Storming Stage.** Instead, concentrate on the constructive resolution of interpersonal conflicts.

Let the jousting and jockeying for position begin.

3. <u>Norming</u> **– By this stage, each team member should have established their (formal & informal) position within the group, and know what is their roles & responsibilities.**

   Hopefully by now; internal conflicts have subsided to the occasional misunderstanding, productivity should have increased significantly when compared to the earlier stages, and cooperation between the team members should have become the norm.

4. <u>Performing</u> **– By this stage, the team should have established (formal & informal) cooperative working relationships between the team members** as they work their way through the Executing Phase of the project's life cycle.

   The team should now be functioning as a complete team; resolving their own internal problems & conflicts, having understood each other's strengths – weakness – values – beliefs – priorities, and with the common goal of successfully completing the project.

   NOTE ❐ **An effective / high performing team would spend the majority of their team development time in the Performing Stage.**

   Whereas, an ineffectual / under-performing team could spend a significant proportion of their team development time bouncing back & forth between the Storming Stage and the Norming Stage.

5. <u>Adjourning</u> **– By this stage, the project is completing and the project team starts to disband.** The team members realize that they will not be working with some or all of these people again, which could result in a feeling of lost mateship, and insecurity about what is next for them (as a group and as individuals).

11.4.3. Team: Cohesive Unity

How do I build team-work and team-cohesion?

Firstly, it must be acknowledged that for small to medium sized organizations and for small to medium size projects there may not be the time nor the budget available for formalized team building activities targeted specifically at "group bonding" and improving "group interactions". That is, it is not feasible to go offsite for workshops & seminars with "Human Relationship" specialists or "Group Dynamics" consultants. Instead, the **team building will come down to the project manager, and the members of the project team**.

This "team building" could be a simple as; Friday afternoon (non-monetary) betting on how regional / national sporting teams will fair that coming weekend, combined with the obligatory Monday morning expert analysis of the results. It could be stirring each other about their love of an iconic brand of muscle car, or the debate over which is the better operating system. A coffee discussion about the latest goings on of some reality TV show, or that blockbuster movie which did not live up to expectations. It could be kicking back after hours as your work "clan" clash swords with Orcs, launch that counter-strike to recapture the battlefield's high ground, a death-match in a derelict space station, or joining forces to holdout against suicide waves of rampaging alien hordes. ... Or, some meaningless banter about the kids, and what they did.

A TEAM THAT PLAYS TOGETHER STAYS TOGETHER.

So, how does the Project Manager build team-work and team-cohesion?

When you were a kid and played Saturday team sports, what made your side a winner, or more specifically what made your team enjoyable to play for?

So, what is different now that you are all grown-up as an adult at work, instead of a kid playing in the park? Though, you as the project manager will now have to act as coach.

The project manager needs to **facilitate a working environment that**:

1) Gives a **common sense of purpose, and emphasizes a commitment & dedication to the common objective** of delivering the project with the agreed [Scope], by the agreed [Time], within the approved [Cost] budget, and with the agreed level of [Quality].

2) Encourages **open, affective, and efficient communications** between the project team members, with the project manager, and with the project's other stakeholders.

3) **Instils trust & mutual respect** between the project team members, with the project manager, and with the project's other stakeholders.

4) Formulates a set of **formal & informal ground rules defining the expected & acceptable behaviour** for the project team members, for the project manager, and for the project's other stakeholders.

5) Establishes **informal & formally accepted ways to constructively resolve internal & external conflicts** between the project team members, with the project manager, and with the project's other stakeholders.

6) Outlines **informal & formally accepted ways that decisions will be made** by the project team members and by the project manager.

7) Encourages the project team **to solve their own problems in a collaborative** manner, before escalating these issues to higher management.

8) Tries to **inspire & motivate** the project team **to achieve higher performances**.

9) Organizes and arranges for **guidance & support from the senior management** of the performing organization.

10) Provides **timely feedback & constructive criticism** on the project team's performance, as well as relaying feedback from the primary-core stakeholders.

11) Ensures that there is a **common identity for the project team**, even providing the project with an identifying name not number (if one has not already been provided).

12) **Encourages a sense of belonging**. Ensures that **all project team members participate**, have a fair go, have their individual input sort, heard, and where relevant accepted.

13) Tries to **co-locate the project team members**. If this is not **physically** possible then tries to arrange for and have **virtual** co-location established by utilizing telecommunications technologies.

14) Ensures that **appropriate rewards & recognition** are obtained for **the project team's good performance**.

15) Tries to provide the project team (and its members) with **work that is challenging**, and provides **opportunities for career advancement & knowledge growth**.

16) Tries to maintain a **friendly, healthy, & sociable atmosphere** for all members of the project team.   *... Including the quiet, the withdrawn, and the newbies.*

17) **Removes those barriers that prevent the team from achieving the project's objectives**.

Lessons Learnt: Miss reading a "Clique Group" as a strong project team.

WARNING: watch out for the "clique group" / "in-crowd" forming amongst the members of the Project Implementation Team.

Comedy movies portray this type of clique group at its extreme, with the 4 or 5 trendy teens who strut their stuff, and make life a living hell for their "lesser" classmates.

In the SDLC world, this "in group" would be composed of 3 to 5 developers / engineers, probably with the team leader / senior implementer as the (male or female) equivalent of the "queen bee". Where this "queen bee" structures the team's operations so that he/she is at the centre, with the "in" team members as the next layer, and the "outer" team members on the fringes. This "queen bee" is also likely to make themselves the sole external representative of the Project Implementation Team, and he/she will have no immediate second in command (else a second with limited power & responsibilities).

In the worst cases, a clique group can:

- Pull in by enticing those developers / engineers that they want for their group, even going as far as directly and/or indirectly undermining other project teams to get that "right" person aboard.   *... Oh, mini "empire building" at others' expense.*

- Push out (by isolating / ostracising) those existing project team members that do not fit-in with the "in" group's perceived ideals. *... Bye-bye to whoever asks why.*

- Reserving the "interesting" / important tasks for their fellow "in" team members, while distributing the "crappy" / "boring" tasks to those "outer" team members.

- And at the extreme; considering the project manager (and senior management) as outside of their "in" group, and hence (in an attempt to protect their "in" group) are prepared to report mistruths when the project's scheduled tasks are not going as well as was planned. Though, this "in" group is quite willing to place blame for the project's failings on those "outer" team members. *... To throw em under a bus.*

Initially, the morale of this "clique group" based Project Implementation Team will appear to be high, but this is a falsehood that is destine to eventually fall apart, because:

1) Those "outer" team members will silently "want out" as they feel that their contributions (and more specifically their potential contributions) are going to waste, that they are being discriminated against by only ever getting the "boring" tasks to do, and that their future career advancement & knowledge growth within the project team and the performing organization is greatly limited / non-existent.

2) Those "in" team members while receiving the benefits of getting the "interesting" tasks to do, will become overloaded & overworked if / when the project falls behind schedule, as the "outer" team members cannot take on more of the heavy burden as they were not (previously allocated) "trained for" those important tasks.

3) The project team's cohesion is predominately via the imposed pecking order.

If all does proceed along well with a relatively productive Project Implementation Team; what happens when this "queen bee" gets promoted into another role, moves onto another project, or leaves the performing organization for "greener pastures"?

When the "queen bee" departs (or is promoted), then most probably this "clique group" will collapse, the project team will quickly unravel as the "in" team members jostle for position in the power vacuum, and the "outer" team members will become under-utilized as there is no "queen bee" to allocate the work and to issue new instructions.

 I would recommend that ... once the project manager realizes that a self-interested "clique group" has formed within the project team:
1) You should take the head position in the daily stand-up meetings.
2) Privately ask individual team members for their own estimates for task progress and for their opinions on the project's health.
3) Hold the "queen bee" (team leader) accountable for delivering each milestone on time ... just as your senior management are holding you responsible for delivering the project successfully.

11.5.   Relationship Monitoring & Control

This is all very interesting stuff, but ...

what does it have to do with monitoring & control?

Well; given that it is this team of people who are the ones that implement the project, and given that monitoring & control is related to the project variables & constraints of [Scope], [Time], [Cost], and [Quality] (that are traditionally used as the determinates of whether a project is deemed a success or a failure), then would it not be wise to monitor & ~~control~~ influence those "human factors" that could greatly affect these determinates of project success.

The monitoring part requires that the project manager keep an eye on:

1) **What is the morale of the project team?**

   - How **happy** are these people to be **working together**?

   - How **happy** are these people to be **working on this particular project**?

   - What is the **level of conflict** between the project team members, are the team members openly arguing for the entire world to see, or is resentment simmering away beneath the surface of civility?

   - Are they happy individuals when compared to their normal state of being, or are they **moody, irritable, grumpy-bums**?

2) **What is the well-being & health of the project team?**

   - Are the team member's taking sick leave unseasonably more than previous years?

   - Do team members look continually under the weather, tired, and/or depressed?

   - Is each team member coming to work at approximately the same time (as per their individual norm) and are they going home at their usual time, or are they coming to work later and/or leaving earlier? ... *Have they become a clock-watcher?*

- Are their lunch-hours taking longer, and/or are their coffee/cigarette breaks getting more frequent or longer in duration?

- Are team members obviously smoking more, drinking more caffeine, coming to work intoxicated or drug influenced? ... *Do they just don't give a damn?*

3) **What is the health of your management, and your senior management?**

- Is there a **disproportionate turnover in project team members** either leaving the performing organization or requesting to be transferred to another project?

- Have any **long-term staff decided to resign** and hence off of the project team, and therefore entirely out of the performing organization?

- **What percentage of staff have left**, irrespective of the number of newcomers?

NOTE   ☐ **The longer that someone has been working on a project and at the performing organization, when he or she decides to leave then a significant amount of essential knowledge & confidential information is walking out the front door with them** (in their head or worse in their bag). Can the project (and can the performing organization) honestly afford to lose such a valuable person, given that it could potentially take weeks / months to find their replacement, and then possibly take several more months for the replacement to come up to speed to be anywhere near as effective as the person who has now departed?

The loss of personnel and the associated loss of knowledge is a valid reason why there should be succession plans in place for everyone in the project team and in the performing organization. This is also why, delegation is such an important activity as it trains ones potential successor.

Does your organization have in place a business continuity strategy for people, or are there several single points of failure?

Therefore, the project manager needs to:

1) **Have some idea of what people are doing**, where they are at, why they are there, and why they are acting & functioning the way that they are.

2) **Know which project team members are overloaded / overbooked**, as well as those team members that are being under-utilized / unproductive.

3) **Know how each project team member is performing** when compared against tangible objectives, provide timely feedback & constructive criticism about their performance, provide positive encouragement, and provide timely rewards & recognition.

4) **Know what necessary skills project team members lack** and organize for them to be either trained or mentored in these skills.

5) **Know about conflicts within the project team** and orchestrate there resolution.

PROJECT TEAM MORALE IS THAT ELUSIVE INGREDIENT THAT CAN GIVE THE PROJECT CONTINUED DRIVE (AND FAITH IN SUCCESS) WHEN THE CURRENT SITUATION AND THE PREVAILING CIRCUMSTANCES HAVE TURNED BLEAK.

Lessons Learnt: "Bingo" signs of a Bad Project Manager.

1) If the project manager always delivers their projects on time, but their **project team has a high attrition rate of personnel** (i.e., sick leave, taking annual leave, absent without leave, and/or resignations) then ...

   "Bingo, a BAD Project Manager".

   Honestly, how can projects continue to be successfully delivered, if no one wants to work for or with this project manager? Consequently, the question should be ... **does the performing organization have a work-place sociopath on their hands?**

2) If the **project** implementation **team** members are **continually working overtime** or having to work weekends to meet the project's delivery milestones then ...

   "Bingo, a BAD Project Manager".

   This is because, there has been a **failure in the estimation** for the project's duration, or there is **uncontrolled scope creep** occurring. Either way; this will mean that the project manager has lost control of the project (when compared to its planned baselines), or the project manager has grossly underestimated tasks.

3) If the project manager rarely asks the project team members for their opinions & suggestions, and when they do it is apparent (by the project manager's verbal language, demeanour, and body language) that **the project manager doesn't** really **want to hear what the project team members have to say** then ...

   "Bingo, a BAD Project Manager".

   That is, if it is apparent that what the **project manager really expects** to see are the project team members nodding their head in agreement like **bobble-head dash ornaments,** or the project manager expects to hear the project team members **parrot off the party line,** then this will be detrimental to the morale of the project team. This is because, the project team members will feel that they are just being patronised to and that this "conversation" is nothing more than a complete waste of their time & energy given that their **opinions & suggestions** will be **ignored** ...

better to just, get on with doing their assigned parts of the project.

"Well that is X minutes of my life that I won't be getting back".

4) If the Project Implementation Team has (developed a case of) **low morale**, has an **apathetic attitude, lacks motivation, communicates poorly**, has **frayed tempers**, is **openly arguing**, and/or shows a **lack of respect** for each other and for others then...

"Bingo, a BAD Project Manager".

This is because; the characteristics of the project team are a reflection of the project manager's (and/or the performing organization's) managerial practices.

5) If the Project Implementation Team has the same members doing the same jobs year after year, with no one changing position then ...

"Bingo, a BAD Project Manager".

This is because; the project manager is **not cultivating** the team members' **career paths** (to the future benefit of the performing organization).

ARE YOU A BAD PROJECT MANAGER?

☒ HIGH ATTRITION RATE OF TEAM PERSONNEL.
☒ TEAM CONTINUALLY WORKING OVERTIME.
☒ TEAM CONTINUALLY DELIVERING LATE.
☒ TEAM OPINIONS + SUGGESTIONS IGNORED.
☒ TEAM MEMBERS HAVE NO CAREER OPTIONS
☒ TEAM HAS AN APATHETIC ATTITUDE.
☒ TEAM COMMUNICATES POORLY.
☒ TEAM HAS FRAYED TEMPERS.
☒ TEAM IS ARGUING OPENLY.
☒ TEAM LACKS RESPECT.
☒ TEAM HAS LOW MORALE.

A PROJECT MANAGER'S ROLE IS NOT TO DO,
BUT RATHER TO MAKE SURE IT GETS DONE
...

ON TIME,
WITHIN BUDGET,
WITH THE AGREED SCOPE, AND
WITH A LEVEL OF QUALITY THAT IS
EXPECTED FOR THE DELIVERABLES
TO SATISFY THE ACCEPTANCE CRITERIA.

MONITORING AND CONTROL IS
WHAT MAKES OR BREAKS A PROJECT,
AND CONSEQUENTIALLY ...
WHAT MAKES OR BRAKES YOUR CAREER
AS A PROJECT MANAGER.

12. MONITORING & CONTROL Phase ... Practical Part 1

Now that the theory behind project monitoring & control has been explained, this chapter (and the following chapters) will concentrate on the practical application of this theory.

NOTE   ❏ **Monitoring & control is what makes or breaks a project, and consequentially what makes or ~~breaks~~ brakes (retards) your career as a project manager**.

    ❏ You are going to be busy as "a one-legged man in a bum-kicking competition", as there are many things that you, as the project manager will need to keep a watchful eye on, many things that need constant adjustment, and there are many things that need to be communicated to the project's stakeholders.

Yep ... one's success as a project manager is not achieved by good luck alone; there is a lot of hard work, persistence, and consistent effort involved. And, just when you think that you're on top of it all, the project's stakeholders will request that changes be made, technical issues will need to be overcome, risks will turn into issues, and the project's situation & the prevailing circumstances will alter. Oh, and by doing such a great job with this project, your bosses will probably decide to give you one or more other projects to manage simultaneously.

AUTHOR'S NOTE

For monitoring & control, there are so many possible techniques, tools, and software applications that could be used; to try to cover all of these would make this book larger than it already is. Also, the monitoring & control means at your disposal is very much dependent on the performing organization that you have been engaged by. Hence, from a tools and software applications perspective this book will limit itself to using one of the most commonly available office productivity suites that you will encounter in small to medium size organizations. That is, a focus on Microsoft Office & MS-Project.

12.1. [Scope] Monitoring & Control

In a practical sense, [Scope] Monitoring & Control [Section 7.1] asks the following questions:

- What is in?

- What is out?

- What is done?

- What is doing?

- What is being added?

- What is being subtracted?

- What is in doubt?

- What is next?

SCOPE

PLUS

QUALITY

EQUALS

CUSTOMER

SATISFACTION

Hence, the practical application of [Scope] Monitoring & Control needs to provide the answers to these questions in a form that can be easily understood & readily utilized by all of the interested-party project's stakeholders.

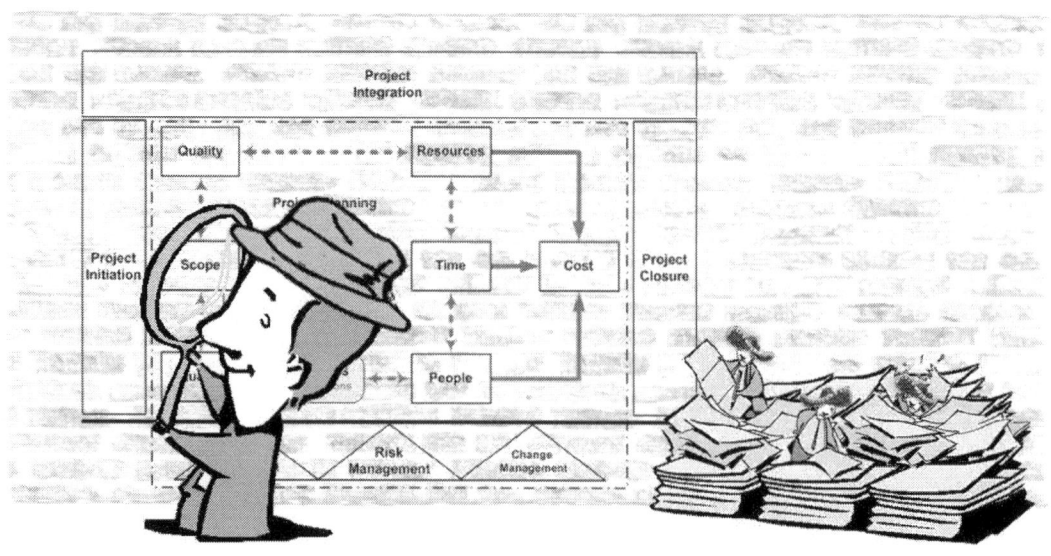

12.1.1. Scope: Requirements Traceability Matrix (RTM)

The following describes using the **Requirements Traceability Matrix (RTM)** as a means to **depict & discover the coverage of the project's approved Detailed Specifications**; *e.g. what agreed to features & functionality have and have not been implemented, and in what releases are these features & functionality to be available.*

Remember that; the requirements are the customer's "wish list", whereas the approved Detailed Specifications is the performing organization's "what will be delivered list".

In the RTM spreadsheet (*e.g. in MS-Excel*) add the following columns:

- An **identifying title** for this particular functionality.

- A **short description** of this particular functionality; make this easy to understand by someone with a basic knowledge of the project.

- **Release Identifier** (or Sprint name) that this particular functionality will be included in a milestone deliverable.

- **Category / Component**; i.e. the grouping of this particular functionality with other related functionality. *For example with the TV set; system input, video output, audio output, user input... etc.*

- (Optional) the **overall priority** of this particular functionality in comparison to the total functionality for this product / application / system. Rank this priority on a 1 – 5 scale where 1 is least important and 5 is most important; i.e. using the same least is 1 and most is 5 scaling that was described previously in [Section 10.1] for Risk Management.

- (Optional) the **release priority** of this particular functionality in comparison to the grouping of functionality to be contained in the associated (same) release of the product / application / system.

- Business Requirements – **BR Identification Number**; i.e. this field will contain the section number or sub-section number from the approved Detailed Specifications for that particular functionality is detailed / defined.
  (Optionally) include the corresponding entry in the Customer Requirements document.

- (Optional) the corresponding Functional Specification – **FS Identification Number**; i.e. the section number(s) or sub-section number(s) in the approved Functional Specification that defines this particular feature / functionality in detail.

- (Optional) the **use-case** or **scenario summation** for this particular functionality.

- (Optional) test case or Test Specification **TS Identification Number**; i.e. the section number or sub-section number in the Test Specifications document that defines how this particular functionality will be verified as compliant with the Acceptance Criteria.

- **Comments** about this particular functionality. There may be two columns for comments; one being for the performing organization, and the other being for the representatives of the customer organization.

- **Status** of this particular functionality; i.e. a text indicator of whether this functionality has been implemented (**DONE**), is currently Work-In-Progress (**WIP**),
  has yet To Be Done (**TBD**), or has been declared as being Out-Of-Scope (**OOS**).
  Alternatively, this field could be replaced or combined with a colour coding system.

Now that the columns of your RTM have been defined, from the spreadsheet application's menu bar choose **View > Freeze Panes** and select **"Freeze Top Row"** for the row that contains all of the column titles. This way the spreadsheet can be scrolled through with the column titles still visible and hence provide easier understanding of the data.

I would make sure that every "finger point-able" element contained in the approved Detailed Specifications is represented by an individual entry in the RTM. This is especially important when the Detailed Specifications is descriptive in content with numerous paragraphs of information instead of bullet-pointed facts & figures.

BR ID No.	Title	Short Description	Category	Overall Priority	Release Priority	Scenario	FS ID No.	Release	TS ID No.	Comments	Status
BR001	channel receiver	receive differentiated signal	System Input	5 (high)	5 (high)		FS001	RES_01	TS011		DONE
BR002	video display	display moving images	Video output	5	5		FS002	RES-01	TS012		DONE
BR003	play audio	play audio sound	Audio output	5	5		FS003	RES-01	TS013		DONE
BR004	channel selector	switch between channels	User input	5	3		FS014	RES-03	TS031		TBD
BR005	volume changer	volume change up & down	User input	3	4		FS004	RES-03	TS032		WIP
BR006	brightness change	brightness change up & down	User input	5	2		FS005	RES-03	TS033		WIP
BR007	contrast changer	contrast change up & down	User input	5	1		FS006	RES-03	TS034		WIP
BR008	Antenna	external antenna connection	System Input	5	3 (medium)		FS015	RES_01	TS010		DONE
BR010	loud speaker	front panel speakers	Audio output	5	4		FS016	RES_02	TS020		DONE
BR011	black & white dis	black & white display	Video output	5	5		FS017	RES_01	TS010		DONE
BR012	coloured display	colour display	Video output	4	5		FS018	RES_04	TS040		TBD
BR013	hand-held remot	user hand-held controls	User input				NA	NA	NA		OOS
BR014	on screen menu	user menus on screen	Video output				NA	NA	NA		OOS
BR015	auxilary port	auxilary input	System Input				NA	NA	NA		OOS
BR015	auxilary port	auxilary output	System Output				NA	NA	NA		OOS

Figure 101: "Candy-striped" Requirements Traceability Matrix (RTM).

As depicted above in [Figure 101] for the RTM spreadsheet, the following suggested colour coding scheme could be used on a per row basis. This chosen colour shading scheme is based on those standard cell colours in MS-Excel 2007-2010.

- **White background** with black text (MS-Excel = normal) – indicates this feature / functionality is **yet to be worked on or included** in any current or previous release; i.e. this feature / functionality has been "**white-listed**".

  o **Black text** – this little part of this feature / functionality is **yet to be examined** or has yet to be fully discussed, let alone agreed to.

  o **Green text** – this little part of this feature / functionality is **good** and there is agreement between the performing organization and the customer over its definition.

  o **Red text** – this little part of this feature / functionality is **bad** and there is not agreement or there is a perceived issue with this little part of this feature / functionality.

- **Black background with white text** (MS-Excel = check cell) – this feature / functionality has been agreed (by the representatives of both the performing and customer organizations) to be **out-of-scope** of this project (or for the current major milestone release); i.e. this feature / functionality has been "**black-listed**".

- **Yellowish / orange background** with brown text (MS-Excel = neutral) – this feature / functionality is **being worked on** for inclusion in the current release.

  o **Brown / black text** – this little part of this feature / functionality is still being worked on for the current release.

  o **Green text** – this little part of this feature / functionality has effectively been completed for the current release.

  o **Red text** - this little part of this feature / functionality is causing a problem.

- **Greenish background** with green text (MS-Excel = good) – this feature / functionality has been **completely finished** (developed & tested) for the current release and is now ready for inclusion in the deliverable.

- **Reddish background** with red text (MS-Excel = bad) – work on this feature / functionality has currently stopped or this is causing **blocking issues** for the current release, and potentially may not be included if these issues cannot be resolved.

- **Greyish background** with dark blue/black text (MS-Excel = output) – this requirement was **completed** in a **previous** release and hence "for all intents and purposes" is **no longer being considered** (other than as part of the regression testing).

I like to have a well structured and logically laid out RTM, with the approved Detailed Specifications replicated in point form (including the "Out-Of-Scope" elements) as I find this very useful to ensuring accurate scope coverage. ... It is also, beneficial for building up the customer representative's confidence in my abilities to competently manage their project to a successful conclusion, and to emphasize that the project team has indeed made real headway with the work being done.

Accepted that; the resultant RTM spreadsheet will look like candy striping, though human beings are visual creatures, hence (from a useability perspective) the end result will depict the coverage of the functionality (or the lack of coverage, as the case may be).

Those persons using this type of visualized RTM will quickly traverse this spreadsheet focusing in on those symbolic colours of current concern to them.

For example; all functionality yet to be assigned to a release (white background), all functionality deemed out-of-scope (black background), all functionality currently being implemented (yellow background), all functionality that has previously been delivered (grey background), all functionality that is ready to be delivered (green background), and all functionality that is causing blocking issues (red background).

Therefore, at today's project update meeting when we need to discuss what the current blocking issues are, we can quickly scroll through the spreadsheet and pick on only those RED rows, instead of like last week when we had to trawl through every row of the entire spreadsheet trying to recall which the blocking issues were.

VISIBILITY IS THE KEY TO SUCCESS, ANYTHING ELSE JUST LEADS TO DIGRESS.

12.1.2. Scope: Work Breakdown Structure (WBS)

The following describes using the **Work Breakdown Structure (WBS)** via MS-PowerPoint or MS-Visio as a visual tool for **depicting the status of each package of work**.

A similar colour-coding scheme could be used for the WBS as was described for the Requirements Traceability Matrix (RTM).

- **White background** – indicates that this work package is **yet to be worked on or included** in any current or previous release; i.e. this work package has been "white-listed".

  - **Black text** – this work package is **yet to be examined** or has yet to be fully discussed.

  - **Green text** – this work package is **good** and there is agreement.

  - **Red text** - this work package is **bad** and there is not agreement or there is a perceived issue with this work package.

- **Black background with white text** – this work package has been agreed (by the representatives of both the performing and customer organizations) to be **out-of-scope** of this project; i.e. this work package has been "black-listed".

- **Yellowish / orange background** – this work package is **being worked on** for inclusion in the current release.

- **Greenish background** – this work package has been **completely finished** (developed and tested) for inclusion in the current release / deliverable.

- **Reddish background** – activity on this work package has currently stopped or this work package is causing **blocking issues** for the current release.

- **Greyish background** – this work package was **completed** in a **previous** release and hence "for all intents and purposes" is **no longer being considered**.

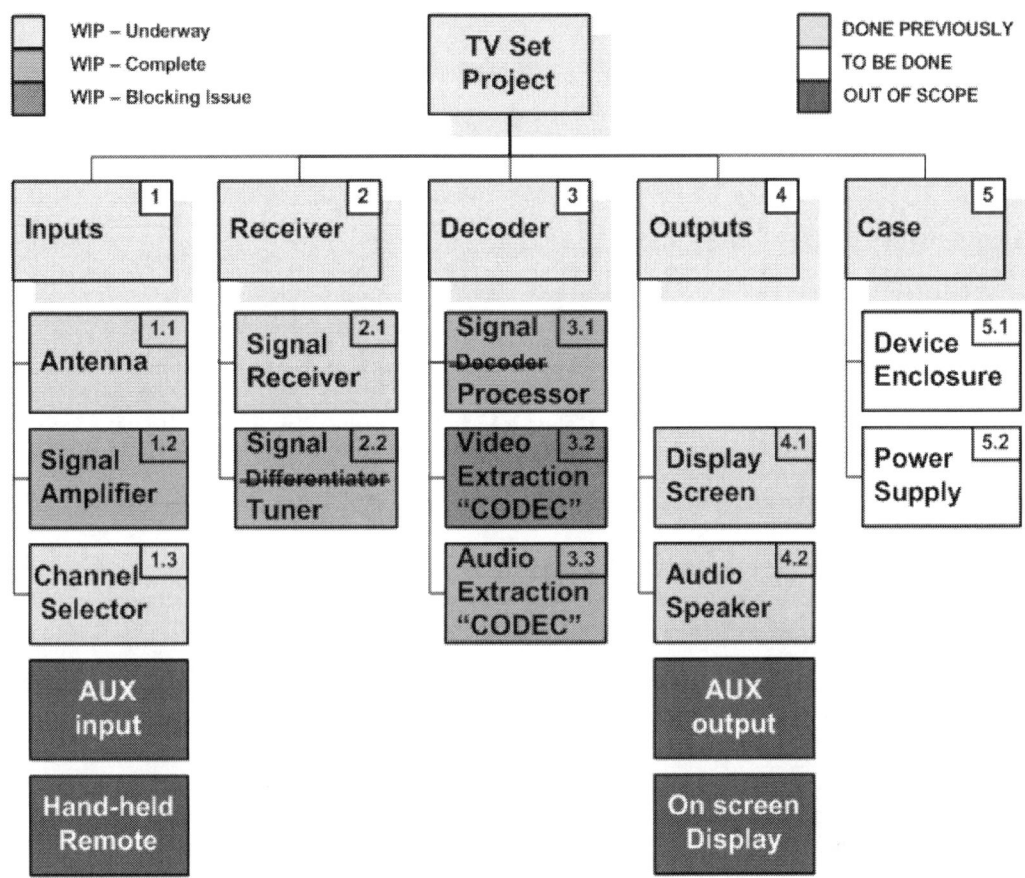

Figure 102: Example "Colour coded" WBS for a TV set development.

Above in [Figure 102] is a "colourized" WBS for a TV set development, where the following are hopefully apparent.

Done:	Antenna, Signal Receiver, Display Screen, & Audio Speaker.
To Be Done:	Device Enclosure, Power Supply.
Work In Progress (completed):	Signal Amplifier, Signal Tuner, Signal Processor, Audio Extraction CODEC.
Work In Progress (underway):	Channel Selector.
Work In Progress (blocking issue):	Video Extraction CODEC.
Out Of Scope:	AUX Input, AUX Output, Hand-Held Remote, On Screen Display.

Another visualization technique for depicting the project's progress is to use a presentation tool such as MS-PowerPoint with a coloured coded WBS on a single slide.

> *For example; Slide 1 – initial WBS when the project or release started,*
>
> *Slide 2 – progress as of last month's meeting, Slide 3 – progress for this month's*
>
> *meeting, Slide 4 – expected progress by next month's meeting.*

With each of these slides in sequential order, then a "page-flipping" animation (like the stick-man cartoons you did in the top corner of your school books) can be created to visually highlight the progress through the project's implementation as a representation of the passing of time or milestones completed; see [Figure 103] below.

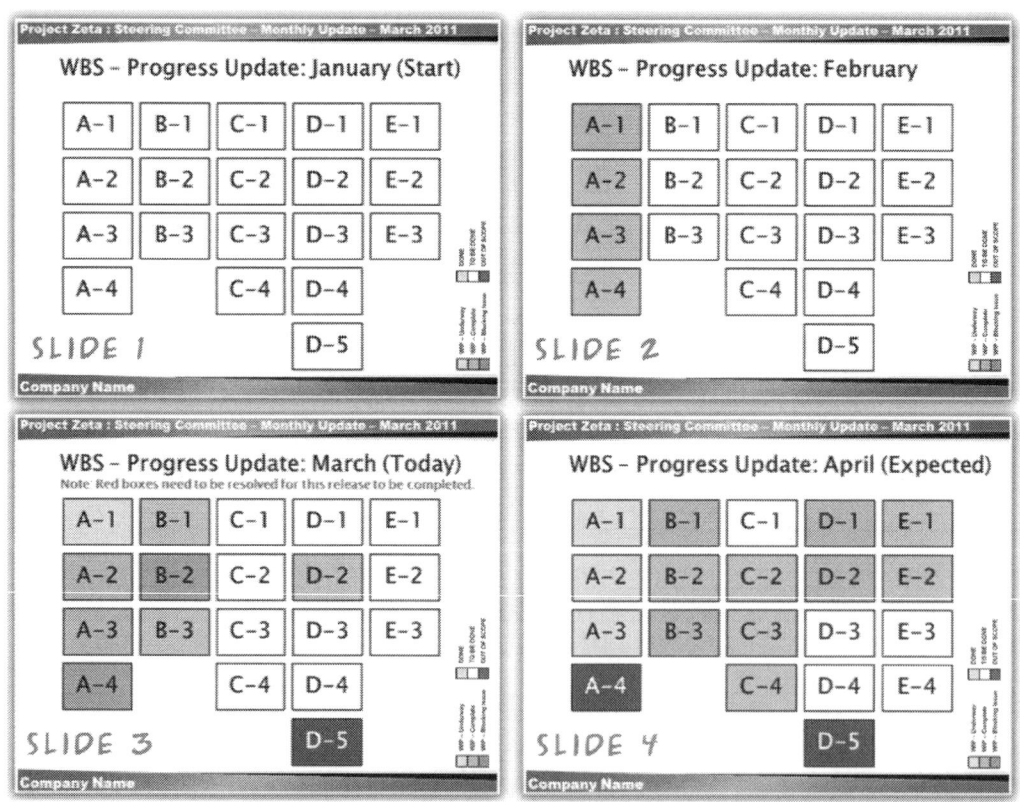

Figure 103: Monthly update of coloured WBS.

Include a note on the slide for any issue that needs to be discussed.

12.2.   [Time] Monitoring & Control

In a practical sense, [Time] Monitoring & Control [Section 7.2] asks the following questions:

- What tasks [Scope] should have been completed by now, or
  how far should these tasks [Scope] have progressed?

- What tasks [Scope], [People], and [Resources] should currently be engaged?

- What tasks [Scope], [People], and [Resources] are currently engaged?

- What tasks [Scope], [People], and [Resources] should be engaged next?

- How far have these [People] and [Resources] progressed through the assigned tasks [Scope] that they are currently engaged with?

- Is the reality of what & when tasks [Scope], [People], and [Resources] are being engaged deviating away from the approved [baselines]?

- How to realign tasks [Scope], [People], and [Resources] with the [baselines]?

- Given the realities of the current situation and the prevailing circumstances does "The Plan" need to be re-baselined?

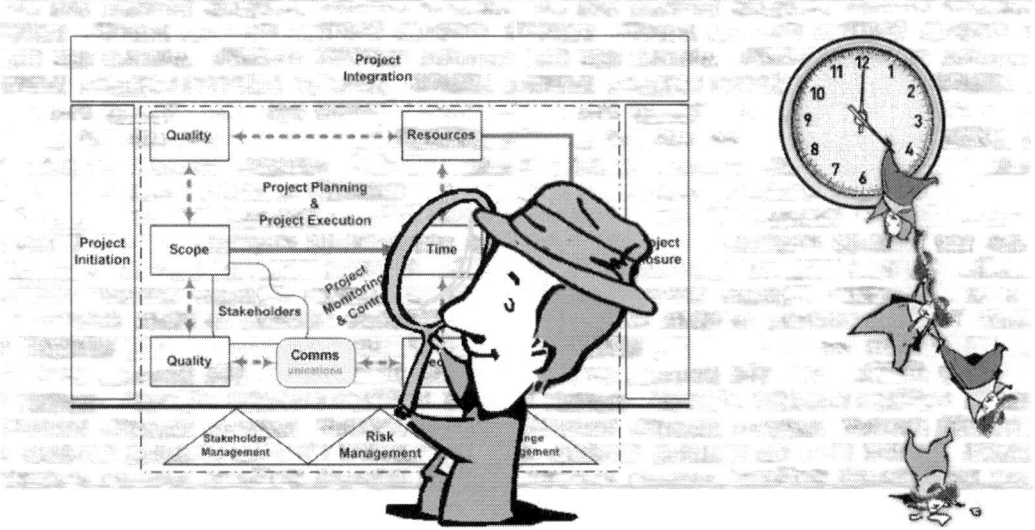

12.2.1. Time: Project Schedule & Gantt Chart

For iterative and waterfall based projects, the **most common way of tracking the relationship of tasks [Scope], [People] and [Resources] with the project's progress over [Time] is via the project schedule; i.e. the Gantt Chart.**

Previously back in [Section 4.3.2]; it was described how to create a project schedule, and how to allocate [People] and [Resources] to that schedule.

Now that the [Time Baseline] Schedule has been revised, reviewed, and approved by the Change Control Board (i.e. the project's primary-core stakeholders), then a copy of this approved [Time Baseline] Schedule will be used for the daily/weekly tracking of the project's progress (i.e. percentage completed).

DO NOT use the originally approved [Time Baseline] Schedule as the editable schedule against which the project is daily/weekly tracked.

K.I.S.S. ... "Keep It Simple, Stupid"

Instead, modify the **"work-in-progress" copy** of the Baseline Schedule to represent the project's daily / weekly progress.

BASELINE SCHEDULE

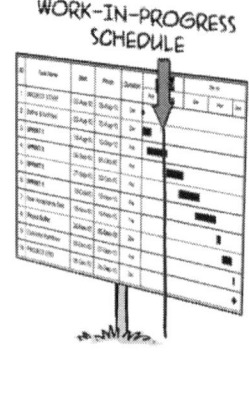

WORK-IN-PROGRESS SCHEDULE

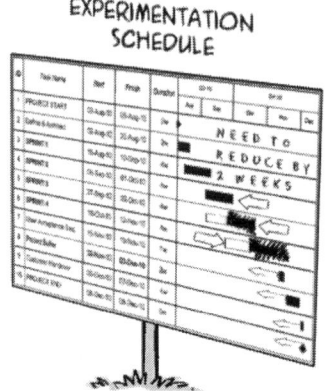

EXPERIMENTATION SCHEDULE

Lessons Learnt: Using different copies of the project schedule for specific purposes.

MS-Project does allow the creation of a Baseline Schedule; but for your own perceived competency as a project manager, do you really want to risk mixing the Work-In-Progress Schedule with the Baseline Schedule?

I personally would not risk it ... given that; not all yearly versions of MS-Project are as effective and defect free as other versions. For example; differences in the undo buffer's size, problems with task numbering when copy-n-pasting several tasks, or even file corruption, etc. While hot fixes and service packs may have resolved such problems, the decision over which version of MS-Project that you will be using may not be of your choosing but rather dictated by the performing organization's Standard Operating Environment (SOE). So keep your scheduling techniques as simple as practically feasible.

It is better to replicate distinct variants of the project's schedule for specific purposes instead of using one schedule file for all purposes. Suggested variants include:

1) **Baseline Schedule** – is the (read-only) version of the project's approved Schedule [Time Baseline] that will be used to pass judgement on the project's real-world progress against the approved "Plan".

2) **Work-In-Progress Schedule** – is the read-write version of the copied Baseline Schedule that is used to track & record the project's real-world progress.

3) **Experimentation Schedule** – is a read-write copy of either the Work-In-Progress Schedule or the Baseline Schedule that is used to experiment with different rearrangements of tasks, durations and resourcing levels to understand the effect on the project of changing circumstances and possible alternate arrangements.

4) **Rolled-Up Summation Schedule** – is the high-level schedule that is used to summarize to the project's primary-core stakeholders how the project is progressing.

However, each variant of the project schedule does not have to be produced with MS-Project, it could also be produced with a presentation tool such as MS-PowerPoint. Though, the Gantt Chart's task bars provide the best representation of the task-time relationship. This is because, the Gantt Chart can visually show when tasks are slipping, as depicted below in [Figure 104]:

- In MS-Project, when a task is entered as being a certain percentage complete then a horizontal black line is drawn along the centre of that task's bar.

- Also MS-Project draws a vertical line for today's date, hence any tasks whose progress bar is not up to or ahead of this date line is evidently behind schedule.

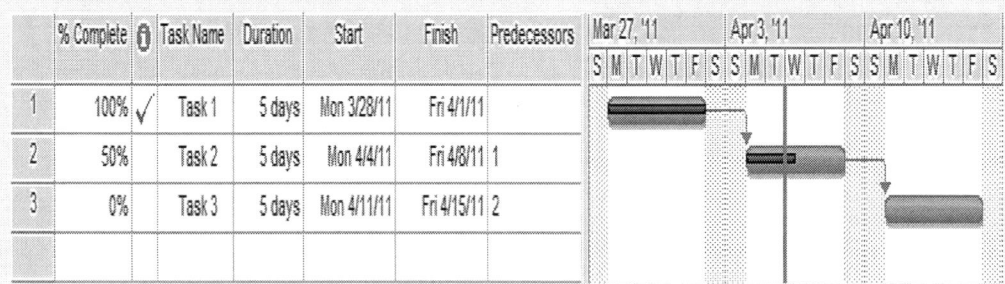

Figure 104: Gantt Chart task bars with progress lines.

Additionally, using a Gantt Chart provides a "psychological advantage" over a spreadsheet because the Gantt Chart (with its task bars and progress lines) gives credibility to the project manager in the eyes of the Project Implementation Team.

Even if this sense of competency is only instilled for a short time.

By being perceived as possessing **"visibility in control", the project manager can leverage off this perception to build the respect & confidence** that the project manager needs from the Project Implementation Team members; see [Section 11.2].

It also indicates to the implementers that the project manager is keeping an eye on & interest in the progress of the work that they are doing. ... So, no slackening off, because I will know soon enough.

How to use the schedule to track the project's progress?

1. Printout that part of the Work-In-Progress Schedule that has to be discussed, or display the schedule onto a screen / white-board so that it can physically be written on. Make sure that you include the task bars representing the tasks for the period of interest.

    NOTE   ☐ **DO NOT undervalue the positive psychological effects of presenting the project's schedule in a physical form** that can be touched and interacted with; i.e. drawn on and notes scribbled down. By being presented in physical form, the listed tasks are transformed from being virtual inside the computer to being physical activities that have to be performed in a given time period.

    This is the same for using the Task Board with agile methods; where the physical act of visibly moving the tasks / story cards across the board on a "publically viewable" wall are significantly more productive than "covertly" shifting text boxes on a computer screen.

2. At least once a week (preferably on a consistent day of the week, e.g. Monday), and possibly at approximately the same time of day, go visit individual members of the Project Implementation Team or the Project Working Group, sit down with them and go through their tasks on the physical paper schedule.

     I would recommend that ... the project manager should be repetitive with the timing of the schedule update visits, because people have a natural tendency to pickup on patterns of behaviour. Hence, in a short time they will adapt to this progress-check "routine".

    Knowing that the schedule update is coming they will prepare in advance for it by thinking about; where are they up to, reasons for why things have not been done yet, how long will it take, and even hurrying them to action on those outstanding tasks.

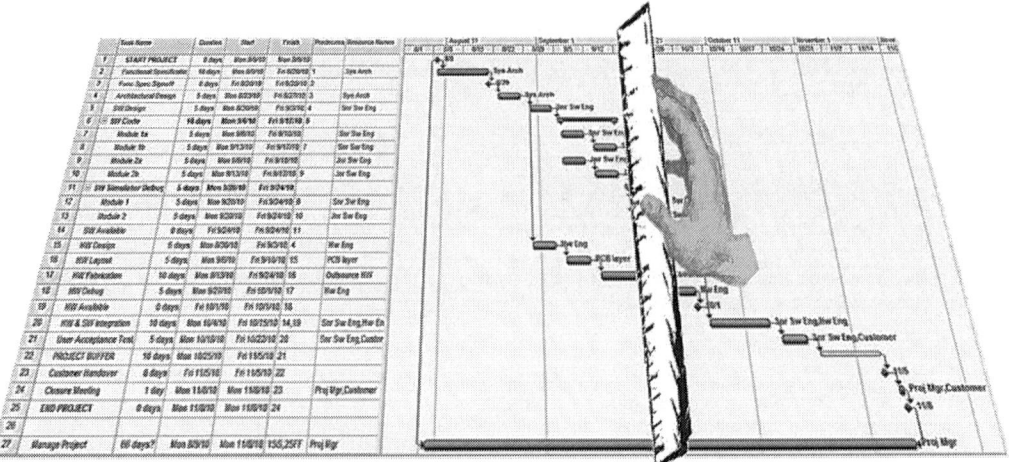

Figure 105: Marking up the Work-In-Progress Schedule.

3. With the person / persons involved seated in a convenient location (i.e. at their desk)
    then spread out in front of them that part of the Work-In-Progress Schedule that is
    relevant to them. With a ruler & pencil in hand, as illustrated above in [Figure 105],
    ask them the following questions:

- **What scheduled tasks have they or their team been working on?**

    It is advantageous to know in advance (i.e. have an approximate guess at
    what their answer will be) and have noted down the task numbers that you expect
    them to answer for. One-way of doing this, is on the paper schedule highlight their
    assigned tasks as follows:

    o   Fluoro PINK – last time this task was behind schedule or
        had experienced a blocking issue.

    o   Fluoro Yellow – this task is scheduled to be worked on during the period that is
        being ~~reviewed~~ discussed.

    o   (Optional) Fluoro GREEN – last time this task was progressing nicely on or
        ahead of schedule.

    This colouring technique will make it easier to select those tasks of interest.

If they are found to be working on tasks that they were not scheduled to be working on (i.e. tasks a lot later in the schedule or tasks that do not appear on the schedule) then enquiries should be made as to "WHY?"

Does the schedule need to be changed to reflect reality? Hence, does Change Management need to be invoked [Section 10.2], or do they have to be steered back on course to do only those tasks that they have been assigned?

- ~~How far have they progressed with the completion of their assigned tasks?~~ **How many days do they still require to complete each assigned task?**

Place the ruler on the paper schedule and line it up with today's date. Any of their tasks that appear in whole on the left side of the ruler should have been completed by now, any tasks that cross under the ruler should be currently worked on, and anything that is right of the ruler probably should not be worked on yet.

- **Write this information down on the paper schedule.**

Back in [Section 4.3.2] it was recommended that the maximum allowable duration for any individual task (before it has to be sub-divided into more tasks) be set as being 5 days. Hence, these five days is equivalent to 100% task complete, so one day would be equal to 20% of the task completed.

Therefore, if they say that they need 2 more days (40%) to complete the task then that means they have only completed 60% of the task.

The use of the day-percentage relationship will mitigate against the common situation where a task remains at 90% complete day after day (and even week after week). If they keep saying, they "need more time" than this means that the task's percentage complete is decreasing, which must raise questions of:

- o What is preventing this task from being completed in the allocated time;
  i.e. are there blocking issues that need to be resolved?

- o Should this task be moved to a more appropriate / opportune moment in the schedule, or transferred to another milestone / release?

- o Does this task have to be dissected into multiple tasks?

Those tasks that are slipping (or have slipped) behind schedule should be highlighted by Fluoro PINK on the paper schedule. When working on the digital Work-In-Progress Schedule, I would suggest that these individual late task entries in the schedule be coloured RED, possibly enlarging their font size & bolding them. The important thing is to make these problem tasks standout from the rest of the scheduled tasks, so that these can be quickly identified and then acted upon.

- Are they **experiencing any blocking issues** that are hindering their progress to completing their assigned tasks on time?

- Are there **any risks to the project** that they perceive will prevent the task's (and the project's) successful completion?

- Are there **any tasks** that they believe could be **better arranged** and thereby occur at a more opportune moment for the betterment of the project?

- Are there **any tasks [Scope], [People], and/or [Resources] availability** that they believe are   **missing** from the schedule?

- Are there **any tasks [Scope], [People], and/or [Resources]** that they believe are **no longer necessary** for the project to be completed successfully?

- Are there **any baseline tasks [Scope], [Time], [Cost], [Quality], [People], and/or [Resources]** that they believe **will need to be adjusted** for the project to be completed successfully?   That is, could a Baseline Change Request be required?

NOTE ❑ Where & whenever possible, involve the project team members **active participation** in the project's scheduling and semi-directly with the decision making for the project. What could help bring their late tasks back on schedule?

The project schedule is only a guide and not gospel

How the project schedule is "idolized" and "utilized" is the big difference between being a "Project Manager" and a "Project Accountant".

The "project accountant" type of manager will expect (even demand) **that the project's schedule be followed as though gospel**. They may consider any deviation (whatsoever) away from "The Plan" as a cardinal sin for which the wrongdoer must be punished or reprimanded. Unfortunately, this misguided belief that, if the schedule is zealously abided by then "all will be fine" is **critically floored because of the simple reality that, things never turn out exactly as was planned**. It is an odds-on certainty that at some point in any project's life; the [People] and/or [Resources] will not be available when these were expected or these will be available in too few numbers or available for too short a duration, and some [Scope] feature / functionality will be a lot more technically difficult than was originally envisioned. Hence, **the "project manager" views the schedule as a guide to the advised path to implement the project; though, understanding that sometimes parts of "The Plan" will have to be adapted, revised, and reapplied to suit the current situation and the prevailing circumstances** that now confront the project.

12.2.2. Time: Task Board & Burn Down Chart

With an **agile** based project, the **Gantt Chart** will **not necessarily** be the primary way of tracking the relationship of task progress to [Time]. Rather, the **Task Board** in [Figure 106] and the **Burn Down Chart** in [Figure 107] could be used for this purpose.

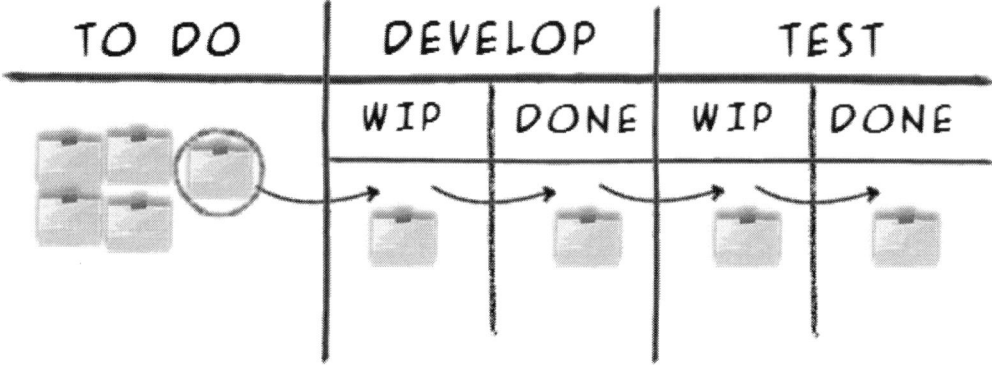

Figure 106: Motion across an agile Task Board.

When an individual task/story passes from left-to-right across the Task Board [Figure 106] to be eventually verified as DONE in testing, then it will reduce the amount of "Work Remaining" and hence it will affect the "Actual Burn Down" line in [Figure 107].

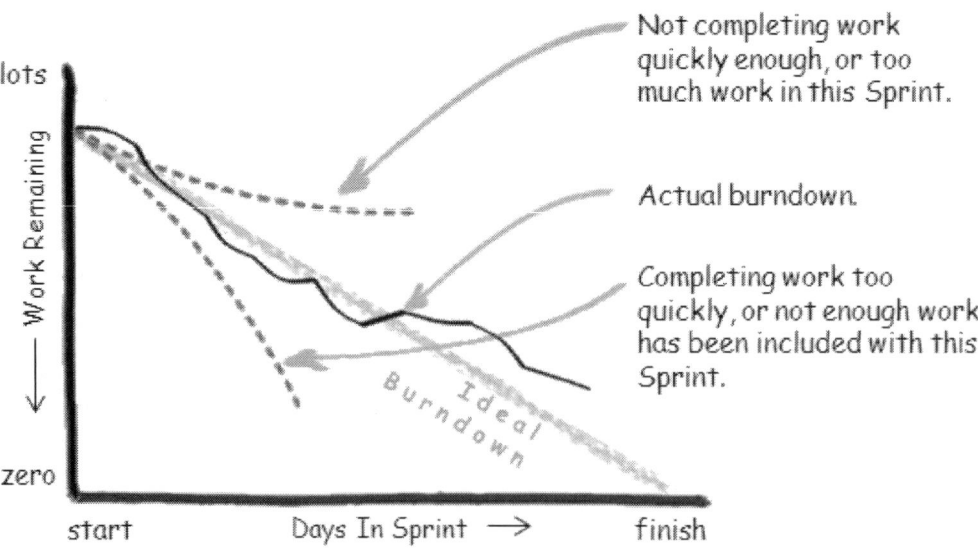

Figure 107: Scrum Burn Down Chart with up-to-date data.

Provided that, the motion of the "Actual Burn Down" line travels downwards around the straight "Ideal Burn Down" line then the project will be progressing along nicely; as illustrated in the first half of the example in [Figure 107].

However, when this "Actual Burn Down" line starts to move away from the "Ideal Burn Down" line (especially movement up towards the right as illustrated in the second half of [Figure 107]), then the current sprint / release has problems with not completing the pool of allotted stories / tasks on time, or supposedly completed tasks are being reopened due to defects being found. ... Alternatively, if a marked dive-down of the line then tasks are being closed a lot quicker than was expected and thus what is the effect on quality?

NOTE ☐ It is essential with the use a Burn Down Charts that each individual Project Implementation Team member **updates** their **story / task status on a daily basis**, or (preferably) immediately after they change their operations on that story / task. ELSE the Burn Down Chart will not accurately depict the true state of the project's progress; hence, be ineffectual as an effective progress tracking tool.

As evident by a Burn Down line that travels relatively horizontally for much of the sprint's duration then dives down towards zero at the end. This kind of motion of the line would clearly indicate that agile best practices are not being followed.

While there are software applications that can plot the Burn Down Chart and replicate **the operations of the Task Board**, I would recommend that ... these **be presented in a physical form** that is "publically viewable" by all within the performing organization.

In a physical form on the wall, this chart and board will be acting as sources of information about the project's progress and hence its achievements. Thereby opening the project up to "public scrutiny" and accessibility, whereas with the software based variants this information would be hidden away.

For more information on agile methodologies, please refer to [Section 7.4].

12.2.3. Time: Timesheets ... NOT!

One could consider timesheets to be a useful tool for the monitoring & control of [Time], however in reality, timesheets are ineffectual for this purpose.

This ineffectiveness is because timesheets are usually submitted on a weekly, fortnightly, or even monthly basis. Consequently, there can be a **considerable time-delay between when the work was actually performed and when the [Time] reporting is processed**. This time-delay can be long enough to limit the opportunity to successfully steer the project clear of an arising issue; *e.g. the over booking of hours and hence labour [Cost] to the project*.

Secondly, **these timesheets do not usually provide the granularity of task details to capture the status of tasks completed**. This is because, the project's job / Cost Account Codes most probably have not been subdivided down to corresponding to individual work packages as was detailed in the Work Breakdown Structure. Hence, it may be difficult based solely on timesheet entries to distinguish which work packages are consuming more effort than was planned for; *e.g. an activity that involved extra [People]*.

Another issue with timesheets is that the **project's job / Cost Account Code that the individual reports against may not correspond to the activities that they spent that time working on**; *e.g. worked on this but charged it against that, or didn't know there was a job code specific to that particular activity so instead used the generic project code that we all use.*

In addition, **the amount of [Time Claimed] on the timesheets may not correspond to the actual [Time Spent]**; *e.g. the project team members may not accurately fill in a timesheet each day, rather at the end of the reporting period they may whitewash the timesheet with the constant number of hours that they believe that they are expected to work (8-8-8-8-8).*

Therefore, **timesheets are not a [Time] Monitoring & Control tool because these do not relate to the "up to the moment" progress being made to complete assigned tasks.**

I would recommend that ... the project manager should keep a close eye on just how much [Time] each project team member is booking against the project's job / [Cost] account codes when compared to the actual amount of effort that they are putting into their assigned project tasks [Scope].

> *For example; did they really do 5 straight days of 7½ hours per day, or was their actual work on the project something like 5-5-7-8-5 hours. That is, instead of doing the 37½ hours that their timesheet claims they only contributed 30 hours towards the project. Hence, there was a 7½ hour deficiency, which translates to a 1 day shortfall in the work performed.*

 I would recommend that ... before inaccurate time reporting becomes an issue, the project manager should remind each member of the project team that they should only book to the project's job codes those hours that they did work on the project's tasks. Also, remind them not to whitewash their timesheet with the daily numbers that they believe that they are expected to expend, as this will make determining the project's true progress highly inaccurate.

PLEASE, ONLY BOOK FOR THE TIME THAT YOU SPENT ON THE PROJECT, AND NOT THE TIME THAT YOU THINK OTHERS EXPECT YOU TO SPEND.

Explain to them that, it is okay if some days they are only able to do a limited number of hours work on the project. Hence, provide them with viable alternative time code(s) to book non-project work against.

 I would recommend that ... it be NOT assumed that each project team member will be working 100% of the time on their assigned tasks. This is because; there are interruptions that will reduce the amount of effort that they can individually put into their assigned tasks, as well as short notice leave (e.g. sick leave). Thus, it is often advisable (especially with senior implementers) that it be planned that each team member will only ever be able to work at a maximum of 70-80% efficiency.

In the following [Section 12.3.5] on [Cost] Monitoring & Control, the relevance of those previous recommendations will be covered in more detail.

NOTE ☐ The project's **Earned Value Measures can take a real hammering if there is a noticeable discrepancy between the timesheet recordings and the actual hours of work effort that have been put in**. This is because; the project will not be progressing as fast as the amount of recorded time indicates it should have.

Whenever it is discerned that there are discrepancies between timesheet entries being submitted and the work that has actually been performed on the project then inquiries should be made as to the causes & reasons for such discrepancies. The reasons could be honest mistakes such as; they used the most appropriate job code that was available to them because they didn't know that they had not been given access to the correct job code, they thought it was project work that they were doing, or it started out as project work that blurred across into another job code's activity (*e.g. fixing a implementation defect ended up entailing the rectification of the generic development environment that is used by all projects*). Once the facts of the situation are known, then where appropriate corrections should be made to the corresponding entries in the timesheet tracking system.

DISCREPANCIES WITH TIMESHEET REPORTING CAN TRANSFORM INTO SIGNIFICANT INFLATION OF THE PROJECT'S COSTS AND RESULT IN DISMAL EARNED VALUE PERFORMANCE MEASURES.

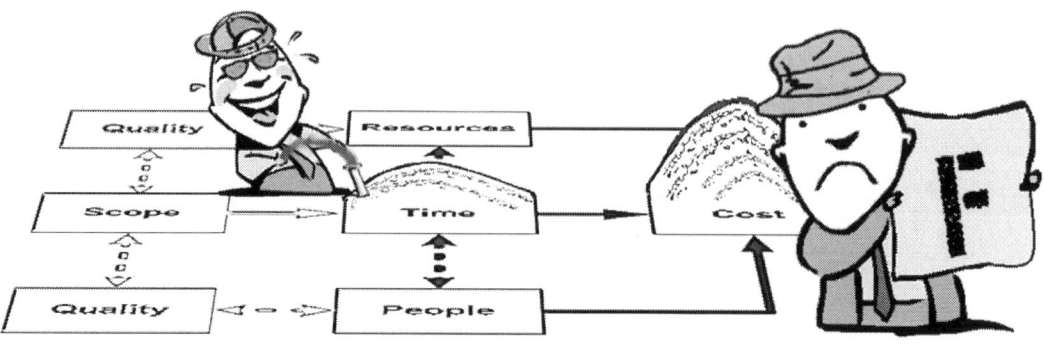

12.3. [Cost] Monitoring & Control

12.3.1. Cost: Earned Value Performance Measures

In a practical sense, [Cost] Monitoring & Control [Section 7.3.1] asks the following:

- **Planned Value (PV)** – is what was the planned spend to date, to get the project to where it was planned to be by now (instead of where it currently is).

- **Actual Cost (AC)** – is what actually has been spent to date (when the measurements were taken), to get the project to where it really is now.

- **Earned Value (EV)** – is what was planned to have been spent to date (when the measurements were taken), to get the project to where it really is now.

- **Cost Variance (CV)** – is how far above or below the [Cost Baseline] budget to get the project to the point where it really is now.

- **Schedule Variance (SV)** – is how far ahead or behind the [Time Baseline] Schedule of where we really are now, against where we planned to be by now.

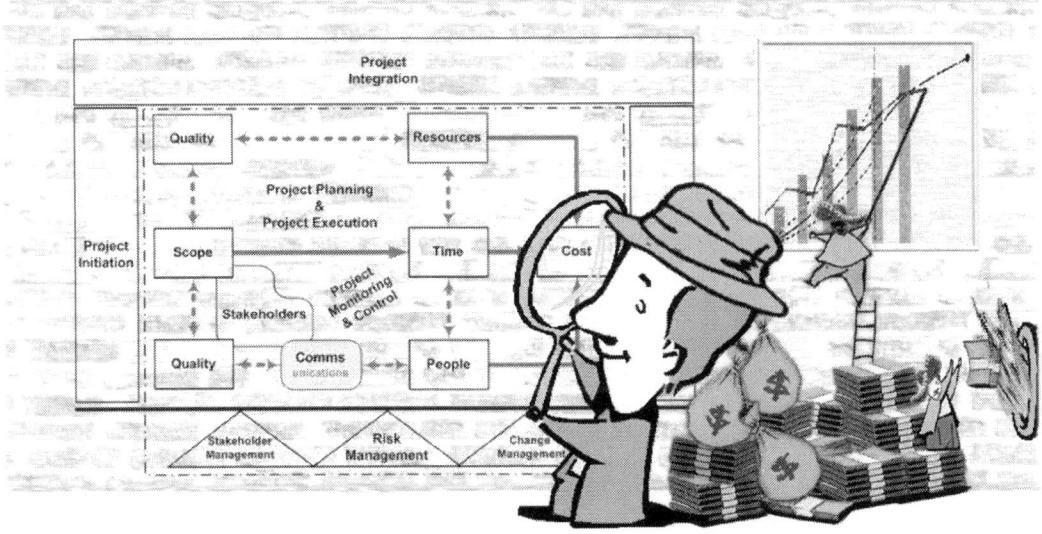

These questions listed so far for [Cost] Monitoring & Control all relate to the Earned Value Performance Measures; i.e. the S-Curves reproduced below in [Figure 71]. Unfortunately, these **cost performance measures judge a project's progress against packages of time and hours of effort spent (NOT specifically against the percentage of tasks completed)**.

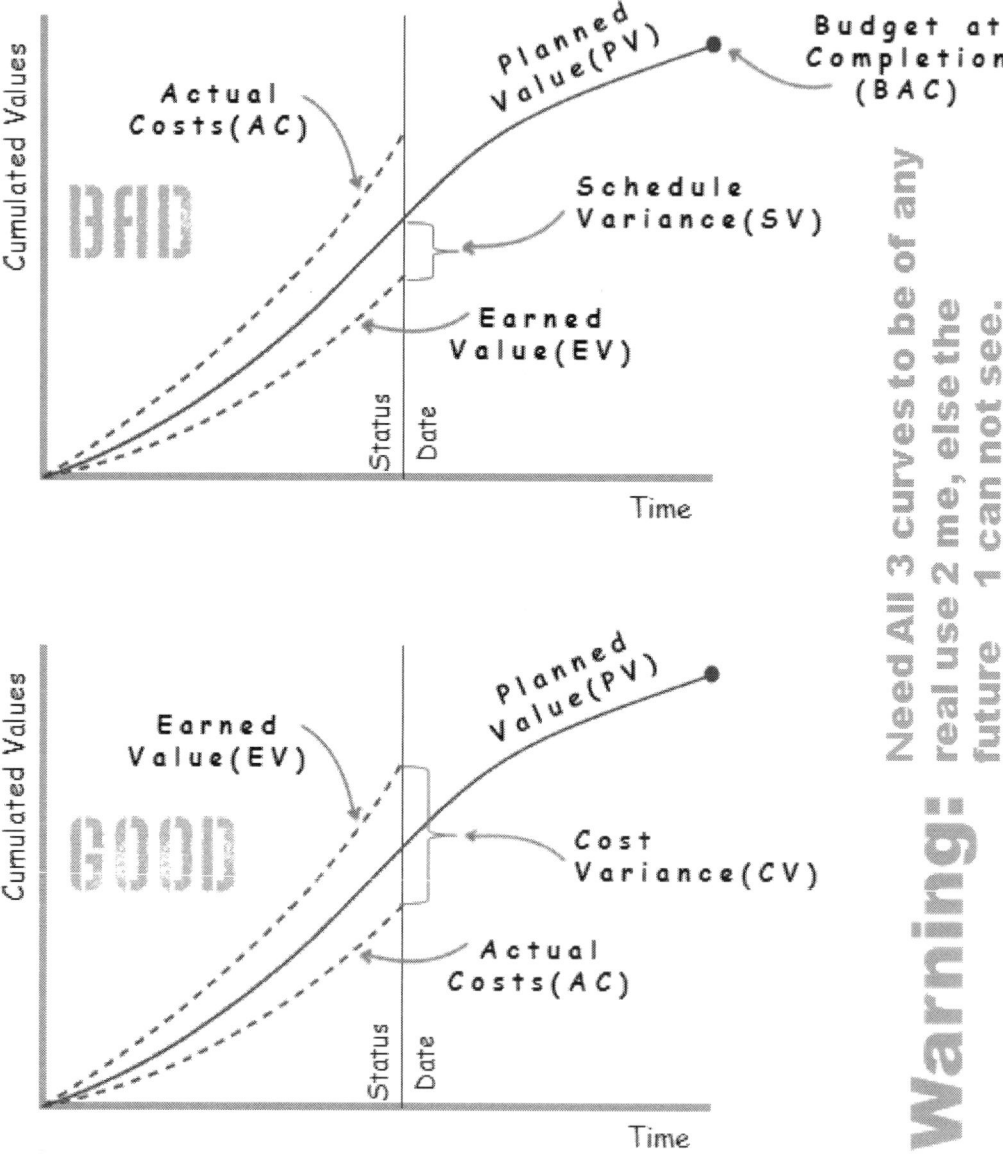

Figure 71: Actual Costs (AC), Earned Value (EV), and Planned Value (PV) versus time mapped as Earned Value Management S-Curves.

These Earned Value Performance Measures are usually expanded upon to answer the following questions related to the completion of tasks:

- **Cost Performance Index (CPI)** – is how good or bad is the project's performance when compared to the [Cost Baseline] budget (in terms of monetary units / labour hours), to get the project to where it really is now?

  > *WARNING: For a fixed-price contract DO NOT present the CPI to the customer's representative, as this would give an insight into what is the performing organization's profit margin related specifically to this project. And, subsequently could be used against the performing organization.*

- **Schedule Performance Index (SPI)** – is how good or bad is the project's performance when compared to the [Time Baseline] schedule (in terms of monetary units / labour hours), where the project really is now, against where it was planned to be by now?

- **Budget At Completion (BAC)** – what is the total Planned Value when the project is finished?

- **Estimate To Complete (EAC)** – What do we have to spend from here on in, to get this project across the finish line?

- **Estimate At Completion (EAC)** – What do we now calculate as the total cost when this project finishes?

Consequently, given the realities of the current situation and the prevailing circumstances, how does one realign the [Real-world Costs] against the [Cost Baseline], or does one need to re-baseline the planned [Costs]? That is:

- Do additional [People] & [Resources] need to be allocated, or

  do the utilization percentages for the existing ones need to be changed?

- Do the scheduled tasks [Scope] need to be refactored, or

  does the schedule [Time] duration need to be compressed? See [Section 7.2.1].

Example Case: Earned Value Performance Measurements in action.

Consider a software development project with seven [People] allocated;

e.g. a project manager come system architect, a system designer, 4x developers, and a tester. Where each of these people had the same "bums on seats" Real Cost rate of $100 per hour with a standard workday of 8 hours.

The project tasks and durations are as follows.

Phase	Task	Duration	Resource
Initiating	Initiation	1 hour	Project Manager
	Conceptualization	5 hours	Project Manager (Architect)
	Work Approval	2 hours	Project Manager
	Requirements Gathering	1 days	Project Manager
Planning	Planning	3 days	Project Manager
	Design	2 days	System Designer
Executing	Development	5 days	Developer 1-4
	Debugging	2 days	Developer 1-3
	Integration	3 days	Developer 1-2
	Verification Testing	2 days	Tester & Developer
Closing	Acceptance Testing	1 day	Tester
	Release / Rollout	1 day	Developer
	Closure & Payment	4 hours	Project Manager

Now create a waterfall project schedule (using Finish-to-Start dependency) and lay these [Scope] tasks out on the [Time] schedule, allocate the appropriate [People] to each of these tasks, and assign [Costs] to each of these people; i.e. utilizing the techniques described in [Section 4.3.2.1].

	ⓘ	Task Name	Duration	Start	Finish	Predeces	Resource Names	Cost
1		☐ PROJECT: COSTCURVE	19.5 days	Mon 1/3/11	Fri 1/28/11			$35,600.00
2		MILESTONE: Project Deliv	0 days	Fri 1/28/11	Fri 1/28/11	24		$0.00
3								
4		Initiation	1 hr	Mon 1/3/11	Mon 1/3/11		Project Manager	$100.00
5		Conceptualization	5 hrs	Mon 1/3/11	Mon 1/3/11	4	Project Manager	$500.00
6		Work Approval	2 hrs	Mon 1/3/11	Mon 1/3/11	5	Project Manager	$200.00
7		Requirements Gathering	1 day	Tue 1/4/11	Tue 1/4/11	6	Project Manager	$800.00
8		Planning	3 days	Wed 1/5/11	Fri 1/7/11	7	Project Manager	$2,400.00
9		Design	2 days	Thu 1/6/11	Fri 1/7/11	7,8FF	System Designer	$1,600.00
10		☐ Development	5 days	Mon 1/10/11	Fri 1/14/11	8,9		$16,000.00
11		Development Part 1	5 days	Mon 1/10/11	Fri 1/14/11		Developer 1	$4,000.00
12		Development Part 2	5 days	Mon 1/10/11	Fri 1/14/11		Developer 2	$4,000.00
13		Development Part 3	5 days	Mon 1/10/11	Fri 1/14/11		Developer 3	$4,000.00
14		Development Part 4	5 days	Mon 1/10/11	Fri 1/14/11		Developer 4	$4,000.00
15		☐ Debugging	2 days	Mon 1/17/11	Tue 1/18/11	11		$4,800.00
16		Debug Part 1 & 2	2 days	Mon 1/17/11	Tue 1/18/11		Developer 1,Developer 2	$3,200.00
17		Debug Part 3 & 4	2 days	Mon 1/17/11	Tue 1/18/11		Developer 3	$1,600.00
18		Integration	3 days	Wed 1/19/11	Fri 1/21/11	16	Developer 3,Developer 1	$4,800.00
19		☐ Verification Testing	2 days	Mon 1/24/11	Tue 1/25/11	18		$2,400.00
20		Test	2 days	Mon 1/24/11	Tue 1/25/11		Tester 1	$1,600.00
21		Defect Fix	1 day	Tue 1/25/11	Tue 1/25/11	20FF	Developer 1	$800.00
22		Acceptance Testing	1 day	Wed 1/26/11	Wed 1/26/11	21	Tester 1	$800.00
23		Release / Rollout	1 day	Thu 1/27/11	Thu 1/27/11	22	Developer 1	$800.00
24		Closure & Payment	0.5 days	Fri 1/28/11	Fri 1/28/11	23	Project Manager	$400.00

Figure 108: Example – development project with [Time] & [Cost] Baselines.

The above schedule will form the project's [Time Baseline] and [Cost Baseline] of; one month's duration and a **Level-Of-Effort (LOE)** value of $35,600.

If these costs per task were combined on a per day basis then the resultant "Daily Cost vs. Time" graph in [Figure 110] would look very similar to the "Cost & Staffing Levels vs. Time" graph illustrated in [Figure 23] back in [Section 2.5].

While adding up the [Costs] on a daily basis; if these [Costs] were also accumulated day after day over the entire life of the project then the resultant curve in [Figure 111] would approximate the theoretical S-Curve for Planned Value (PV) and Earned Value (EV). Similarly, the accumulated weekly [Costs] would result in a weekly expenditure / cash outflow as depicted in [Figure 112].

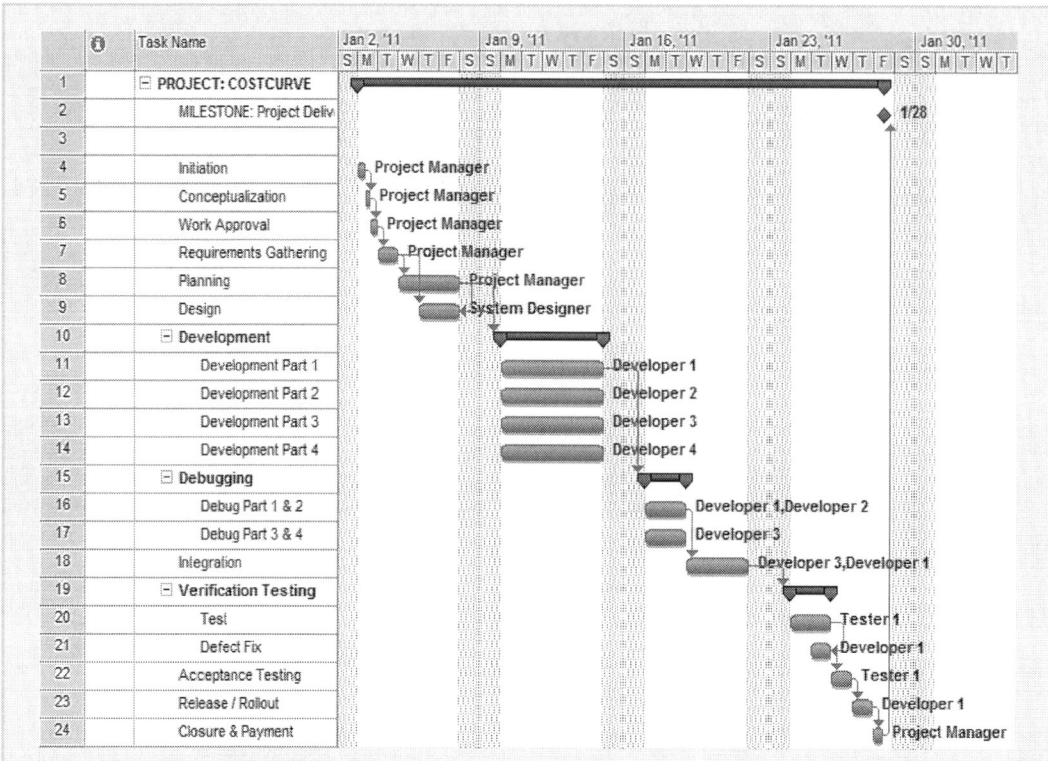

Figure 109: Example – development project with baseline task bars.

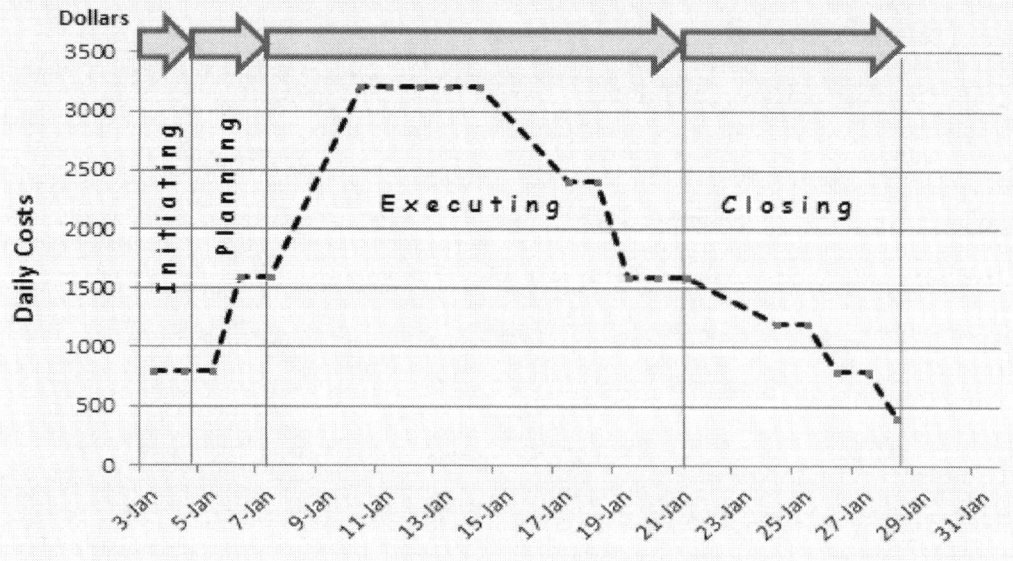

Figure 110: Example – Daily Costs vs. Time.

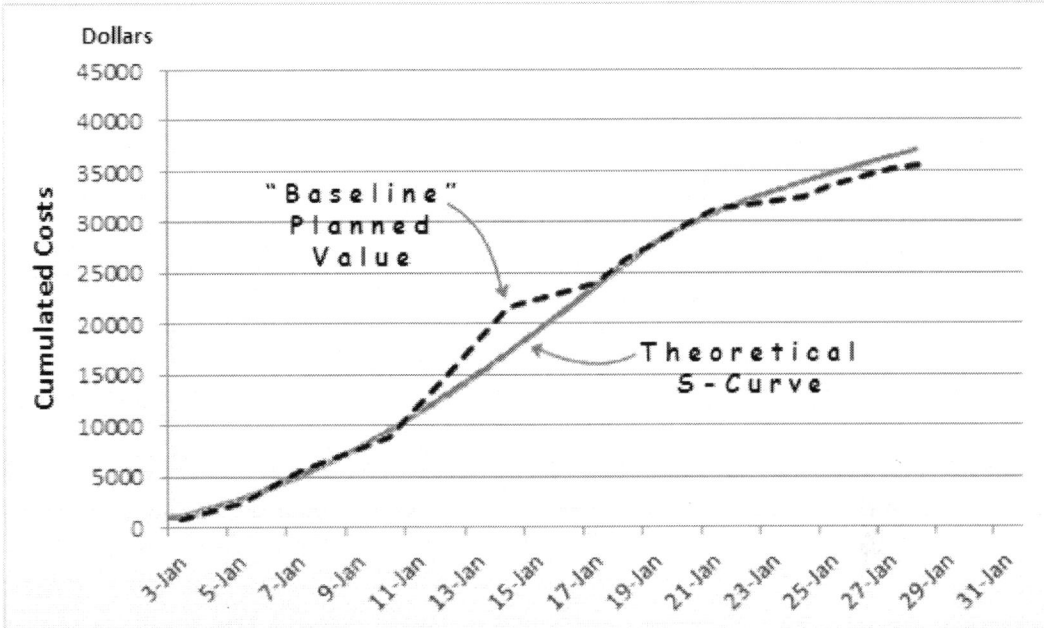

Figure 111: Example – comparison of an actual Accumulated Cost vs. Time compared to the theoretical S-Curve.

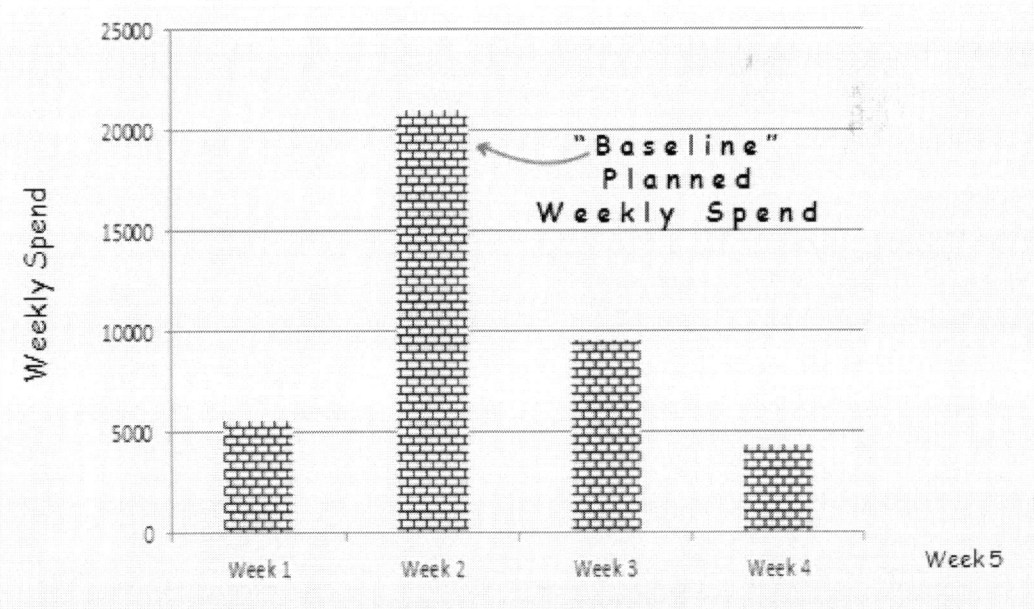

Figure 112: Example – baseline weekly spend (cash outflow).

Once the project is underway the Execution Phase runs into a few snags and subsequently the project's development is dragged out by "a coupla days"; i.e. tasks 11 through 14 all have increased from 5 days to 7 days in duration which is longer than was planned for and presented in the Baseline Schedule.

Hence, the Work-In-Progress Schedule [Figure 113] and [Figure 114] looks worse than the original Baseline Schedule [Figure 108] and [Figure 109].

	❶	Task Name	Duration	Start	Finish	Predeces	Resource Names	Cost
1		⊟ PROJECT: COSTCURVE	21.5 days	Sat 1/1/11	Tue 2/1/11			$42,000.00
2		MILESTONE: Project Deliv	0 days	Sat 1/1/11	Sat 1/1/11			$0.00
3								
4		Initiation	1 hr	Mon 1/3/11	Mon 1/3/11		Project Manager	$100.00
5		Conceptualization	5 hrs	Mon 1/3/11	Mon 1/3/11	4	Project Manager	$500.00
6		Work Approval	2 hrs	Mon 1/3/11	Mon 1/3/11	5	Project Manager	$200.00
7		Requirements Gathering	1 day	Tue 1/4/11	Tue 1/4/11	6	Project Manager	$800.00
8		Planning	3 days	Wed 1/5/11	Fri 1/7/11	7	Project Manager	$2,400.00
9		Design	2 days	Thu 1/6/11	Fri 1/7/11	7,8FF	System Designer	$1,600.00
10		⊟ Development	7 days	Mon 1/10/11	Tue 1/18/11	8,9		$22,400.00
11	📋	Development Part 1	7 days	Mon 1/10/11	Tue 1/18/11		Developer 1	$5,600.00
12	📋	Development Part 2	7 days	Mon 1/10/11	Tue 1/18/11		Developer 2	$5,600.00
13	📋	Development Part 3	7 days	Mon 1/10/11	Tue 1/18/11		Developer 3	$5,600.00
14	📋	Development Part 4	7 days	Mon 1/10/11	Tue 1/18/11		Developer 4	$5,600.00
15		⊟ Debugging	2 days	Wed 1/19/11	Thu 1/20/11			$4,800.00
16		Debug Part 1 & 2	2 days	Wed 1/19/11	Thu 1/20/11	11,12	Developer 1,Developer 2	$3,200.00
17		Debug Part 3 & 4	2 days	Wed 1/19/11	Thu 1/20/11	13,14	Developer 3	$1,600.00
18		Integration	3 days	Fri 1/21/11	Tue 1/25/11	15	Developer 3,Developer 1	$4,800.00
19		⊟ Verification Testing	2 days	Wed 1/26/11	Thu 1/27/11	18		$2,400.00
20		Test	2 days	Wed 1/26/11	Thu 1/27/11		Tester 1	$1,600.00
21		Defect Fix	1 day	Thu 1/27/11	Thu 1/27/11	20FF	Developer 1	$800.00
22		Acceptance Testing	1 day	Fri 1/28/11	Fri 1/28/11	21	Tester 1	$800.00
23		Release / Rollout	1 day	Mon 1/31/11	Mon 1/31/11	22	Developer 1	$800.00
24		Closure & Payment	0.5 days	Tue 2/1/11	Tue 2/1/11	23	Project Manager	$400.00

Figure 113: Example – project schedule delayed by 2 days.

NOTICE   how in [Figure 113] when it is compared to [Figure 108] the project's total [Costs] has increased from $35,600 to $42,000 and the project's total [Time] duration has increased from 19.5 days to 21.5 days.

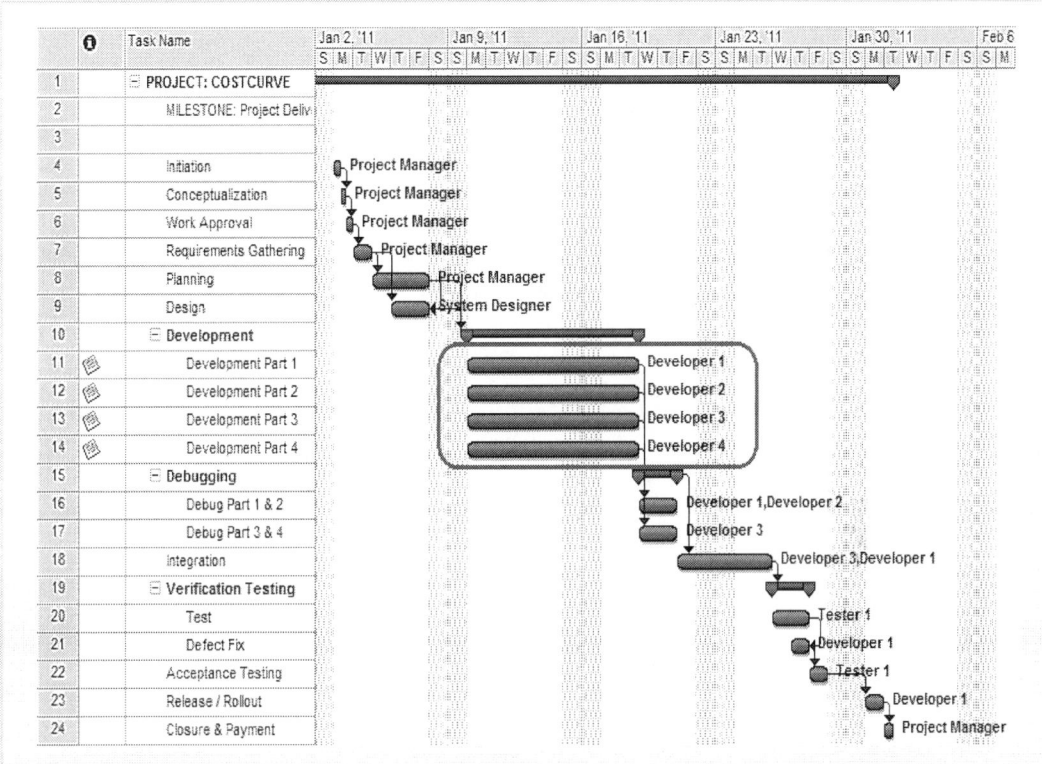

Figure 114: Example – effect on the schedule's task bars of a 2 day delay.

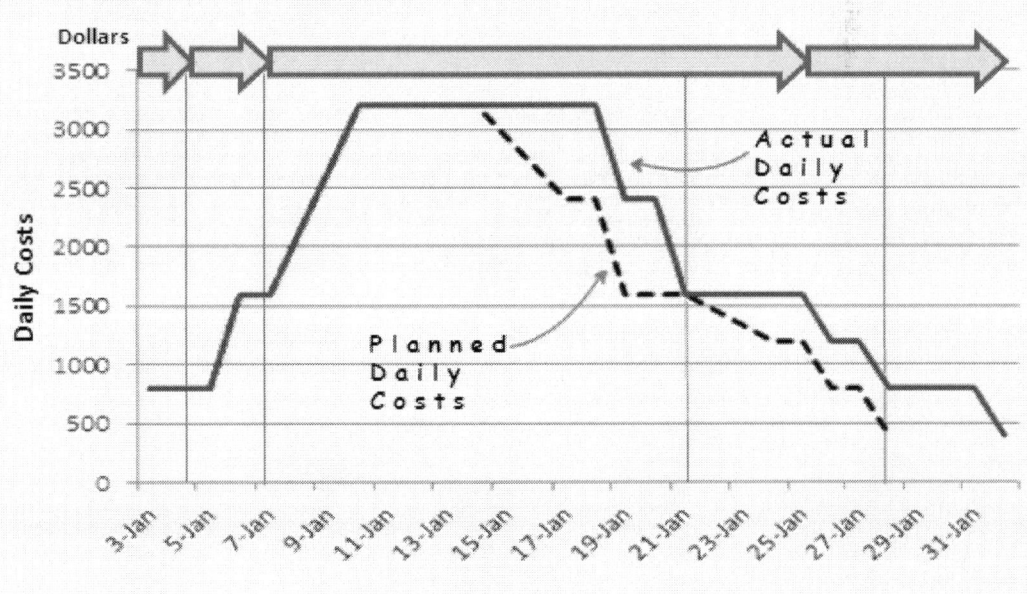

Figure 115: Example – Daily Costs vs. Time with a 2 day delay.

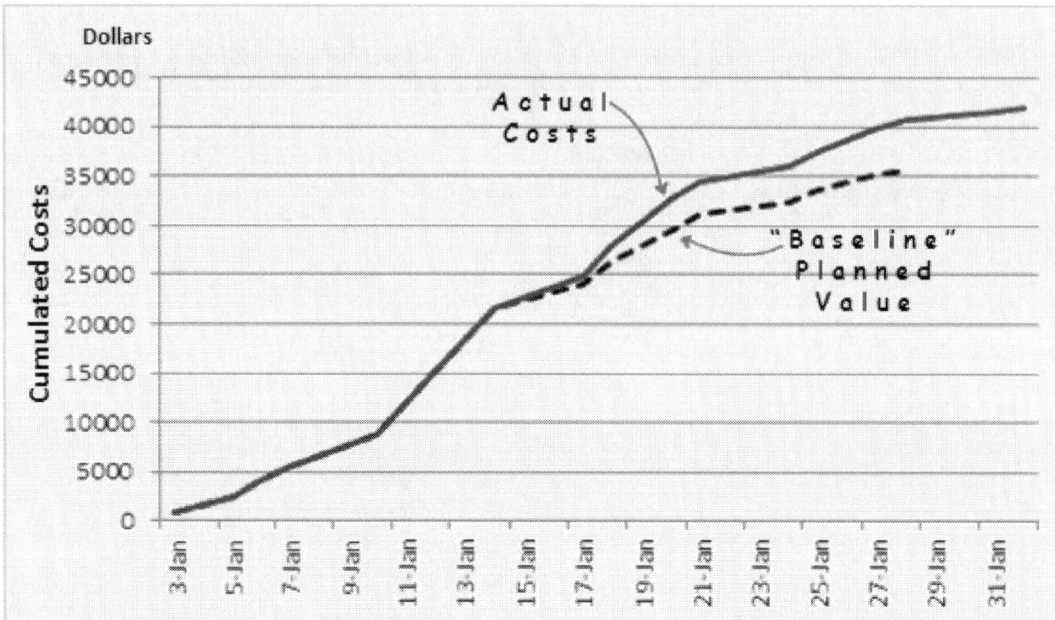

Figure 116: Example – the actual Accumulated-Cost vs. Time with a 2 day delay.

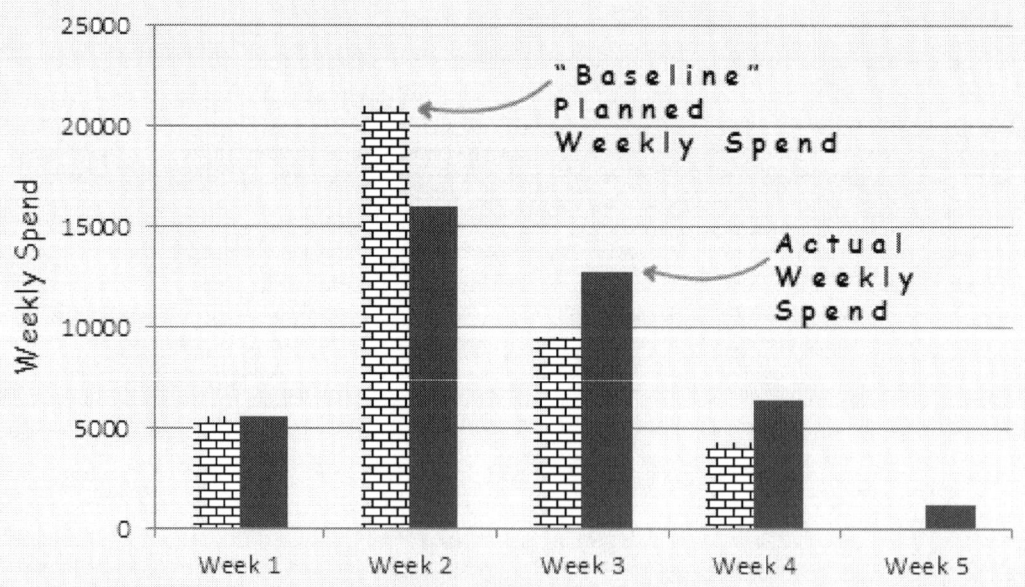

Figure 117: Example – baseline weekly spend compared to the actual weekly spend due to a 2 day delay.

So, the project was delivered a few days late and that meant it was $6,400 over budget ... big deal.

Well yes, in this simple SDLC project there is only a $6,400 difference to the bottom line which may not seem to be much of a "big deal", but:

- What if, this was for a lot larger project and there were many tasks affected. How about a magnitude increase of another one or two zeros at the end of that cost figure; i.e. $64,000 or $640,000 or $6,400,000. ... *Ouch, that's gotta hurt.*

- What if, the contract with the customer stipulated that the project must be delivered by a specific end date else **liquidated damages** (inline with the contract's **penalty clauses**) will be enacted, where the performing organization will be **paying compensation** to the customer organization for every day the deliverable is late.

- What if, those extra few days of work in the 5th week of [Figure 117] had to be paid for by the performing organization's **own cash reserves** and not by the customer.

- What if, there was a "time sensitive" partial payment point on delivering the project on time, and due to this delay the performing organization subsequently experienced a **negative cash flow crisis**, which thus pushed the organization into insolvency (of not being able to pay its outstanding debts when due).

Similarly, this example was only for a couple of days delay, but:

- What if, this delay was for a month. What about for a lot larger project where the accumulated delays for multiple tasks could extend out the project by several months. Hence, these additional costs could quickly snowball in to a sizable problem for the performing organization. ... *AKa, a Vampire-sucking project.*

Please refer to [Section 9.1.1] for an explanation of the "Potential Cost" to the performing organization of a project running late and/or the project being over budget.

12.3.2. Cost: Percentage Complete Measures

I have a fundamental issue with equating earned value with time spent where the project is "up to today", as this is simply a date reference and not real progress. IMHO, earned value should reflect the percentage complete as denoted by the progress bar inside each task bar in the Gantt Chart schedule.

Okay then, other formulas for calculating Earned Value (EV) and Planned Value (PV):

EV = BAC x Actual % Complete ... where BAC is Budget At Completion.

PV = BAC x Planned % Complete

Let's consider a 5 day project that consists of 4x tasks and has 4x people allocated to work independently on each of these tasks.

	% Complete	Task Name	Duration	Start	Finish	Apr 3, '11 S M T W T F S
1	20%	Task 1	5 days	Mon 4/4/11	Fri 4/8/11	
2	60%	Task 2	5 days	Mon 4/4/11	Fri 4/8/11	
3	40%	Task 3	5 days	Mon 4/4/11	Fri 4/8/11	
4	80%	Task 4	5 days	Mon 4/4/11	Fri 4/8/11	

Figure 118: Example percentage complete task bar analysis.

The allocated budget for this project is $1,000 / day / person.

Budget At Completion = $1,000 x 5 (days) x 4 (people) = $20,000

That is, each of these project tasks has a budget of $5,000 to complete.

Given that it is the conclusion of the third day of this five day project, then each of these tasks (according to the schedule) are planned to be 60% complete;

PV = BAC x Planned % Complete = $20,000 x 60% = $12,000

However, the reality is that the implementation of the 1st task is a bit more difficult than was envisioned and this task is only 20% complete, the 2nd task is progressing nicely at 60% complete, the 3rd task is a little harder than expected at 40% complete, and the 4th task turned out to be a breeze and is now 80% complete.

Mathematically, at the conclusion of the third day, the Earned Value is:

EV = BAC x Actual % Complete

(Task 1) EV1 = $5,000 x 20% = $1,000

(Task 2) EV2 = $5,000 x 60% = $3,000

(Task 3) EV3 = $5,000 x 40% = $2,000

(Task 4) EV4 = $5,000 x 80% = $4,000

EV = EV1   + EV2   + EV3   + EV4

EV = $1,000 + $3,000 + $2,000 + $4,000

EV = $10,000

Therefore, in this example, the project is in a bit of trouble because at the current status date, the Earned Value ($10,000) is less than the Planned Value ($12,000).

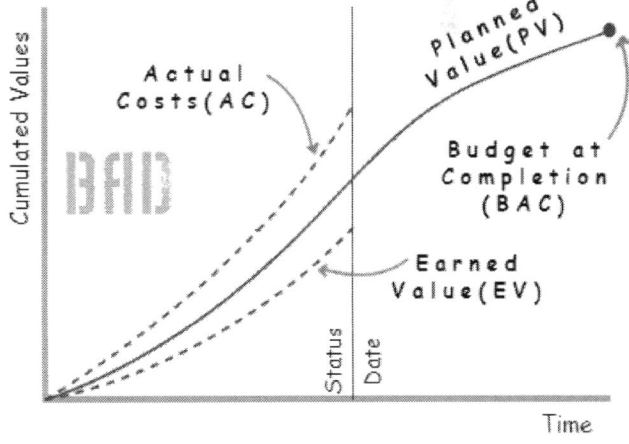

12.3.3. Cost: Performance Indexes

These techniques for Earned Value Performance Measures (in the previous Sections [12.3.1] and [12.3.2]) can be combined together to provide another way of monitoring & controlling the project's progress.

Consider the following SDLC project that has been broken up into the stages of; Specification, Design, Development, Verification / Test, and Release. This particular project has an underline End-Date, which happens to give a project duration of 10 weeks. Each one of these stages is planned (expected) to require a certain number of hour's effort per week. This project also has a fixed monetary value per hour worked, where each hour worked is worth N dollars, N bars of gold-pressed latinum, N peanuts, or N whatever's.

	Specification		Design		Development			Verification / Test		Release
Week	1	2	3	4	5	6	7	8	9	10
Planned Hours of Effort	5	10	20	40	60	60	60	40	25	7

If these "Planned Hours of Effort" per week (i.e. **Planned Value – PV**) were accumulated over the duration of the project, then this would equal the "planned" total hours to conduct the project (i.e. **Budget At Completion – BAC**).

	Specification		Design		Development			Verification / Test		Release
Week	1	2	3	4	5	6	7	8	9	10
Planned Hours of Effort	5	10	20	40	60	60	60	40	25	7
Accumulated Planned Hours	5	15	35	75	135	195	255	295	320	327

Based on the weekly "Accumulated Planned Hours" it can be determined via the Budget At Completion (e.g. 327 hours) what is the **"Planned Percentage Complete"** per week.

Planned % Complete = PV / BAC

	Specification		Design		Development			Verification / Test		Release
Week	1	2	3	4	5	6	7	8	9	10
Planned Hours of Effort	5	10	20	40	60	60	60	40	25	7
Accumulated Planned Hours	5	15	35	75	135	195	255	295	320	327
Planned % Complete	2%	5%	11%	23%	41%	60%	78%	90%	98%	100%

The resultant table above describes the **Budget Cost of Work Scheduled – BCWS**.

Now that this project is underway and work is being done, then each week the project team members can diligently submit their timesheets (i.e. "Actual Hours Worked") against the project's job code.

	Specification		Design		Development			Verification / Test		Release
Week	1	2	3	4	5	6	7	8	9	10
Actual Hours Worked	3	12	14	52	61	37	79	57	12	17

If these "Actual Hours Worked" per week (i.e. **Actual Cost – AC**) were accumulated over the duration of the project then this would equal the "actual" total hours to conduct the project (i.e. **Estimate At Completion – EAC**).

	Specification		Design		Development			Verification / Test		Release
Week	1	2	3	4	5	6	7	8	9	10
Actual Hours Worked	3	12	14	52	61	37	79	57	12	17
Accumulated Actual Hours	3	15	29	81	142	179	258	315	327	344

Based on the weekly "Accumulated Actual Hours" it can be determined via the Estimate At Completion (e.g. 344 hours) what is the **"Actual Percentage Complete"** per week.

Actual % Complete = AV / EAC

	Specification		Design		Development			Verification / Test		Release
Week	1	2	3	4	5	6	7	8	9	10
Actual Hours Worked	3	12	14	52	61	37	79	57	12	17
Accumulated Actual Hours	3	15	29	81	142	179	258	315	327	344
Actual % Complete	1%	4%	8%	24%	41%	52%	75%	92%	95%	100%

The resultant table above describes the **Actual Cost of Work Performed – ACWP**.

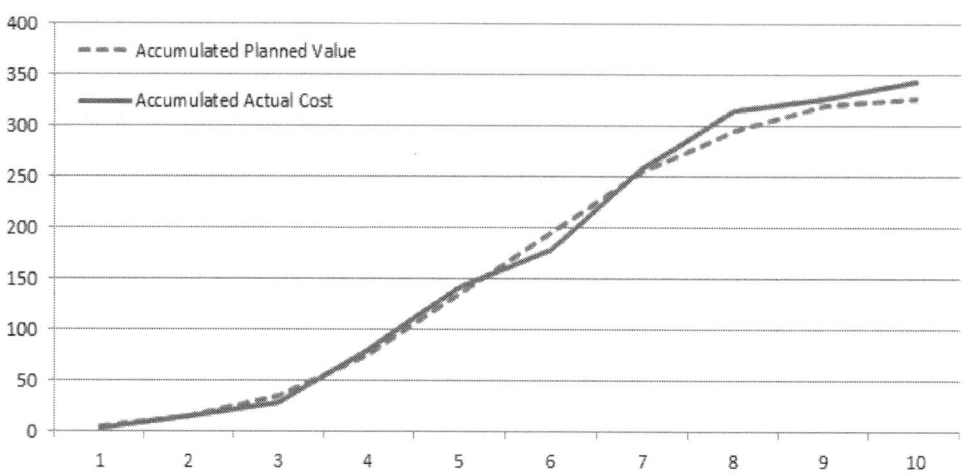

Figure 119: Accumulated Planned Value compared to the Accumulated Actual Cost.

NOTICE how the above plot in [Figure 119] for the Accumulated Planned Value (Hours) compared to the Accumulated Actual Cost (Hours) corresponds to the shape of the theoretical Earned Value S-Curve.

This is all fine & dandy, but... the Actual % Complete presented so far has been calculated based on the mathematical relationships, and is not based on the "true" percentage of work that has been completed.

This "true" percentage of work completed would be extracted from the weekly update to the project's Work-In-Progress Schedule. Look back at [Figure 105] in [Section 12.2.1], where a ruler was used to mark-off on the schedule to determine what has been really completed.

For this example, the "Real % Complete" doesn't match those values previously calculated. Instead, the "real" Actual % Complete as derived from the project's Work-In-Progress Schedule happens to be as follows.

	Specification		Design		Development			Verification / Test		Release
Week	1	2	3	4	5	6	7	8	9	10
"Real" % Complete	2%	7%	10%	20%	35%	45%	85%	90%	95%	100%

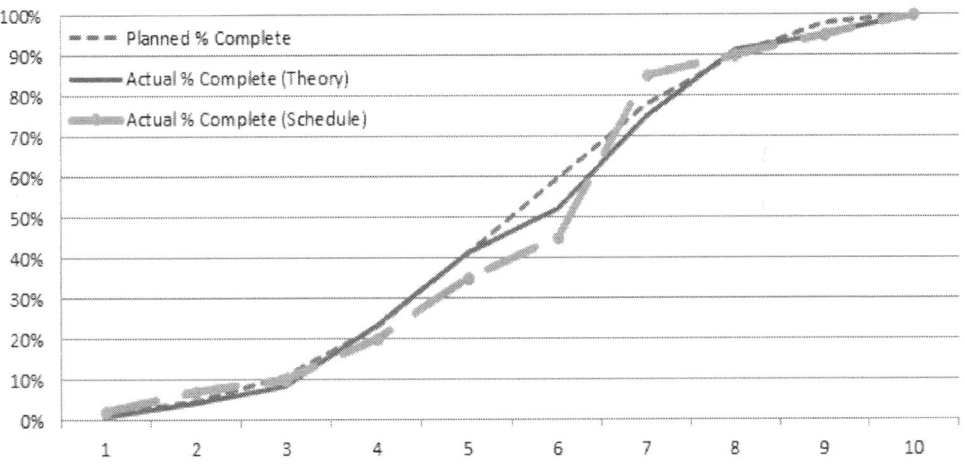

Figure 120: Schedule derived Actual % Complete compared to Planned % Complete.

Using the "real" Actual % Complete, then the **Earned Value – EV** can be derived.

EV = BAC x Actual % Complete

	Specification		Design		Development			Verification / Test		Release
Week	1	2	3	4	5	6	7	8	9	10
"Real" % Complete	2%	7%	10%	20%	35%	45%	85%	90%	95%	100%
Earned Value	6.54	22.89	32.7	65.4	114.45	147.15	277.95	294.3	310.65	327

NOTICE how the Earned Value at the end of the project is equal to 327 (i.e. the BAC) and not 344 (i.e. the EAC). This is because, the Earned Value maps where the project currently is (i.e. the "real" Actual % Complete) against the equivalent point in the planned baseline.

This previous table describes the **Budget Cost of Work Performed – BCWP**.

Now that the accumulative Earned Value – EV, the accumulative Planned Value – PV, and the accumulative Actual Cost – AC have all been determined, then the **Schedule Performance Index – SPI** for [Time] and the **Cost Performance Index – CPI** for [Costs] can be calculated on a weekly basis.

$$SPI = (EV_{accumulative}) / (PV_{accumulative})$$

$$CPI = (EV_{accumulative}) / (AC_{accumulative})$$

	Specification		Design		Development			Verification / Test		Release
Week	1	2	3	4	5	6	7	8	9	10
Earned Value	6.54	22.89	32.7	65.4	114.45	147.15	277.95	294.3	310.65	327
Accumulated Planned Value	5	15	35	75	135	195	255	295	320	327
Accumulated Actual Cost	3	15	29	81	142	179	258	315	327	344
Schedule Performance Index	1.31	1.53	0.93	0.87	0.85	0.75	1.09	1.00	0.97	1.00
Cost Performance Index	2.18	1.53	1.13	0.81	0.81	0.82	1.08	0.93	0.95	0.95

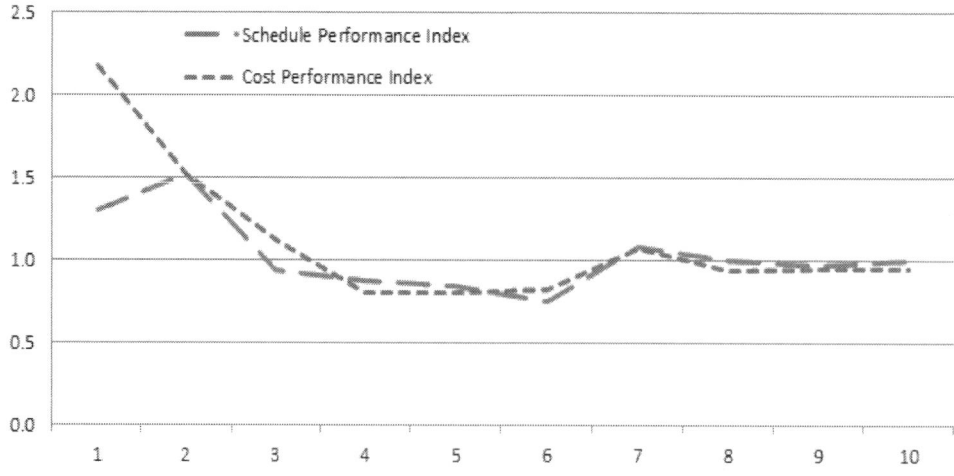

Figure 121: Schedule Performance Index (SPI) and Cost Performance Index (CPI).

		Specification		Design	Development				Verification / Test		Release		
		Week											
		1	2	3	4	5	6	7	8	9	10	Total	
Budget Cost of Work Scheduled													
Planned Value	PV	5	10	20	40	60	60	60	40	25	7	327	Budget At Completion (BAC)
BCWS — Accumulated Planned Value		5	15	35	75	135	195	255	295	320	327	·	
Planned % Complete		2%	5%	11%	23%	41%	60%	78%	90%	98%	100%	·	
Actual Cost of Work Performed													
Actual Cost	AC	3	12	14	52	61	37	79	57	12	17	344	Estimate At Completion (EAC)
ACWP — Accumulated Actual Cost		3	15	29	81	142	179	258	315	327	344	·	
Actual % Complete (Theory)		1%	4%	8%	24%	41%	52%	75%	92%	95%	100%	·	
Budget Cost of Work Performed													
Actual % Complete (Schedule)		2%	7%	10%	20%	35%	45%	85%	90%	95%	100%		
BCWP — Earned Value (accumulated)	EV	6.5	22.9	32.7	65.4	114.5	147.2	278.0	294.3	310.7	327.0		EV = BAC x Actual % Complete (Schedule)
Performance Indexes													
Schedule Performance Index	SPI	1.31	1.53	0.93	0.87	0.85	0.75	1.09	1.00	0.97	1.00		SPI = (EV acc) / (PV acc)
Cost Performance Index	CPI	2.18	1.53	1.13	0.81	0.81	0.82	1.08	0.93	0.95	0.95		CPI = (EV acc) / (AC acc)
Traffic Light / Key Performance Indicators													
Schedule Performance Index	SPI	GREEN	GREEN	GREEN	AMBER	AMBER	RED	GREEN	GREEN	GREEN	GREEN		Upper limit (lose) = None / Upper Limit (tight) = None / Lower Limit (tight) = 0.90 / Lower Limit (lose) = 0.80
Cost Performance Index	CPI	RED	RED	RED	RED	RED	RED	AMBER	AMBER	GREEN	GREEN		Upper limit (lose) = 1.10 / Upper Limit (tight) = 1.05 / Lower Limit (tight) = 0.95 / Lower Limit (lose) = 0.90

Figure 122: Performance Indexes and Traffic Light KPIs.

NOTE ❒ The **SPI** may be **GREATER than OR LESS than ONE** depending on how early or late the project is delivered. Whereas, the **CPI** may be **GREATER than OR LESS than ONE** depending on how **"profitable"** an endeavour is the project.

12.3.4. Cost: Key Performance Indicators

For the previous table in [Figure 122], each week a project dashboard for all of the projects currently being undertaken by the performing organization / business unit would be presented to senior management. This below example dashboard [Figure 123] of Key Performance Indicators (KPI) would tabulate each project's CPI and SPI as well as the ongoing trends for both CPI and SPI, overlaid with a red-yellow-green traffic light indicator of each project's respective good-indifferent-bad progress toward completion.

INDICATOR	PROJECT ALFA	PROJECT BRAVO	PROJECT CHARLIE	PROJECT DELTA
PROJECT MANAGER	PAUL M.	JOHN L.	GEORGE H.	RINGO S.
CPI	1.03	0.97	0.91	0.22
SPI	0.99	0.98	0.94	0.85
CPI TREND	⇨	⇨	⬊	⇨
SPI TREND	⬊	⬈	⬊	⬈
MILESTONE	GREEN	GREEN	YELLOW	RED

Figure 123: Project Dashboard.

Back in [Section 7.3.1] it was stated that:

- For the Schedule Performance Index (**SPI**) a value of less than ONE (**< 1.0**) indicates that **less progress** had been made **than was planned for**. Whereas, a **SPI** value greater than ONE (**> 1.0**) indicates that **more progress** had been **made than was planned for**.

- For the Cost Performance Index (**CPI**) a value of less than ONE (**< 1.0**) indicates a **cost overrun** for the **work performed to date**. Whereas, a **CPI** value greater than ONE (**> 1.0**) indicates a **cost underrun for the work performed** so far.

For a fixed-price contract DO NOT report or present the CPI to the customer's representative; as to do so would give an insight into what is the performing organization's profit margin for this project.

If the CPI is much greater than ONE then this could / would be interpreted by the customer's representative as. ... *"Hay sucker, we are charging you a lot more than what this job really should have cost".* This realization would in turn result in an acrimonious relationship between the customer and performing organizations.

In reality, it is not a tragedy if the SPI and/or the CPI are not exactly at 1.0 each week when these performance index values are calculated; as some weeks these index values will be independently lower or higher than 1.0. However, what is important is how much higher or lower these index values are, and **what are the continual weekly "trends"** for these index values; i.e. is the **movement of each index on the rise or on the fall** compared to 1.

For the purpose of monitoring & control, an adjustable value range window is used. *For example; a relative loose window of 0.90 to 1.10, or a tighter window of 0.95 to 1.05.* **As long as the current index values are within the prescribed "Performance Index Window" then all is well,** but outside this window then all is not good; see [Figure 124].

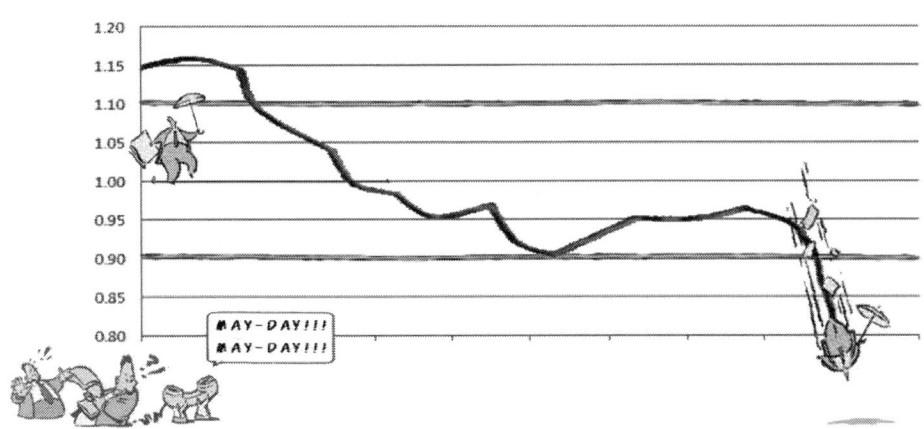

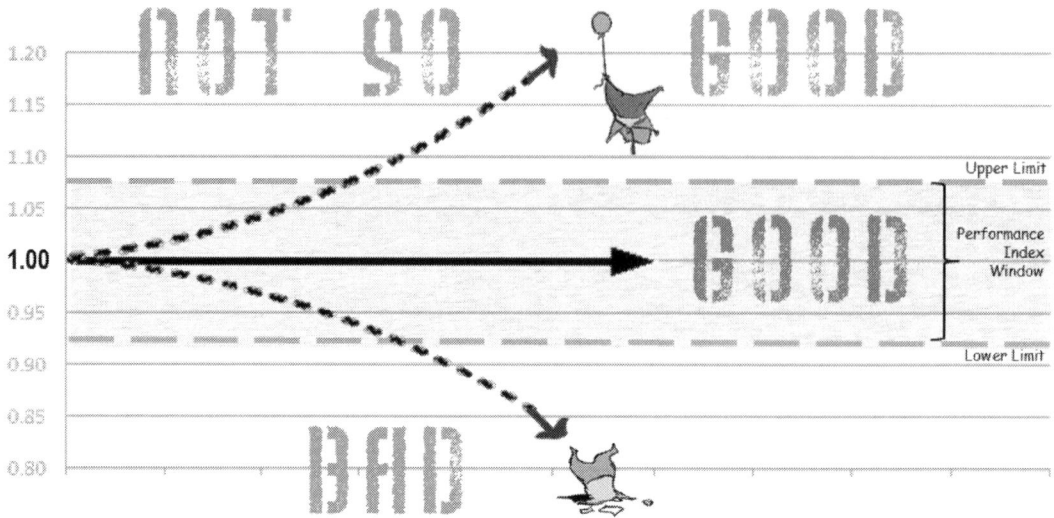

Figure 124: Performance Index Window with upper & lower ranges.

If the current value of the **SPI exceeds the upper limit** of the prescribed "Schedule Performance Index Window" **then possibly the schedule has too much buffering "fat" included, or** it could also mean that the **[People] & [Resources]** are being **worked excessively** to get more done than what was originally planned for.

Conversely, if the **SPI** is **below the lower limit** of the Schedule Performance Index Window **then the percentage of work completed is falling behind** that which was planned for.

If the current value of the **CPI exceeds the upper limit** of the prescribed "Cost Performance Index Window" **then possibly the budget has been over estimated "inflated", or** it could also mean that the **[People] & [Resources]** are being **underutilized**, and/or the expected amount of [Time] is not being booked against the project's job code.

Conversely, if the **CPI** is **below the lower limit** of the Cost Performance Index Window **then the cost of the work completed is exceeding** that which was planned for, and/or there is non-related work being booked against this project's job code.

If at present, **both the SPI and the CPI** were with**in the "Good" zone** of their respective (independent) Performance Index Windows **then** it would be reasonable to assume that there are **currently no performance problems** with this project's progress.

NOTE   ❏ **For the first couple of weeks, DO NOT get into a panic if either the SPI or the CPI values are outside their respective Performance Index Windows**, because not enough time has passed and not enough work effort has been accumulated to provide any meaningful data about the true progress of the project.

> *For example; if it was planned that for the first week's tasking 20 hours of effort would be required, but instead 24 hours was needed to finish this work then that is a 20% difference or a CPI = EV/AC = 20/24 = 0.83.*

However, what is **of interest is the general direction (trend) of each performance index over a few weeks.** ... *Do these start homing in on 1 ?*

❏ **The Performance Index Windows used for the SPI and the CPI values DO NOT necessarily have to have exactly the same limits**. Depending on what is more important, schedule progress or costs spend then either the SPI or the CPI window could be tighter or looser than the other **performance** index's window.

❏ **The Performance Index Window does not necessarily have to have an upper limit**, for it could be decreed that as long as the value is greater than 1.0 then all is good with that **performance** index's aspect of the project.

❏ However, **there must always be a lower limit to the Performance Index Window**, and what value this lower limit is set at is **dependent on the risk tolerance** / risk appetite of the primary-core stakeholders; [Section 10.1.1].

With the **performance index values** being calculated each week then these values could be **affiliated with traffic light indicators** based on a comparison with the tight Performance Index Window and the loose Performance Index Window; [Figure 122] and [Figure 123].

> *For example; values in the range 0.95 to 1.05 are designated as GREEN, values in the range 0.90 to 0.95 and 1.05 to 1.10 are AMBER, and values less than 0.90 or greater than 1.10 are RED. Alternatively, the scheme could have the Performance Index Window having any value above 1.0 as being GREEN.*

12.3.5.   Cost: Via Other Performance Measures

I tend not to use these earned value calculations for projects of a relatively short duration or for projects of low budgetary value.

There is a simple way of doing this performance monitoring; back in [Section 7.3] after the Earned Value Performance Measurements were presented there was a simplified cost formula provided, as shown below.

$$\text{COST}_{\text{TO DATE}} = (\text{ PEOPLE}_{\text{RATES}} \times \text{TIME}_{\text{USED}})$$
$$+$$
$$(\text{RESOURCE}_{\text{UNITS}} \times \text{AMOUNT}_{\text{USED}})$$
$$+$$
$$(\text{FIXED COSTS}_{\text{TO DATE}})$$

At any moment during the project's progress the equation parameters that are usually known upfront are; the "PEOPLE rates" as monetary values (*e.g. $100 / hour*), the "RESOURCE units" monetary values (*e.g. $1000 / tonne, $1000 / day*), and the "FIXED COSTS" (*e.g. $100 / day overhead cost*). Therefore, if the Cost-To-Date (i.e. **Burn-Rate**) can be kept within a predefined allocation then all is fine with this project.

However, some of these parameters are not always so readily known, for example; the amount of [Time] that each [Person] has worked on the project for the reporting period, and the amount of each [Resource] that has been used during the reporting period. This is because the duration of the reporting period is often weeks to a month, hence making these parameters inaccessible for daily and weekly monitoring & control purposes.

Comment: The reasons why earned value calculations may not be appropriate for short-term monitoring & control.

As stated on the previous page, there are three underlying parameters whose characteristics make cost performance measurements ineffectual for [Cost] Monitoring & Control:

1) **TIME used** – people's time that they have worked on the project.

2) **AMOUNT used** – resource units that have been used on the project.

3) **TO DATE** – the reporting duration for the period being analysed.

Oh, the first parameter is easy; you get the 'TIME used' from the timesheets?

Well actually, it is not that easy because as stated back in [Section 12.2.3] timesheets have a few problems:

- It is not unusual for project team members to **whitewash the project's timesheet** (job code / Cost Account Code) with a constant number of hours that they are expected to work each day; *e.g. 8-8-8-8-8 hours*.

  This kind of behaviour will inflate the project's expenditure and corrupt accurate recording of how much time was really spent.

- It is not unusual for project team members to be **tardy with updating and submission of their timesheets**; providing numbers that kind of sound right to them, rather than factually related to the time that they actually spent working on specific tasks. This inaccuracy gets worse the longer the period between the performing of the tasks and when the time expended is recorded for those tasks.

- Timesheets are usually required to be submitted on a weekly, fortnightly, or even monthly basis and not daily; hence, timesheets cannot be used for the daily monitoring & control of the project's progress.

- Once the timesheets have been submitted these have to be approved, these then will go into the performing organization's administrative systems to be processed. Therefore, adding even more delay before the time-cost reporting information will be available for analysis to help with the project's [Cost] Monitoring & Control.

- The timesheets don't usually have the granularity of activity identification to enable the detection of specific tasks that are going to cause problems for the project.

- Access to these administrative systems that churns through the accumulated timesheet information may be limited solely to admin / accounting staff that will disseminate such project information at specific days of the month or fortnightly.

 **Watch what is being charged to the project's job / account codes.** It is not uncommon for some project team members to use their current project's job code as a **catchall** on their timesheet for everything they did during the day **including work not related to the project**. Additionally, as the project is nearing completion or has practically been completed, **watch out for illegitimate time being recorded against the project's job codes.**

WARNING: the over inflating of hours that were actually worked on a project (e.g. overstating or by pushing across hours from another project), and the charging of resources not used on the project is ... "fraudulent" ... and potentially such fraudulent actions, if exposed by the customer organization (or by third party auditors), could result in legal problems for the performing organization. Let alone this exposed fraud being very detrimental to the relationship between the performing organization and the customer organization (as well as their industrial "friends"). ... Where, reputation is everything to an organisation's future survival.

For the second parameter you get the 'AMOUNT used' from the suppliers' invoices?

Well actually, it is not that easy because:

- A representative of the supplier / vendor will firstly have to accumulate the information on the amount of their supplied [Resources] that has been used. This may be done on a monthly or quarterly basis; meaning that there is an inherent delay before this true costing information is available. Moreover, this is only from one supplier; there are probably several others involved with the project.

- Once this usage information has been obtained then the supplier / vendor's financial accounts group will have to produce the invoice. This will most probably be done as a batch with other invoices, hence produced on a monthly or quarterly basis ... meaning delays, delays, delays before the information is usable.

- Once the invoice is produced, it then has to be sent from the supplier / vendor's organization to the performing organization ... meaning more delays.

- At the performing organization, the invoice will be sorted, distributed to appropriate people to authorize its payment ... that is even more delays.

- Finally, the invoice will hopefully be paid.

Though, what is the financial accounts group's policy on paying invoices; do they pay it immediately, after 30 days, or do they use up the entire payment period before paying.

QUESTION: When does the performing organization recognize the cost of the resource? At the time when the resource is received & signed for by a representative of the performing organization, or when the invoice is received from the supplier, or when the invoice is paid by the performing organization?

For the third parameter of 'TO DATE' for the reporting period?

As described back in [Section 6.4], there are three distinct reporting cycles which relate to three distinct levels of management hierarchy; hence, there are three corresponding windows for monitoring & control:

1) **Daily / weekly** – for the Project Implementation Team.

2) **Weekly / fortnightly** – for the Project Working Group.

3) **Fortnightly / monthly** – for the Project Steering Committee.

As illustrated next in [Figure 125], speaking "honestly", the members of the Project Implementation Team have little to no interest in the project's [Costs]; rather, they are concentrating on getting their assigned tasks [Scope] done in the allotted [Time] as best they can [Quality].

The members of the Project Working Group are primarily concerned with coordinating the works of the Project Implementation Team, where the [Costs] starts to be of some concern to them. However, the Project Working Group is primarily concerned with balancing [People] & [Resources] versus [Scope] & [Time] for a [Quality] return.

That leaves the members of the Project Steering Committee who happen to be the ones who traditionally are most concerned about the [Costs] & [Time], especially during the monthly executive management meeting when they review those active projects.

Therefore, given that the Project Steering Committee only meets on a monthly basis (maybe fortnightly), and that the performing organization's financial-accounts group usually produce the costing information on a similar monthly (fortnightly) basis; thus, these cost performance measurements are ineffectual for the daily monitoring & control of the project.

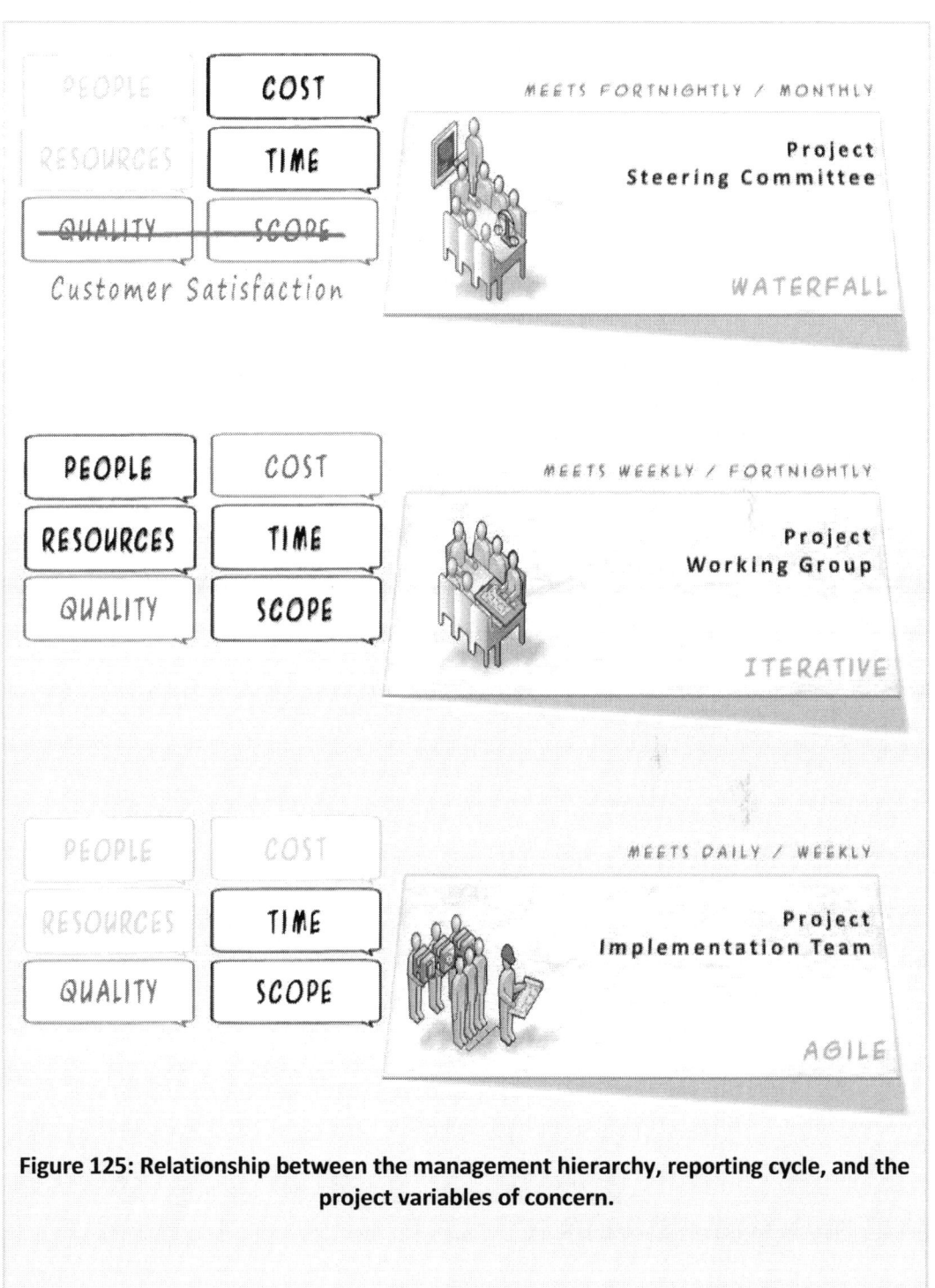

Figure 125: Relationship between the management hierarchy, reporting cycle, and the project variables of concern.

How are costs monitored & controlled on a daily / weekly basis?

Back in [Section 7.3], after the simplified cost formula was presented it was stated that:

> "As a novice project manager, you will most probably only be responsible for the project's [Scope], [Time], [People], [Resources] and not the [Cost]".

The reasoning behind this is that; the project's [Scope] into [Time], [People], & [Resources] has a direct flow-on effect on the project's [Cost]. The project's agreed [Scope], [Time], [People], & [Resources] can be monitored & controlled on a daily / weekly basis; refer to [Figure 68] and [Figure 70]. Therefore, if the project's [Scope] were delivered on [Time] using only the [People] & [Resources] that were allocated then in all probability the project would be delivered within the allocated [Cost] budget.

Hence, **from a daily / weekly monitoring & control perspective; take care of the [Scope], [Time], [People], [Resources], [Quality], and subsequently the [Cost] will take care of itself.**

PROJECT MANAGEMENT ... THE ART OF BALANCING THE BUDGET WHILE SIMULTANEOUSLY BALANCING THE EFFECTIVE AND EFFICIENT IMPLEMENTATION OF THE PROJECT.

13. MONITORING & CONTROL Phase ... Practical Part 2

13.1. [Quality] Monitoring & Control

In a practical sense, [Quality] Monitoring & Control [Section 8.1] asks the following questions:

- Are the project's tangible / measurable deliverables correct when compared to the agreed [Scope] and the agreed Acceptance Criteria?

- Is there an alignment between the customers and the other stakeholders' perceptions & expectations when compared to what is in the approved Detailed Specifications?

- Do the project's deliverables satisfy the needs for which these were intended?

- Are the appropriate Quality Assurance and Quality Control processes & procedures being followed?

- Is [Quality] being considered as today's issue, or is [Quality] mistakenly being put off as tomorrow's problem?

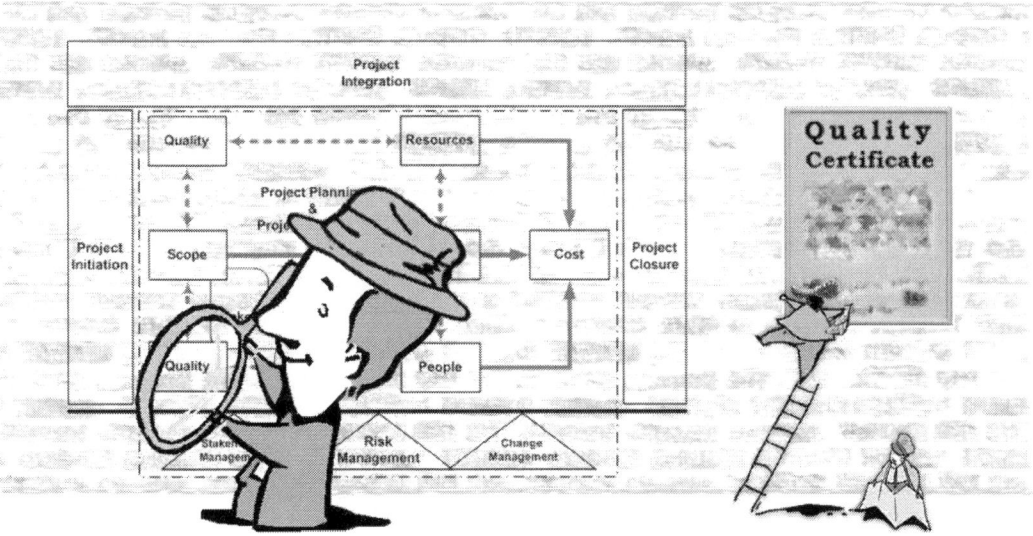

13.1.1. Quality: Verification, Validation & Testing (VV&T)

AUTHOR'S NOTE

This section is by no means a comprehensive statement about Verification, Validation & Test (VV&T) as this would constitute a book inside a book; however, what is provided here should be reasonable to the understanding of testing for SDLC projects.

The first thing to note about testing is that, there is not one specific set of test methodology that is commonly used, as these will vary greatly based on the performing organization's processes & procedures, the customer organization's requirements & expectations, and on the relevant industry (and market segment) standards & practices.

Secondly, the names and acronyms used to identify each grouping of the tests conducted will vary between organizations and across industries. Hence, the acronyms and names used in this book have been chosen for descriptive purposes, rather than for a specific industry; i.e. as used in defense, telecommunications, or banking & finance.

	Test Type	Tested By	Description
1.	**Debug Testing**	Developer	Testing undertaken during the implementation of the task. That is, as the code is being written; *e.g. 1 = 1.*
2.	**Unit Testing**	Developer	Testing of a completed individual "unit" of functionality that is isolated from other units. That is, the smallest "building block" of testable / usable source code, or a portion of circuitry; *e.g. 1 + 1 = 2.*
3.	**Module / Component Testing**	Developers	Testing of multiple units of associated functionality that aggregate to a larger component of functionality; *e.g. (1 + 1) x (4 + 3) = 14.*

4.	**Integration** Testing	Developers & (Testers)	Testing of the completed modules which either combine together to form one deliverable or one system that interfaces with each other. That is, testing that the individually completed deliverable / system do conform to agreed Acceptance Criteria (or a subset of Criteria).
5.	**System** Testing	Testers & (Developers)	Testing that the completed deliverables / systems are interoperable with each other and interoperable with existing applications & systems. Also, testing that the completed deliverables / systems does conform to the agreed Acceptance Criteria. It is advisable during system testing to privately "dry run" through all of the proceeding FAT test cases, prior to the customer's involvement.
6.	**FAT** – ~~Factory~~ *Functional* Acceptance Testing	Tester & (Customer)	Testing to verify that every in-scope feature & functionality described in the approved Detailed Specifications has been successfully implemented in accordance with the agreed Acceptance Criteria. That is, ensuring that everything relevant in the Requirements Traceability Matrix has been accepted by the customer as PASSED.
7.	**SAT** – ~~Site~~ *Satisfaction* Acceptance Testing	Tester & Customer	(re) Testing of those features & functionality found to be non-conforming during the FAT and verifying that it is "satisfactory" to provide the deliverables to the customer, or that it has been deployed "satisfactorily" to the customer's site.
8.	**UAT** – ~~User~~ *Usability* Acceptance Testing	Customer & (Tester)	Testing that the deliverables "in situ" (operation from the intended deployment location) is functioning as expected and hence that the deliverable / system is "fit for use".
9.	*Warranty & In-Service Support*	*End User*	*Continual usage of the application / system by the end user under real-world conditions for the duration of the warrantee period and any subsequent In-Service Support contract.*

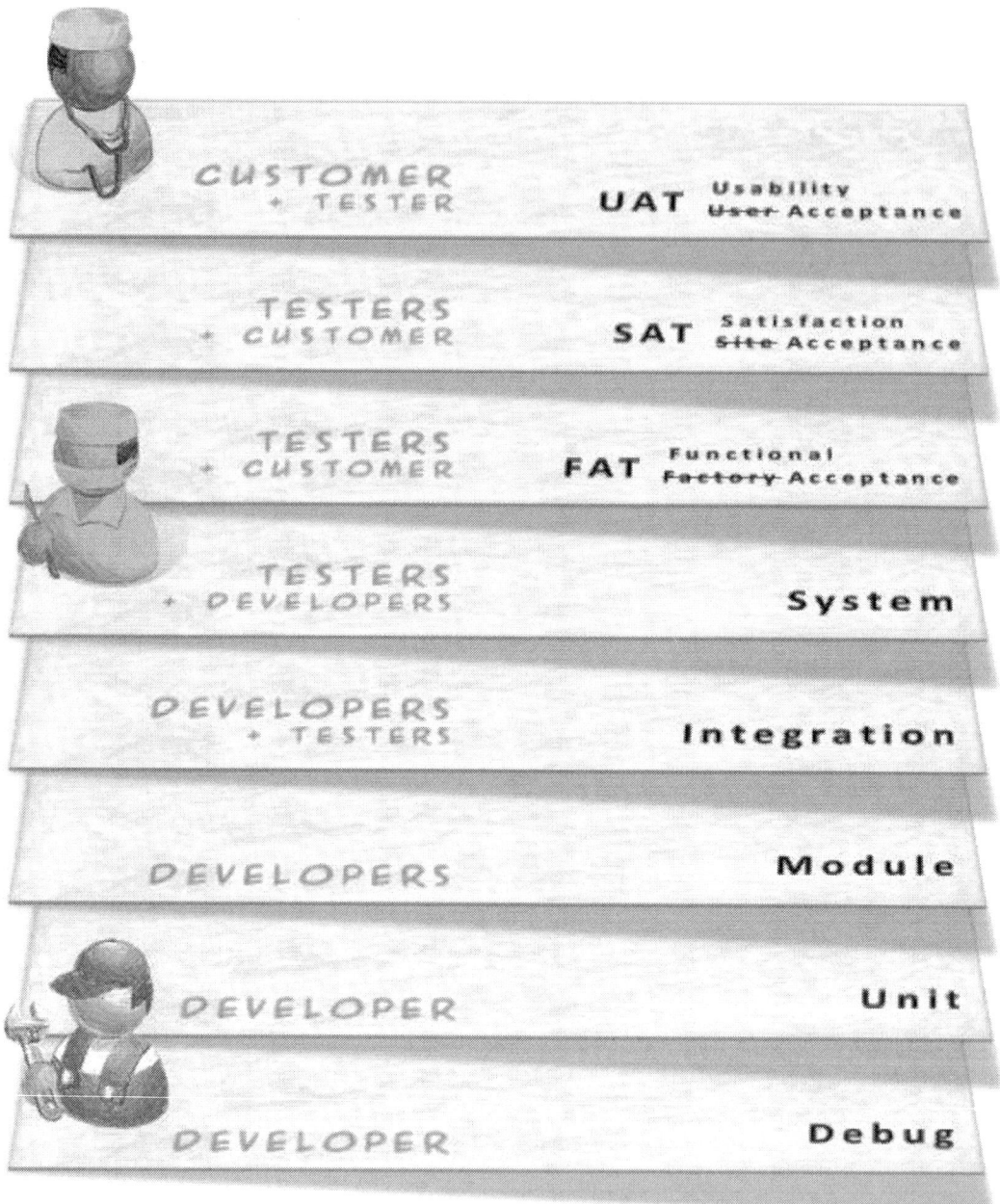

Figure 126: Hierarchy of testing and the groups involved.

NOTE ❑ Each suite of these tests builds upon the successful outcome of predecessor tests. However, there may be more or less than what is shown here; hence, find out which suite of tests are appropriate for the project that is being managed.

Quality Assurance & Quality Control Aspects Of Testing

❖ **Quality Assurance** group of tests deals with **ensuring** that the deliverables **will conform** to the approved Detailed Specifications.

❖ **Quality Control** group of tests deals with **validating** that the deliverables **do conform** to the agreed Acceptance Criteria.

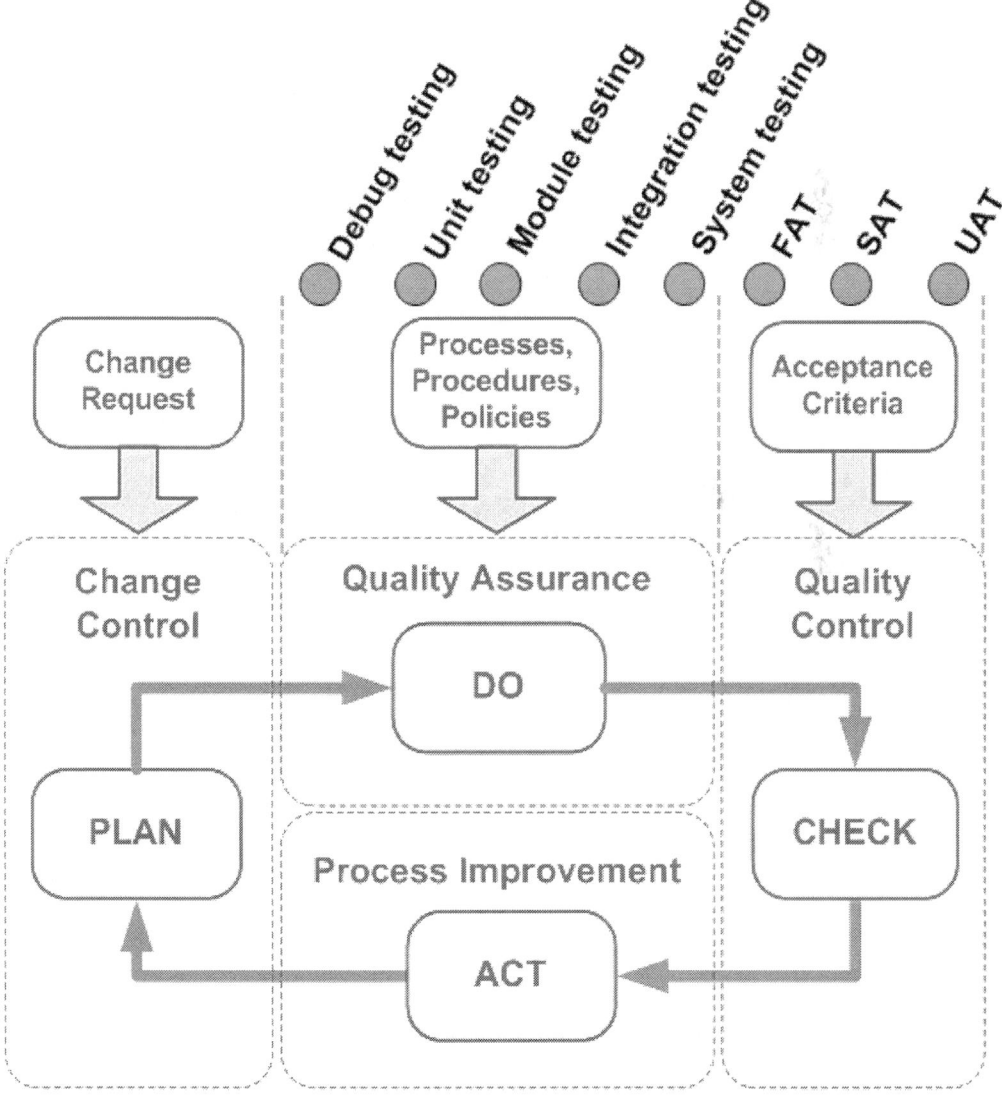

Figure 127: Relationship of testing to Quality Assurance and Quality Control.

These Quality Assurance & Quality Control tests suites also have differences with how readily accessible are the internal operations of the application / system during testing:

- **White Box Testing – are Quality Assurance tests that** are undertaken predominately by developers where these persons have **access to the internal structure of the application** / system; i.e. access to the source code, data structures, and interface protocols. Thus, they can get down to the nitty-gritty of what is going on inside the box and determine exactly what is going wrong; *e.g. debugging, unit testing, module testing, integration testing, and system testing*.

- **Black Box Testing – are Quality Control tests that are** undertaken predominately by testers (and the customer) where these persons **do not** have **access to the internal structure of the application** / system nor do they have knowledge of its internal workings. Black Box testing would include; FAT, SAT, and UAT acceptance testing.

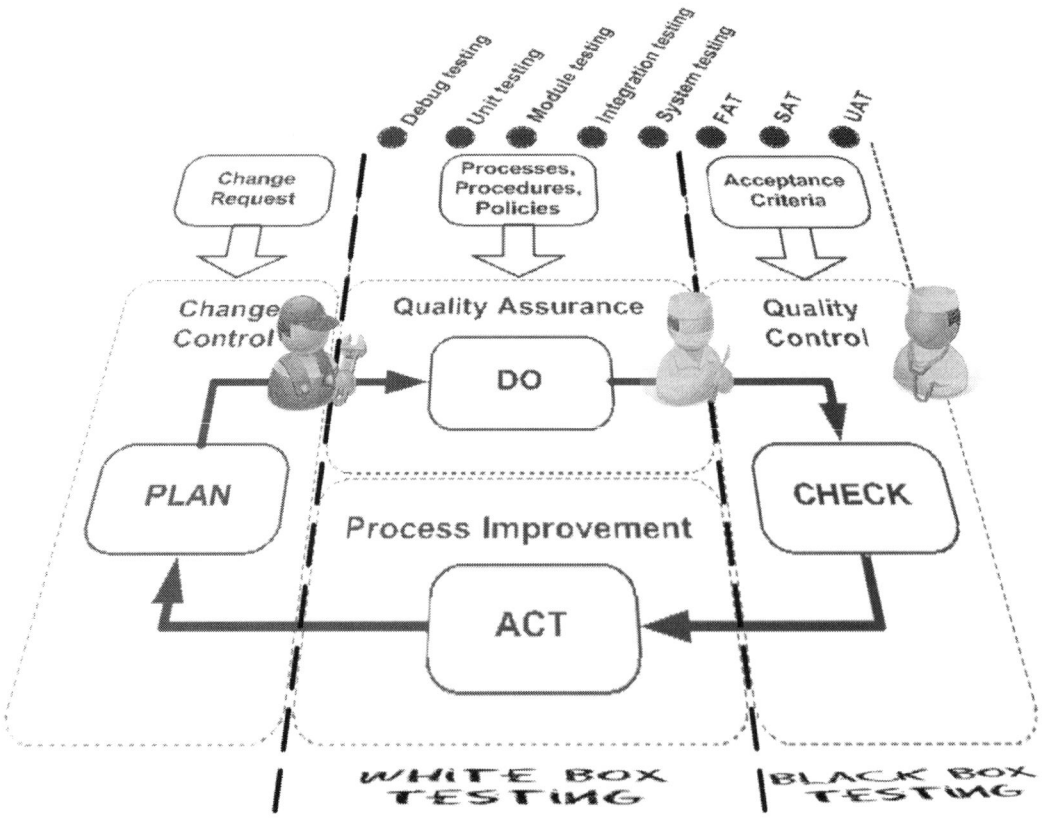

Each suite of tests illustrated previously in [Figure 126] and [Figure 127] form a "Gate" to the next suite of tests. That is, if the implemented functionality does not meet (or exceed) the threshold of acceptability for the current suite of tests, then this implemented functionality will not proceed onwards until it does. As illustrated below in [Figure 128], the defective implementation will be passed backwards to be rectified, or else a Change Request will have to be raised to redefine how that functionality is to operate.

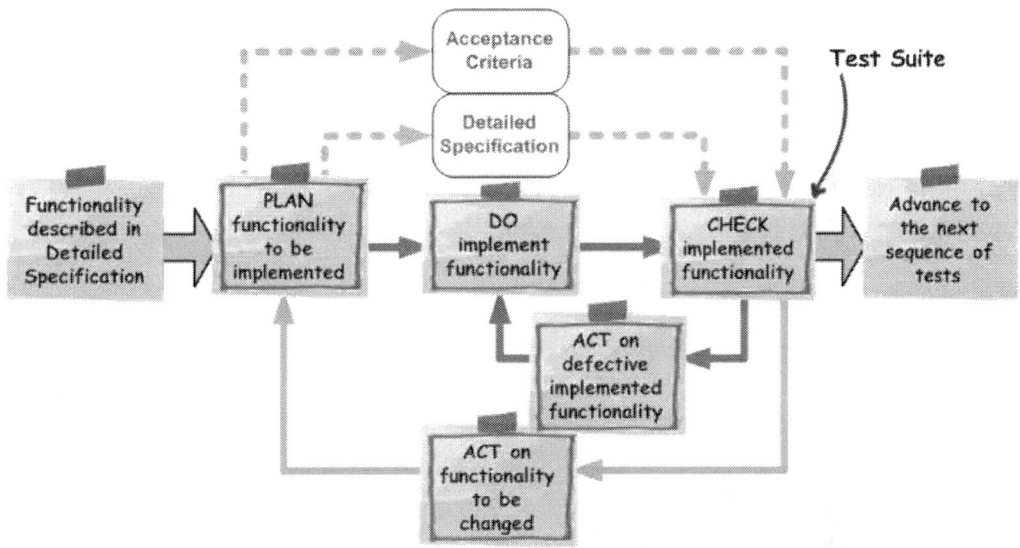

Figure 128: Iterative nature of testing of the implemented functionality.

NOTE   ❑ For the **Quality Control** suite of testing (i.e. FAT, SAT, and UAT), **whoever is undertaking these tests should be "independent" of the implementers**, else there could be potential for the delivered application / system to not be "objectively examined" and hence not be as stringently evaluated / scrutinized.

❑ Secondly, **whoever does the Quality Control suite of testing must have:**

(1) the **jurisdiction to FAIL** the application / system, if the delivered features & functionality does not comply, and

(2) the **authorization to push back** on the application / system being passed onwards to the next Gate / stage of release.

Other Types Of Testing

- **Regression Testing** – By the Integration & System testing stages the application / system should be in a stable enough condition to include testing targeted towards verifying that what previously worked still does (even after new features and changed functionality has been included). These regression tests can be automated to save the testers and the developers from having to go back and test everything again and again. Though, the point of **automated regression testing is NOT to verify that the application / system functions as described in the approved Detailed Specifications**; rather, the **purpose of automated regression testing is to verify that the application / system still functions as it did previously when the test scripts were written**.

 Automated regression testing cannot detect conformance to the specifications; rather, automated regression testing can only detect changes in functionality to what it was previously when the regression tests were last conducted.

There is no point in running automated regression tests while the targeted portions of the application / system is still being extensively developed. This is because, the tester will spend a significant amount of their time re-modifying test scripts that were broken by the nightly source-code build. Hence, automated regression tests need to be targeted at relatively stable portions of the application / system.

- **Smoke Testing** – is a quick test to **confirm whether the most basic functionality of the application / system is working;** hence, to extrapolate whether it is worthwhile continuing with more stringent testing. *For example; does this test mobile phone power-on and does its operating system boot-up, or does it go up in smoke & flames?*

- **Sanity Testing** – is non-exhaustive testing to **verify that the application / system's essential functionality is still working as expected**; i.e. not verifying every feature and function. *For example; can the user of this test mobile phone make and receive calls, but not whether the user can modify a playlist on the phone's music player.*

- **Load Testing** – is using testing tools to burden the application / system with a continuous stream of inputs & outputs that **replicate the highest usage** that the application / system is expected to be able to handle. In addition, use load testing to determine at what point the application / system's performance starts to degrade, and thereby determine what is the **maximum sustainable recommended load** that can be supported.

- **Stress Testing** – is using testing tools to **heavily burden the application / system** with a continuous stream of inputs & outputs that **replicate excessive usage that is expected to exceed the application / system's specified capabilities**; thereby, gauging how & when it will fail. *For example; a web-server application that is designed to accept 10 thousand requests per hour is instead bombarded with 100 thousand requests per hour.*

 I would recommend that … when conducting **Load & Stress testing, your project team replicate the customer's application/system's configuration & resource availability**; i.e. the amount of RAM, hard-disk space, CPU power, bandwidth, and even environmental constraints such as temperature & humidity.

Different usage parameters could result in significantly different outcomes and hence impact on the customer's satisfaction.

Was that new operating system really that bad, or did those numerous disappointed end-users try to run it on under-spec hardware that was not up to the intended purpose?

- **Alpha Testing** – is testing where a limited number of selected / volunteer end users use a partially or mostly completed version of the application / system. Observations are made to determine potential changes to the design & specifications required due to semi real-world usage, additionally obtaining end-user feedback, and usage analytics. That is, end users have an uncanny ability to find & stumble upon ways to utilize the application / system that were not previously contemplated.

- **Beta Testing** – is testing where a reasonable number of selected / volunteer end users use what is essentially the completed application / system. Thereby determining any remaining defects or problems that have to be resolved prior to the "go-live" public release of the application / system. ... *Also, obtains broader audience feedback.*

- **Release Candidate Testing** – is the final testing where selected / volunteer end users (for all intents & purposes) use the completed application/system. Thereby, determining prior to "general availability" if there are any problems with the deployment process, and are there any high & medium severity defects remaining?

There are other types of testing that can be conducted; though, as the project manager you might want to consider including additional "testing" which enables the implementers to demonstrate their efforts & capabilities to other members of staff and specifically to senior management. Consider such "testing" as:

- **"Come Show Me"** – where each implementer (or a group of implementers) gets their management to "checkout" what they have been working on. This only has to be the occasional demonstration, but sufficient to show what progress has been achieved.

- **"Peer Group Demonstration"** – where at the completion of a group of tasks / module an implementer (or a group of implementers) demonstrates to their fellow team members (and/or to the local staff) what they have been working on.

> This type of informal in-house "testing" puts a small amount of pressure on the project team members to complete their tasks with quality, and to then perform in front of their peers.
> This form of testing can be used to prepare them for the day when they may have to present to customers.
>
> Secondly, this form of testing exposes their efforts to the rest of the organization and especially to senior management.
>
> I would also recommend that ... this form of testing be used as a constructive lead into Friday afternoon drinks & nibbles with the subsequent informal question & answer discussion.

13.1.2. Quality: Acceptance Testing

NOTE ❏ Prior to the acceptance test plans & procedures being written, let alone before any acceptance testing is conducted, it is **essential that both** the representatives from the **performing & customer organizations have agreed on what are the Acceptance Criteria**.

❏ Once the **Acceptance Test plans & procedures** have been written, prior to any acceptance testing being conducted, these documents must be **reviewed & approved by both** the representatives from the **performing & customer organizations**. This is especially important for the "Functional" Acceptance Tests.

❏ **Representatives from both the performing & customer organizations need to be present during these acceptance tests** (definitely for the SAT and at least for the FAT of the "1st production" unit).

❏ The **FAT** should be **undertaken in a regimented manner and thereby ensure that all of the included functionality is tested in a reproducible way**.
The "Satisfaction" Acceptance Testing (SAT) maybe a little more flexible; though, the "Usability" Acceptance Testing (UAT) can be an ad-hoc affair that is only bounded by the limitations specified in the accompanying Release Notes.

I would highly recommend that ... a **dry-run** / dress-rehearsal be conducted **prior to** the occurrence of **the "official" FAT**.
This dry-run should be conducted in privacy by the Project Implementation Team away from the prying eyes of the representatives of the customer organization and also from view of the senior management of the performing organization.

Remember that, perception is a significant contributor to the project being deemed a success; hence, "practice makes perfect", and (more specifically) practice uncovers those obvious mistakes and silly defects that would degrade the perception of a quality deliverable.

Customer Participation in the Acceptance Testing

The reason that it is so important for representatives from both the performing and customer organizations to be present and (preferably) actively involved with the acceptance testing is because:

1) The **FAT is the last feasible and affordable opportunity to resolve any issues** and differences of opinion over whether non-conforming functionality constitutes a defect or is a change request. See [Section 10.2].

2) Provides formal written acknowledgement by both parties that the project's [Scope] has been covered and delivered satisfactorily; i.e. the acceptance of the "Satisfaction" ~~Site~~ Acceptance Test (SAT). Where the **formal sign-off is essential for limiting the potential for possible future disagreements over** non-conforming or missing functionality that is discovered later on; i.e. the "political" resolution of **latent defects**.

3) **When the customer signs off on the deliverables then the warranty period starts ticking down**. However, with no signed off acceptance of the deliverables the problem is, "when does the obligated warranty period begin and when does it end"?
   Under such circumstances, it is possible that the warranty provided by a third party vendor/supplier could expire before the performing organization's obligation to the customer organization has concluded. Potentially the performing organization could end-up having to cover (out of its own pocket) those claims lodged against an expired third party warranty. This is especially true for hardware elements of the deliverables, such as computer components; e.g. *motherboards, memory, CPU, hard disks*.

   So, do you have any spare parts stashed away, for a rainy day?

4) Limits the potential for future warranty claims for non-conforming functionality, because (depending on the contractual agreements and the regional statutory requirements) **after the warranty period has expired, any functionality issues encountered would thereby constitute a Change Request and not a Defect Report**.

Establishment of Test Documents, Processes & Procedures

It is essential that the following be undertaken prior to commencing the corresponding acceptance testing:

1. The **test scenarios** are established.

2. The **test cases** are determined per scenario.

3. These test cases are **planned** out **and resourced** accordingly.

4. These test scenarios and test cases are **documented** in the Acceptance Test Plan.

5. This test documentation is **reviewed & approved** by representatives from the performing organization (and possibly from the customer organization).

6. This test documentation (and the included test cases and scenarios) are "frozen" as the **test baseline** for the corresponding milestone release.

It is highly beneficial that once the Detailed Specifications have been agreed to (and while the Project Implementation Team is in the process of implementing the deliverables) then the Acceptance Test plans & procedures should be written by those persons who will be involved with the conduct of the acceptance tests (with the assistance / recommendations / review by relevant Subject Matter Experts).

Testing Coverage Matrix (TCM)

While these test documents (processes & procedures) are being prepared, the Testing Coverage Matrix (TCM) should be created / updated so that it will be possible to indicate when each feature has been verified as functioning correctly or incorrectly. Subsequently, when the acceptance testing is underway, the **TCM would be marked up to represent the PASS : FAIL of each test**. Additionally, the TCM might include references to any; associated Defect Reports generated, an indicator of the type of failure, the perceived severity of the defect, and the perceived priority for rectifying this failure. The TCM may also include **references to any waivers** if these were **granted for specifically failed tests**; i.e. the test failed but this failure has been agreed to be acceptable under the current circumstances.

13.1.3. Quality: Defect Reporting

Defects Reported Due to Differences of Perspective

During the acceptance testing (especially **during the FAT**), **the customer's representatives** in addition to confirming that the deliverables satisfy their organization's stated "needs", they **may also try to confirm that the deliverables meet their "wants" & "expectations"**. This can come about because; it may be only now while using the deliverables that they obtain a clearer understanding of what they are really trying to achieve.

What can also occur during the acceptance testing is that, there is a mismatch between what the customer representative "interpreted" as being described in the Customer Requirements, and that which was agreed to in the approved Detailed Specifications. Consequently, **due to these misunderstandings & misinterpretations, the implemented functionality may conform exactly to the definition included in the approved Detailed Specifications, yet this functionality does not correspond to the purpose for which the customer had intended**.

Both of these above factors could mean that the customer's representative submits dubious defects, some of which may be [Scope] changes in disguise; see [Section 10.2].

If the project manager and the Project Implementation Team do not catch these [Scope] changes that are submitted as Defect Reports, then the project's duration could drag-on, until the project is no longer a profitable venture for the performing organization due to the project's haemorrhaging in [Time] & [Cost]; see [Section 7.1].

It is of the utmost importance that the **project manager keeps on guard against [Scope] changes and [Scope] creep that are disguised as defects**. Therefore, the project manager should **plan for a reasonable amount of [Time]** during and after the completion of these acceptance tests to allow selected members of the project team (and the project manager) **to analyse & compare the reported defects against the approved Detailed Specifications**.

And, definitely not compare against the Customer Requirements.

When comparing those received Defect Reports against the approved Detailed Specifications, also check whether any of these Defect Reports are for the same issues as any existing Defect Reports. ... This is because; every duplicate Defect Report will inflate the Defects Found Count and thereby worsen the perception of quality.

After conducting this Defect Report analysis, the project manager will have to sit down with the customer's representatives and decide / debate whether certain defects are in fact scope changes and should consequently be handled via the Change Request process. **This defect review process is when the skills of the project manager as a good communicator / negotiator / conflict resolver will come to the fore.**

Defects Reported as a Quantifiable Measure of Quality

For application development projects, **the easiest way** for the customer's primary-core stakeholders **to quantifiably measure the quality of the project's deliverables is via the number of defects reported**. Where a defect is a non-conformance with the agreed Acceptance Criteria, and features & functionality that is not as defined in the ~~Customer Requirements~~ approved Detailed Specifications; see [Figure 129] and [Figure 128].

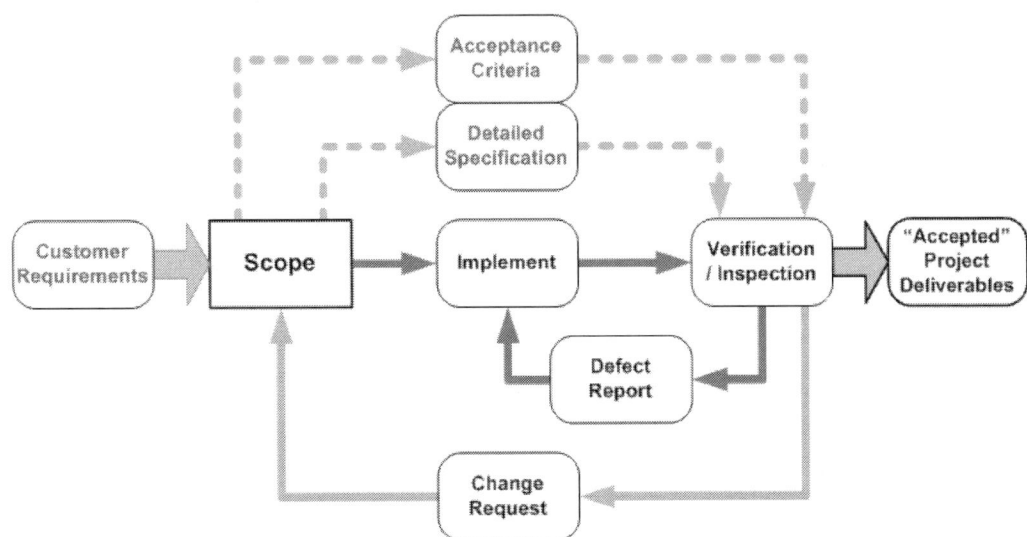

Figure 129: Project deliverables verification & Defect Reporting.

However, **without an alignment of** both the performing and customer organizations' **understanding & interpretation of the functionality**, what are **the deliverables,** and what will be included in **each release / milestone, then ... a significant number of (perceived) Defect Reports are going to be generated by the customer representative**. As a result, the customer representative will probably perceive the project's deliverables as having poor quality.

NOTICE how in the preceding paragraph the line is, has *"a significant number of (perceived) Defect Reports"* and not *"a significant number of (real) defects"*.

This is because for many an application development project, the first opportunity that the customer has to judge the quality of the project's deliverables is during the acceptance testing. During this testing, the customer's representative will be submitting their Defect Reports; if the deliverable is found to have a (single) major crash-bug that causes the application to fail dramatically then this does not necessarily have a significant negative impact on their perception of the project's quality. *... But, multiple crash bugs will.*

$$MISUNDERSTANDINGS \atop MISINTERPRETATIONS \quad = \quad HIGH\ NUMBER\ OF \atop DEFECT\ REPORTS$$

$$\textbf{PERCEPTION OF} \atop \textbf{POOR QUALITY} \quad = \quad \textbf{HIGH NUMBER OF} \atop \textbf{DEFECT REPORTS}$$

What does start to build a negative perception of quality is if the customer's representative is entering umpteen Defect Reports for the day. In which case, it is **not necessarily the severity of the defects that influences the customer representative's perception of poor quality; rather, it is the quantity of defects detected during testing**.

The reasoning behind this impression is that, one (software) crash-bug has the potential to be quickly fixed and thus has the promise that testing can soon resume. Whereas, numerous defects will require some time to fix each one, and hence in the meantime the customer representative will have to continue using this deficient deliverable which will impair their testing and will not be enjoyable (to the point of being downright frustrating).

However, if there are found to be several major defects with the deliverable then it will definitely be considered as having poor quality. Similarly, if the crash-bug takes an unreasonable amount of time to fix, or crash-bug after crash-bug is encountered after each fix is rolled-out then the deliverable will also be considered as having poor quality.

Interestingly, the customer's primary-core stakeholders (i.e. their senior management) would be judging the deliverable's quality based on; the number of "reported" defects, and the feedback & impressions that they receive from their testing representatives. That is, these customer's primary-core stakeholders may not necessarily base their judgment on their own personal experiences with using the project deliverables.

Back in [Section 1.3] it was stated that,

> "*The determination of whether a project is deemed a success or a failure is based entirely on the expectations and perspectives of the*" stakeholders.

Hence, **it is essential for the project's deemed success that Defect Reporting and defect resolution be undertaken methodically, analytically, and supervised**, so as to reduce the number of spurious defects being reported. Do this by answering the following questions:

1) What should the deliverables be compared against? ... *the Detailed Specifications*

2) How should Defect Reports be recorded? ... *in a Defect Tracking System*

3) How should Defect Reports be categorized, prioritized, & severitized? ... *consistently*

4) How should defects be resolved? ... *in an orchestrated manner*

5) How should defects be verified as resolved? ... *also in an orchestrated manner*

 I would recommend that ... some form of defect tracking be always used. Whether this is paper forms, a spreadsheet, a simple database, or a dedicated Defect Tracking System. Irrespective of the methodology, without some form of defect tracking, the project's [Quality] Monitoring & Control will be ineffectual.

DO NOT skimp on using an appropriate Defect Tracking System.

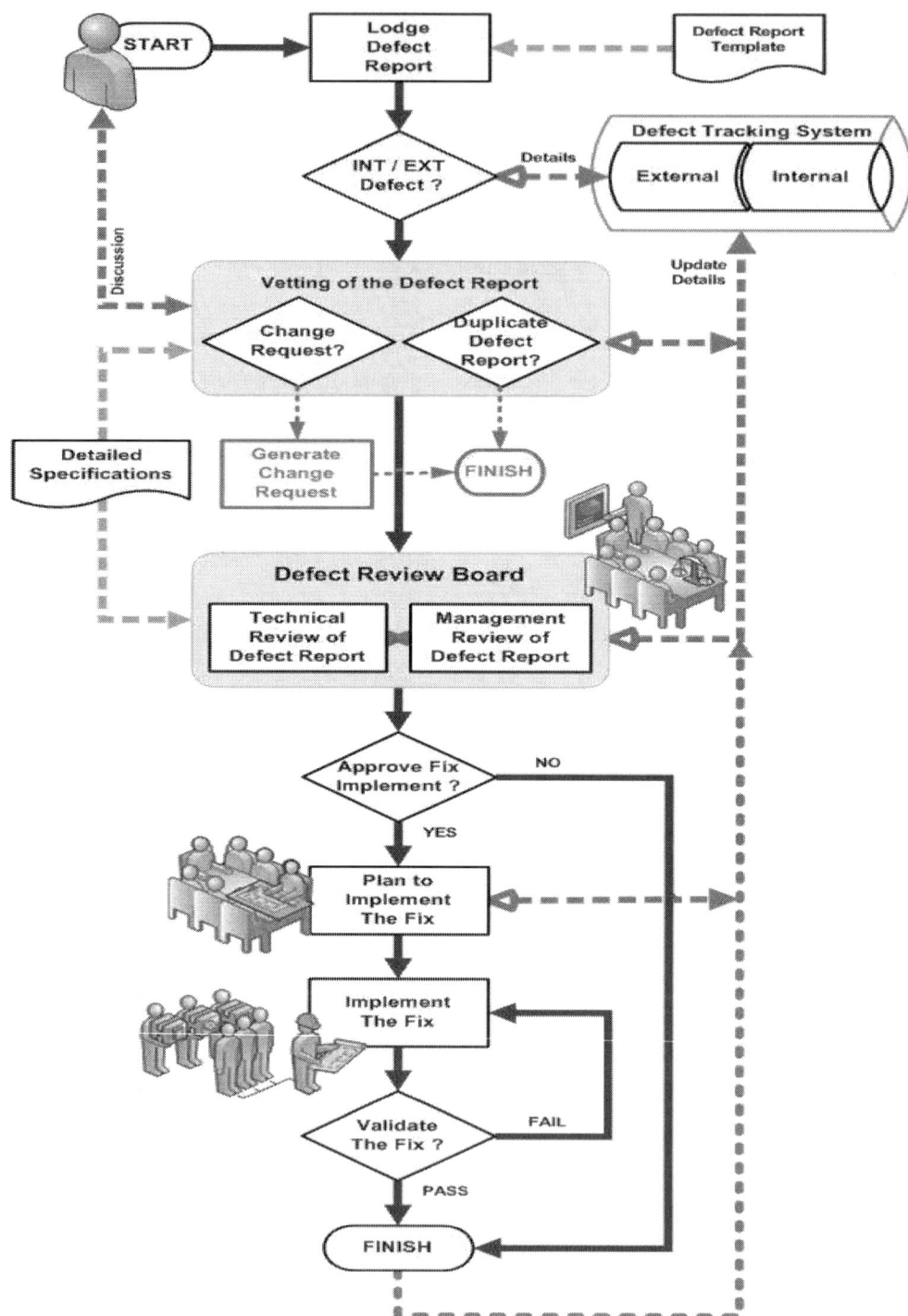

Figure 89: A simplified defect processing workflow.

Separation of Internal & External Defect Tracking

For the project's deliverable to be perceived as having high quality, it is beneficial to separate into two distinct Defect Reporting areas (i.e. issues / Defect Tracking Systems) those defects found "internally" by the performing organization and those defects that were found "externally" by the customer.

The reasoning behind the separation of the groupings of Defect Reports is because:

- **Those defects found externally will relate directly to the customer's perception of the deliverable's "fitness for use" and its conformances with the specifications.** Whereas, those defects found internally could include non-functional issues such as non-conformance with coding standards, deviations from policies & procedures, etc.

- **From a customer's** primary-core stakeholders' **perspective** (i.e. their senior management), **the measure of the quality of the deliverables is proportional to the count of the defects that have been found, and how many of these defects remain unresolved**. Hence, it would be detrimental to pollute this quantifiable measure of quality by inflating the numbers with internally reported defects.

Severity Grading and the Prioritization of Defect Resolution

There is a distinction between the priority of a defect and its severity.

- ❖ **Severity – is how much impact** the defect has on the application / system.

- ❖ **Priority** – is an indication to the implementers of **how soon this particular defect has to be resolved when compared to the other defects that are still open.**

Grade the defects based on the level of impact / severity that each defect has on the operability of the project's deliverables. Use a severity grading as follows:

1. **Critical** – is where that particular defect renders the **deliverable** as either **completely inoperable or** major parts are **unusable, or** where **data is lost or corrupted**, and there is **no reasonable workaround**. *For example; a crash bug.*

2. **High** – is where that particular defect renders the **deliverable** as **unusable or is not fit for use**, but there is **a reasonable workaround available**.

3. **Medium** – is where that particular defect is for **some feature / functionality of the deliverable** that is **not operating as specified** and this deficiency is causing difficulties, but the **rest of the deliverable is usable** and there is **no loss or corruption of data**.

4. **Low** – is where that particular defect is for a **minor non-conformance** with a part of the deliverable but this **does not impede that deliverable's fitness for use**.

NOTE   ❑ **Just because a defect has a low severity grading does not mean that it can't have a high priority.** There may be some low severity defects found early in the project's life where for reasons of perception it would be better to fix these sooner rather than later; as to let these defects linger on would form a constant reminder to the customer representatives of the issues with the deliverables.

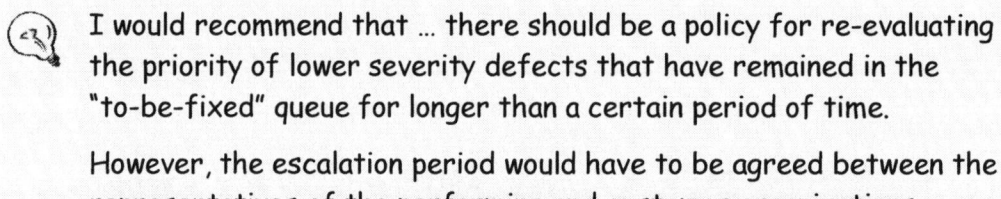

I would recommend that ... there should be a policy for re-evaluating the priority of lower severity defects that have remained in the "to-be-fixed" queue for longer than a certain period of time.

However, the escalation period would have to be agreed between the representatives of the performing and customer organizations.

I would recommend that ... all of the found issues be quickly looked at prior to being submitted as a defect, because these issues maybe for functionality that had not been included in the release that is being tested. Also, gives the opportunity to suggest an appropriate severity grading of the defect, instead of just marking each one as "Critical".

If your customer's representative insists on grading every defect that they find as 'Critical' then generally you are left with no option other than to have your own person there at the testing location to intervene (and haggle) before the defect is registered.

Thereby, your person can debate whether it is or is not a defect and to argue for each defect to be assigned a more appropriate grading.

13.1.4. Quality: Defect Resolution

If you originally come from an implementer's background then you are probably very proficient and knowledgeable at resolving individual defects; however, what you have probably never given much thought to is **the "politics" of resolving defects**:

1. Resolve ~~as soon as possible~~ *immediately* those defects (problems) that **embarrass or humiliate the customer organization**. That is, those defects which are publically viewable by the customer's customers. *E.g. crash bugs and out-of-service notices.*

2. Fix as soon as possible those defects with the **highest priority**, then severity & priority. This may necessitate fixing a 'low severity but high priority' defect before a 'high severity but low priority' defect.   ... *Perception can overrule factuality.*

3. Fix those defects / issues **causing several other defects**; i.e. fixing this one issue will resolve or partially resolve several other defects or eliminate the apparent potential for additional defects to occur in that affected area.

4. Fix those defects that the customer's representative **keeps noticing**; *"oh, that bug is still there"*. Even if this defect was of a 'Low' severity and 'Low' priority, from the perception of quality and evidence that the customer is being heard, it is advisable to, *"just fix the damn thing and move on"*.

5. Try to keep the rate that the 'Critical' and 'High' graded defects are being **fixed in pace with the number of defects that are being reported;** i.e. evidence that the project is making progress and is still under control.

 I would recommend that ... incomplete code should not be accessible during testing by the customer's representative.

Hence, stub-out access to such incomplete code and present some form of "Under Construction" indicator, such as a "To-Be-Done" message. Whatever method is chosen, make sure that the implementation retreats gracefully from that incomplete operation.

Presentation of Defect Resolution

From the perspective of improved communications with the project's stakeholders, it may be highly beneficial to depict the number of open / outstanding defects per grading classification against the project's time.

Recall back in [Section 7.4] when the agile Burn Down Chart in [Figure 76] was used to illustrate the progress through the work remaining in the number of days left in the sprint. Well, a similar idea can be used for plotting the progress through the closing of open defects; as illustrated below in [Figure 130]. However, instead of plotting a single line, there would be multiple overlaid plots for 'Critical', 'High', 'Medium', and 'Low' grades of defects.

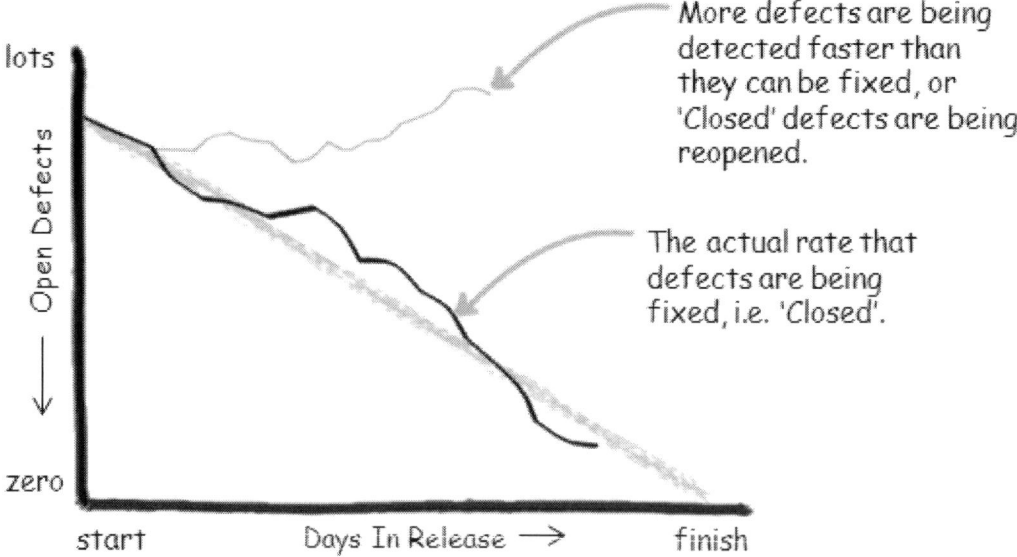

Figure 130: Defect Resolution Burn Down Chart.

NOTE   ❑ As with most of monitoring & control ... one of the keys to success is the "visibility of that which must be addressed", and here, it is the visibility of the count of open defects being reduced down to zero. However, if these defects are not fixed properly then these defects are just going to be re-opened to increase the count; consequently, rising up the Defect Resolution Burn Down Chart.

13.1.5. Quality: Evidence, Standards, Processes & Procedures

Objective Quality Evidence (OQE)

Once each suite of tests and acceptance testing has been performed, it will be necessary to accumulate all of the **Objective Quality Evidence (OQE)** that **proves** that **these tests were completed successfully**. This evidence would also include any **waivers** that were **granted for non-conforming functionality** during that associated suite of tests.

This accumulated OQE would then be presented to the performing organization's **Quality Authority** (and any chosen representatives of the senior management) for review, and to accept or reject the outcomes of these acceptance tests.

This **reviewed & approved OQE** would then form the **"artefacts of proof" that the project's deliverables are ready to be handed over** to the customer organization.

Additionally, it may be a contracted stipulation of the project that selected extracts of this OQE be provided to the relevant representatives of the customer organization, so that this particular OQE can be inputted into their own Quality Assurance processes and Quality Control procedures.

NOTE   ❏ **What constitutes the necessary Objective Quality Evidence will be dependent on what industry (and market segment) the performing organization and the customer organization are engaged.**

        ❏ National and/or state **regulatory requirements and statutory authorities** will also **prescribe what OQE is necessary** so that the deliverable will be allowed to operate in their domain.

        [] *Have no stamped OQE, then do not pass GO. ... As this OQE may form a component of the deliverables necessary to activate the next partial payment point (as per the contracted milestone list).*

The following is some of the Objective Quality Evidence that the project may be required to provide; however, the following may or may not be required for your specific project.

- **Assembly Traceability Record (ATTR)** – lists and describes the (major) **itemized components that went into constructing the physical unit** (i.e. "inside the box" or sub-box). An ATTR would be generated per unit produced and signed by the person / team-leader who produced that unit. *E.g. serial numbers of the CPU and motherboard.*

- **Inspection & Test Record (ITR)** – the documented **results of the inspection of the built physical deliverable** (i.e. "the box"). Signed by the team-leader or individual that conducted the inspection, to signify that the produced individual unit does conform to the approved product baseline. Hence, this particular box is physically "the same" as the original box (First-Of-Type) that was approved for delivery to the customer.

- **Functionality / Factory Release Test (FRT)** – the documented **results of the "in-house" system testing that was conducted to verify that the deliverable is functioning** as per the performing organization's understanding and interpretation of the **approved Detailed Specifications**. The FRT can be considered as recording the results of the dry-run of the ~~Factory~~ "Functional" Acceptance Testing (FAT). The FRT would be signed by the team-leader or individual that conducted the testing.

- **Acceptance Test Record (ATR)** – the documented results of the ~~Factory~~ "Functional" Acceptance Testing (**FAT**), ~~Site~~ "Satisfaction" Acceptance Testing (**SAT**), and/or the ~~User~~ "Usability" Acceptance Testing (**UAT**). Each ATR produced would be signed by the team-leader or individual that conducted the testing, the supervisor of the testing or the project manager, and the customer's representative (if they happened to be involved with the acceptance testing).

- **Deviation** – written **permission for a departure** from the approved processes, procedures, and/or specifications used to produce the deliverable.

- **Waiver** – written **permission for a non-conformance** so that the deliverable may be accepted.

- **Concession** – a "special case" approval form signed off by the representatives of the performing organization (*e.g. the project manager and the quality authority*). This concession **grants the release to the customer of a deliverable** (with or without repair) **that is known to have specific non-conformances to the approved baseline**; i.e. either waivers and/or deviations are in effect for this particular delivered unit.

  Concessions are usually time limited (to the current special case) and are granted because; while the non-conformance is not technically a match with the approved baseline, the non-conformance does have the characteristics of the approved baseline. The Concession would not necessarily last for the entire duration of the project; either, the cause of the deviation and/or waiver would be resolved for units produced thereafter, or an **Engineering Change Proposal (ECP)** would be generated to have the approved product baseline modified so that such a Concession would not be required for any future builds of the deliverable. *For example; increasing the storage capacity of the installed hard-disk, because that original model (and its smaller storage capacity) are no longer available from suppliers due to technological obsolescence.*

- **Declaration Of Conformity (DECL) / Certificate of Conformance (COC)** – a formal document signed off by the authorized representatives of the performing organization as commemorating that the product unit being delivered does conform to the approved product baseline and the product unit has passed all of the associated Acceptance Criteria tests (as was evident by the previously mentioned ATRs Acceptance Test Records, ITRs Inspection & Test Records, Concessions & Waivers).

  That is, the authorized quality representatives of the performing organization are saying, *"With hand-on-heart, we the signatories stand behind our product and are proud to present this project deliverable to our honourable customer"*.

- **Third Party Certifications** – the formalized documented outcome of an audit of the performing organization and/or of specific project deliverables undertaken by an externally independent, impartial, accredited third party. The purpose of the certification audit is to evaluate whether the performing organization and/or the project deliverable under examination does comply with a given sanctioned standard or standards; *e.g. ISO standards*.

The non-obtaining of this Third Party Certification could mean that the performing organization is not eligible for consideration as an acceptable project participant, and/or that the project deliverable is not eligible for a specific usage. Without the necessary Third Party Certification, then the customer organization must consider the legal ramifications of engaging the performing organization or using that uncertified project deliverable. *For example; would you consider having your new home built by a builder who was not legally certified? Similarly, would you honestly consider buying an existing home that did not have local government records indicating that the house was constructed in accordance with the statutory building regulations?*

- **Licenses** – Similar to certification, a license gives **official (statutory / legal) permission** for the performing organization and/or the project deliverable to do a particular something. *For example; the equipment is licensed to be connected to the national telecommunications network infrastructure.*

- **Audit Report** – objective appraisal & evidence compiled by either a first party (internal to the performing organization), a second party (internal to the customer organization), or by a third party (external to both of these organizations).

This audit report would **detail the findings of comparing the performing organization and/or specific project deliverables against a pre-establish set of Audit Criteria** which determines whether this criteria has or has not been met. *For example; ISO 9001 certificate of compliance / accreditation audits.*

I would recommend that ... the project manager **ensure that sufficient Objective Quality Evidence be produced for the project so that the deliverables and the associated processes & procedures involved with producing those deliverables would successfully pass scrutineering** (such as with a quality audit).

Does my project really require this much Objective Quality Evidence, as I have never had to produce this amount of bureaucratic paperwork before?

Yes and No ... **The amount of "traceability records"** (or bureaucratic paperwork as some would describe it) **that is required per project is very much dependent on the expectations of the project's core-primary stakeholders, the associated statutory authorities, the industrial norms, and the agreed quality standards that must be complied with in order for the project deliverables to be accepted.**

For some organizations / industries then a minimal amount of OQE is acceptable, but for other organizations / industries then a sizeable amount of OQE is necessary.

IMHO, if someone could lose their life (or lose the shirt off their back) due to the use of this deliverable then you can bet that a sizeable amount of OQE will probably be required.

NOTE ❑ **This Objective Quality Evidence could prove to be vital later on, if the project happens to be scrutinized and/or the performing organization is audited for quality certification, or due to legal action brought against the organization.**

Therefore, the trick is for the project manager to come to a quick realization of just how much Objective Quality Evidence is required for the project and the project's deliverables to satisfactorily comply with the expectations of the project's core-primary stakeholders' and to abide by those relevant industrial / statutory regulations.

There is **no point turning out the perfect operational deliverable**, if the customer is unable to accept the deliverable (or is not legally allowed to use the deliverable) **without all of the associated paperwork** (objective quality evidence) having been **correctly signed, sealed, and delivered**. *... in triplicate, in the correct format.*

Blindly Following Quality Standards, Processes & Procedures

Whatever industry the project is being undertaken for, there most probably will be some set of quality standards (*e.g. the ISO 9000 family*), and a collection of quality processes & procedures that the project and/or its deliverables must comply with.

While these "quality formalities" are very much relevant, **the zealous pursuit of complying with quality standards, and processes & procedures does not necessarily guarantee that the project will be successful when it comes to producing "quality deliverables".**

Recall back in [Section 4.3.4] it was stated that:

> *"The true measure of quality is analogous to, the taste of the cake and not the processes used to bake the cake".*

It is essential to comply with those quality standards, processes & procedures that are "mandatory" for the project deliverables to be formally accepted.

As for **all of the other quality formalities**, these **are "negotiable"** based on the project's current situation and the prevailing circumstances.

DO NOT blindly follow quality standards, processes & procedures without firstly questioning whether these are "really applicable" to the project due to; that project's size, scale, duration, deliverables, and the performing organization and the customer organization's current situation and the prevailing circumstances.

Because **following quality standards, processes & procedures to the nth degree (without relevant justification) could be detrimental to the project' success** by; significantly adding to the project's [Time] duration, the number of [People] & [Resources] required, which intern could impact on the project's [Cost] and hence the project's profitability. ... *and, needlessly frustrating the implementers.*

NOTE ❑ **The decision as to which quality standards, processes & procedures are "mandatory" and those that are "negotiable" will potentially introduce risk to the project**. So, think about this choice carefully and choose wisely, as an incorrect decision could result in a lot of retrofitting rework (just when the project should be finishing, or should have already finished).

Always remember that; **the purpose of the Quality Assurance / Control processes & procedures is to assist in achieving the project's objectives.** ... And, **not dictating how to achieve these project objectives**.

That is, the project's objective is to produce the deliverables, and (most probably) not the production of copious amounts of documentation. Unfortunately, inappropriate forms of Quality Assurance and Quality Control can result in the generation of a disproportionate amount of documentation (and associated paperwork) which does not really result in the production of quality project deliverables.

As stated back in [Section 2.7.1]:

THE QUANTITY AND QUALITY OF DOCUMENTATION APPLIED TO A PROJECT SHOULD BE PROPORTIONAL TO THE SIZE, SCALE, AND SIGNIFICANCE OF THE PROJECT.

Misalignment of Quality Processes & Procedures

Sometimes during the project's life, you (and more specifically the Project Implementation Team members) will feel that the quality processes & procedures have become burdensome and seem to be hindering / delaying the project's timely delivery.

This **negative view of Quality Assurance and Quality Control as a burden can arise due to a misalignment between the quality processes & procedures that have been "dictated as necessary" and that "which are necessary" for the successful delivery of the project**.

This misalignment of perspectives on Quality Assurance and Quality Control can result because:

1) Those people who wrote (and authorized) the quality processes & procedures have a **different industrial background** to those people who are implementing the project.

   For example; the Quality Authority comes from a mass-production hardware manufacturing industry, but the project is being implemented by persons from the bespoke software development industry.

2) While those people who wrote (and authorized) the Quality Assurance and Quality Control processes & procedures have the same industrial background as those people who are implementing the project; the quality authorities have extensive experience with a waterfall **development methodology** whereas the implementers are using agile techniques.

   For example; the Quality Authority sees the production of the project deliverables occurring via a system of gate releases with in-triplicate sign-off at the conclusion of each release, and hence the Quality Assurance/ Control processes & procedures replicate this methodology. However, the Project Implementation Team is being hamstrung by this and subsequently cannot respond to the dynamic nature of this project and the customer representative's various requests.

In my humble opinion; for **Quality Assurance / Control processes & procedures** to be seen as beneficial, these **must relate to the realities of the project**'s current situation and the prevailing circumstances, and how those persons that are bounded by these processes & procedures actually do their work (in the real-world).

Hence, if the relevant sections of the Quality Assurance / Control processes & procedures cannot be explained as a hand-drawn sketch on a single A4 sheet of paper, then maybe these processes & procedures are too complicated for their supposed purpose.
So don't be surprised when these processes & procedures are quietly sidestepped by those persons who are restrained by these burdens.

There is also another difference in perspectives on Quality Assurance / Control to consider.

3) The members of the Project Implementation Team are often solely focused on completing their assigned tasks [Scope] within [Time]; as a result, occasionally a corner or two will be cut and thus the project's [Quality] could suffer. However, the project's success is based on the relationship of [Scope], [Time], [Cost], to [Quality]; therefore, it falls to the Project Manager to **keep the Project Team's mind on the bigger issue of "customer satisfaction"**.

In conclusion, **the project's management** (i.e. the members of the Project Working Group and the Project Steering Committee) **have to be responsible for ensuring that only those relevant quality standards, processes & procedures are being applied when producing the project's deliverables**.

 **Quality is NOT the sole domain of the Quality Control Group** (*e.g. Verification Validation & Test team*) **and the quality authority; rather, [Quality] is the domain of the entire project team** (i.e. the Project Steering Committee, the Project Working Group, and the Project Implementation Team). *... IMHO, quality is equally the responsibility of the generals as it is the foot soldiers.*

Reviewer's Comment ... *Quality isn't meant to be a burden*

I think that at some points in that chapter, you gave a slightly negative view of quality. The other side of the coin is that, quality processes & procedures are in place because these do enforce quality, and help in avoiding the shortcuts that can lead to poor quality deliverables.

Okay, that last part might have come across as a bit "negative" on quality processes & procedures, and I do agree that quality processes & procedures do help avoid and limit the effect of poor quality deliverables... as stated back in [Section 8.1]:

> *"DO NOT underestimate the cost of poor quality and the effect that it can have on the performing organization's financial bottom-line".*

QUALITY ASSURANCE + QUALITY CONTROL
THE NEW SHERIFF IN TOWN ...
UPHOLDING LAW + ORDER WITHIN
THE PROJECT TERRITORY.

14. MONITORING & CONTROL Phase ... Practical Part 3

14.1. [People] Monitoring & Control

In a practical sense, [People] Monitoring & Control [Section 9.1] asks the following questions:

- What can I do to help "my people" perform at their best?

- Is there anything that disrupts their day or stands in their way?

- Are they satisfied that they are contributing and do they feel valued?

- Do I need to change their priorities to align with the project's objectives?

Remember that, without the [People] the project is not going to be implemented let alone be delivered with the agreed [Scope], on [Time], within [Cost] budget, and to the level of [Quality] that was agreed to be acceptable. Yet, it is surprising how many performing organizations don't place more importance & concern on this [People] aspect.

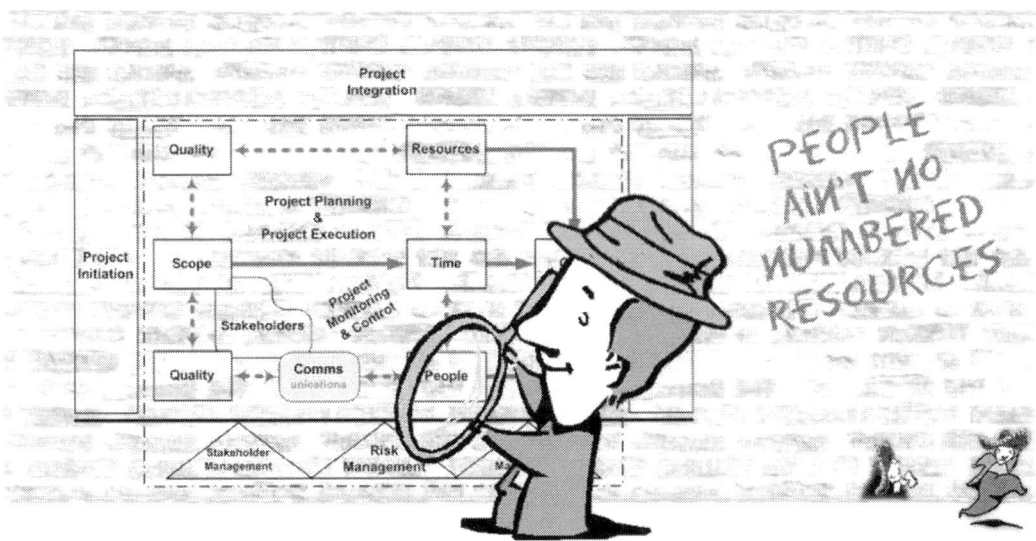

⊠ [People] have to be considered as an important project element and not just treated as numbered [Resources].

Way back in [Section 1.4] it was stated that;

> "People should NOT be grouped together with Resources (i.e. inanimate objects), because a successful project requires the right number of people with the appropriate skill sets" and "the People performance aspect will change dynamically throughout the life of the project, whereas the characteristics of the inanimate Resource performance aspect will remain relatively constant."

"Project Accountant" types of managers generally place an emphasis on the "hard skills" and are more likely to consider people simply as "interchangeable resources" that can be easily exchanged / bought / let go, as though these people are components in a machine.

Thinking back to the road-trip analogy that was introduced in [Section 6.1], these people are analogous to tyres on the car. Though, what happens if there is no viable spare tyre (person) available at that moment in time when it is so desperately needed?

The vehicle (project) is going to be stuck there in the middle-of-nowhere, while the driver and passengers wait for roadside assistance to arrive. All the while, simmering over how all hell is going to break lose because the necessary precautions where not taken to get home in time (i.e. complete the project on time & within budget).

War Story … Our project is screwed due to people mechanisation.

Our project is now "screwed" because another long-term senior engineer resigned today. Senior management are in a panic, trying to encourage that particular engineer to stay (at least to the end of the project, while privately shoring up those other key engineers). But, anyone who was paying attention would have realized that this situation has been in the making for some time, given theirs (and others) dissatisfaction with the mechanistic structure that has been imposed by this senior management.

The project manager has to be a "soft-skilled" manager of interpersonal relationships.

As was detailed in [Section 11.2] and [Section 11.3], it will be the project manager who will have to deal with these persons via the manager's capabilities with interpersonal communications, negotiating & influencing, leading & motivating, problem solving & conflict resolving, and showing empathy & concern.

NOTE ☐ **DO NOT make the mistake of thinking that the responsibility for dealing with the humanities aspect** of the project's people **can be just palmed off onto the Human Resources department**. It is very important that the project manager personally monitor & actively control the humanities aspect of the project team.

14.1.1. People: The Bounds of Performance

How often have you heard the line?

DON'T TELL ME HOW HARD YOU'RE WORKING ... RATHER TELL ME HOW MUCH YOU'VE GOT DONE.

And that perennial favourite,

WORK SMARTER ... NOT HARDER.

IMHO, what a load of narrow-minded rubbish.

I totally agree to disagree with these adages ... this is "a load of non-constructive management double-speak". In reality, the majority of implementers [People] such as engineers, programmers, and testers will do what is required to get the job done. As more often than not for these types of implementer people, work transforms into a personal quest to resolve the technical problems that confront them and the self-satisfaction of building something for which they can proudly say, "I made this", "we achieved that".

Hence, these types of implementer people don't usually shirk work or buck-the-system (unless they are bored or see no future for themselves with that performing organization); rather, they learn to play by those unwritten rules of their team and work within the boundaries decreed by their employer.

It is the project manager's responsibility to:

1) **Help** senior management to **define the boundaries of [People] performance.**

2) **"Empower" each project team member with the authority to get done what they are responsible for achieving**.

3) **Remove those barriers that prevent these [People] from** doing their jobs to the best of their abilities and thereby enabling them to **perform optimally.**

4) It is the **project manager's own behaviour,** performance, planning & organizational skills, plus their leadership style that will also contribute to **influencing the performance of these [People]**.

 For more information on team building and teamwork, refer to [Section 11.4].

THE PROJECT MANAGER SHOULD REMOVE THE BARRIERS AND OPEN THE DOORS THAT WILL ENABLE THE PROJECT TEAM TO MAKE FURTHER PROGRESS.

Just think back to when you were an implementer, did you ever have your manager:

- **Explain** (to a level of detail that you understood) **what was expected of you?**

  OR, did you only find that out later on, when you didn't score as well as you expected during your yearly performance review?

- **Define** (to a level of detail that you understood) **what you had to do and what you had to deliver?**

  OR, did you only find this out later on when they scolded you for not doing "what they told you to do"?

- **Complemented you when you did a good job?**

  OR, was the only time your manager "communicated" with you was to yell at you when you didn't perform as they expected (when "you stuffed up")?

- **Take the time to ask you how your day was going, and show some general interest in you as a human being?**

  OR, did they just hideaway in their corner office / cubical fortress only reappearing to issue decrees and to berate those who came under their stern gaze?

- **Concerned with the number of hours overtime that you were putting in and how this maybe affecting your home life?**

  OR, were they only concerned with meeting some delivery date deadline, which after you had put in all that effort (and overtime work) turned out to be not so important?

Now that you are the project manager, are you going to repeat their bad characteristics as though you are bound to "do unto others as was done unto you"?

OR, are you going to strive to be different to your predecessors?

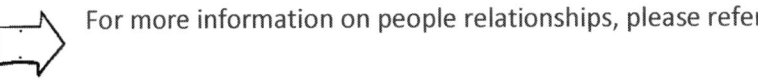 For more information on people relationships, please refer to [Section 11.5].

14.1.2. People: Handicapping Performance

This section could be filled with the bad elements of those facets about project management and team building, as described previously in [Chapter 11]. However, instead of covering all of those elements again, this section will focus on one specific characteristic of management and on an associated implementation structure that can hamper the performance of the project team and limit the achieving of project success.

Micromanagement

Micromanagement is that misguided belief held by some managers (and/or organizations) that to get the best performance out of their ~~people~~ *human resources* they must ~~manage~~ *control* every aspect of their workers' daily lives via rigidly defined hierarchical work structures and draconian work practices, processes & procedures.

Reviewer's Comment ... why micromanagement proliferates.

One possible reason for micromanagement could be that the project manager / senior management have invested a significant amount of time & effort fastidiously planning every aspect of the project. Thus, it could be very hard for them to allow others to come along and potentially muck-up their well thought out plans (given that no one else knows the project anywhere near as well as they do). Subsequently, for the micromanager the slightest sign of worker self-autonomy could be perceived as a direct threat to "The Plan" and its success. Therefore, any dissentient thought or effort must be terminated with extreme prejudice.

Overzealous micromanagement can also be systemic within an organization, originating from the senior management and proliferating on downwards. Prolonged organization wide micromanagement can eventually result in a workplace that is predominately composed of "Drone" persona workers and "yes men" middle management, as all those persons with some nous & savvy eventually leave, having given up the fight to "try to make a difference".

However, micromanagement may be appropriate for certain industries, activities, and situations; *e.g. production lines, new employee training, and crisis resolution*.

But ... for an SDLC project, a tendency towards micromanagement will (more often than not) hinder the project team's performance and can even result in poor productivity, mediocre quality, and missed project milestones.

Let alone foster poor team morale, and stifle initiative.

IT IS HARD TO THINK OUTSIDE THE BOX WHEN ONE IS BOUND UP INSIDE ONE.

The micromanager or the micromanaging organization (with their Theory X nature) can be highly restrictive with the delegation of work and the allocation of responsibility to their "workers"; even going as far as to silo team members into discretely demarcated job roles (then locking them into these defined roles via bureaucratic controls).

Assembly-Line Tasking

Due to the micromanager's innate desire to "continually maintain control over everything", the micromanager will often turn the project activities into sequences of assembly-line tasks that they will allocate to the workers. These workers will be restrained to only perform a limited set of ordered tasks, and may not be allowed to proceed to the next set of tasks until the manager has confirmed the outcomes of their completed tasks and (personally) given permission to proceed, or the worker has to wait for the manager to issue new instructions as to what each worker is to do next.

While this assembly-line process may (for a short-term) generate productivity benefits from a "project accountant" perspective, what this also does is transforms each worker into a brain-dead zombie who has become "neutered to taking the initiative".

Hence, **for an SDLC project where the current situation and the prevailing circumstances can change dynamically, micromanagement can result in a project and a project team that is TOO SLOW to "RECOGNIZE, RE-ASSESS, REVISE, AND RE-APPLY".**

Human Machines ... *You are just another cog in the machine.*

An associated problem with micromanagement and assembly-line tasking is that, by siloing the works into highly specified and demarcated taskings then "these workers" are being treated as "disempowered" human machines. **By being considered as mere human machines this effectively renders those** ~~workers~~ **[People] with intelligence / creativity / talent as almost superfluous** *("anyone for a lobotomy")*, it also **prevents the cross-fertilization of ideas & knowledge** from one encapsulated group to the next, and it **limits the feeling of "ownership"** of the whole by those persons working on a portion of the project. All of this can have a very **demoralizing effect on** the project team, and can **contribute to growing job dissatisfaction** (potentially contagious dissatisfaction within the performing organization) because individuals perceive that they are **not doing what they would consider "meaningful work", let alone not doing "stimulating work".**

A common scenario for experiencing regrettable **problems** with micromanagement and assembly-line tasking is **when a mechanistic waterfall methodology is mandated onto a project where the majority of the project team members are used to working autonomously or semi-autonomously** (such as with iterative and agile developments). By imposing the siloing of the work into demarcated tasks, some project team members (especially those junior and non-subject-matter-expert members, and even the senior members) are going to end up working on tasks that they consider as **menial** or **mind-numbing jobs that effectively eliminates their opportunity to expand their experiences & responsibilities;** thereby, **decreasing** their **"job satisfaction"** and **lack of "self-esteem"**. As a result, they may seriously be contemplating resigning, hoping to move onto other organizations where their skills & experiences will be better suited and more importantly, "appreciated" (let alone rewarded).

Yep, under this type of Theory X organizational reign there is limited opportunity to contribute outside of the assigned box and much of your skills & previous experiences will go to waste. ... So, don't be surprised when several good staff decide, "stuff this, I am out of here, you Jerks".

 **Micromanagement is bad for the Project Manager's future** (let alone for the performing organization); because, micromanagement **takes up a considerable amount of time** that could be better utilized elsewhere. Secondly, micromanagement **results in non-self-reliant project team members**; consequently, what will happen if the project manager is away for a period of time, will these "workers" stop working effectively because they are not sure how to constructively & productively manage their own time or are they rendered incapable of resolving problems by themselves (given how disempowered they are)?

In addition, because of this siloing of activities when a task is in the grey-area between demarcated zones then things can fall between the cracks. However, neither group of "workers" knows how to take unassigned responsibility, believing that the other group will be ordered to take responsibility for that grey-area task. Consequently, the project can become deadlocked as no one takes responsibility for anything outside of their little box.

PEOPLE WHO ARE BEATEN INTO SUBMISSION TO ONLY DO WHAT THEY ARE TOLD, SOON FORGET HOW TO THINK FOR THEMSELVES, AND ARE HESITANT TO SPEAK UP EVEN WHEN THERE IS SOMETHING THAT THEY REALLY SHOULD SAY.

14.1.3. People: Judging Performance

Before the [People] can be successfully monitored & controlled it will be necessary to define how to judge each project team member's performance.

These judgements must be based on:

1) **Tangible, meaningful, and constructive measures that contribute to achieving the project's objectives** and NOT based on a set of arbitrary criteria (i.e. figures thought up by some "accountant" type). **These measures must take into consideration the realities of the project's situation and the prevailing circumstances, as well as based on a reasonable understanding of the project(s) being undertaken.**

Inappropriate performance measures for SDLC projects would include a scheme based on the number of features implemented per day (i.e. the burn down velocity), the number of lines of code written per day, the number of defects closed per day, the number of defects reported per week, the number of XYZ somethings per month.

> *For example: the development teams' new performance schemes are based on the amount of lines of compiled code produced per day. The senior programmers are not happy about this situation because they argue that this is encouraging "a room full of monkeys" coding that has resulted in software applications that are "bloated, resource-hogging, performance-dogs". They claim this fall in quality is all because the new performance measures have discouraged the production of well thought-out, efficient, and effective programming.*

> *For example: a support & maintenance team's new performance scheme is based on exceeding a set target of how many Defect Reports a person closes per day. Within weeks, animosity has built up among members of the once harmonious team because some members have been reserving / checking-out multiple Defect Reports that appear to be relatively easy to resolve, and some members are prematurely closing cases before these have been properly concluded.*

 **DO NOT attach performance judgements to purely quantitative measures because quality suffers BADLY in the pursuit of quantity over substance.**

This is because quantitative measures DO NOT encourage intelligent, efficient, and effective work practices.

2) **Measures that the person being judged actually has some amount of control over.** That is, they are relatively the masters of their own fate as their own efforts (and that of their team) directly influences the outcomes of their performance review.

Else, **these people will give up trying if they perceive that the measure of their performance is completely out of their own hands.**

For example: their salary's "at risk" component is determined 25% by their own performance, 25% by their department's performance, 25% by their business unit's performance, and the remaining 25% by the performing organization's results. While from an "accountant's" perspective these may sound reasonable judgement parameters, from the people's perspective they see themselves as having only a 1 in 4 chance (maybe 2 in 4) of influencing whether they obtain their bonus or not. As a result, if the business situation worsens and these external parameters don't look good then they may question whether their additional effort is worthwhile given that they figure they have "bugger all chance" of collecting their bonus.

IMHO, this is why bonuses (and performance based stock options) are losing their effectiveness as a motivating tool in non-high-flyer industries, because given the economic conditions (after the Global Financial Crisis and European Debt Crisis) such bonuses can no longer be reasonably counted on, ... no matter how hard you worked nor how good your personal performance results were for the period being judged.

 **DO NOT attach monetary bonuses to performance factors outside the control of the person being judged.**

This is because their morale and effort will decline when they realize that no matter how hard they work and how well they perform, it won't change the outcome of whether they obtain their bonus or not.

3) **Measures that are relevant to the role that the person being judged has been required to fulfil** and NOT the role that was defined in their job description (or roles common to the performing organization's core business). This is because; **the project's work situation and the prevailing circumstances may require some team members to step out of these defined confines and to perform alternate roles**.

For example; My career started out as a software coder, but somehow it deviated into being a system tester, customer support, web developer, IT administrator, business analyst, document writer, project manager,... and all because the project's success was dependent on someone immediately stepping into each of those roles.

Here is a laugh ... I once had a performance review where the function of the job role that I was fulfilling did not match the majority of the questions on the review form. What made things worse was that the review form did not consider the possibility of any such deviations outside the business industry's norms, and hence there were no fields on the questionnaire for "not applicable". Therefore, as specified on the form I could only be graded a 1 out of 5 for each of the irrelevant questions. Though somehow, I did get a really nice pay rise off such a low performance score.

4) **Judgments that are based on the person's actual performance and NOT based on their conformance to prescribed norms.** How a person dresses, speaks, behaves, and their unique quirkiness can influence the judgment of their performance. Similarly, their gender, age, ethnicity, and apparent religion can be negatively held (covertly) against them when evaluating what they add to the performing organization.

For example; a long haired, long bearded, sandal wearing, alternative-living hippy or the well groomed, suit wearing, polished shoes, conservative business professional. If these two persons were being interviewed for identical roles at a banking organization and a research & development organization then which is more likely to get which role, or more specifically who would not be considered for which role based purely on their appearance?

Well, the same goes for the judgement on the performance of their daily work activities.

5) **Judgments that are based on their actual performance and NOT based on the perception of the hours of effort that they are putting in.** Differing work & home life circumstances will mean that some of the project team members will need to work different hours to the norm. Some members may need to start earlier and finish earlier because they have to pick up their kids after school, whereas others may decide to skip the rush-hour traffic-jams and hence arrive later at work and go home later.

You may not think that the hours when a person starts and finishes work could affect their perceived effort, but it can. *... Especially with a Theory X type manager.*

For example; if one of your team members leaves work each afternoon at exactly 3:30 PM, how do you honestly perceive their efforts?
If they happen to start work at 7:30 AM in the morning when you also start, then you probably will perceive them as working just as hard as you do; but, if you don't start work until 9:00 AM then you possibly will perceive them as not working as hard as others who work similar hours to you (as you pack-up in the dark of night).

What if a team member gets in 2 hours earlier and another gets in only 15 minutes before the manager, they will subconsciously be perceived as both starting at about the same time. Similarly, for team members who finish 15 minutes or several hours after the manager goes home, they both could be perceived as working late. That is, those team members whose work hours deviate noticeably from the performing organization's core business hours risk being perceived as not putting in the same level of effort as those others whose work times do conform to the norms.

I once had a manager who was on my case because I would come into work at quarter past 9 in the morning, instead of being at my desk coding at 9am on the dot. He was not the least bit interested that public transport only stopped near their office every hour, nor that I was still coding away long after he had driven home.

Yet somehow, in his mind those 15 minutes in the morning were more important than those additional hours that I worked each day.

6) **Management styles of Theory X and Theory Y can greatly influence the judgement of a team member's performance**. Some team members may require more of a Theory X style of defining all of the tasks that they are to perform; whereas, other team members' performance may thrive when given a freehand by the Theory Y manager. Consequently, the best performer for one project manager could be another project manager's worst performer, and all due to differences in management styles ... and possibly due to **clashes of personalities** (and in a worst case due to discrimination).

MICRO-MANAGER
THE LITTLE DICTATOR
OF THE PROJECT REALM

14.1.4. People: Reviewing Performance

AUTHOR'S NOTE

There is a rather large caveat when it comes to doing performance reviews, ... most performing organizations will already have some defined ways (and templates) for reviewing their staffing performance; hence, what follows in this section is often constrained by pre-existing determinates.

However with that stated, the following should provide you with some food-for-thought when you are undertaking the performance review of your project team members, and hopefully this will balance those purely "accounting" measures.

In addition, you never know; one day in the future you may be the one who decides how performance reviews are to be conducted in your organization.

When it comes to reviewing performance, there is more to it than just judging each person's **technical skills** and their project delivery accomplishments for the time period being evaluated. As thought should also be given to how that individual **personally interacted** with other people inside and outside of the project team and within the performing organization. That is, how are their "inter-personal" / "social" skills?

Simply put, **there is not much benefit in having a technically brilliant team member if no one else is prepared to work with them** (or worse, people refuse to / begrudge participating with that team because of that person). Do they make others feel uncomfortable and ill at ease, or are they a catalyst for disharmony within and around the project team?

THERE IS NOT MUCH LONG TERM BENEFIT IN HAVING A TECHNICALLY BRILLIANT TEAM MEMBER, IF NO ONE ELSE IS PREPARED TO WORK WITH THEM

A GENIUS MAY COME UP WITH THE BRILLIANT IDEAS BUT IT IS A TEAM OF MUNDANES WHO TRANSFORM THESE IDEAS INTO REALITY

Dual Axis Performance Grading Scheme

What is required is some technique to **place inter-personal skills on the same judgemental plain as technical skills**.

Back in [Section 10.1.1] the Risk Matrix was presented and a similar arrangement was used for stakeholder analysis in [Section 10.3.3].

So, why not use this same matrix idea to analyse people's performance?

Consider a 5x5 grid as presented in [Figure 131], where the x-axis represents the technical skills of the person being judged, and the y-axis represents their inter-personal skills.

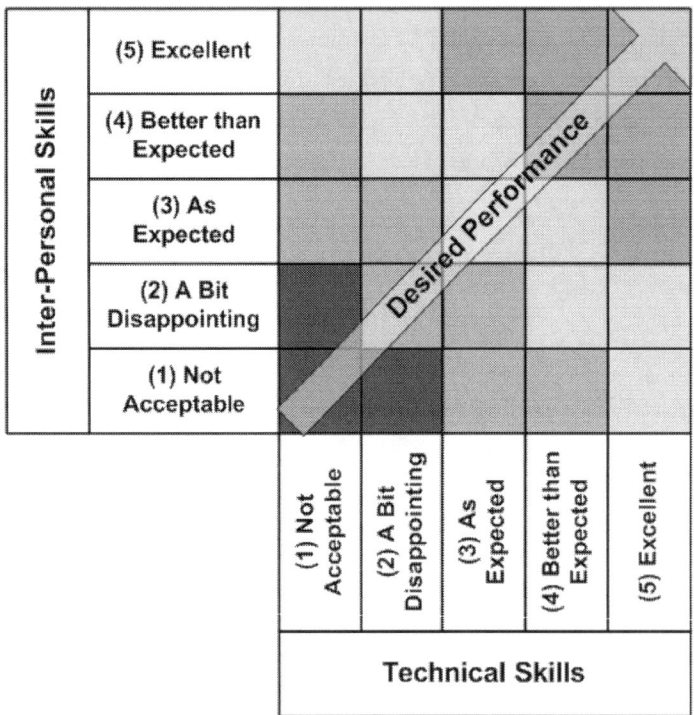

Though, with the desired outcome at the top right instead of the bottom left.

As with most grading schemes, the scoring goes from lowest to highest; *e.g. they have a technical skill level of very low to very high.*

Figure 131: Dual axis performance review matrix.

PERFORMANCE REVIEWING IS SERIOUS BUSINESS, NOT AN ANNUAL HASSLE.

Though, let's have another think about this grading scheme ...

If I got 4 out of 5 for doing certain activities last year, when I am reviewed this year and my performance level is exactly the same as it was last year, then should I still be graded as a 4 out of 5 or should it be downgraded to a 3 out of 5 given that I now have had a year's more experience at my job?

How about an alternative grading scheme where the evaluations are more personalized:

1) **Not Acceptable** – for the period of concern your performance (with respect to the technical or inter-personal aspect) was well below what we expected from you; i.e. *"we need to talk"*.

2) **A Bit Disappointing** – for the period your performance (in this aspect) was less than what we expected from you; i.e. *"I am sure you can do better next time"*.

3) **As Expected** – for the period your performance (in this aspect) was exactly what we expected from you ... given your relevant experience and the prevailing circumstances you faced; i.e. *"keep up the good work"*.

4) **Better than Expected** – for the period your performance (in this aspect) was a bit of a surprise for us ... given your relevant experience and the prevailing circumstances you faced; i.e. *"that was a pretty good effort"*.

5) **Excellent** – for the period your performance (in this aspect) was more than we could have imagined ... given your relevant experience and the prevailing circumstances you faced; i.e. *"take a bow, the drinks are on me"*.

The important things to note about this dual axis performance-grading scheme are:

- **The grading for the technical skills and inter-personal skills are independent of each other**; i.e. these are NOT a combined "score" but rather a visual positioning on a grid.

- The scheme is **NOT about comparing one person against another person as some form of competition. Rather, the scheme is all about comparing an individual against their own performance for this period versus their previous periods performances, as** well as comparing them against what was realistically expected from them, ... given their relevant experience and the prevailing circumstances that they faced.

- **A person who is highly productive (when compared to their compatriots) can still position poorly on the grid; if they happen to be not working anywhere near to their known technical capabilities, or if they happen to be exhibiting some very undesirable inter-personal behaviour.** Inversely, an individual who has relatively low productivity (when compared to their compatriots) can still position well on the grid, because they are continually improving technically and their attitude is desirable & beneficial to the project team and the organization as a whole.

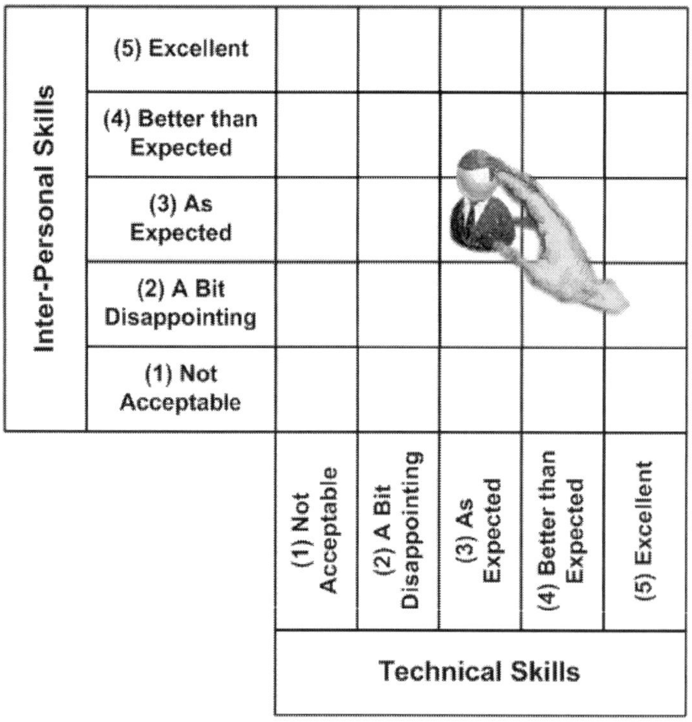

For the dual axis performance review matrix, the expected norm is the centre / middle of the grid and not the top right of the grid.

To position in the top right of the grid can only be described as an astonishing performance (both technically and inter-personally).

Figure 132: Use of a people dual axis performance review matrix.

With this dual axis **performance grading scheme the emphasis is on self-improvement, continued growth within the performing organization;** as a result, the person is being rewarded for such efforts and desired behaviour that contributes to the whole.

Therefore, the aim of the person being reviewed is to strive to be positioned in the top-right corner of the grid and definitely not be positioned in the bottom-left corner of the grid; for an example see [Figure 132].

The problem with more traditional grading schemes is that the longer that someone has been in the organization or the longer that they have been performing the same role then it is probable that after a while they can become complacent and even expectant of receiving a certain grading at each performance review. Consequently, they may "rest on their laurels". This expectation of receiving their "normal" review score can make them very moody if they happen not to get their "usual" grading (and not receive its "usual" associated monetary bonus). This associated bonus would no longer be directly related to their performance; rather, it is considered as part of their regular salary package.

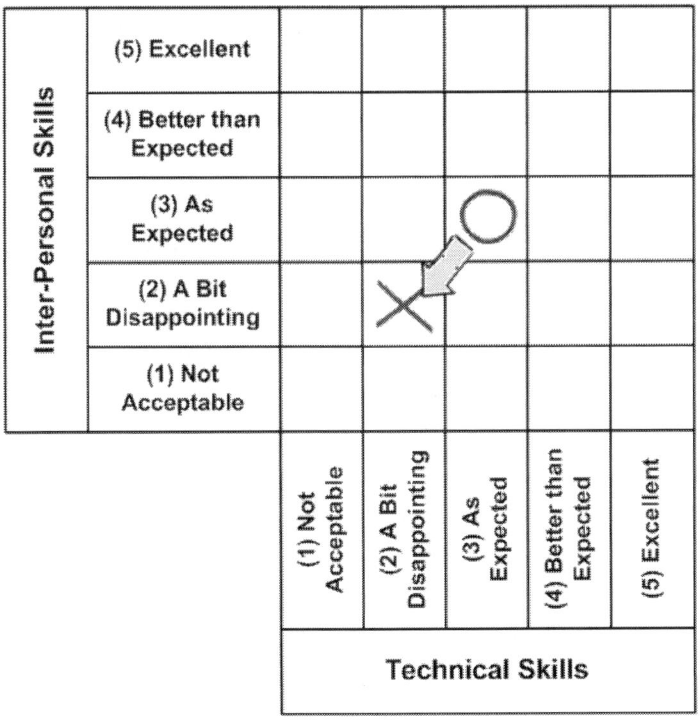

Figure 133: Performance review matrix used to represent current progress.

The 'X' indicates where a continuation of their current performance will probably be graded at the next review.

An acquaintance of mine would get angry when they did not get four and five at each year's performance review. But the thing is, they had done nothing new to improve their work-skills let alone seek internal promotion.

Honestly, what is better for the performing organization... someone who is improving all of the time, or someone who has plateaued and potentially is no longer providing beneficial stimulus into the organization? *Mentoring is also beneficial to the organization.*

Additionally with this dual axis performance grading scheme, future information about the person being judged could be included; i.e. the direction that their performance is heading. That is, an indicator that if their current performance continues as it is today then this is what they will most probably be rated next time, as depicted by the cross in [Figure 133]. This technique of representing directionality can be quite affective for reinforcing desirable behaviour and for discouraging not so desirable behaviour.

> *For example; a junior engineer who for ¾ of the review period has technically performed exceptionally well, but in the last few months has lost focus and has become a little overconfident (to the point of being arrogantly "cocky"). While a "quite chat" will go some way to help correcting this undesirable behaviour, the technique presented here will visually reinforce the message that the continuation of such behaviour will have a negative impact on their performance review (and thus directly affect their associated pay review).*

> *Similarly, a senior developer with a technical superiority complex was shocked to discover that his interpersonal attitude had a negative impact on his review.*

I would recommend that ... prior to the performance review meeting being conducted, the person being reviewed should fill out their own copy of the performance review form and they should put a marker where they believe they would be positioned on the grid.

During the review meeting, the manager's version of the form and grid should be compared with the individual's version. Any major differences of opinion should then be discussed, and thereby any differences in expectation & perception of what is required from them should be explained by their manager.

Remember that, while the top-right of the grid is the desired result, it is the center / middle of the grid that is the expected norm.

Performance Reviews and the Bell-Curve of Death

If such a technique as the dual axis performance-grading scheme is used, then **DO NOT** pollute its purpose by trying to **superimpose a bell-curve** on to it.

A **bell-curve** will totally **undermine the principle idea** of the scheme **to promote "self-improvement" and "continued growth"**. Instead, the bell-curve will **replace it with arbitrary competitive judgements** where people are either winners or losers.

Similarly for a pyramid grading scheme where a limited few employees are on the top level, where some more employees are in the middle, and the rest are on the bottom.

But, *I thought competition was good to have?*

While competition may be appropriate for groups such as sales, it is highly detrimental to the evolution of teamwork within the project team; this is because, competition emphasizes self-interest over collaboration.

> For example; *helping you could mean that you position higher on the review curve which will mean that I could be pushed down the curve and out of the "to be rewarded" quadrant, or by taking the time to help you I am limiting my own time to push myself up the curve.*

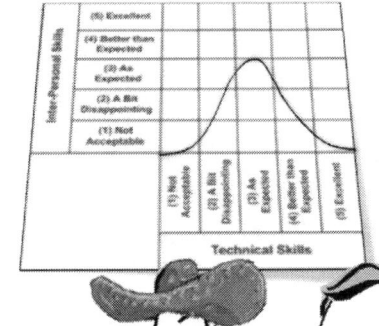

POLICY DICTATES THAT WE HAVE TO HAVE SOMEONE IN THE BOTTOM PART OF THE CURVE AND YOUR NUMBER WAS DRAWN. IT IS NOTHING PERSONAL... IT IS ONLY BUSINESS.

Example Case: the bell-curve bastardization of a good idea.

Once upon a time, I worked at a small "adaptive" organization whose new international owner imposed a similar dual axis performance grading scheme to the one described in this section. However, they used a ridiculously small grading scale of 1 to 3, and it was mandated that a certain portion of the local staff had to be in the bottom percentile of the bell-curve, ... and these "low achieving" people were to be "let go" (as though these local "resources" could be so easily replaced). This for a team oriented development department where "retained people and knowledge is golden"; but alas, under this bastardized scheme these "low achievers" were to be terminated just to comply with the dictated bell-curve rules. This was a self-destructive scheme that outraged the local staff and irritated the local management because, next year was it expected that some of this year's "good" people would be moved into the "bad" area of the grid to be sacrificed to appease the bell-curve's gods. To implement such a bell-curve performance measurement scheme as this would only result in a churning of people [Resources] and demoralize the development group with the loss of colleagues / friends.

IMHO, the churning of ~~people~~ "resources" is a complete waste of time, money, and is highly detrimental to overall productivity.

Let's face facts, the new person who comes on board (irrespective of how much experience & skills they have) will most probably not be anywhere near as productive as the person who left. This is simply because the new person has to learn the organization's ways, obtain an understanding of the project / projects, and integrate into the community that makes up the project team and the performing organization. And, unless you want the new person's participation to be incompetent then someone else is going to have to show them the ropes; as a result, this mentor or mentors will have their own productivity restricted during the period of time that is required to successfully assimilate the new person into the organization.

14.1.5. People: Accepting the Performance Review

Now that the performance judgements have been determined and the performance review materials have been prepared, the next step is to conduct the one-on-one performance review meetings and (hopefully) have the reviewee respond positively to the reviewer's well thought out feedback.

Consider the following scenario;

In your project team, you have two senior developers who happen to have the same amount of relevant experience and very similar skillsets, they both face the same work circumstances, they are both performing the same type of tasks at about the same proficiency, and they both appear to have the same state of well-being. You have judiciously prepared each of their performance reviews using a traditional grading scheme of scoring them on a scale from 1 to 10, and as you suspected it is calculated that they both received the same scores of 6.5 out of 10. You write up very similar review forms for both of them.

This afternoon, you conduct their individual performance review meetings using the exact same presentation style, and you expected both reviews to turnout in very similar manner.

The first one of these interviews goes exactly to plan with the team member accepting the review with what appears to be a pleasant smile, a shake of hands, and they comment, "see you the same time next year".

The second interview doesn't go off so well with the other team member evidently not being too impressed with what you have to say nor with the whole interview process, as their body-language, tone of voice, and spoken words make it very clear that they consider this "whole thing to be a charade", "a complete waste of time" that "could have been better spent doing project work".

Confused by the marked difference in their reactions, you seriously wonder if one of your essential team members will be handing in their resignation tomorrow morning, and with that, you come to the realization that this would definitely "throw a spanner in the project's works".

Why was this performance review received so badly?

Let's be honest with ourselves ... if you received 6.5 out of 10 would you consider this to be a good, bad, indifferent, or an excellent grading?

Umm, the topic being graded would influence my opinion and my subsequent reaction. If I got 6 out of 10 for a test on, "the mathematics of a quantum singularity in multidimensional space" then I would be happy with this result.

However, if this was for the equivalent of a 5th grader maths exam then I would be very disappointed and maybe even a bit peeved off, ... and especially if I had spent what I considered to be an unreasonable amount of my time preparing for that test, ... and if I had expected to definitely do a lot better than that grading indicated, ... and if I knew that getting a score like that would mean that I was not going to get the bonus that I was counting on then I would be very "dis-a-point-ED" to outright frustrated.

So the answer to this question of why the same scored performance reviews may not have been received in the same manner can be due to the reviewee's self-expectation, their perception of the "Return On Investment" for their efforts, their work & home life balance, and whatever else is riding on the outcome of that review. ... In addition, their personality traits, and even their cultural / ethnic & community background can be influencing factors.

What? ... I now have to be a psychologist, as well as being a project manager.

Think about it, most of those previously listed reasons are probably obvious when **considered from the reviewee's perspective**:

- **Self-expectation …**

  Deep down we all have a realistic idea of what our effort is probably worth. If the grading for this effort comes back better than our inner-self honestly expected then we are generally happy (even if externally we may make a show of pleasure or displeasure at the results).

  > *For example; an exam that you didn't do so well on, but you feel that on that day and at that time you honestly could not have demonstrated your knowledge & understanding of that topic any better than you did. Therefore, if you just pass with a score of 51 out of 100 then you would gladly accept that result.*

- **Perception of the "Return On Investment" …**

  With all endeavours where time & effort is required to be spent (whether it be work, study, sports, or relationships), you will have to give up partaking in something else or at least reduce your interests in other things for the interim. When the results from that endeavour do not balance with what has been surrendered then it is only natural to question whether it was all worth it (sometimes regretfully so), and even to become angry with the party(s) who put you in that situation.

  > *For example; team members who have continually been expected to do additional work outside of their contracted norm to achieve some supposedly important deadline (which eventuates into nothing), or they receive no "thank you" and no sign of gratitude from the project manager (let alone from the performing organization); i.e. no Time-In-Lieu, no financial compensation, or no rewards.*

  Yeah, I once worked for this partnership that seriously expected everyone to put in a ridiculous number of billable hours, else you got this dirty look from the owners as though you were robbing them blind by only doing your contracted hours per week.

These owners just did not get it, from our perspective. Honestly, what was in it for us "the hired-help" when the owners were the only ones who were going to reap the benefits from the endeavour. It was not like they were offering us share options in their business, yet they expected us to invest so much of our unpaid time into their business without some form of compensation (nor show of gratitude).

- **Work versus home life balance ...**

Everyone has to balance his or her work life versus home life, and weigh up which is more important to him or her (at that particular point in his or her live). It may seem essential to the performing organization to complete that project by that specific deadline date, but is it personally that important to them when compared to what outside of work could be lost or be execrably damaged by doing so.

In addition, everyone does not live and work in isolation. What is happening in the ~~worker's~~ person's home life can and will eventually affect their work life and vice versa. If the situation happens to be bad at home then this can affect their emotional state at work. Similarly, a bad work life can exert pressure on their home life, which can eventually result in an unpleasant death-spiral of misery at home and at work, where something will finally have to give.

- **What else is riding on the outcome ...** *just keeping up with the cost of living*

If the person being reviewed is dependent on obtaining that performance bonus or at least receiving an inflation-matching pay rise, then difficulties could arrive if these are not forthcoming.

Pay raises and bonuses need to take into consideration the economic realities for both the employee and the performing organization. If year after year the employee's pay is not keeping up with the cost of living then eventually (irrespective of whether they like their job or not) they will be forced to find work elsewhere, hoping to at least be paid at the going market rate. This is especially true for employees who have been at the organization for some time. If their pay level is not keeping up with that of the new hires, then they could become very disgruntled with the organization for evidently taking advantage of their "loyalty".

- **I did all and more than what was expected & required of me ...**

How do you think they will feel and react if they believe that they have been consistently performing as required (or better) to activate the receipt of their bonus, but due to factors beyond their control, their efforts have been deemed null-n-void?

NOTE ☐ **Performance based bonuses must be realistically obtainable**; else the intended recipients will give up trying if they perceive that "*there is no way in hell that I am ever going to achieve that*".

- **Their personality traits ...**

Every one reacts to feedback and "criticism" in different ways. While some persons may see it as constructive information, others may take it as a personal affront.

What the hell is this, SIX AND A HALF OUT OF TEN!

You have to be $#@% kidding ... You $#@% JERK!

- **Cultural, ethnic, and community background ...**

Today's workplace is a multicultural environment with peoples from diverse cultural, ethnic, and community backgrounds. And, every one of these groups will have a slightly different perspective on what differentiates "A Winner" from "A Loser", and what are acceptable grading scores for performance.

Think about your own personal background; what do you, your family and friends consider as the equivalent grading scores for good, bad, average, and excellent. Some cultures, ethnicities, and community groups could consider 6-7 out of 10 to be a reasonably good result, whereas others would consider anything less than an 8 out of 10 (that is 4 out of 5 gold stars) to be shameful.

So, should I just give them appeasing appraisals?

No of course not, but a grading scheme such as the dual axis performance review, as described in this section, can have a **distinct advantage over competitive mechanistic rankings** as used by some organizations (aka bell-curve metrics).

- With the dual axis performance review scheme, it is **all relative to the individual and not a comparison to others**. Consider that, the longer someone has been in a specific role within the organization then they would be normalizing at a personal performance rating of 3 out of 5. Where 3 out of 5 translates to a desired result of, *"you have performed exactly to the standard that we expected from you, given your relevant experience and the prevailing circumstances you faced"*.

  Conversely, with a competitive mechanistic scheme, the longer that someone has been in a specific role within the organization then generally the higher will be there ranking. As a result, it is possible for senior / experienced team members to "rest on their laurels" and cruise along with the expectation of automatically receiving at least 7-8 out of 10 just because they have more relevant experience than their fellow junior / less experienced workmates.

- With the dual axis performance review scheme, **the focus is on highlighting personal extra-ordinary performance**, given that a good result would normalize at 3 out of 5. Therefore, if a member of your project team puts in extra-ordinary performance then they would receive a 4 or 5 out of 5 (even if their performance was not necessarily in the same league as their more experienced workmates). Whereas, with a competitive mechanistic scheme their extra-ordinary performance may only rank as a 5 out of 10.

- With the dual axis performance review scheme, **the inter-personal aspect can be used to highlight the effort that they have been putting in**. So while they may not have performed as well as others or not as technically sufficient, their efforts & attitude (for positively aiding in driving the performing organization forwards) can be officially acknowledged & potentially rewarded.

- Additionally, with the dual axis performance review scheme, it can be emphasized to a person (who always technically performs well) that **to improve their overall personal performance rating they will need to consider their inter-personal characteristics** and consider how they interact with others.

This dual axis performance review scheme kind of
sounds a bit scared to hurt someone's feelings.

Remember that, these performance reviews are for work activities and not to obtain some university accreditation. Unlike with university, the work relationship between the organization and "the worker" does not necessarily conclude at the end of the semester / term or at the end of the unit / course. Often with SDLC based performing organizations, the relationship between the organization, the management and the worker will continue from one project to the next (potentially for many years to come).

Therefore, **the performance review technique utilized by SDLC based performing organizations should ideally aim to; encourage & strengthen the inter-personal relationships, encourage & acknowledge the inter-personal as well as the technical performance, and to highlight to the ~~worker~~ person when they are not performing as well as was expected of them.** That is, **the emphasis should be on the ongoing mutually beneficial relationship between the performing organization and the individual person.**

Bi-directional Performance Reviews

So far in your career, have you ever been given the opportunity to review your manager's performance? Most probably not, as this **bottom-up review** is a bit of a radical concept. Though, **when the manager's performance is reviewed by the people that they manage,** the results can be very interesting. With the team members anonymously passing their individual reviews of their manager to the manager's manager (indirectly via the human resources department), then the manager's own strengths & weaknesses can be examined (which may not always be so evident to senior management).

> *For example; it may come to light that – the manager has excellent people harmonizing skills, the manager lacks understanding of the project's underlying technology, the manager has a tendency to micromanage, or the manager's negative personality traits which are not apparent to senior management (such as bullying). For more information on the characteristics of a manager, refer to [Chapter 11].*

The most interesting thing that can occur with bottom-up reviewing is that, the people being managed start to take performance reviews a lot more seriously and often they no longer consider performance reviews as a "complete waste of time".

On those rare occasions when I've been at an organization that has had the courage to do bottom-up reviews, I've found it interesting that the lower levels of the organization don't see it as an opportunity for payback. Rather, most of them provided valuable feedback on how their manager could improve their performance to the benefit of the organization and the people they managed.

Reviewer's Comment ... Suggested reading.

This section reminded me of a very good book:

\#   "Peopleware, Productive Projects and Teams" 2nd edition 1999
     by Tom DeMarco & Timothy Lister.

14.2. [Resources] Monitoring & Control

Introduction

This book makes a clear distinction between inanimate [Resources] and those living & breathing [People] resources. Please refer back to [Section 1.4] for the reasoning why.

NOTE   ❑ [People] resources also include those contractors & subcontractors who work on the project on a continual or semi-continual basis as though they were regular members of the project team. Whereas, ...

❑ [Resources] includes componentry/goods, *e.g. servers, sensors, integrated circuitry, hard disks, memory, motherboards, chassis cases, labels, display screens, antennas, connectors, etc.* [Resources] also includes "black box" services; *e.g. external manufacture, warehousing, internet access, website hosting, database hosting, and temporary contractors*. That is, [Resources] includes both goods and services.

Based on this book's differentiation between [People] and [Resources], it should not be a surprise to find out that this section on practical [Resources] Monitoring & Control will concentrate specifically on those inanimate [Resources] and not the [People] aspect. Thus, this section will focus on [Resource] procurement, processing, and dispatch management.

However, as this book's author I had better come clean and let you in on a deep-dark secret. ... for much of my SDLC career, my interactions with inanimate [Resources] have been limited to sending off a wish list of those resources that the project required (including the dates by when these were needed) then sometime later the "resource-fairy" would magically materialize these resources for the project to utilize.

Then   *Welcome newbie, to resource procurement, processing,*
    SHAZAM   *and dispatch.*

Overview

In a practical sense, [Resources] Monitoring & Control [Section 9.1] asks the following questions:

- What things do I need?

- When do I need these things?

- From whom do I get these things?

- Where do I require these things to be, and
  how will these things get there?

- When will these things arrive?

- What will I do with these things once I have them in my possession?

- When do I need to get the deliverable things out the door?

- How should these deliverable things be packaged before these leave?

- How are these deliverable things going to get to where these things have to go?

- Exactly to whom and to where do these deliverable things need to go?

14.2.1. Resource: Procurement & Processing

AUTHOR'S NOTE

There is a major caveat when it comes to resource procurement; your performing organization probably already has some defined processes & procedures related to procurements. Hence, what is contained in this section could be constrained by these pre-existing formalized guidelines. However, what follows should provide you with some food-for-thought when you are undertaking resource procurement, and especially when you are planning & budgeting for this aspect of your project.

14.2.1.1. Procurement: Sources

Resources can be acquired from three generalized directions:

1) internally to the performing organization,

2) internally to the customer organization, and

3) externally to both the performing and customer organizations.

Procuring Resources Internally

1) **Borrowing** – is the ad hoc acquiring of the required resource (for a relatively short duration) from another department or project team while that party doesn't need to use that particular resource. The success of this type of procurement will be dependent on the project manager's ability to negotiate and influence others to temporarily make such resources available. Please refer to [Section 11.3] for more information about the relevant characteristics of the project manager.

2) **Negotiating** – is the acquiring of the required resource (for a relatively extended duration) from another department or project team by coming to an agreement based on some form of exchange or the promise of a future exchange which will benefit both parties. Again, please refer to [Section 11.3] for more information on getting others to do what you need them to do.

3) **Chain-Of-Command request** – is the acquiring of the required resource from another department or project team by going through formalized management channels that hopefully will instruct the relevant department or other project teams to relinquish the necessary resources for the specific period of time.

NOTE ☐ The **procurement of resources** internally to the performing organization **will require the project manager to utilize their interpersonal skills of "networking"**.

> As an aside, it is amazing what can be achieved for your project's success by just taking the time to build up friendly working relationships with other people within the organization. This is especially true for working relationships with your peer project managers and with people in different positions. Though, this is not limited to management type people, as there can be significant advantages to keeping on the good side of the office coordinator, the store person, network admins, and project teams.

 IMHO, you will improve the prospects for your project's success by treating others with respect, dignity, and generally being a "nice guy".

Procuring Resources From The Customer

Procuring resources from the customer organization is effectively having the required resource "**lent**" to the performing organization specifically for the purpose of conducting the project. Obtaining the necessary resources this way will require the establishment of a good working relationship between the project manager and the customer's corresponding manager and with management higher up in both organizations' hierarchy.

This type of resource procurement requires a reasonable amount of rapport between the management at both the performing and customer organizations.

IMHO, when building this type of interpersonal working relationship with the customer, it is important to remember that, it is not advisable to try to create too "familiar" a relationship, because you are still representing the performing organization and hence what you do, say, and how you act (even in jest) will reflect back on your employer. ... And, what if the contractual relationship sours, you will need to represent the best interests of your employer.

NOTE ❑ When resources are procured from the customer organization then these items should be clearly marked as not being owned by the performing organization (therefore not your asset and not available for communal usage), and these items should be entered into some form of **Third Party Property Register (TPP)** which would detail;

- Which customer organization owns this particular item?
- Who in the performing organization is responsible for this item?
- When was this item received from the customer organization and in what operational condition was it in when received?
- When, where, and to who in the customer organization is this item to be returned?
- Who in the performing organization is currently in possession of this item?

Procuring Resources Externally

Procuring resources from third party organizations, this type of procurement involves the following phases:

1. Planning the procurement,

2. Conducting the procurement,

3. Administration of the procurement, and

4. Closing out the procurement.

Hmm, interesting how this list reads like the phases of a project.

14.2.1.2. Procurement: Planning

Prior to deciding on which vendor/supplier to procure the resources from, the following decisions need to be made:

1) **What has to be procured?**

2) **When are these goods/services required?**

3) Is it **better to buy, build, short-term hire, or long-term lease** these resources?

4) Should these goods/services be **sourced internally or externally** to the performing organization (and customer organization)?

5) How will these goods/services be **defined and scoped**?

   That is, what are the specifications of the goods required or the Statement Of Work for the services to be performed?

> I would recommend that ... before you start communicating with vendors/suppliers, you should understand what you want, know what you are talking about, and have a reasonable understanding of the subject matter of concern. Because this insight will put you in a better negotiating position than coming across as a complete newbie who could so easily be taken advantage of.

6) How will the competing **vendors/suppliers** be **differentiated** so that the most appropriate one can be logically chosen? That is, what will be the **Selection Criteria for comparing the potential vendors/suppliers** with each other and with the rest of the market place? Use techniques such as; the vendor/supplier's **qualifications**, **screening** via compliance with the **mandatory requirements**, **weighted scoring** of all of the vendors/suppliers performance attributes, and **comparison of** vendor/supplier's offering **against independent estimates & expert opinions**.

Having a noted down Selection Criteria is essential to judging the various vendors/suppliers based on their relevant merits and thereby not be swayed by the razzle-dazzle of a salesperson who "talks a very good game" even though their offering is only second rate.

7) How will the **vendors/suppliers** be **identified**, and

how will they be approached to gauge their level of interest?

Will this contact be via an invitation for the vendors/suppliers to apply to provide the goods/services; i.e. Request For Information (RFI), Request For Proposal (RFP), Request For Tender (RFT), Request For Quotation (RFQ), or Rough Order of Magnitude (ROM). Alternatively, will a representative of the performing organization solicit a quotation from the vendors/suppliers over the phone or via email?

- **RFI – Request For Information** is a request for the vendors/suppliers to provide information about themselves and the goods/services that they can provide; so that they can be evaluated to determine whether they would be capable of delivering the required goods/services. This is analogous to placing a job advertisement in the paper online and then reviewing the resumes and cover letters that are received in response from the potential candidates.

- **RFP – Request For Proposal** is a private sector request for the vendors/suppliers to submit a freeform (or semi freeform) proposal outlining how they would provide the required goods/services and at what approximate cost & duration.

- **RFT – Request For Tender** is a public sector version of the RFP.

  The RFT has a rigid formal structure that the vendor/supplier's response must conform to in order to be taken into consideration for the tender. This conformity to structure is imposed to ensure that the competing vendors/suppliers can be appraised in a fair & open manner; thereby, reducing the potential for the misappropriation of public funds (aka to limit corruption).

- **RFQ – Request For Quotation** is a (semi-formal or formal) written request (including via letter, email, or fax) to the vendors/suppliers asking them to submit a quotation that will specifically address the requested specifications & quantity of the required goods/services including accurate cost & duration statements.

The RFQ does contain some amount of details specifying the requirements, and hence it would be expected that the vendor/supplier would put some reasonable amount of effort into determining the quotation (taking into consideration such things as; the amount of materials that would be used, the person hours & labour rates, and packaging & freight). This quotation would be returned in a formally written form with a time period stated on how long this quotation is valid. This quotation could be based on variations of the following;

- **Fixed Price** – the amount to be paid is not going to change from that stated in the quotation irrespective of whether the actual costs incurred are lesser or greater. A modification being **Fixed Price with Economic Price Adjustments**.

- **Cost Reimbursement** – the amount paid is the costs incurred plus some additional percentage or fixed margin. Could also add Performance Incentives.

- **Time & Materials** – the amount to be paid is based on the charge-out rate for each person involved plus the sell-price of the materials used.

- **ROM – Rough Order of Magnitude** is a verbal or written request of the vendors/suppliers to provide an "in the ball-park" guesstimation (**within at least a 20 – 25% range**) of the [Costs], [Time], [People], and [Resources] involved. Unlike the RFQ, **the ROM is not a binding obligation by the vendor/supplier**; rather, provided purely for evaluation purposes of deciding whether it is worth further investigation into the procurement via this vendor/supplier, or determining whether it is even worthwhile considering procuring that type of goods/services.

 I would recommend that ... when providing a ROM, it doesn't have to be accurate to the exact monetary value or work hours, so round the estimated values up to whole values of 10s – 100s. However, when providing a quotation then randomize the final digit or digits of the estimate to give the appearance that some real thought has been put into determining the quote; e.g. 107 hours instead of 100 hours.

Procurement Guidelines

Have you read through your organization's procurement guidelines?

Though, sometimes there are a few things that these guidelines forget to mention:

1) **There may be no option on who the chosen vendor/supplier is due to pre-existing formal arrangements** between the vendors/suppliers and the performing organization. *For example;*

   - *Support & maintenance contracts or Service Level Agreements (SLA).*

   - *Preferred vendor/supplier status which has associated financial advantages on price and discounting on bulk procurements across the entire performing organization.*

   Thus, the procurement strategies and partnerships related to the project may have been established well in advance by those higher up in the performing organization's management hierarchy.

   Sometimes your project is stuck with which vendors/suppliers can be used. ... To use an alternative may require some damn good excuse.

2) **There may be only one appropriate choice for the vendor/supplier** due to;

   - **Pre-existing proprietary knowledge** of the performing organization's applications / systems (or that of the customer organization's applications / systems) that are involved with the project. Hence, this vendor/supplier possesses **exclusive knowledge or intellectual property that cannot be obtained elsewhere.**

   Therefore, you may have no other realistic choice but to pay whatever fee the vendor/supplier is demanding.

   Do the words, "being held to ransom" come to mind, because under such circumstances it can feel as though that vendor/supplier does have the project over the proverbial barrel with respect to Time, Cost, and "when I am darn ready".

- There is a **contractual requirement** imposed by the customer organization to use a specific vendor/supplier for the project.

- There may be **no other realistic choice due to the limited number of competitors in that market** (in your geographic proximity), or the market accessible to your organization may have become **monopolized by one predominate vendor/supplier**.

3) **There is a pre-approved list of preferred & trusted vendors/suppliers** which has been established based on;

- The goodwill of previous successful relationships with a particular vendor/supplier.

- The favourable financial Terms & Conditions.

    We had this problem the other day; there was a regional vendor who was the only recognizable supplier of a specific component that was required by the project. I obtained a quotation from that particular vendor, I put in my Purchase Request to be approved by the accounts department, then a few days later when the resultant Purchase Order did not arrive for me to sign, I went in search of why. It turned out that for some unknown reason that particular vendor had been "black-listed" by the accounts department and hence the purchasing system would not accept my Purchase Request. Seems that there was an issue between our accounts department and the vendor's over the payment Terms & Conditions.

4) Even though there are alternatives available, **a particular vendor/supplier is so well recognized by their reputation** for the quality of their goods/services that only a "brave fool" would risk choosing someone else.

    Do you want to take the risk of some very probing questions from your senior management if your choice of a non-recognized vendor goes all wrong?

14.2.1.3. Procurement: Conduct

NOTE ❑ **The procurement of the [Resources] required by the project have to occur well in advance of when these [Resources] will be deployed,** as any such delay in [Time] could mean the late delivery & subsequent failure of the project.

❑ Under certain circumstances such as with [Time] constraints, **it may be necessary to initiate the procurement of these [Resources] as early as possible,** potentially while still in the project's Planning Phase.
This is **essential for "long lead time" items that have a direct effect on the project's critical path and the associated project milestone dates.**

 I would recommend that ... during the Planning Phase you determine as accurately as possible the project's resourcing needs, the real costs involved (including supporting infrastructure costs) and lead times for procuring these resources, because to do this later on during the Executing Phase could result in some "nasty surprises" with overruns for [Time] and/or the [Cost] budget.

Let us assume that for this project, it has been decided to procure the required goods/services externally to the performing organization (and externally to the customer organization). The next step is to; communicate with each of the potential vendors/suppliers (preferably with at least three), evaluate their responses/quotations, select the winning vendor/supplier, and to contract the winning vendor/supplier so as to get their own processes rolling for the timely delivery of the goods/services.

QUESTION: Does the performing organization that you work for have formalized procurement procedures that the project manager (possibly as the signatory on Purchase Orders) must be aware of, and are these procedures to be rigidly adhered to when acquiring goods & services?

For the flowchart of the procurement process being described in the remainder of this section, please refer to [Figure 134] a few pages on from here.

External Procurement Pre-Considerations

But wait, before diving off into the procurement activity there are a few other things that will have to be considered before deciding on the vendor/supplier.

1) What is the **common "market availability"** of these goods/services being procured? Because this availability will affect the price, the delivery lead time, and the quantities that can be ordered. This needs to be known now, as there may not be quantities available later on for any additional orders. ... *Will this be an End Of Life Buy?*

2) What is the **going "market rate"** for these goods/services being procured? Because there could be significant variations in price ranges between vendors/suppliers, and who has the cheapest price today may not necessarily be the cheapest tomorrow. Though, also consider how much extra the vendors/suppliers are charging for shipping & handling, what is and is not included in the purchase price; *e.g. warranty periods, level of support, number & type of licenses*. When these "optional extras" are added to the equation, then do these price differences start to make more comparative sense?

3) Are there **alternative parts** suitable to replace those ones that are to be acquired; i.e. are there other parts that have the same **"Form, Fit, and Function" (FFF)**? Even though these items may have the same generalized Form, Fit, and Function these could in fact have very different operational ranges; e.g. humidity, pressure, temperature, Meantime Between Failures (MTBF), and serviceability period.

4) Do the items that are required have **alternate identification numbers** that are used by various suppliers; i.e. **exact same item only with different Product Codes**? This is possibly due to this particular item being sourced from different manufacturers, targeted at different end-user markets, or intended for different geographic regions.

5) Is the item to be procured a **proprietary / custom-build** that can only be sourced from one solitary supplier? Hence, the project would be constrained by that supplier's lead times and prices. Subsequently, would it be **wiser to** redesign the project deliverable's solution / architecture to be able to **utilize a readily available / generic item**?

6) Can the items that are required be **described/identified in alternate ways**?

   Does the market/supplier use different terminology?

   This is important when the vendor/supplier doesn't speak the same first language.

7) What is the required items **availability in the project's implementation location**?

   As some items may not have been made available to the project's specific country or geographic region. Alternatively, these items could have significantly different pricing between geographic regions possibly due to import & export tariffs, and subsidies.

8) Are **import and/or export licenses** required to be obtained from relevant national & international authorities for the use of these items, and/or for these items inclusion in the project's deliverables; *e.g. prohibited materials, national / international trade embargoes, and national security.*

9) Does the customer organization have **contractual requirements that local distributors / manufacturers are to be used in preference to foreign-based ones**; i.e. requirements for a certain **percentage of local content** to be included in the product delivered? Hence, does **"Made In Country X"** carry significant weight when selecting vendors / suppliers? This could be a major consideration on how the project deliverables are treated by the customer's national authorities.

10) What are the **Terms & Conditions** and **licensing restrictions** imposed by the vendor / supplier? Are there stipulations on use; is there a remote access (backdoor) capability to disable the provided application if the terms of the **end-user license** are breached?

11) Would dealing with a local distributor provide the performing / customer organization with "**customer support advantages**" when compared to an overseas provider?

   For example; if an overseas vendor/supplier is used but technical assistance is required, then would there be several hours delay while waiting for a response to those support calls? This situation is especially bad when the vendor/supplier is on the other side of the world and their business hours are completely out of sync. There could potentially be 24 to 48 hours delay on the turnaround for support calls.

12) How does the **warranty** of the overseas supplier compare to that of the local distributor? That is, if something goes wrong with the items that were acquired, is there some form of recompense or is it a case of "buyer beware" you're on your own.

 **Be very careful if the price or availability** of the items to be procured is **too good to be true**, because these items could in factuality be **"grey market" or "counterfeit" parts**.

Grey Market Parts – may look & function the same as the "real thing", but there could be significant differences in operating specifications; *e.g. these could be scrapped batches that did not consistently perform within a specified operating range*. Hence, while the resultant products that use these grey market parts may very well function correctly for some of the time these products' overall performance will be poor when compared to the design specifications and when compared to the First-Of-Type unit that was (built with the proper components and) certified for end-user usage.

Counterfeit Parts – is something masquerading as the parts that were specified. Unlike the grey market parts these counterfeit items do not come anywhere near the operating specifications. Hence, every unit produced with these counterfeit parts will fail or function significantly worse than the design specifications. At best, resulting in many production-line failures, at worst litigation by end-user groups affected by the product failures.

NOTE ☐ The reason for such caution about grey market and counterfeit parts is that by the time it is discovered that the project has been stung by these items there could already be a significant amount of damage done (both financially and reputational) to the performing organization and the customer organization.

For example; product recalls, copious amounts of warranty repairs, numerous product returns & refunds, storage & eventual disposal of unusable & unsellable stock, market/media/public ridicule, customer complaints to statutory authorities, legal cases, and class-action suits.

External Procurement Process Flow

As illustrated in [Figure 134] on the next page, the procurement process is as follows:

1. It has been decided that it is necessary to acquire certain goods/services externally to both the performing and customer organizations.

   - **Determine the specifications** of the goods/services to be acquired, ensure that these are clearly defined so that there will be minimal opportunity for the wrong goods/services to be delivered.

   - **Establish the Selection Criteria** based on both the specifications of the goods/services and the performing organization's procurement guidelines.

2. **Contact the appropriate vendors/suppliers** to gauge their level of interest in doing business, and to request that each of the selected vendors/suppliers provide a quotation (plus their Terms & Conditions).

3. Once the quotations have been received from the competing vendors/suppliers then **evaluate these quotations against the Selection Criteria and the specifications**.

   - If there are any questions, misunderstandings, or misinterpretations then contact the relevant vendors/suppliers to have them adjust their quotations accordingly.

4. Once **the most appropriate vendor/supplier** has been **selected** (given the current situation and the prevailing circumstances) then **generate a Purchase Request** to seek financial approval to proceed with the procurement of these resources.

5. If the **Purchase Request has been approved** by senior management (or by the finance department) then a **Purchase Order will be generated**, signed off by the appropriate representative of the performing organization, and then sent to the selected vendor/supplier as an agreement to proceed with the provisioning of the specified goods/service. If the Purchase Order was not approved then the process may need to go back to the beginning and re-evaluate the goods/services to be acquired. ... *Need to consider the effect on the organization's cash flow.*

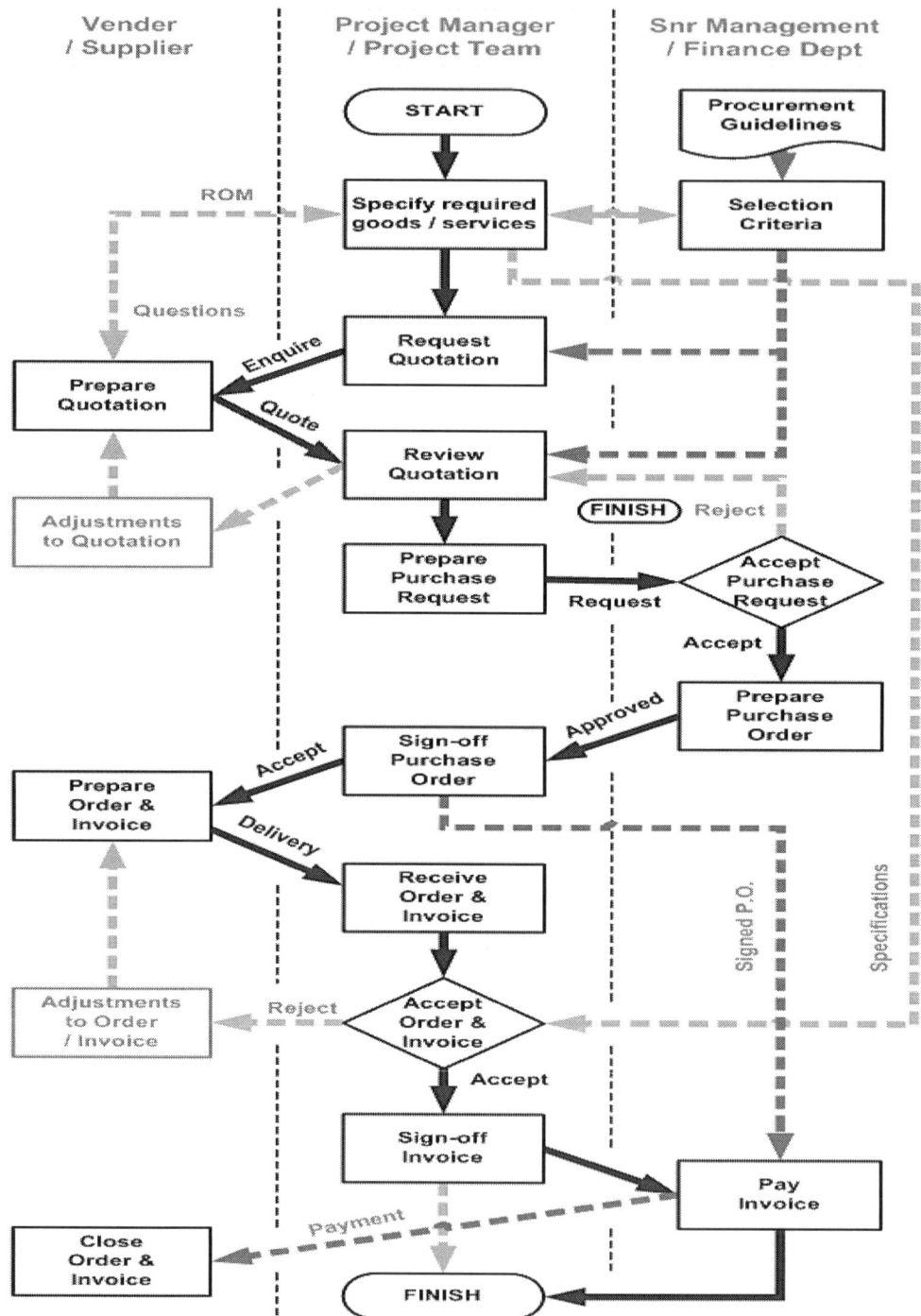

Figure 134: Process flow for the external procurement of resources.

6. On receiving the Purchase Order, the selected vendor/supplier would initiate their own procedures to acquire the goods/services, and then either forward these goods to the performing organization or arrange for the services to be conducted.

 I would recommend that ... the project manager should confirm with the selected vendor/supplier that they have received the Purchase Order, and also enquire on approximately how long it will take to **fulfil the order** (even if this information was previously provided on the quotation).

7. At this point the payment for these goods/services will either be deferred to a later date (*e.g. payment via received invoice*) or immediately (*e.g. from petty cash or on the company credit card*).

- **Deferred Payments** would be used for goods/services that will directly **feed into the project and thus appear against the project's Cost Account Code**.
  The cost of these goods/services would be incorporated as an itemized element of the price that is charged to the customer. Hence, **a solid "paper trail" will be essential for the project's financial accounting and auditing purposes.**

- **Immediate Payment** (such as via petty cash) would be **used for low value consumable items, and items obtained face-to-face.** Though immediate payment (such as via company credit cards) is increasingly being used for online transactions. The cost of these goods/services may or may not be incorporated as an itemized element of the price that is charged to the customer. These costs may alternatively be included as part of the performing organization's Operating Costs.

 Before you start making these types of immediate payments, I would highly recommend that ... as the project manager, you firstly find out what you are allowed and not allowed to procure this way, what are your financial limits, and **what are your authorization limitations.**

14.2.1.4. Procurement: Logistics

Continuing with the previous sequence of procurement steps, as illustrated in [Figure 134].

8. As the project manager, it is in yours (and the project's) best interest to **maintain good working relationships & communications with each of the involved vendors / suppliers** who are fulfilling the resource procurements, as you do not want any nasty surprises of finding out too late that some of the goods/services won't be available when these [Resources] are needed or these will arrive a lot sooner then was planned.

    By establishing good working relationships with vendors/suppliers then hopefully, enough forewarning of resourcing issues and storage concerns will be received to enable alternate arrangements to be put in motion, so as to be ready when the goods/services do arrive.

9. **Keep the organization's stores person / warehouse manager informed** of when these [Resource] items are expected to arrive and under what conditions will these items have to be stored. By giving them plenty of forewarning then hopefully this will give them the opportunity to make sufficient space available for the project's [Resources].

    How often have you witnessed project resources shoved under people's desks or in the corner of an office, while a perfectly good storage room goes to waste with materials that have not seen the light of day in years? Alternatively, the warehouse with stacks of new items accumulated around the entrance's roller-door while archived materials (aka "junk") occupies secure positions inside.

10. **Keep an eye on the intended location** where these soon to be arriving **[Resource] items are to be stored**, as you don't want to end up having a heap of "pristine" boxes delivered only to find out that there is absolutely nowhere appropriate to put these or there is no one or equipment available to receive these items.

This "lack of storage preparation" situation can lead to other problems such as:

- **Damage & devaluation** of the [Resource] items due to inappropriate storage conditions; *e.g. water or moisture damage, excessive light damage, weathering, dirt & dust, overspray from nearby activities, misplaced items, etc.*

  Consider the scenario where the plan (and associated [Cost] budget constraint) is to re-use those delivery boxes (that the large [Resource] items arrived from the supplier), as the customer delivery packaging. However due to inappropriate storage, these delivery boxes have been damaged and are no longer acceptable for presenting to the paying customer. Thus, the plan to reuse these boxes is no longer viable. Subsequently, replacement delivery boxes will have to be procured specifically for packaging the deliverables, which will introduce an additional [Cost] burden and procurement [Time] delays to the project.

- With no appropriate **secure location** available to store these [Resource] items then there is potential for opportunistic passers-by to take advantage of the situation and do some unauthorized "five-fingered" procurement of their own.

 **DO NOT underestimate the costs and logistics involved with warehousing & inventory control;** because, when the Real Cost of these are billed against the project's Cost Account Code then there could be a shocking impact on the project's profit margin.

14.2.1.5. Procurement: Closure

As the ordered goods/services start to arrive from the various vendors/suppliers,

do not just accept these [Resource] items without having:

11. **Securely stored these [Resource] items** in a known holding location that is dedicated

    to the receipt of items for this particular project.

 **DO NOT "squirrel away" [Resource] items in "secret stashes"** because if the person who controls that stash is away then these stashed resources could be unavailable when required.

---

12. **Check that each [Resource] item** received exactly **matches those specified** on the

    Purchase Order that was sent to the vendor/supplier.

    - Check that the quantity of each item received exactly matches the **numbers listed**

      on the Purchase Order and also matches the **numbers listed on the invoice**.

    - Check that **what was ordered** and that which **was received** are in fact what **was**

      wanted, because what was specified may not necessarily be what was thought to

      be needed; i.e. differences in understanding of the terminology used and hence

      these items may have to be returned or exchanged for the correct ones.

    - Depending on the quantity of items received check each item or **randomly sample**

      a few of the items (at least three differently positioned items from each shipping

      box, i.e. top – middle – bottom / front – middle – end) so as to ensure that **these**

      items do conform to the specifications that were listed on the Purchase Order.

      For example; having ordered 20 cans of mango-yellow paint, as the last

      batch of project deliverables roll off the production line, it is discovered

      that the last few cans contained baby-poop yellow (even though all of

      these cans were labelled as mango-yellow), and hence these units have to

      be stripped back and repainted (once the correct paint is acquired).

NOTE ☐ **Contact the vendor/supplier immediately** if there are any discrepancies, as you will want **to resolve these issues** a.s.a.p., before these result in difficulties for the project; i.e. arrange for exchanges, returns, refunds, and even consider alternate sourcing.

13. **Check that the invoice matches what [Resource] items were delivered**. Note down any discrepancies (for or against), and before these become real problems, contact the vendor/supplier to rectify these discrepancies.

14. If the performing organization's procurement processes allowed for the ordering of goods & services without a Purchase Order being generated then **check the invoice against the vendor/supplier quotation**. If there are any inconsistencies then immediately contact the vendor/supplier, as differences between the invoiced value and the quoted price could result in the project being over budget; hence, the project's manager could have some explaining to do.

15. **Sign-off on the invoice only if all** of the [Resource] items have been delivered in the **correct quantity, quality** and with the **correct monetary values**.

DO NOT sign-off on the acceptance of the invoiced deliverables until it has been confirmed that these agree with the approved Purchase Order.

DO NOT approve the payment of the invoice until the contractual agreement with the vendor/supplier has been satisfactorily completed.

If it was your hard-earned money on the line, would you honestly be happy to receive these items and hand over your cash? I did not think so. ... So, why should you when acting on behalf of your employer. You have a "Stewardship of Responsibility" when you are spending someone else's money.

Be very aware of the procurement of goods/services that have payment milestones **around the End Of Financial Year** (EOFY). Because, if an invoice arrives or is paid at the start of the next financial year then it could result in the complex situation of having the Cost Account Code that corresponds to the Purchase Order (and its associated invoice) having been closed out.

With the accounting books for the last financial year having been closed then how will this outstanding invoice be paid?

In addition, will there be funds in the new financial year's budget to cover this outstanding invoice?

16. **Clearly label the received [Resource] items** with:

- The name of the person in the performing organization **who is responsible for these items**; *e.g. the project manager*.

- The **project** that these **items belong to**.

- **Identify the components & quantities** that are contained inside each package; e.g. 21x motherboards, 42x memory modules, 20x CPUs.

17. Update the project **documentation** that records these items as having been received.

BOTCHED RESOURCE PROCUREMENTS CAN GRAVELY AFFECT THE PROJECT'S SCOPE, TIME, COST, AND QUALITY. ... Often the project manager has to clean up this mess.

 For external resource procurement, I would recommend that ... it is best to **deal with any potential problems with these items ASAP**, because if there happens to be an issue then hopefully this can be resolved before these items cause any major [Time] delays or impose unexpected [Costs] on the project.

14.2.2. Resource: Production & Packaging

DO NOT underestimate the [Cost] & [Time] involved with production, packaging, and dispatch ... as these can very quickly erode away the project's profit margin.

Now that all of the material [Resources] have arrived, the project team can get down to work and use these items to produce the project's deliverables.

Unfortunately, it is not always that simple; while, the procurement of these [Resources] for the Executing Phase may well be completed, what about:

- When the product is ready to go into multi-unit production?

- When the product is ready to be delivered to the customer?

There is still more [Resource] procurements that have to occur.

Example Case: a box inside a box, inside another box or few.

Back in [Section 4.3] the scenario of 'No-Television-Land' was used to demonstrate the use of the Work Breakdown Structure (WBS). Well, several hundred pages later the project team has finally completed the development of this ground breaking product (which they affectionately refer to as "the box" or "TV" for short).

The first batches of these hand-built "TV boxes" are now ready to be shipped to their eagerly waiting customers. But, these rare 'TV boxes' cannot just be put onto the back of a truck or dropped into a cargo-hold without some form of protection. Hence, each one of these 'TV boxes' will have to be put inside its own protective '**product box**'.

As the packers place the 'TV box' inside its 'product box', it is observed that the 'TV box' is free to move around and often hits up against the sides of its protective 'product box'. Fortunately, a smart engineer has had the brilliant idea to use contour moulded foam to bookend each 'TV box' so that it fits snuggly inside the 'product box'.

Oh great, the marketing people have just seen the nice blank sides of the 'product box' and they have concluded that this is a perfect '**Point Of Sale**' (**POS**) location to put the company logo along with the product branding. Well, this self-promotion ends the opportunity to use '**Off The Shelf**' (**OTS**) mass-produced boxes, instead custom 'product boxes' will have to be procured. This will in turn mean that these 'product boxes' will have their own specifications, supplier, associated lead-times, handling, and storage needs in readiness to be used. As a result, these 'product boxes' will require their own little procurement process to ensure that these are available '**Just In Time**' (**JIT**) for the finished 'TV boxes' to go into. Similarly, a reasonable quantity of the contour bookend moulded foam will also have to be ready to go, which is more inventory to manage.

Oops, almost forgot that each 'TV box' is going to require its own dust cover prior to being put inside its 'product box'. Then into each 'product box' has to go the user manual, the remote control, batteries for the remote, the antenna cable, plus a copy of any certificates / licensing, the warranty information, and any promotional materials.

Though, you don't want too much invested with packaging held in storage waiting to be used, as these non-utilized packing items cost money to procure, which is money that could have been used elsewhere. Then there is also the cost of the storage facilities, the expense of logistics, and the cost of insuring these items while they are being stored.

Okay, now we're ready to ship these 'product boxes' to market. But wait, the freight company just informed us that, to ship these individual 'product boxes' is going to cost significantly more in handling fees than if several of these 'product boxes' were shipped together inside a larger '**shipping box**'. Oh great, you can bet those pesky sales & marketing people won't miss an opportunity to put the company logo and product branding on the side of each one of these 'shipping boxes'. Well that will be another batch of custom boxes that will have to be procured and inventoried.

As demonstrated in this example case, just because the development of the deliverable units has neared completion doesn't mean that there is not a significant amount of effort still remaining; related to **production, packaging, and shipping the project's deliverables to the customer**. This **is akin to a mini-project** related specifically to **getting the product out the front-door** of the performing organization and into the customer / end-user hands.

If the project happens to produce a product destine for the retail market (*e.g. hardware or software media in a box*) then there is several more boxes and materials to go inside each of the boxes described previously in the example case. Each of these packaging components will need to be procured and inventoried prior to scheduled use.

1) The product; *e.g. a mobile phone.*

2) The '**Point Of Sale**' (**POS**) product box; *e.g. the glossy box with the product image on the front and potentially different branding depending on which telecoms service provider the device is being sold through.*

3) The packaging to secure the product inside the POS product box.

4) Any associated accessories to accompany the product; *e.g. battery, power supply adaptor, USB to PC connector cable, and headphones.*

5) Any associated paper materials; *e.g. user manual, quick reference guide, regional warranty information, and POS promotional information.*

6) Then there is the retailer box, which contains multiple product boxes for efficient distribution to the retailers.

7) The wholesaler's box, which contains multiple retailer boxes for efficient distribution to the wholesalers.

8) The shipping box, which contains multiple wholesaler's boxes for the efficient transport by the freight companies to different geographic regions.

9) The consignment / palette, which contains multiple shipping boxes for efficient national and international distribution.

10) Any consumables, which happen to be used during the packaging and distribution processes; such as glue, tape, anti-static shielded bags, dust covers, anti-moisture bags, etc. ... *Oh, and then there are itemized stickers and labels for each of these boxes and those associated container boxes.*

As demonstrated by the previous list, many additional [Resources] will have to be procured prior to the product being delivered to its intended customer / end-user.

Just as there would be a '**Bill Of Materials**' (**BOM**) **detailing all of the components that are required (and need to be procured)** prior to producing the deliverable product, there would also be a similar BOM related specifically to **packaging & shipping**.

For small organizations and/or for small to medium size projects the list of **packaging resources** required is **often "forgotten" to be included as part of the product's 'Manufacturing Data Package'** (**MDP**). Similarly, production consumables are often left off the BOM. *And, guess who often has to do the last minute run-around to coordinate securing these items; i.e. the project manager. Oh, and WERE THESE PACKAGING & SHIPPING ITEMS BUDGETED FOR IN THE PLAN? ... Umm, NOPE!!!*

JUST HOW MANY 'RUSSIAN DOLL' BOXES ARE REQUIRED TO GET THE PRODUCT INTO THE CUSTOMER'S HANDS?

14.2.3. Resource: Packaging & Dispatch

Finally, the project's deliverables / product are complete, and it is all nicely packaged up ready to be delivered to the customer / end-user.

 **DO NOT underestimate the impact that the product's packaging & dispatch can have on** the success of the project; whether that be on the **timely delivery to the customer or on the financial bottom line**.

 I would recommend that ... a contingency should be put into any quote involving packaging & freight, because several months later at the time of shipping, there can be a significant difference between the quoted price and the actual cost to be paid.

Lessons learnt from my real-world experiences with packaging & dispatch.

There are several aspects to packaging & dispatch that need to be considered in order to reduce the [Cost] & [Time] burden on the project: *... Better, design these in.*

1) **Handling & Containment** – does the package's contents require any specialized handling and containment? *For example; is it fragile, is it temperature – pressure – humidity sensitive, does it contain toxic or hazardous materials, does it contain sensitive or classified information?* **Any specialized requirements will add to the [Cost]** of the dispatch, **as well as effecting the [Time] required for delivery**, and the **means available to transport** the deliverable to the customer.

   Was this [Cost] & [Time] included in the project's baselines?

2) **Stacking** – can the package deliverable have other packages placed on top of it or does it need to reside by itself? **If it cannot be stacked then** this will mean that its presence will limit the quantity of other packages that the freight carrier can transport with this package and hence **the greater [Cost]** that will be incurred.

3) **Volumetric Weight** – is based on an assumed weight for a specific volume, or the actual weight of the package; whichever is the greater will be used.

For example; would you prefer to carry, 100kgs of bricks or 100kgs of feathers? While most people would quickly choose the feathers; however, from the freight carrier's perspective then the bricks would definitely be preferred because of its smaller volumetric size, which means that it is much easier to handle then all of those feathers. Secondly, its smaller volumetric size means that it gives more room for the courier to transport other customers' packages. Therefore, it is advisable to **ensure that appropriately sized packaging is used**, else the project could incur a considerable amount of [Costs] for transporting empty space.

4) **Physical Extremities** – when the package passes through the freight carrier's handling systems, the package will most probably be weighed and its physical extremities will be determined. If a part of the package is protruding then the package's physical extremities will be scanned as being greater than the package actually is. Hence, this increase in the **package's extremities** could be enough to push the delivery cost up into a **higher volumetric-weight pricing bracket.**

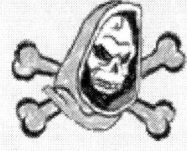 **When obtaining a quotation for freight, try to be as accurate as possible with the physical dimensions & weight of the packages** to be transported; because, a few units of measurement out and the package could be put into a different volumetric weight class and subsequently be invoiced at a significant different amount to that which the project's [Cost Baseline] budgeted for.

---

5) **Recycling of packaging materials** – given today's economic conditions and environmental concerns, it is often beneficial to recycle packaging. However, if a box is going to be reused then ensure that any old stickers with address details and barcodes are completely removed or are completely blacked out; because, these

old barcodes and addresses could cause confusion for the freight company and potentially cause [Time] delays to the delivery. Similarly, **confirm that the recipient's delivery details are correct** as these maybe different to the customer's details.

Yep, a package that gets misplaced and arrives a few weeks / months late can have a significant impact on not meeting the project's milestones. ... And, subsequently doesn't look good for you know who.

6) **For When** – different times of the year will incur different cost rates and possibly have different delivery durations. *For example; during high demand periods such as Christmas, during the wet season, during winter months. ... End of Financial Year boundaries.*

7) **By When** – the less time that is available to deliver the package (i.e. the more urgent) then the more costly will be its delivery. *For example; by overnight express or regular delivery, by land or by air.*

8) **Where to and where from** – the pickup location and the delivery address will affect what options are available for handling & transporting the package, plus the scale of the costs incurred during the transportation. *For example; the cost of delivering the package to a remote location in the middle of nowhere is going to be a lot more than when delivering the same package to a metropolis transport-hub.*

9) **Current Circumstances** – the current circumstances when the package is to be delivered could be significantly different to when the cost of freight was originally budgeted for. *For example; there has been a steep change in fuel prices, a change in the competitiveness of the freight market, a change in the economy, and changes in the monetary exchange rates.*

10) **Import & export licenses, custom clearances, import & export taxes, duties** ... can all add to the [Cost] and [Time] while introducing complexity to the delivery process.

The project manager had better plan all of those considerations into the project from the start, else expect the project to have a rocky finish.

Reviewer's Comment ... "Jack Of All Trades" PM.

I am not sure that this section on resources feels right, because the inclusion of packaging & dispatch goes beyond project management into production and logistics coordination.

Hmm, have you forgotten your younger days as a project manager, when you worked for that **small to medium sized organization** with their small to medium sized projects, when you as the **project manager needed to be a "jack of all trades"** (coverall goalkeeper)?

During the project's life, much of the project manager's focus will be on the Planning Phase and the Monitoring & Control Phase. Unfortunately, it is during the Executing Phase when certain aspects of the project are inadvertently overlooked until the project's deliverables are about to be handed over. Consequently, it can fall to the project manager to find the person (or rather be the person who has) to execute the missing "End Game" activities, such as how the project deliverables will be packaged & dispatched to the customer organization / end-user. ... And, all of this needs to be scheduled & budgeted for.

NOTE   ❏ These **forgotten "End Game" activities can greatly affect the project's chances of being a success by over running the [Time] and inflating the [Cost].**

Additionally as the project manager, you should be very aware of **how the product is presented (including its delivery packaging) as this will directly reflect the [Quality] image of the project deliverable and subsequently that of the performing organization.** That is, if the presentation & packaging looks second-rate then this will induce a poor perception of both the project deliverable and the performing organization.

PACKAGING AND DISPATCH COME TO TYPIFY THE PERFORMING ORGANIZATION'S QUALITY IMAGE, AND HENCE INFLUENCES THE CUSTOMER'S PERCEPTION OF ACCEPT-ABILITY AND EXPECT-ABILITY.

15. MONITORING & CONTROL Phase ... Practical Part 4

15.1. Risks & Issues, Stakeholder, and Change Management (R.I.S.C.)

Introduction

What I originally planned for the practical monitoring & control of Risk Management, Change Management, and Stakeholder Management was to write three independent sections specifically about each one of these aspects. However, while I was writing the theory in [Chapter 10] I mashed up the cartoon below and I feel that this is a better representation of "a day in the life" of a project manager when it comes to dealing with these three aspects. That is, **from a practical perspective of Risk Management, Change Management, and Stakeholder Management these blur together into one complex relationship**, or more specifically, you cannot deal with one of these aspects without having to deal with the other two aspects at the same time.

Overview

Given at the time when this section of the book was originally written, I was working in the defence industry where abbreviations and acronyms are the order of the day, I thought I had better come up with a four-letter acronym to represent this complex relationship between Risk Management, Change Management, and Stakeholder Management.

R.I.S.C.   [R]isk [I]ssues [S]takeholders [C]hange

Umm, where does the Issues part come into the acronym?

Well, to paraphrase [Section 4.3.5] ... a 'Risk' is something that could hurt the project in the future; whereas, an 'Issue' is something that is hurting the project right now or will most probably be hurting the project in a relatively short time from now.

The purpose of R.I.S.C. Management is to orchestrate those activities, interactions, relationships that will prevent, mitigate, avoid, or eliminate "bad things" happening to the project.

Alternatively, the purpose of R.I.S.C. Management is to ensure that the project achieves its business objectives of delivering the agreed [Scope], by the agreed dates [Time], within the agreed budget [Cost], and to an agreed level of customer satisfaction [Quality].

Please refer back to [Section 1.3] to the four traditional determinants used to decide whether a project is considered a success or a failure.

Though, is it really such a surprise for Risks & Issues Management, Change Management, and Stakeholder Management to be handled together. ... And, to then be associated with these traditional determinants of project success?

R.I.S.C. Management is what happens when everyone had planned for something else to happen, but it ain't gonna happen that way now, so you better get down to dealing with the realities of what it is now.

Cyclic Nature of R.I.S.C. Management

Back in [Section 10.2], it was stated that;

> *"Any change can inadvertently introduce undesirable side effects, and*
>
> *hence introduce some form of risk to the project being successfully completed".*

What [Figure 135] below is depicting is the mutually binding relationship between
Risk & Issues Management, Change Management, and Stakeholder Management.

**Figure 135: the cycle of interaction between the management of Risk, Change, and
Stakeholders.**

That is, risks & issues (and the response to such risks & issues) will necessitate changes to
the project's baselines and these risks & issues will affect the project's stakeholders.

While conversely, changes to the project's "Triple Constraints" will introduce risks to
the project being completed successfully, and these changes will also simultaneously affect
the project's stakeholders.

Likewise, the project's stakeholders could require changes to be made to the project's baselines and each of these stakeholder's individual needs, wants, concerns, expectations, and perspectives will influence whether they and others consider the project to be a success or a failure; see [Section 10.3.2].

To put it another way …

Stakeholder Management is about managing those risks associated with the stakeholders, based on those stakeholders' power, interest, position, uncertainty, influence, and impact; see [Section 10.3.3]. Where, the failure to successfully manage these stakeholders will result in some amount of damage to the project's chances of being successful (or being considered a success), and thereby interfering with the performing organization being able to achieve its business objectives.

Change Management is about managing those risks related to changes to the project's baselines, more specifically, [Scope Baseline] changes made with or without authorization. Where, the failure to manage these changes successfully will cause the project to haemorrhage financially [Cost] and duration-wise [Time] due to the amount of rework, over-work / under-work [People] & [Resources] that could result from these changes, and not forgetting the impact that these changes can have on the [Quality] of the project's deliverables.

Similarity of the R.I.S.C. Management Processes

While you are looking back at the theory in [Chapter 10], also have a look at [Figure 85], [Figure 87], and [Figure 90] which have been reproduced on the next couple of pages. Notice how all three of these management processes have very-very similar structured flows as summarized in [Figure 136].

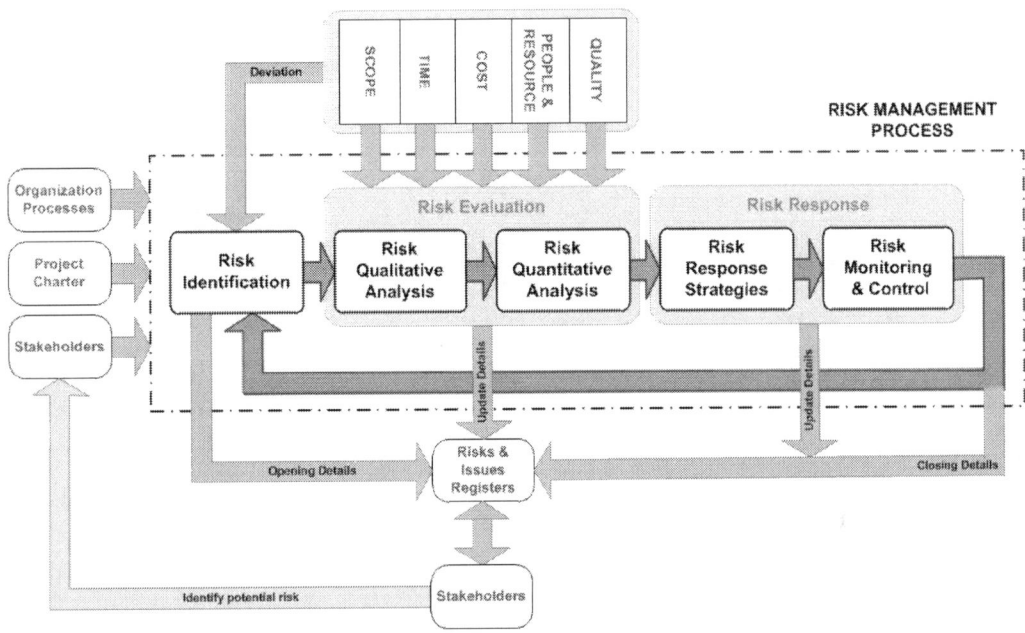

Figure 85: Risk Management Process.

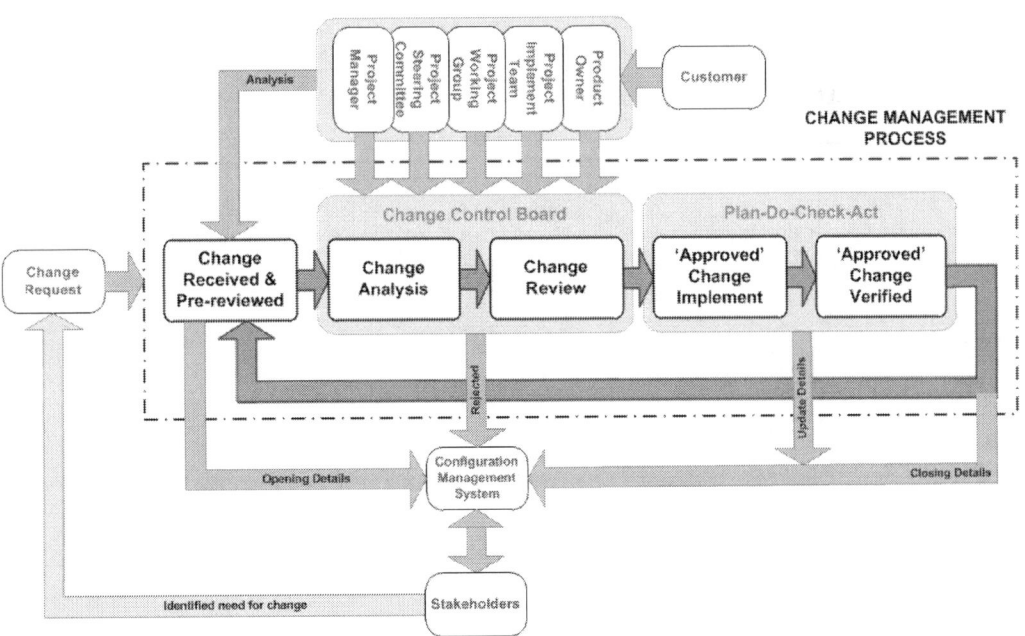

Figure 87: Change Management Process.

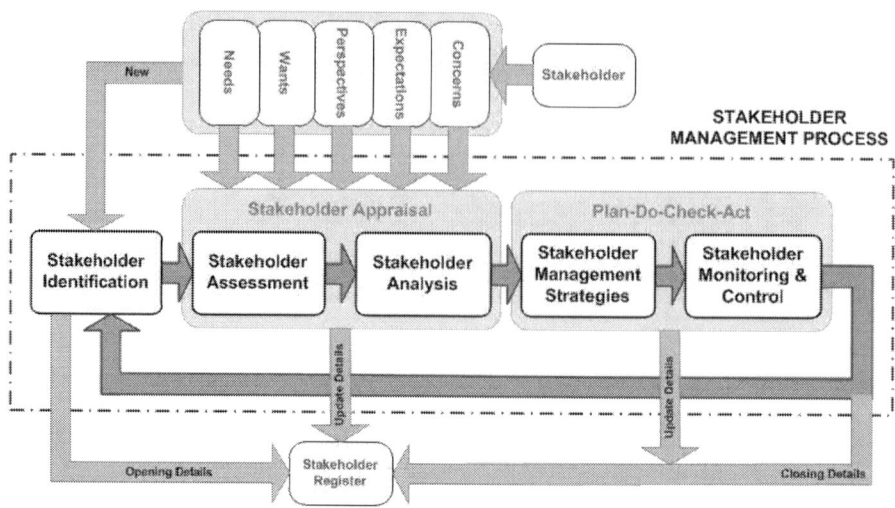

Figure 90: Stakeholder Management Process.

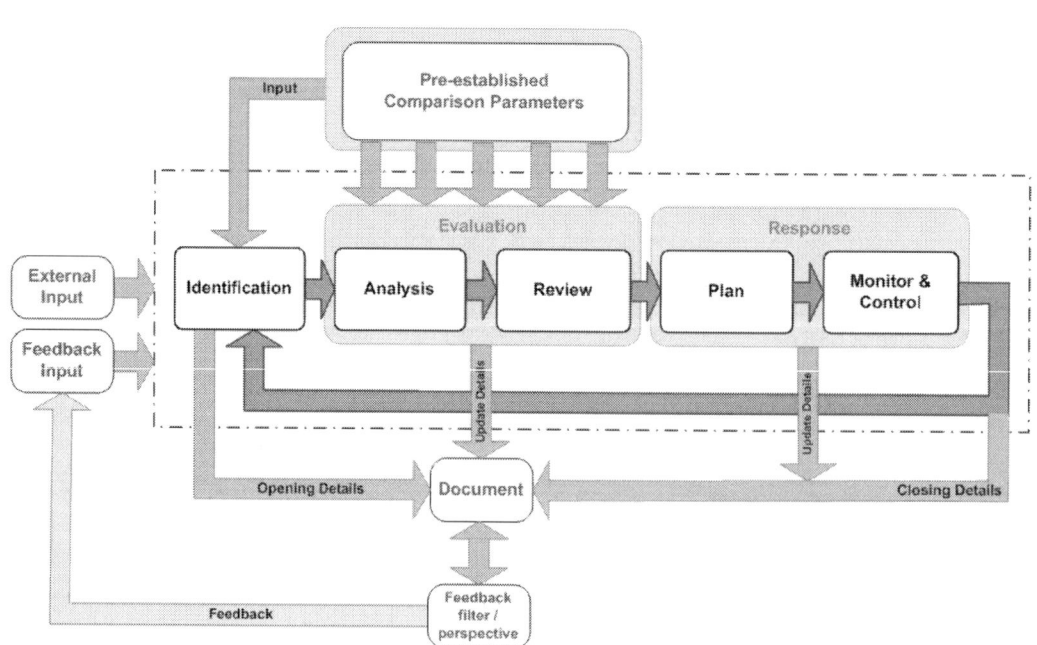

Figure 136: R.I.S.C. Management Process.

As summarized in [Figure 136], these three management process models all have the following in common:

1. **Identification** – Identify the risks (issues) / changes / stakeholders associated with the project.

2. **Analysis** – qualitatively & quantitatively analyse these identified risks (issues) / changes / stakeholders effect on the project.

3. **Review** – judge each of these identified risks (issues) / changes / stakeholders in comparison to the other risks (issues) / changes / stakeholders, and determine the overall impact / affect when it comes to the project as a whole.

4. **Plan** – the responses & strategies to be applied to these identified risks (issues) / changes / stakeholders.

5. **Monitor & Control** – manage the interaction of the chosen response implementation with these identified risks (issues) / changes / stakeholders.

6. **Document** – for each risk (issue) / change / stakeholder record the findings, the decisions made, the strategies to be implemented, those actions that were implemented, and the outcomes of the current pass through the management process.

7. **Feedback** – take what is "now known" about the risk (issue) / change / stakeholder and feed this back to the beginning of the management process to be combined with any new inputs, then iteratively pass all of this through the management process again & again until the matter is resolved or the project has concluded.

But, *there is no actual executing in these models?*

This is because as the project manager, the concern is no longer on the "doing"; rather, the emphasis is on "ensuring it gets done". Hence, the project manager's focus is on the planning, coordinating, and monitoring & control of the implementation.

15.1.1. The Rules of R.I.S.C. Management

As with any process, there has to be a set of rules to be applied so as to ensure that things run smoothly.

1st Rule … DO NOT ignore the RISC

To ignore the RISC is at the project's and the project manager's own peril.

Ignored RISCs do not usually just fade away; rather, the negative aspects of an ignored RISC will continue to grow until it has **"snowballed" into a major problem that can no longer be ignored.**

LEFT TO THEMSELVES, THINGS TEND TO DETERIORATE AND ACCELERATE.

By the time that the ignored RISC has become an apparent problem, it will probably have already affected the satisfaction of some of the project stakeholders. Subsequently, because of this RISC, the project stakeholders (i.e. the primary-core stakeholders) will be demanding that changes are made to the project. … And in the worst case, they may be demanding changes to how and by whom the project is being managed.

2nd Rule ... DO NOT hide the existence of the RISC

As soon as possible, communicate about the RISC; even if the acknowledgement of this RISC tarnishes the perceived prowess of the project's management.

 IMHO, the customer does not (and should not) need to know about every single RISC associated with the project. However, the other stakeholders such as members of the project team (i.e. the Project Steering Committee, the Project Working Group, and the Project Implementation Team) will definitely need to know of this RISC so that they can take the appropriate actions and make the appropriate arrangements to deal with this RISC.

If the RISC affects or impacts on any of the project constraints such as; the [Scope Baseline], [Time Baseline], [Cost Baseline], [Quality Baseline], [Resources Baseline], and [People Baseline] then this RISC has to be communicated to the relevant project stakeholders within the performing organization. Additionally, the customer's representative may need to be notified of significant RISCs, as a Baseline Change Request (BCR) process may need to be undertaken. Therefore, be prepared to devote time to communicating with the relevant project stakeholders about the situation and how the RISC will be prevented, mitigated, avoided, transferred, or shared; see [Section 10.1.3].

3rd Rule ... DO NOT be overly pessimistic or optimistic

Be realistic about the significance of the RISC.

You may have heard the following proverbs; "to make a mountain out of a molehill", "a storm in a teacup", and the obscure "the mountain that gave birth to a mouse".

All of these proverbs are symbolic to R.I.S.C. Management:

- **DO NOT overstate the significances of a RISC**, because the stakeholders involved with the project may well think, "Here we go again, an exaggeration of the situation". This could result in the project manager destroying their credibility, and hence potentially suffering the fate of "the boy who cried wolf".

- **DO NOT understate the significances of a RISC**, because the stakeholders involved with the project may well think; "if this RISC is not that significant then I've got plenty of time remaining to deal with it", or "not a serious RISC !!, are you joking !!".

Therefore, a continual tendency to be either overly pessimistic or optimistic about RISCs is not beneficial to the project's success, as it does not present **a realistic representation of the RISCs** to the project's stakeholders. Hence, this skewing the truth of the situation does not **result in an appropriate level of action, response, and concern** by the relevant project participants.

4th Rule ... Assign responsibility and ownership of the RISC

Without **responsibility & ownership for a specific RISC having been assigned to a certain person or the leader of a group** then everyone involved with that RISC could assume that it's someone else's problem, and hence the RISC "ball could be dropped", "fall between the cracks", or "roll behind the couch".

 I would recommend that ... even when responsibility for the RISC has been assigned to someone else, the project's manager should still presume partial ownership. Keep a watchful eye on the RISC; because, if it all goes horribly wrong then (more often than not) the project's primary-core stakeholders will still consider the project manager as the one who is ultimately responsible for the RISC (and not the persons who were assigned to deal with the RISC).

Therefore, choose wisely to whom responsibility for the RISC is transferred. ... Are you honestly confident in their capability to take care of this RISC?

5th Rule ... not capable of resolving every single RISC

The project manager is not capable of personally resolving every single RISC; hence, other persons will have to be involved with doing something about the RISC. That is, to maximize the effectiveness of the R.I.S.C. Management strategies and RISC resolution implementation, it is essential to involve those people who have enough power & influence in the performing organization and/or within the customer organization to be able to do something about the RISC (whether it be personally resolving the RISC or by coordinating others to resolve the RISC). Therefore, assigning responsibility & authority to someone who is more capable of dealing with that specific RISC.

 A new project manager (not wanting to be perceived as ineffectual) may try to fix RISCs by oneself, or they may not notify the primary-core stakeholders about the RISC in a vain hope of dealing with that RISC before it becomes evident to all. This RISC self-burdening is potentially a dangerous and futile strategy by the project manager.

The reality is; given the current situation and the prevailing circumstances, the project manager may not have the power, position, or influence to successfully resolve the RISC. Thus, **it is essential that the project manager accept that they are not always capable of controlling RISCs by oneself.** Similarly, **the project manager has to be prepared to deliver bad news about the RISC to the project stakeholders.**

15.2. Risks & Issues Management

15.2.1. Risks & Issues Register, Risk Matrix

Back in [Section 10.1.1] the Risks & Issues Register was introduced, and it was described how the Risk Matrix (and the Risk Breakdown Structure) could be used to analyse the project's risks & issues. However, in the real-world of small to medium size projects and with small to medium size performing organizations, these risks & issues management tools are not always utilized on a continually active basis. Rather, these risks & issues management tools may only be used for the formalized tracking of those risks & issues that are visible and/or known to both the performing and customer organizations.

That is from a practical perspective, the Risks & Issues Register may primarily be used as a traceability record (for project auditing purposes); whereas, other means will be used to actively manage the project's risks & issues.

15.2.2. Risks & Issues recorded via the Project Meeting Minutes & Actions

Instead of an active Risks & Issues Register, possibly the action points listed in the minutes (summation email) of the regular project meeting will be used to track the project's risks & issues. Where, each minuted action has someone or some group assigned responsibility for dealing with a specific risk or issue, and at the next project meeting this minuted action will be recalled to report the progress on resolving that risk or issue.

I would have to agree with that observation, because many a time I have seen the project's current risks & issues being handled via the ongoing actions in the minutes of the weekly / fortnightly project meeting; rather than being actively tracked via entries in the Risks & Issues Register.

If the Risks & Issues Register is being used then most probably these will be utilized to formally record those project-meeting concerns of some significance, and to chronicle those resultant actions that have been decided upon. Additionally, the Risks & Issues Register provides a beneficial location to incorporate some logical analysis of the priority for resolving each of these risks & issues.

Umm, previously there was all this information about
the Risk Management process, *yet this section reads like it is*
being done via a different process?

No, the Risk Management Process in [Section 10.1] is not being undertaken in a completely different way. Rather, the Risk Management Process is being incorporated into the **weekly / fortnightly project meeting**, which happens to be **the perfect opportunity to undertake the management of risks & issues.**

The regular project meeting often **contains the quorum of people** necessary to identify the risks & issues, evaluate each risk & issue, propose viable strategies to resolve the risks & issues, and often has present the people who have the power & authority to have the appropriate risk & issue responses implemented.

I would recommend that ... the project manager should volunteer to take down and distribute the minutes for all meetings related to the project.

 I would also recommend that ... when politically feasible, DO NOT let the customer's representative be the one who writes the project's minutes. This is because, it is not unheard of for some unscrupulous stakeholders to utilize the project's minutes as an opportunity to ~~record~~ rewrite the project's history to the betterment of their own interests. Hence, as soon as possible always read over any distributed minutes related to the project. Immediately (but politely) highlight to those that were present at the meeting, any inconsistencies that you perceive between the contents of the minutes and what you recall as having occurred and outline / state what you believe was agreed to (and not agreed to) during that meeting.

 I would also recommend that ... after any conversations / telephone calls with the project's primary-stakeholders (i.e. the customer's representative) that a summation email be sent to reiterate what are perceived to be the facts and the outcomes of that discussion.

HE WHO WRITES THE MINUTES,
INTERPRETS THE EVENTS THAT OCCURRED,
AND
~~RECORDS~~ REWRITES THE PROJECT'S HISTORY.

15.2.3. Risks & Issues information exchanged via Emails

Rather than detailing the risks & issues in the Risks & Issues Register, possibly a significant amount of the information pertaining to the risks & issues are contained within email exchanges between the involved project stakeholders. In addition, emails in general will be used to communicate to all of those persons (directly & indirectly) affected by the risk / issue. Likewise, emails and verbal discussions will be used to coordinate with those persons who can help to analyse, strategize, and resolve the particular risk / issue.

I would recommend that ... your work email's mailboxes should be highly structured with individual folders for each project so that there is a clear and encapsulated body of information.

Inside each parent folder create subfolders per project for:

- **Received from Customer** – to store all of those topic relevant emails received from the customer organization.

- **Sent to Customer** – to store all of those topic relevant emails transmitted either directly or CC to the customer organization by anyone inside the performing organization (either sent by you personally or that you were CC on).

- **Received from 3rd Party** – to store all of those topic relevant emails received from third parties (such as vendors, suppliers, and transient contractors), that do not involve the customer organization (i.e. the customer is not in the TO or CC lists).

- **Sent to 3rd Party** – to store all of those topic relevant emails sent to third parties by anyone inside the performing organization (either sent by you or that you were CC on), that do not involve the customer (i.e. not in the TO or CC lists).

- **Internal** – to store all of those topic relevant emails that were sent and received exclusively to & from persons inside the performing organization (and from 'pseudo staff' contractors attached to the project on a permanent basis).

15.2.4. Risks & Issues Management Conclusion

Irrespective of which risks & issues management tools (formal and/or informal) that are being used for the project, from a practical perspective the key things are:

1) Any individual <u>risk/issue should not be ignored</u>, but rather its existence has to be <u>acknowledged</u>; *e.g. the risk/issue (i.e. action) has to be covered during each weekly / fortnightly project meeting until that risk/issue is either resolved or no longer relevant.*

2) Each risk/issue should be entered into <u>some form of project record</u>; *e.g. the minutes of the regular project meeting, and/or the Risks & Issues Register.*

3) <u>Responsibility & authority</u> for dealing with each risk/issue (i.e. action) should be <u>assigned</u> to a specific person or the leader of a specific group; *e.g. as an assigned action in the regular project meeting minutes, and/or as an entry in the Risks & Issues Register (that has been communicated to the assignee).*

4) The details of each risk/issue and the status of the resolution of that risk/issue should be <u>communicated</u> to all parties affected by and involved with resolving that risk/issue; *e.g. via the minutes of the regular project meeting, with emails, through person to person discussion, and/or entries in the Risks & Issues Register.*

5) Irrespective of whether or not the project manager has been assigned responsibility & authority for dealing with a specific risk/issue, **the project manager should keep a <u>watchful eye</u> on all of the project's risks & issues, and the progress towards mutually acceptable resolutions of these risks & issues.**

 I would recommend that ... each risk/issue (i.e. action) be included in the minutes of the regular project meeting, indicating such things as: (1) Date when the action was raised, (2) a short one-sentence description of the action, (3) the person assigned responsibility and authority for resolving that action, (4) the date by when that action is expected / required to be resolved, (5) the status of the action when these minutes were taken down [open, closed, paused, unknown].

15.3. Stakeholder Management

Back in [Section 10.3] the Stakeholder Register was described, and the various Stakeholder Matrices were detailed as ways to analyse each project stakeholder's power, interest, position, uncertainty, influence, and impact.

However in the real-world, the project manager only has a limited amount of time during each work day to deal with all of the different aspects of the project; let alone time to deal with the many stakeholder needs, wants, concerns, expectations, and perceptions (henceforth to be referred to as "requests"). Consequently, these stakeholder analysis techniques, documented stakeholder information, and prepared Stakeholder Management strategies ~~could~~ often fall by the wayside after the project's Planning Phase is completed, or alternatively these formalized Stakeholder Management tools may never be utilized.

That is for practical purposes, **Stakeholder Management is often done on an ad-hoc basis**, where those stakeholders who have the most recent and loudest "requests" are the predominate focus of the Stakeholder Management activities.

IT IS THE SQUEAKY WHEEL THAT GETS THE GREASE ... UNFORTUNATELY, IT MAY NOT BE THE WHEEL THAT REALLY REQUIRES IMMEDIATE ATTENTION.

How to manage the stakeholders if the stakeholder management tools are not going to be utilized?

When it comes to the practical management of the project's stakeholders there are a couple of facets that need to be understood:

(1) **Perceptions & Deceptions**,

(2) **Respect, Humility & Intelligence**,

(3) **Actions & Distractions**.

15.3.1.   Stakeholder: Perceptions & Deceptions

The project stakeholders (especially those in the customer organization) have to perceive that the project manager (and the rest of the project team) are:

- **In charge & in control of the project ...**

  While the project manager may be able to quietly go about the business of getting things done (by interacting & communicating with the appropriate persons), the project manager has to be **"visibly seen" as the one who orchestrated the project's activities and is delivering what they said they would do, by when they said it would be done**.

  For example; the project manager should be the one who:

  o   *oversees & distributes the project management documentation (i.e. project status reports, project plans & schedules, the Detailed Specifications),*

  o   *maintains the project's tracking information (i.e. the Requirements Traceability Matrix, the Risks & Issues Register),*

  o   *chairs the project's meetings, sends out the project meeting agendas, takes down & distributes the project meeting minutes,*

  o   *facilitates discussions about project risks & issues, and gets the resolution moving along,*

  o   *communicates / emails the summation of decisions made for actions & tasks to be undertaken, and*

  o   *coordinates the [People] and the [Resources] for the project.*

  Hence, **from the stakeholder's perspective, the project manager should be perceived as the "public face" of the project**. However, In-Charge & In-Control does not limit itself to project coordination, but also with how the project manager personally acts & reacts to the project's changing situation and the prevailing circumstances.

- **Calm & in self-control ...**

  The project manager has to be **visually perceived as unflappable, collected,** and
  not panicked by the project's changing situation and the prevailing circumstances.
  While in reality the project manager maybe paddling like crazy to catch up with what is
  going on with the project, and having to rush to get the situation back under control;
  however, from the stakeholders' perspectives, the project manager should appear
  unruffled as he / she methodically & logically works their way through project matters.

PERCEPTION AND DECEPTION ... AN ILLUSION INTENDED TO PROPAGATE A BELIEF THAT IS NOT COMPLETELY THE TRUTH.

- **Master of the project's domain ...**

The project manager has to be **perceived as knowing what the project is about**, and **knowing what is going on with the project**. As listed previously in [Section 11.2]:

1) The project manager has to **understand the relevant principles & concepts** behind project management and the SDLC methodology being used by the Project Implementation Team.

2) Is **aware of the relevant details about the project being implemented**. Such as, the architecture overview of the product / system / application, and the technologies & processes involved with the project.

   However, not necessarily knowing how to technically hands-on implement the project themselves.

3) Is **aware of what is going on around the project**, within the project team, within the performing organization (*e.g. resource availability, other projects' major milestones*), and knows what is going on with the customer organization.

NOTE ☐ Just as "you cannot hope to manage that which you do not know or reasonably understand"; **you cannot expect to manage those stakeholders who do not believe that you know what you are doing.**

- **The 'GO TO' person for project related things ...**

The project manager has to be **thought of as the first Point Of Contact (POC) for finding things out about the project, and considered the "go to person" for that project**; i.e. the "Mr. Gets It Done", and the "Ms. Makes It So".

Subsequently, the project manager has to be prepared to listen to the stakeholder's requests, politely declining those inappropriate & unrealistic requests, to act on those relevant requests, and to then continually communicate the outcomes of the resultant activities until the associated request(s) has been resolved or is no longer relevant.

- **Confident and trustworthy ...**

Based on past performances and current appearances, the project manager (and the project team) have to encourage confidence and trust from the customer's representatives (the customer's equivalent of the project manager and technical peers). This is because, the more confidence and trust that the customer's representatives have in the performing organization's capabilities to successfully get the right things done, then potentially the less future effort will be required by the performing organization's project manager (and project team members) to deal with that stakeholder's requests.

That is, the primary-core stakeholder will not feel the necessity to continually "supervise" the performing organization; hence, the primary-core stakeholder will not feel the necessity to continually reiterate their "requests" (concerns, needs, wants, expectations, and perceptions) ... again and again.

Umm, why did you replace stakeholder needs, wants, concerns, expectations, and perceptions with stakeholder "requests"?

The reason that stakeholder needs, wants, concerns, expectations, and perceptions were replaced with stakeholder "requests" (other than to save having to type all of this every time) is because **as the project manager you can choose to; ignore a stakeholder's request, or decide to deal with the request at a later time, or immediately deal with the request**. As the project manager, the choice is yours; but, as with every choice in life there is some risk associated with that choice, hence choose wisely.

Else you could find that the project snowballs into a mass of stakeholder related risks & issues.

15.3.2. Stakeholder: Show Respect, Humility, and Intelligence

When meeting with stakeholders (and especially with the customer representatives) then an appropriate level of respect has to be shown towards the relationship.

- **Equivalent managerial ranks present ...**

  If a member of the customer organization is coming to visit the performing organization's premises (for a non-regular meeting), or alternatively a member of the performing organization is going to meet with the customer then it is appropriate for an equivalent level or higher level member of the performing organization's management team to be present during this meeting (at least initially or in conclusion of the meeting). The reason for equivalent managerial ranks being present is because, the meeting's discussion could deviate away from those project meeting's agenda topics and onto strategic / contractual matters.

- **Intermediary between the customer and subcontractors ...**

  If there is to be a meeting between the customer organization and a subcontractor then a representative of the primary contractor (i.e. the performing organization) should also be present, so as to ensure that there is no resultant confusion and misunderstandings about the project's deliverables, milestones, progress, arrangements, remunerations, obligations, commitments ... etc. That is, you do not want the subcontractor to state or agree to things that are not in the best interest of the performing organization, or to undermine / refocus the relationship between the performing organization and the customer organization.

- **Small talk enabled ...**

  If a senior member of the customer organization invites you for tea or coffee then go. Even if you know that the conversation will be a one-sided spiel and will not directly contribute to the current project's success; this "get together" could form the basis for further working relationships between both organizations.

- **Dress appropriately for the occasion ...**

Dress ready to achieve the business objectives that are the purpose of the get together. If it is a meeting with the customer's representatives and they will be wearing suit & tie then the performing organization's representatives should also dress similarly. However, if the purpose of the get together is to perform a technical work activity (such as installing equipment) then dress befitting the primary activity.

In the realm of SDLC projects, when appearing before the customer it is not such a good idea for you (and/or your colleagues) to be wearing an overpriced outfit, donning lots of bling, and/or turning up in an extravagant vehicle; aka over "dressed for success". Because, the customer could start to wonder whether their money is being misdirected towards paying for the performing organization's "visual show", rather than providing them with value for their money "go". Additionally, a sense of humility is important if the project has gotten into a state of bother.

Hence, for SDLC projects dress to present a persona of efficiency & effectiveness.

PRESENT RESPECT, HUMILITY AND INTELLIGENCE NOT GLOATING ONE'S SELF IMPORTANCE

15.3.3. Stakeholder: Actions & Distractions

From a practical perspective, **the project manager's actions will result in the satisfaction of the project stakeholder's requests**. That is, those actions that give these stakeholders; what they need, when they need it, the way that they need it, in a form that they can use.

During the project's life, the various project stakeholders are going to make many and varied requests of the project manager. Although the project manager may think that they can multi-task sufficiently to deal satisfactorily with all of these stakeholder requests, the truth is that he/she won't be able to deal with all of these stakeholders at the same time.

Thus for reasons of self-preservation, the project manager has to:

- **Be proficient at Time Management ...**

  The only truly effective multi-tasking that the project manager can hope to achieve is the pseudo multi-tasking of doing one activity (or part of an activity) at any one time – for a reasonable duration of time – before switching to the next highest priority activity. That is, the project manager should deal with one activity (i.e. a stakeholder request) and work on that activity until it has been taken to a point where:

  1) that activity **cannot effectively be worked on anymore** because there is something that is hindering its effective progress, or

  2) that activity is **dependent on some other activity** that has to be completed or partially completed first (*e.g. a quotation received from a third party*), or

  3) that activity has **received enough** of the project manager's **attention** for today (*e.g. a couple of hours effort*), or

  4) that activity is of a **lower priority** than a higher priority stakeholder request that has just been received.

  Unfortunately, what can result when the project manager tries to deal with multiple activities all at the same time is the jumping from one activity to the next without having effectively completed anything to a stakeholder satisfying level.

Attempting to deal with multiple activities all at the same time can result in project management inefficiencies and a lack of productivity because time is wasted churning between tasks. Where, **task churning** is when the current activity has to be put aside, then the next activity is picked up, then time is required to figure out what was going on with that activity, then that activity is also put aside for a higher priority activity, then another and another activity until the project manager is no longer sure what is going on with all of these activities, and **subsequently ends up "firefighting" these activities as problems arise.**

This task churning can also happen to Project Implementation Teams, when project management reassign team members from incomplete task to incomplete task, and "bouncing" them from one ongoing project to another project. Effectively inducing a form of project dementia where "I know I was doing something, but I just cannot remember what it was, let alone where, I was up to when I was last here".

- **Effective Constraint & Restraint Management ...**

The project manager has to develop that elusive art of sufficiently servicing all of those relevant stakeholder requests without losing focus on and control over the project's [Scope], [Time], [Cost], and [Quality]; i.e. maintaining control over those traditional determinates of whether the project is deemed a success or a failure.

NOTE   ❏ By the Executing Phase these four project variables will usually have become project constraints due to; the sign-off of the approved Detailed Specifications [Scope Baseline], the schedule [Time Baseline] having been accepted / milestoned, the project's budget [Cost Baseline] having been allocated, and the Acceptance Criteria [Quality Baseline] having been defined.

As the project manager (and the members of the project team) try to service every stakeholder request, the project manager may feel that much of their day (and that of the affected project team members) is taken up with dealing with these stakeholder requests instead of getting on with the project's technical implementation.

Therefore, **before** the project manager (and the affected members of the project team) **undertake any action in response to a stakeholder request**, they should stop for a moment and question; **"does acting on this stakeholder's request** (now or later) **contribute to the project's overall progress towards the successful completion of the project?"**

If the answer to this question is "**NO** it does not directly contribute", **then it may be prudent to restrain from undertaking that action,** and then **enquire as to its necessity.**

It is very easy to try to **satisfy every stakeholder request** without these actions effectively contributing towards the production of the project's deliverables, and thereby **may no longer be working within the boundaries of the agreed project constraints.**

However, depending on whom the primary-core stakeholders are and what is their attitude & demeanour, **the project manager (and the members of the project team) can sometimes find themselves having no "political" option other than to do this non-project-advancing activity specifically to appease this all-powerful & highly influential project stakeholder.**

As stated previously, the determination of whether a project is deemed a success or a failure is based entirely on the expectations, perspectives, and opinions of the project's primary-core stakeholders. ... *Sometimes it is better to appease, than.*

- **Effective Time Record Management ...**

  As the project manager, you could find that (due to the amount of stakeholder requests being handled) the recorded 'project management' hours are in fact closer to a third of the total project hours reported thus far.

  Yes, customer handholding can consume a lot of time and effort.

However, excessive 'project management' hours are not going to look good for the project (even if these hours were necessary for the project's successful completion) as this can be interpreted as a sign of a lack of control over the project.

Thus, the next question that the project manager should ask is; "**was the undertaken activity really 'project management' or was it an extension of an implementation activity?**" That is, just how much of that time recorded against the 'project management' job code was related to technical & semi-technical activities such as engineering support, architectural design, document review, or acceptance testing of the deliverables? In addition, project team members may use the 'project management' Cost Account Code as a catch all for tasks that were not specifically assigned to them. Consequently, when the project's (weekly / monthly) timesheet report is generated, it may come as a bit of a shock to find out that there has been a disproportionate amount of time booked to 'project management'.

Yep, so much for that approved project plan that had the 'project management' hours as being only 10-15% of the project's total people hour allotment.

 I would recommend that ... those activities that were not really 'project management' be recorded against an appropriate Cost Account Code, such as 'technical authority' and 'business analyses'. Also, it may be politically wise to write-off some of those excessive 'project management' hours against the project's other tasks such as 'support' and even 'design' & 'development'.

Therefore, in addition to 'project management', it is advisable to have additional job codes for the situation when the project management is servicing "technical" related stakeholder requests that were not planned for as scheduled tasks in the project's [Time Baseline] schedule.

For SDLC projects, I usually include a "Business Analyst & Technical Authority" task of 10-15% of the implementers' total allotted time.

- **Prioritize the order in which these stakeholder requests are dealt with ...**

A previous bullet point emphasized that not all of the stakeholder requests can be dealt with simultaneously; in fact, some of these stakeholder requests should not be addressed at all (let alone immediately).

 There are not enough hours during the work day to **satisfy every stakeholder request**; so **DO NOT try to,** else the project manager (and/or project team) could suffer from overworked burn-out, as well as unnecessarily impacting on the project's [Time] & [Cost].

The project manager has to order these stakeholder requests using some logical and **consistent means of prioritization** based on:

1) Those requests that have an actual **imminent time constraint on response**.

2) Those requests from **stakeholders who possess great power – interests – position – uncertainty – influence – impact over the project's successful outcome**. That is, requests from those stakeholders that appear in the top right hand corner of the Stakeholder Matrices depicted in [Section 10.3.3].

3) Those requests from primary-core **stakeholders who make the direct decision over whether the project is considered a success or a failure**.

4) Those requests from **stakeholders who have been waiting** sometime for some form of response.

 I would recommend using a High – Medium - Low grading scheme for stakeholder requests where;

1. **High** - are stakeholder requests that have the **greatest importance** (i.e. power, influence, impact) **AND** the **greatest urgency** (i.e. shortest necessitated response times).

2. **Medium** – are stakeholder requests that have **great importance OR urgency, but not both** at the same time.

3. **Low** - are stakeholder requests that have the **lowest importance AND** the **lowest urgency**.

Reviewer's Comment ... Tennis ball stakeholder management.

Personally, when it comes to managing stakeholders, I often use what I refer to as, "tennis ball" management, where the objective is to answer the stakeholder's request swiftly & sufficiently to put the next action of that stakeholder's request back onto their side of the court. Thereby, buying time to enable myself (and the project team) to focus on those remaining tasks necessary to get the project to a successful completion.

This can involve a tactic of managing to the next milestone / deliverable.

- **Deal with stakeholder requests in an intelligent & timely manner ...**

  The previous points have emphasised how it is important to deal with the relevant stakeholder requests in a prioritized, effective, and multitasking fashion; but just as important is to **deal with these relevant stakeholder requests well timed and well informed.** That is, to give these stakeholders what they need, when they need it, in a form that they can use.

  So, before the project manager gets the urge to act on a stakeholder's request, they should find out the following:

  1) **What are the latest facts pertaining to that stakeholder?**
     For example; what is the current business situation and what are the prevailing circumstances (are they pushed for time, are they running out of budget).

  2) **What is their demeanour towards the performing organization** (are they not too impressed with the effort so far).

  3) **What are the up-to-date details of their request** and not just some hearsay, and not what where the facts about this request some time ago?

  4) **What are the actions / solutions** that the performing organization intends to undertake **in order to respond to this request** and what will be the response?

Example Case: project manager who engages mouth too soon has a lot to explain.

You have just stepped out to buy lunch from the local café and while you are waiting in the queue your mobile phone rings.

You respond; *"YOHO MAN, WHAT'S UP"*

... Your 1st mistake. ... A non-professional demeanour during work hours.

You forgot that your personal mobile phone number was included in the signature block of your work email.

It is not your best friend calling at lunchtime to tell you whether they scored those tickets to tonight's big game; rather it is Nigel, your customer's representative and he sounds rushed (or is that flustered) and it sounds like he is in traffic.

Nigel asks about how long it will take to deliver Project MINOW.

You are now only a couple of steps away from being served at the lunch counter.

As you quickly place your lunch order, you tell Nigel "9 weeks" because that is what you recall from when you last looked at that project's schedule.

... This is your 2nd mistake. ... As those facts in the mind, do not necessarily match the facts as these stand, especially when distracted.

Nigel hurriedly responds with thanks and hangs up the phone.

On arriving back at the office with your chicken-n-egg roll in hand.

You sit down in front of your workstation and pop open Project MINOW's schedule.

"OH SHIRT"... as red sauce drips down the front of your clean white shirt.

That was a 9 weeks work effort ... but you completely forgot about the 2 weeks resource unavailability delay that was mentioned late last Friday afternoon.

"OH SHIRT"... you quickly call Nigel back, but his phone goes straight to voice mail.

Later in the afternoon, Nigel calls you back in response to your voice-mail message. You inform him of the fact that the work effort was 9 weeks but the duration has increased to 11 weeks due to resourcing lead-time delays.

Nigel despondently answers with; "*YOU'RE KILLING ME HERE!*"

It turned out that Nigel needed that information ASAP to provide to his senior management so that they could plan the work streams for the next two financial quarters. Unfortunately, that 2-week delay for Project MINOW will mean that the project's deliverables won't be available in time to make the project's assigned go-live test window. Hence, the flow on effect that Project MINOW's deliverable will have to wait an additional 2 months until after the strategically important Project SUPERMAX goes through its go-live release testing, i.e. Project MINOW will "miss the boat".

Evidently, this forgotten resource delay will be putting your customer's representative, Nigel into a tight spot with his own senior management. Likewise, this delay will also put the squeeze on you as the project manager because that resultant 2 months wait before Project MINOW's deliverables can be acceptance tested will mean that the performing organization will not be paid (until the final sign-off by the customer). Consequently, this 2 months delay will result in cash flow problems for the performing organization due to progress payments being associated with that customer sign-off.

Thus, the moral of this little story is ... it only took a few seconds to answer the stakeholder's request, but it would have been significantly wiser to have said,

"*I WILL GET BACK TO YOU IN A COUPLE OF MINUTES*"

When the correct facts are known then formulate a response. Therefore, unless you are so sure of what you are about to say and are prepared to stake your career on it, then ...

DELAY BEFORE YOU SAY, ELSE SOMEONE MAY VERY WELL HAVE TO PAY, FOR THE CONSEQUENCES.

Now this example case is not proposing that the project manager should leave the customer's representative on permanent hold while the project manager eventually gets around to responding to that stakeholder request; i.e. not leaving it for a few days or weeks. Rather, **all stakeholder requests should be replied to within a respectful amount of time**. Even if that response is only to say that, more time will be required to formulate an appropriate response.

TO SAY, "NOTHING HAS CHANGED AT PRESENT" IS A LOT BETTER THAN SAYING NOTHING AT ALL.

Recall back in [Section 10.3.4] how continuous bi-directional communications was stated as one of the important facets to Stakeholder Management. In addition, given that everyone likes to feel special and cared for; then a good way to achieve stakeholder satisfaction is to keep these stakeholders informed about what is and what is not going on (as it relates to them).

- **Be more Proactive rather than solely Reactive ...**

When dealing with these stakeholders, the project manager could operate reactively or proactively. That is, the project manager could choose a **reactive / "defensive"** management style of only responding to stakeholder requests once these have been initiated, or the project manager could try to **proactively / "pre-emptively"** deal with some of those potential stakeholder requests before these have been asked.

> *For example; it is coming up to the end of the financial year, so maybe it would be beneficial to provide the primary-core stakeholders with a synopsis of what has and has not been achieved on the project during the current financial quarter.*

Admittedly, dealing with the project's stakeholders **reactive**ly **is** a lot **easier** than dealing with them proactively. This is because to be **proactive requires the project manager to have a lot better understanding of each stakeholder's needs, wants,** concerns, expectations, and perspective ... as well as a good understanding of the relationship between each stakeholder's power – interest – position – uncertainty – influence – impact; see the stakeholder analysis in [Section 10.3.3].

However, it is not that hard to partially pre-empt some of these stakeholder requests, given that **for all SDLC projects there will be several stakeholder requests that will be inherently common** from one project to the next, for example:

1) **What are the current states of the four traditional project constraints used to judge a project's progress / Success : Failure?**

   o **How much** of the project's **[Scope Baseline] has been covered so far**, and **how much [Scope] remains** to be covered by the current release and for the remaining releases?

   o **How much** of the **[Time Baseline] has been covered so far**, how much billable [Time] has been recorded against the project's Cost Account Code, and **how much [Time] is required to complete** the remainder of the current release and to complete the remaining releases?

   o **How much [Cost Baseline] has been spent so far**, how much is it expected to **[Cost] to complete** the remainder of the current release and to complete the remaining releases?

   o How many defects have been detected thus far, and **what is the perceived [Quality] of the project deliverables**?

2) **What are the current states of the other project constraints/variables?**

   o **What [People] have been utilized** to get the project to where it is now, and **what [People] will be required when, where, and how many to complete** the remainder of the current release and to complete the remaining releases?

   o **What [Resources] have been used** to get the project to where it is now, and **what [Resources] will be required when, where, and how many to complete** the remainder of the current release and to complete the remaining releases?

3) **What are the RISCs involved with the project?**

- o **What risks & issues currently confront the project**, and how will these be communicated, responded to, and possibly resolved?

- o **Which stakeholders** (other than the one for whom this question is intended to be answered) has **requests on** this project and/or has a change in their power–interest–position–uncertainty–influence-impact that will affect this project?

- o **What baseline changes** to the project does this stakeholder need to be made aware of and/or will be required to act upon?

If you were that stakeholder then what would you need, want, and expect to be told? Hence, as the project manager, if you tried to answer these questions before the stakeholders have to ask (or have to chase up the answers to these questions) then you will be perceived as a highly effective & efficient project manager.

Lessons Learnt: Kudos earned by being prepared to roll up one's sleeves.

There will be a time when a member of the Project Implementation Team, the Project Working Group, or even the customer's representative will make a request of the project manager. Some of these requests are going to seem "trivial"; however, these "trivial" matters are going to provide the perfect opportunity for the project manager to quickly roll-up their sleeves and just do that "trivial" thing themselves (instead of instructing / interrupting someone else in the project team to do it). This type of "getting it done" attitude demonstrated by the project manager can build a significant amount of kudos and respect from that requesting stakeholder (and from those team members who were not interrupted by this "trivial" request).

However, the trick here is to "serve" the project's stakeholders and not become "subservient" to those stakeholders' every whim. Basically, if you as the project manager only need to take a relatively short amount of time to resolve that matter then *"G.S.D. – GET STUFF DONE"* and keep the project moving with no big show of it.

- **Go back to the basics to maintain control …**

For the management of day-to-day project matters, the use of a diary / notebook can prove to be highly beneficial.

I find that when I am stressed; if I write lists of all of those tasks that I have to do, then sort these into prioritized order - I get back into a mindset of being able to control the situation. As, too many tasks in my head and not written down causes me problems.

You could also use a 'To-Do' list on your mobile phone / tablet, or reminders in your computer's calendar application (or the performing organization may use a dedicated workflow software application) to manage many of the project's activities.

Whichever tool is at your disposal or whatever technique you decide to use, there are a few things that you might want to note down:

1) **What are the currently outstanding To-Do actions?**

   In an easily accessible common location so that all of these activities can be seen together, put down each of these actions as a list of simple sentences.

2) **What is the priority** for completing each of these To-Do actions; i.e. low, medium, high?

3) **When does each To-Do action need to be completed by**?

   For example; "to be completed by close of business Thursday 9th Sept", or "to be sent first thing on Monday morning 12th Sept".

4) **What are the closure criteria** for completing (*e.g. what is required to be handed over*), postponing, scrapping that action?

5) **When will you allocate some time for working on each action?**

   It maybe that you will try to knock-off some of these actions in the morning before most of the Project Implementation Team members have arrived at work.

But, do not honestly expect to spend all day working through your 'To-Do' list because the project manager role is to support the Project Implementation Team and deal with the project's stakeholders, and not to lock themselves away to concentrate on their list.

That is, **the project manager has to deal with continual interruptions and with spontaneous requests that will have to be reacted to in a timely fashion**.

The thing to remember is that, **the project manager does not personally have to do these actions all by themselves**. Rather the project manager orchestrates getting these actions done in a timely & efficient manner. Hence, the **'To-Do' list should be used as a cross-off reminder of what currently has to be dealt with**.

Sometimes the most effective solution is the most simplest / primitive technique such as with pen & paper.

AN ACTION A DAY
KEEPS THE ANGRY STAKEHOLDERS AWAY

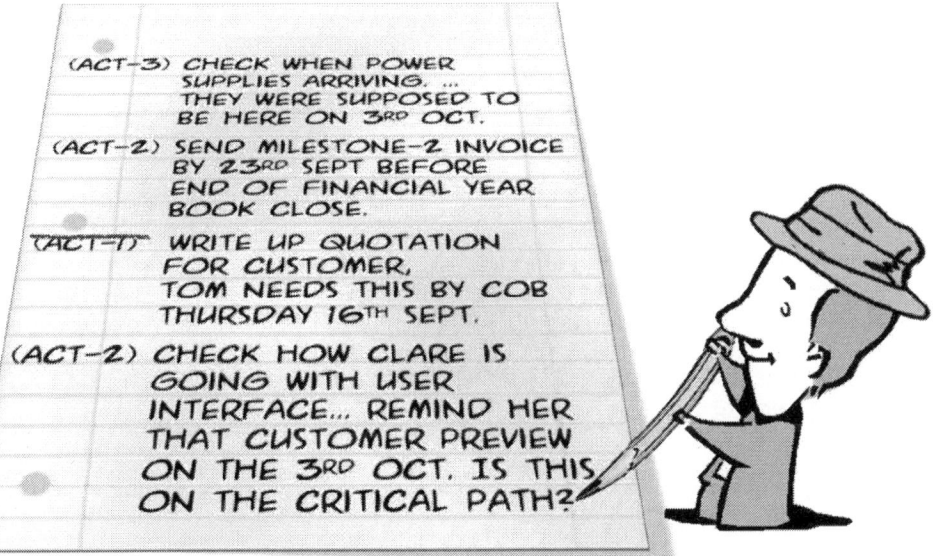

15.4. Change Management

Back in [Section 10.2] the Change Management workflow in [Figure 88] ... which happens to have been reproduced a couple of pages on from here ... and the Change Control Board were described. However, in the real-world of small sized performing organizations and with small projects there may not be such **formalized Change Management in place**. As a result, it may be necessary for the project manager to coordinate a "pseudo" Change Management Process, and for the project manager to take on the role as one-half of the Change Control Board opposite the representative of the customer organization.

Yep, with some projects and with some performing organizations there is no formalized Change Management Process in place.

You may find, that the project's [Scope Baseline] was never **formally agreed to** because the **Detailed Specifications** had not been **signed off**. What complicates this situation even further is that there is no **specific point in time when a [Scope Baseline] was declared**. These oversights will enable [Scope] changes to be made ad-hoc in response to the project's ever-changing situation & prevailing circumstances. And, to make matters worse, these [Scope] changes will be expressed verbally with the occasional written reiteration.

As a consequence of the above; scope creep and even scope shrinkage will occur virtually unchecked. But later on during the acceptance testing of the project's deliverables there will be some dissension between the performing organization and the customer's representatives over whether the provided functionality is correct or not. Where the interpretation of the "correct" functionality will be based on the individual's personal understanding of what was supposed to have been included within the project's [Scope]. These "disagreements" over [Scope] could eventually turn into a "political" and financial wrangle over whether such-n-such constitutes a contract deviation or is simply a clarification of pre-existing ~~requirements~~ specifications.

So, **how can one honestly expect the project to be delivered successfully when there is no relevant Change Management Process in place?**

 **The lack of a formalized Change Management Process can be a catalyst for problems during the later phases of the project's life cycle**. Combine this with an unsigned Detailed Specifications, and a non "frozen" [Scope Baseline] for each release, then the project is likely to culminate in an argument over what was and was not agreed to, and who is right and who is obviously wrong.

What exasperates the situation is when there is a long time between delivery releases (i.e. as with a waterfall life cycle).   In such a situation, scope creep and scope shrinkage could occur without being detected until a significant way into the project's life. At which point, the necessary corrective actions for these scope deviations could result in a significant amount of rework; thereby, having a catastrophic ripple effect on the project's [Time] & [Cost], and hence on the project being deemed a success.

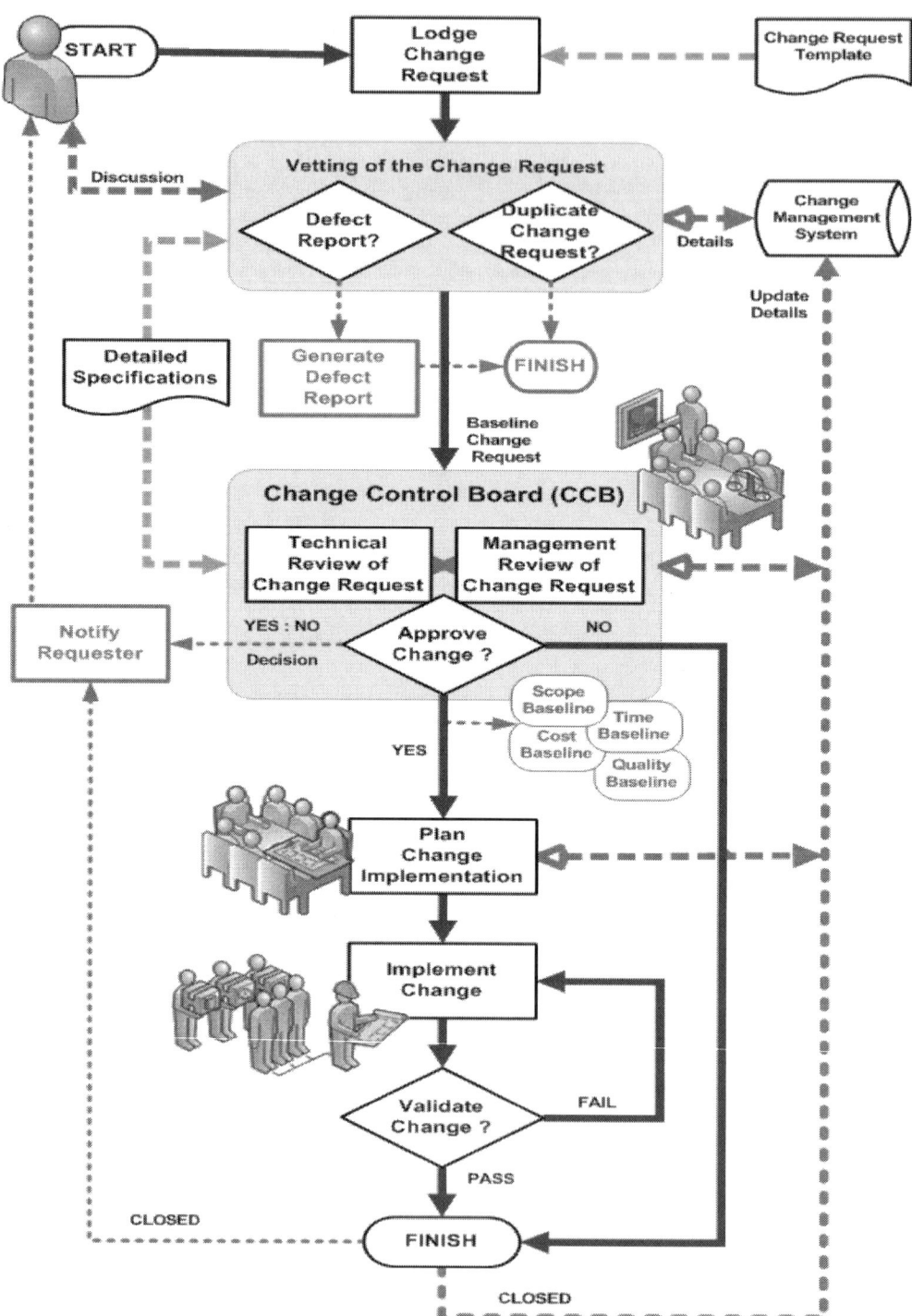

Figure 88: A simplified [Baseline] Change Request workflow.

How to manage changes if the project manager and the customer representative are the only members of the Change Control Board, and the project manager has to coordinate a "pseudo" change management process?

When it comes to the practical management of change in small organizations and for small projects then there are a couple of ways of looking at it:

1) "She'll be right mate" … "Err, no she won't".

2) "My project is my small business".

15.4.1. "She'll be right, mate" … "Err, no she won't"

As an Australian, we have many a colloquialism and one of these is the classic;

"SHE'LL BE RIGHT, MATE"

Generally, this expression will be used by those less educated Aussies who have an attitude of, "I will worry about it tomorrow if and when it goes wrong".

However, as the project manager your success and perceived benefit to the project cannot be based on having such a laissez-faire attitude towards life.

Do you honestly expect to leave the Project Implementation Team to make whatever changes they deem necessary to complete the project?

While the Project Implementation Team do have a vested interest in finishing the project, they do not have the same self-interest as the project manager has for ensuring that the project is completed within the boundaries of the traditional [Scope] – [Time] – [Cost] – [Quality] determinants that judge the project as a success (or as a failure).

Secondly, many a project manager has found out that while they were not looking an implementer or few has (with the best of intentions) deviated away from the [Scope Baseline] (i.e. the approved Detailed Specifications) and instead implemented what they thought the customer had intended, what they believed is "obviously how it is supposed to be", or what the customer representative personally told them they wanted.

Unfortunately, these **well-intentioned implementer changes do result in deviations in the agreed [Scope], [Time], [Cost] and [Quality].**

HOW DOES A PROJECT GET INTO REAL TROUBLE ?

... ONE EXTRA DAY AT A TIME,
... ONE EXTRA DOLLAR AT A TIME,
... ONE EXTRA PERSON AT A TIME,
... ONE EXTRA RESOURCE AT A TIME,
... ONE UNSOLICITED AND UNCOMPENSATED CHANGE AT A TIME

Thirdly, remember that someone (i.e. the project sponsor) has invested a considerable amount of time, money, people, and resources (as well as their business reputation and possibly even their business's future) into the successful outcome of this project. However, if the implementers and/or the customer representative have interposed unsolicited changes to the project then eventually this someone (i.e. the primary-core stakeholder) is going to have to bear the burden of undertaking these changes that were not agreed to.

 A laissez-faire attitude of "she'll be right, mate" just don't cut it in the real-world of project management, as this is a dereliction of a project manager's duties.

Neither does project management via telepathy and osmosis. Successful project management comes via pro-active bi-directional communications with all involved stakeholders.

15.4.2. "My project is my small business"

Imagine that your project is your self-funded business and that the Project Implementation Team members are the employees of your business.

As the proprietor of this small business, are you going to work for "free"?

Of course you're not going to choose to work for "free"; you're a professional organization that has to find the cash flow to pay the bills and pay its staff on time (every time).

As the project manager / proprietor, **whenever** you allow **a change to be implemented without some form of compensation being agreed to (in advance) then in essence** you are committing your business to **working for "free"**.

Thus, **even if your project / performing organization does not have a formalized Change Management Process it is still in "your business's self-interest" to undertake some form of (Baseline) Change Management**.

Example Case: What do you think I am, a charity?

Imagine that your self-funded business is an automotive paint shop.

A customer has come in requesting that you paint their car a specific blue colour.

Based on a calculated $6500 Real Cost, you quote the customer a Sale Price of $10,000 for a two week job including; stripping the existing paint, prepping the surface prior to painting, applying the undercoat, painting the car with the specified blue colour, then a few topcoats of clear, and also all of the rubbing down and buffing between layers.

A few days later, the customer comes in and asks for the job to have a few extra paint layers to give their car a richer depth of colour. You say, "no problem" as the car is still at the priming stage and painting won't start for a few days.

A few days after that, the customer again comes in and this time requests a few extra layers of clear topcoat to give their car a luscious shine. You again say, "no problem" as the car is just about to finish the first paint layer.

The car is now completely finished; your people have put in their worksheets for the job including the extra paint layers and clear coats that all up added three extra days to the job's billable time. You draw up the customer's invoice for $13,000; i.e. approximately a thousand dollars a day.

The customer comes in all bouncy and happy at the sight of their beautifully painted car. You present them the invoice; they baulk at the amount and then whinge that they only have $10,000 for the job and that you didn't explain that those extras would add more to the price. The customer's companion then comes over and chips in;

"but I wanted a different colour blue".

Now lets add some "what if" scenarios to this case study.

- What if, instead of painting the customer's car the specified blue colour it was accidently painted red due to a clerical mix-up. Won't the customer be within their rights to demand that their car be stripped back and re-painted in the correct colour, without them being charged anymore than what was originally quoted ($10K) to do the job properly in the first place (which doesn't even cover the $6500 x 2 = $13K Real Cost expended)?

- What if, due to the urgent necessity to get their car back the customer is time limited and hence they have to take their car back even though it has been painted the wrong colour; should the customer still be charged the original quoted amount, or rather a discounted amount, or does the customer get the paint job for free?

- What if, instead of painting the customer's car in the correct blue colour it was painted another shade of blue (that was as close a match as the painter could achieve) but it still was definitely a different colour to that which was specified?

- What if, the "craftsman" assigned to undertake the job took it upon themselves to include a few extra layers of paint and clear coat because the car "just didn't feel right"; should the customer have to pay for this extra work effort and materials consumed even though these were not requested by the customer nor were these instructed by the paint shop owner?

- What if, while the paint shop owner was away the customer came into the shop and asked one of the workers to include a metallic flecks to the paint to which the worker off-the-cuff replied, "ya mate she'll be right" that the customer understood as an acceptance of their change request but this was not done or would cost more?

How about, if the customer decided to send a representative to pickup the car and to pay for the paint job on their behalf.

- What if, the customer's representative accepted the work and paid for the job, but a few days later the customer comes in to complain that the car was not painted the "right" colour or that the amount paid was more than what was quoted?

- What if, the customer's representative decides to reject the paint job and insists on paying a reduced amount because in their opinion "the car was not painted correctly" even though it had been painted exactly as was originally quoted to the customer?

These additional "what if" scenarios to this example case are the types of situations that a small business owner has to deal with on a daily basis, while trying to run a profitable venture.

What makes these scenarios any different to that faced by a project manager for any SDLC project? Given that:

- The customer is going to ask for changes to be made to the project's specifications and to the deliverables.

- Mistakes and misunderstandings are going to occur.

- An implementer or few are going to occasionally stray outside the scope boundary.

- Defects will occur and these will have to be rectified.

- The customer or their representative is going to question the price that they have been charged and they will disagree on whether they should or should not accept the deliverables.

As the **"small business owner"** (i.e. the project manager) **it would be in your self-interest to have a consistent way of handling these Change Requests when there is no formalized Change Management Process in place**.

Another way of considering Change Management other than by the formalized process outline in [Section 10.2] could be as follows:

> **Request – Investigate – Quote – Present – Negotiate – Amend – Accept/Reject ...**
>
> **Address – Deliver – Judge – Negotiate – Correct – Accept/Reject ... Invoice – End.**

How would this sequence be appropriate change management for my project?

Imagine that the project you are managing is your own small business; *e.g. you are the equivalent of the automotive paint shop, a contract house painter, plumber, electrician, carpenter, or landscaper.*

Example Case: Change management as a small business operation.

1. A customer contacts your business and asks for work to be done for them, or an existing customer asks for some modification to be made to a job that you are currently doing for them; i.e. a **REQUEST**.

2. As the small business owner you **INVESTIGATE** the work opportunity.

   - You listen to what the customer has to say.

   - You write down some notes about the work that the customer is wanting done.

   - You evaluate the work location and work objectives.

   - You take down some measurements.

   - You collect any additional information that you deem necessary to successfully complete the job.

3. Out of sight of the customer, you open up your secretive worksheet / spreadsheet and calculate the **QUOTE**, taking into consideration the following factors:

   - **How much material [Resources]** will you require to complete this job? You may need to call up your local suppliers to enquire about the availability of these items, what is the [Cost], and what is the lead [Time] on delivery.

     With these details in hand you include a markup percentages to obtain the **"Sell Price"** that will be presented in the quotation to the customer. Also need to add on some safety margins to the delivery dates to cover the lead-time for each of these items.

   - **How many workers [People]** will you require to complete this job? Taking into consideration; who will be available to do the work, when will they be available, how much [Time] will they require to do the work, and what [Cost] is each one of them charging for their services.

   - You will also need to include some amount of **[Cost] for your own [Time] managing this job**.

- Oh, and lets not forget to include some amount for the **[Cost] of your business's operating overheads**.

- With all of these details in hand you can then include a **Profitability Margin** as well as some **Safety Margin** for the job's duration.

4. You **PRESENT** your quotation to the customer.

5. The customer looks over the quotation; they give it some thought, doing their own calculations on the [Time] & [Cost] as well as what they perceive will be the delivered [Scope], and their envisioned satisfaction with the expected [Quality].

6. The customer asks a few pertinent questions which you try to answer as best you can. Then the customer will probably want to **NEGOTIATE** on the price and/or the duration of the work. You come to an agreement with your customer on what is involved with the job and discuss what can be left out or additionally included for an amended price & duration.

7. Based on the outcomes of these negotiations, you include the appropriate **AMENDMENTS** to the quotation. Though, you **need to take into consideration what is necessary for this job to remain a viable venture for your business while satisfying your customer's expectations**.

8. The customer either **ACCEPTS or REJECTS** the quotation and informs you of whether they wish to proceed with the work. Only when the quotation is accepted, do you then arrange when would be the best opportunity to start the job.

9. With the job details having been finalized, you organize for the relevant [People], [Resources], and [Cost] budget, to be in place ready to start the job.

10. With the job now rolling along, you watch to ensure that **only the agreed [Scope] is being ADDRESSED**, that the **[Costs] are not getting out of hand**, that the **[Time] is not slipping away**, and that the **expected [Quality] is going to be satisfied**.

11. As the work nears completion you check that the **DELIVER**able is ready to go and

that no [Scope] has been overlooked. You ensure that all of the [People] work time is recorded, and that all of the [Resources] used have been accounted for.

12. You organize for the customer to come and inspect the work.

13. The customer looks over the work; they make their own **JUDGE**ment as to whether the agreed [Scope] has met with their "expected level of" [Quality] when taking into consideration the [Time] taken and the [Cost] involved.

14. The customer asks a few pertinent questions about the [Scope], [Time], [Cost], and [Quality] which you try to answer as best you can. Then the customer may **NEGOTIATE** for any corrective actions they feel are necessary. You come to an agreement with your customer on what [Scope] changes / corrections are necessary and what the associated compensation will be and to whom this compensation will be directed.

15. Based on these negotiations with the customer, you consider the possible **CORRECTION**s to the [Scope] and [Cost] so that your small business will still be a viable venture while satisfying the customer's expectations.

16. The customer either **ACCEPTS** or **REJECTS** the [Scope] of work delivered.
If the customer decides to reject the deliverable then some corrective actions may need to be undertaken, or the customer may take the deliverable "as-is" with some consideration given to compensation for either party involved. In a worse case it may fall to a third party (i.e. legal) to try to resolve the differences and disagreements.

17. If the customer accepts (or takes) the deliverable then you draw up an **INVOICE** for the customer to pay.

18. With the customer having (or intending) to pay, then if you have not already done so, you pay your [People] and your [Resource] suppliers. Though these [People] will most probably be paid on an ongoing basis; *e.g. weekly, fortnightly, monthly, or when their invoice is received.*

19. With the work completed you **close-out / END** that job, and move on to worrying about the next job.

Having looked through these example cases would it not be beneficial to consider the project that you are managing as a strategic job for your own small business.

If you don't manage your small business (project) in a sensible & profitable way then you will go out of business; i.e. the project's primary-core stakeholders / your senior management will relieve you of your position, show you the door... "liquidate your assets".

THE CUSTOMER IS NOT THE ONLY PRIMARY STAKEHOLDER OF THE PROJECT... THERE IS ALSO THE PERFORMING ORGANIZATION'S SENIOR MANAGEMENT WHO WILL BE CONCERNED WITH THE PROFITABILITY OF THE ENDEAVOUR.

NEED TO TAKE INTO CONSIDERATION WHAT IS NECESSARY FOR THIS ENDEAVOUR TO REMAIN A VIABLE VENTURE WHILE SATISFYING YOUR CUSTOMER'S EXPECTATIONS.

15.5. Key Performance Indicators (KPI)

For a performing organization with multiple projects and programs (all on the go at the same time), then the senior management of the organization (i.e. the primary-core stakeholders) don't have the available time to "get down amongst the weeds" and understand the inner workings of each project. Instead, what these primary-core stakeholders need is a **standardized and regimented indicators** (such as with a stock market board or ticker tape) which **summarizes the key information about the performance progress of each of the projects in their domain of responsibility.**

The collection of these **"Key Performance Indicators" (KPI)** on a **project dashboard** would thereby enable these primary-core stakeholders to have a **quick scan across the various projects and notice the warning-signs** (*e.g. stoplights*) of those individual projects that currently require their immediate attention or at least necessitate their further investigation. Please refer to [Figure 122] and [Figure 123], back in [Section 12.3.3].

Thereby, enabling these primary-core stakeholders to quickly meet the challenges of the project's current situation and "hopefully" enable them to make sound decisions in enough time to ensure the continued success of the project and the survival of the performing organization.

Subsequently, each of these Key Performance Indicators needs to be composed of quantifiable measures of a particular aspect of the project that can be used to consistently evaluate (and be historically compared) on a regular basis.

NOTE ❏ These **Key Performance Indicators (KPI) must be based on tangible and measureable objectives** such as; advancement towards specific milestone dates [Time], the current budgetary spend [Cost], the [People] & [Resources] utilization, the defect rate & clearance rates [Quality] ... etc.

❏ **KPIs cannot be based on opinion, conjecture, nor rumours.**

The thing about Key Performance Indicators (KPIs) is that **there is no one prescribed set of KPIs that must be used for the project** ... well not quite true, as the performing organization and/or the customer organization may have stipulated certain suites of KPIs that must be used for all projects under their domain.

Possible KPIs may include but not limited to the following:

- **Scorecards** for the coverage of the Requirements Traceability Matrix and/or the Work Breakdown Structure in [Section 12.1].

- The **current velocity** of the schedule and/or agile **sprint burn down**; see [Section 12.2].

- Earned Value Performance Measures, percentage complete measures, and/or performance indexes; see [Section 12.3].

- **Defect** reporting **rates**, defect resolution & closure rates, the quantities grading and the categories of the defects, the amount of Change Requests, and/or "scorecards" for the PASS : FAIL of the acceptance testing; see [Section 13.1].

- **People utilization rates**, performance **grading**, and even the amounts of unsolicited leave taken; see [Section 14.1].

- **Resource utilization rates**, production & shipping numbers; see [Section 14.2].

- **Numbers of risk & issues** raised / open / closed, stakeholder satisfaction survey numbers; see [Section 15].

As the list above demonstrates, **almost anything that can be consistently measured can be used as a Key Performance Indicator**.

THE TRICK WITH SELECTING KPIs ARE CHOOSING THOSE THAT ARE THE MOST EFFECTIVE AT UNDERSTANDING THE PROJECT'S CURRENT SITUATION GIVEN THE PREVAILING CIRCUMSTANCES.

Reviewer's Comment ... To accelerate and/or decelerate.

Projects (and performing organizations) don't exist in a situational & circumstantial vacuum; sometime a project will need to accelerate or decelerate towards re-baselined milestones due to marketplace / industrial / environmental / customer factors outside of the performing organization's control.

For example; a necessity to get to market earlier than was originally planned & scheduled, so as to have a chance to be a competitive player in an exponentially growing market, and thereby not be perceived as a "Johnny come lately" jumper onto the latest trend.

For example; a need to slowdown (halt) the project because the customer will not have the necessary infrastructure ready in a reasonable form to even contemplate being able to accept the completed deliverable. ... So in the interim, it would be better to redeploy that project's People and Resources onto other projects.

War Story ... The current situation & the prevailing circumstances.

Our company (and alas my project) just got a major kick in the guts, as the Executive Management have just now internally acknowledged that the company is experiencing a cash flow crisis due to the financial exchange source of our regular cash stream having recently been declared insolvent, thereby taking a sizeable portion of our cash reserves down with them.

Subsequently, until the next major cash injection of government grants and pending the sale of some of the owners' fixed assets, we will need to scale back our spending and re-scope all of our projects to stay within our financial means. ... While simultaneously trying to produce a revenue generating deliverable, a lot sooner than was originally planned for.

UNLESS YOU HAVE HIT A WALL AT FULL SPEED, THEN EVERYONE WILL STILL WANT THE PROJECT TO BE A SUCCESS.
SO, STAY ON COURSE AND MAKE IT WORK, BECAUSE EVEN A WRECKED PROJECT CAN BE SALVAGED IN SOME FORM.

16. Project Rescue & Project Recovery

16.1. Introduction

The project's Implementation Team has been working hard and the project's monitoring & control has been diligently performed, yet the Key Performance Indicators (**KPI**s) **have changed to RED or the performance trends are on a continual downwards slide, thereby signifying that the project is in trouble**.

With a red warning light, I have a chance to fix a problem, especially if it was amber for a while before. However, with no warning lights displayed (or with warning lights artificially set on GREEN), I could unknowingly plough into the wall at full speed.

Do you as the project's manager, mix metaphors and "turn into the skid" ...

Do you, "put peddle to the metal and power out of trouble" ...

Do you, "put your head between your legs and kiss your assets goodbye" ...

Do you, "bury your head in the sand and hope that it just goes away" ...

Do you, "sweep it under the carpet and promise to deal with it another day" ...

Do you, "hide it in the cupboard and hope that no one saw it go in there" ... or

Do you, "move it somewhere else and let some other sucker take the blame"?

Thinking back to the road trip analogy; if you were teaching your son or daughter to drive then would not the most useful piece of advice that you could give be;
"when you're in trouble ... back off the throttle, regain control, and only then smoothly reapply the power, or pull over to the side for a safe stop".

16.2. Overview

Project Rescue is emergency actions undertaken to save the project from failure due to suddenly changing circumstances. That is, analogous to a sharp jerk of the steering wheel to swerve to miss an unseen child who chased a wayward ball out onto the road. Whereas,

Project Recovery is planned actions undertaken to mitigate a pending project failure after having lost control. That is, analogous to digging the car out of a bog after skidding off the road due to that ball-chasing child.

With an appropriately timed and competently executed Project Rescue & Project Recovery, it may be possible to salvage the situation, and return the project to some resemblance of normality. Thereby, limiting the likelihood that the primary-core stakeholders will deem the project as a failure; as per the traditional determinants of [Scope], [Time], [Cost], and [Quality] defined back in [Section 1.3].

This chapter on **Project Rescue & Project Recovery is essentially a crash-course in the application of project management**; because, in order to save the project from pending failure then simultaneously all of the various process models of project management (that have previously been covered in this book) will have to be meshed together in a well-timed and competently executed manner.

That is, the Project Rescue & Project Recovery effort will involve the following:

1) The "**Project Variable Star**" model of [Scope] – [Time] – [Cost] – [Quality] – [People] – [Resources] that was introduced in [Section 1.5].

2) The "**PLAN – DO – CHECK – ACT**" model that was introduced in [Section 2.6.1].

3) The "**R.I.S.C. Management**" of the amalgamation of Risk & Issue Management, Stakeholder Management, and Change Management that was introduced in [Section 15.1.1].

And, an as yet to be explained process model;

4) A **"Four Re's"** model of Recognize – Reassess – Revise – Reapply that will be introduced in the proceeding [Section 16.3].

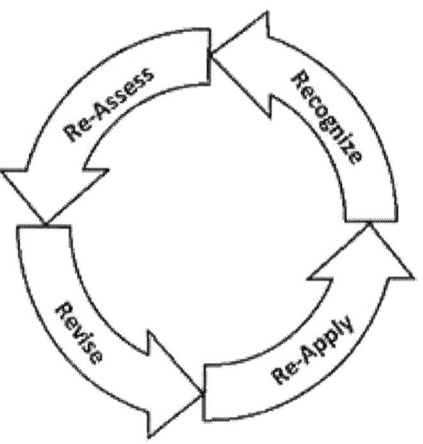

Conceptually, these various project management models will be applied simultaneously and iteratively as though overlayed on semi-synchronous spinning discs; see [Figure 137] below.

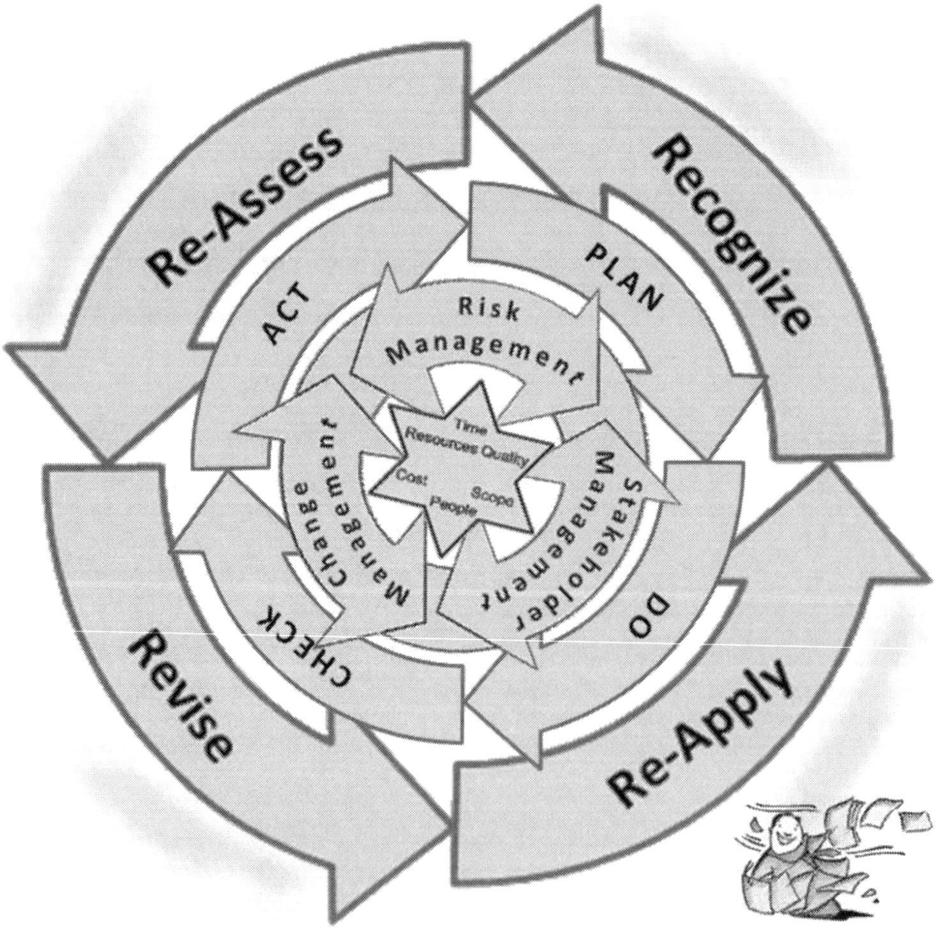

Figure 137: Interaction of the various project management models related to Project Rescue & Project Recovery.

16.3. Steps for Project Rescue

1 RECOGNIZE

Before anything can be done to resolve the problem, it must be **realized that a problem exists and that the problem is occurring** (is about to occur, or has occurred).   That is, someone must accept & acknowledge that the actual performance / result does not compare well with the expected / planned outcome; *e.g. spotting that a Key Performance Indicator has turned RED or has been AMBER for some time now.*

2 RE-ASSESS

Before being able to do something "appropriate" to resolve the problem one must; **evaluate** exactly **what happened**, determine **the cause** of the differences between what was expected to have happened and what actually happened. Recall from [Section 11.2], **"You cannot hope to manage that which you do not know or reasonably understand"**.

3 REVISE

Now that one has an understanding of what went on and what went wrong then in sequential order; **decide whether it is necessary to do something** about it, decide whether something has to be done **straight away or can it wait** for a more opportune moment, decide **what exactly has to be done**, decide **what is** the newly **expected** performance, and **revise the expected outcomes**.

4 RE-APPLY

Implement that which was decided to be done.

As reproduced below in [Figure 57] and explained previously in [Section 5.3],

the Executing Phase can be represented as a "PLAN – DO – CHECK – ACT" process model.

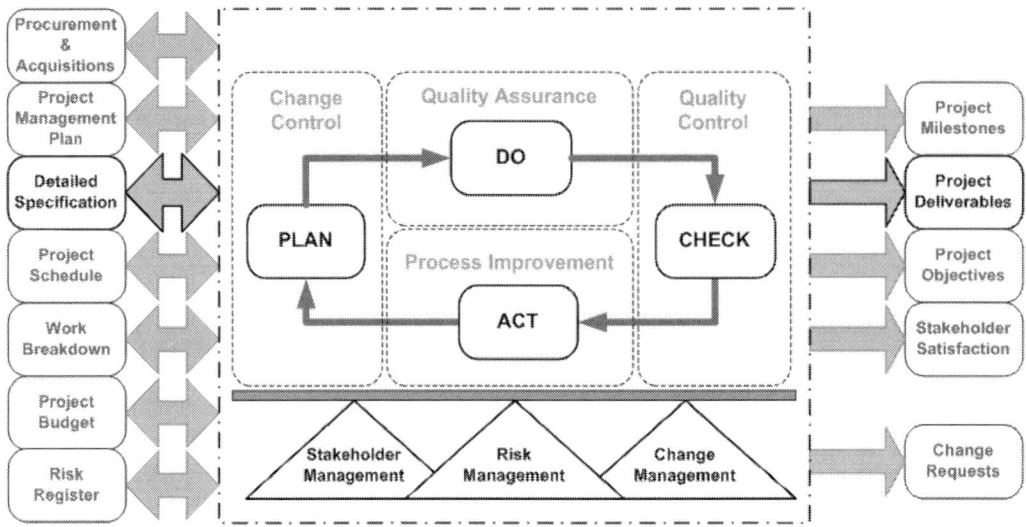

Figure 57: The Executing Phase as a "PLAN – DO – CHECK – ACT" process model with associated inputs and outputs.

It is during the Executing Phase when the project is most likely to get into trouble and

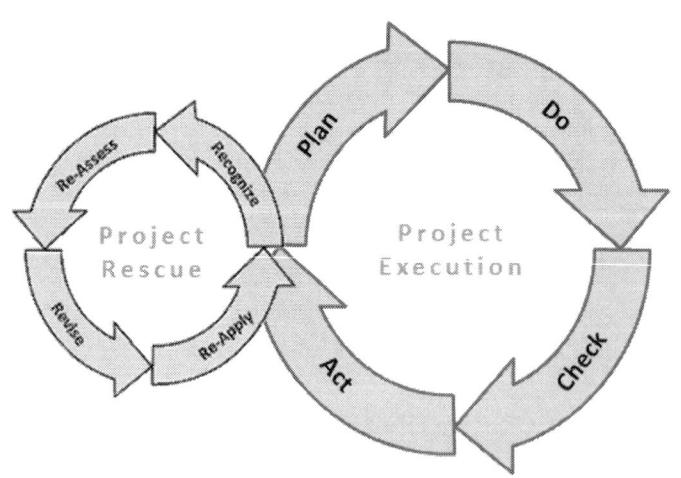

hence require some form of rescuing. Subsequently, Project Rescue is that companion gear that silently meshes with the project's Executing Phase and keeps it moving in the right direction.

Figure 138: Relationship of Project Rescue and the Executing Phase.

Ah, making sure things do not get out of whack.

16.4. Steps for Project Recovery

1 RECOGNIZE THE FACTS THAT THE PROJECT IS IN TROUBLE

Realize when the Key Performance Indicators (KPIs) are exhibiting signs that the project has deviated away from those agreed baselines that were established during the Planning Phase. That is, **detect when the project is having difficulty maintaining balance between the project's constraints and the project variables**. See [Section 1.6].

2 ACCEPT THE FACTS THAT THE PROJECT IS IN TROUBLE

That is, face the reality that failure is happening, has happened, or is imminently about to happen. **DO NOT ignore the facts** in the misguided belief that, "this can't happen to us", because "it really is happening to us" so get use to the idea that **this problem will have to be dealt with now and not tomorrow**. The quicker it is accepted that this problem is not going to sort itself out and somehow self-correct, then the sooner will be the response and subsequently the better will be the chances of recovering the project.

3 ACKNOWLEDGE THE PROBLEM TO THOSE WHO NEED TO KNOW

There is no point hiding the reality that the project is in trouble or "sugar-coating" the facts to make things appear disproportionately better than things really are.

While it may not be "politically" welcoming to be the bearer of bad tidings by **laying the project's true predicament out before the Project Steering Committee**, to not acknowledge the project's problems will only result in negative consequences (potentially financial penalties and reputational damage) for the performing organization.

With the known facts now in hand, the Project Steering Committee should decide when is an appropriate time to inform the representative of the customer organization.

NOTE ❒ **The decision to acknowledge to the customer's representative that the project is currently in a negative predicament is NOT for the project manager to decide**. This acknowledgement decision is a "strategic decision" that must be made by the Project Steering Committee in consultation with the Project Working Group and the project manager; see [Section 6.4.1].

❒ Only the Project Steering Committee should decide the "orientation" of the information that shall be communicated to the customer's representative.

Usually ... before informing the customer's representative of the project's current troubled predicament the Project Steering Committee will want to:

1) try to mount some form of **Project Rescue** "to save the day" ahead of time,

2) have some form of "authorize-able" **Recovery Plan** drawn up and ready to go,

3) have some idea of what they are going to do if the Recovery Plan does not resolve the problem; i.e. **a Plan-B and maybe a Plan-C**.

 As the project manager, **DO NOT decide to take it upon yourself to inform the customer's representative of the project's current negative situation without having firstly conferred with members of the Project Steering Committee (and the Project Working Group).**

To inform the customer's representative of the project's problems without prior consultation and approval can create undesirable tension (and sometimes unnecessary or disproportionate tensions) between the management hierarchies at the customer organization and the performing organization.

Additionally, this information can needlessly strain the working relationship between the customer's representative and the project manager. Because, the moment that the customer's representative is informed of the project's problems then that places them in a similar dilemma of how do they present these bad tidings to their own senior management, and subsequently how do they go about orchestrating the resolution of this problem?

BETTER TO FIRSTLY CHECK THE FACTS BEFORE YOU AND THEY OVER REACT

Yep, definitely do not want to "cry wolf", if in hindsight the current problem could have been kept "behind closed doors" and "resolved in-house".

4 BACK OFF THE THROTTLE AND TRY TO REGAIN CONTROL

There is no point continuing at full speed because the project's current predicament

makes it apparent that these activities are not working as well as was expected.

Might as well throttle back, think the whole situation through, and

then "**work the problem**" to regain control.

Yep ... back off, do not be afraid to brake, but do not slam on the brakes
as a first reaction because you could very well lose complete control of
the situation.

5 DO NOT BE IN SUCH A RUSH TO PERFORM C.P.R.

There can be an overwhelming desire to perform C.P.R. [Section 7.2.1] of throwing

whatever re-assignable (C)ash, (P)eople, and/or (R)esources at the problem in an attempt

to overwhelm and mitigate the failing project. Yet, often these reassignments are

not targeted specifically at the critical path tasks; rather, scattergun fired at the project,

and thereby unduly increasing the [Cost] and the complexity of the failing project.

Furthermore, there are other unfortunate side-effects of this C.P.R. approach:

1) The re-assigned Cash, People, and/or Resources have to come from somewhere

   else; often this is either from another project(s), and/or from the performing

   organization's own reserves. Consequently, other projects are negatively impacted

   upon, usually in [Time] because these other projects no longer have available the

   same quantities of Cash, People, and/or Resources that were previously assigned to

   be utilized.

 **Before mounting that project rescue and the subsequent project recovery ... STOP for a moment** and give **serious consider**ation as to whether such an undertaking could end up dragging down other projects and even possibly sinking the performing organization.

2) The performing organization may have to "play catch-up" and "fire-fight" the "knock-on effect" of having redirected much needed Cash, People, and/or Resources away from other projects towards the rescue of this failing project. Consequently, one or more of those other projects may now need to be bailed out.

3) This "firefighting" redirection could become contagious within the performing organization and subsequently cause a form of "debt crisis" where there is extensive borrowing from other projects to fix the current failing project; i.e. "robbing Peter to pay Paul". Thereby, transferring the failing condition from the rescued project to one or more of the other projects that sacrificed its Cash, People, and/or Resources; i.e. the satisfying of Paul has now also put Peter into debt. In the worst case, this can go as far as resulting in the performing organization spiralling to its own demise.

THE DROWNING PROJECT THAT PULLS DOWN ALL OTHERS THAT COME TOO CLOSE.

 6 GET A GRIP ON REALITY AND FACE THE FACTUALITY

In order to be able to recover the project, it must be understood what exactly went wrong to result in the project's current troubled predicament?

Therefore review; what was happening, what was the current situation pertaining to the project and the encompassing circumstances at the time leading up to the failure, and subsequently **identify the cause(s) of the project's failing**:

- Was it due to **misunderstandings or misinterpretations of the customer's** needs, wants, concerns, expectations, and perspectives?

- Was it due to **underestimating** the scale of the project?

- Was it due to **misconstrued requirements** or **ambiguous specifications**?

- Was it due to **unforeseen technical difficulties**?

- Was it due to the **unavailability** of the necessary **[People] and/or [Resources]** when these were planned / expected to participate?

- Was it due to **poor project team performance**?

- Was it due to **mismanagement**?

- Was it due to **circumstances beyond the project's control**?

- Was it due simply too **bad-luck**?

Now, **make a clear statement about the identified cause(s) of the project's problems**.

This statement of the facts will most probably be as an update to the members of the Project Steering Committee, though it will definitely need to be presented to the other members of the Project Working Group (and when appropriate, to the Project Implementation Team).

7 A PLAN, A PLAN, A RESCUE PLAN

Based on the information collected so far, as it currently stands is it apparent **what things can be done right now** to rescue this failing project?

Can these things be successfully **done without the need to involve the customer's representative**?

Is it possible to **come up with a reasonable & sensible sounding Recovery Plan** that when presented to the customer's representative won't result in them panicking about the current state of the project?

NOTE ☐ At this stage, **no commitment has been made by the Project Steering Committee to inform the customer's representative of the project's problems**, because there may still be opportunities to rescue the project (behind the scene) and therefore **not needlessly traumatising the customer's representative**.

Recall back in [Section 13.1.3] that,

HALF THE BATTLE FOR PROJECT SUCCESS IS DEALING WITH WHAT HAS TO BE ADDRESSED.

8 THE RECOVERY WILL REQUIRE CHANGES TO THOSE EXISTING BASELINES

Hooray, it is now apparent that the project can still be saved.

In order for the project manager and the Project Working Group **to be able to come up with a viable Recovery Plan** (or alternate recovery plans), **then one or more of the existing project baselines** (i.e. the project constraints) **are going to have to be changed** and possibly changed dramatically.

These changes will most probably mean that either **the traditional determinant baselines of [Scope], [Time], and/or [Cost] will have to be restructured; else [Quality] will bear a disproportionate burden** as the project continues to fail.

In order to have a chance of recovering the project, **it must be accepted that those existing baselines are now unrealistic to achieve. To continue to insist on the conformance with all of those existing project constraints / baselines is a recipe for disaster**, and to continue to place faith in these baselines being achieved is opening oneself up to major disappointments and ripple-effect problems.

Therefore, **the project ~~constraints~~ variables of [Scope], [Time], [Cost], [Quality], [People], and [Resources] will all need to be reassessed, and subsequently "re-baselined" if the project is to have any chance of being successfully delivered.**

That is, the undertaking of an accelerated retrospective Initiating Phase and a circumspect Planning Phase, as described in [Chapter 3] and [Chapter 4], with consideration given to the specifics of the project's current situation and the prevailing circumstances.

Firstly, the members of the Project Steering Committee must be conferred with in order to establish whether, there **are any project constraints that the performing organization needs to self-impose on the project**. These self-imposed project constraints are due to the reality that, most probably by the time of the Executing Phase the performing organization has made other commitments and has other obligations that limit the flexibility of the performing organization when it comes to redefining these project variables. Therefore, are there any self-imposed restrictions on:

1) **What [Time] remains** to complete the project?

2) **What [Cost] budget is available** to complete the project?

3) **What [People] are available** to complete the project?

4) **What [Resources] are available** to complete the project?

With these self-imposed project constraints having been defined, then based on the existing relationship with the customer organization and that customer's known needs, wants, concerns, expectations, and perspectives then determine what other prior project constraints are and are not feasible to be adjusted:

- **What [Scope]** is 'a must have', 'a would be nice to have', 'a could do without', and 'a can do without'? That is, **prioritize** all of the outstanding / incomplete [Scope].

- **What [Time]** must the project definitely be delivered by or ELSE?

- **What [Cost]** budget must the project stay under to remain viable?

- **What [Quality]** acceptance criteria must the project satisfy in order for the deliverables and associated artefacts to be accepted?

- **What [People]** will be available to undertake the remainder of the project? Additionally, what skilled [People] will be required to implement the selected [Scope]?

- **What [Resources]** will be available for the remainder of the project? Additionally, what quantities and grade of these [Resources] will be required to implement the selected [Scope]?

9

TO DO OR NOT TO DO, NOW THAT IS THE QUESTION

Based on the information obtained from the previous steps:

- ☑ **Can this project still be "successfully" completed**; either 'as is' or in some modified form?

- ☑ **Does undertaking this project still make sense** to the performing organization by providing value to the business and to its owners / shareholders?

- ☑ **Would this project still be considered a worthwhile endeavour** for the customer organization when the desired outcomes and the perceived benefits are taken into consideration ... in light of the project's current situation and given the prevailing circumstances?

Therefore, should the **proposed recommendation** to the project's primary-core stakeholders (i.e. the project sponsor / customer's representative) be that the project should:

a) **proceed as it currently is,**

b) **revise the project constraints,**

c) **recover only a portion of the project,**

d) **restart the project as a "do-over",** or

e) **terminate the project**?

The proposed recommendations (along with the draft Recovery Plans) should be laid out before **the Project Steering Committee,** so that they can **pass judgement** on the **direction of the project's recovery or its planned demise**.

10 AGREE ON THE TERMS OF ACTION OR NEGOTIATE THE TERMS OF SURRENDER

With the Project Steering Committee having decided on what the performing organization should do next (i.e. "to do or not to do") then the customer's representative should be notified of the project's current predicament and informed about what are the performing organization's intensions to recover the situation.

 I would recommend that ... the project manager should volunteer to inform the customer's representative, as this will demonstrate that they are taking responsibility for the project's current predicament and also taking responsibility for recovering the situation.

The project manager and the customer's representative(s), as well as appropriate members of the Project Working Group and selective members of the Project Steering Committee need to **sit down together and work out** the details of the "**Terms Of Action**" (i.e. **what conceptually is to be done**), or come to an **amicable resolution** on how to terminate the project (or that portion of the project).

NOTE   ❏ It is essential that these "**Terms of Action**" be agreed to and accepted prior to the establishment of the Project Recovery Plan (or the Termination Plan).

Thus, the "**Terms Of Action**" is the equivalent of t**he Project Charter for the project's recovery**, though the Terms Of Action may simply be an email exchange that culminates in an acceptance by both organizations of the actions agreed / approved to be taken.

If **negotiating the "Terms Of Surrender"** then the discussion **must involve members of the Project Steering Committee**, as they will need to **deal with any litigation, liquidated damages,** and **performance guarantees**. That is, these project termination negotiations may exceed the responsibilities of the involved project manager.

11 THE PLAN, THE PLAN, THE RECOVERY PLAN

With the customer organization and the performing organization having reached an agreement on the "Terms Of Action", then the next step is to do an accelerated pass through a mini Planning Phase as outlined previously in [Chapter 4]:

1) **What portions** of the existing project's planning **can be reused**?

2) **What is the new scope boundary** of this project?

   That is, what **[Scope] can be realistically included** given any customer and self-imposed constraints on [Time], [Cost], [People], and/or [Resources]?

3) **What is the priority for implement**ing these selected [Scope] items?

   May I suggest that, you look at the Agile topic in [Section 7.4] as this could be beneficial when determining the Scope to be included. Though, do not go changing the existing implementation method (just go borrow some ideas) as changing the existing SDLC implementation method can cause other problems, and can give the project stakeholders something convenient to blame if the project does not improve or gets itself into trouble again.

   However, when a project gets into trouble, many an organization will go rigid waterfall and try to micromanage their way clear of the problem ... potentially strangling the project team's already declining morale.

4) Based on the revised project [Scope] then **what portions of the existing implementation can be reused**?

   Definitely discuss this over with the members of the Project Implementation Team as they may come up with some ingenious solutions to recover the situation.

5) Based on the [Scope] and the work that can be reused then **redo the project's schedule to determine realistic [Times] for the revised project milestones**.

The members of the Project Team will need to be involved with providing realistic estimates for the work to be undertaken; else, they may not **"buy-in"** on the revised project schedule. Such a lack of buy in with the recovery plan can result in another project failure (soon after the "rebooted" project commences).

6) Incorporate into this **revised schedule the allocated [People] and [Resources]**.

Ensure that other managers are consulted on the re-assignment of [People] & [Resources], because you do not want to be inadvertently sinking other projects (and BAU activities) while trying to recover this particular problem project.

7) Determine the **revised [Costs]** for this problem project.

Rework & revise all of the above until a realistic & reasonable Project Recovery Plan is achieved; i.e. one that is acceptable to the customer's representative, the Project Steering Committee, the Project Working Group, and the Project Implementation Team.

DO NOT forgo (necessary) **documentation and/or testing due to the need to speed up the project's recovery, as this will only result in future failings of the project** (especially during acceptance testing and the receipt of the deliverables).

12 NEGOTIATE THE REVISED PROJECT PLAN

Now that the internally approved Project Recovery Plan has been produced, it is time to sit down with the customer's representative(s) and negotiate on the proposed Project Recovery Plan, its subsequent acceptance, and its eventual implementation.

This interaction is often a **compromise negotiation** held over a few meetings, during which both sides work out exactly what they "can and cannot live without". As a result of these negotiations, the project manager and the Project Working Group may need to go back to an earlier step, reassess, and revise the Project Recovery Plan.

Hopefully, some agreement can be reached in a reasonable amount of time.

NOTE   ☐ **DO NOT immediately commit to the acceptance of the Project Recovery Plan**, because it is highly likely that the customer's representative(s) will want other reevaluated changes to be incorporated into the revised project baselines.

The customer's representative(s) will probably try to retain more of those pre-existing project constraints / baselines than was agreed to during the discussion of the "Terms Of Action".

The project manager will need to have at hand the "Terms Of Action", and thereby remind the customer's representative(s) of what was agreed to in principle. Also, the project manager will need to point out what can and cannot be realistically achieved given the constraints imposed by the project's current situation and the prevailing circumstances (while privately keeping in mind those self-imposed project constraints).

Any revisions during this negotiation will **need to** be **internally reviewed before being agreed upon. To commit to such changes without re-assessing the potential impact** on the revised plan **is just setting the project up for a future failure.**

---

13 REACH AN AGREEMENT TO COMMIT TO THE REVISED PROJECT PLAN

Once the representatives of the customer and performing organizations have come to a mutual agreement on the revised Project Recovery Plan, then a **Baseline Change Request (BCR)** document will be **signed-off to signify the authorization of the revised baselines**.

That is, authorization for the agreed revised project schedule [Time Baseline], the agreed revised project budget [Cost Baseline], the agreed revised Detailed Specifications [Scope Baseline], the agreed revised Acceptance Criteria [Quality Baseline], and/or (if set as project constraints then) the agreed allocation of [People Baseline] & [Resources Baseline], **will henceforth supersede (invalidate) any pre-existing baselines.**

 **The fact that there are now approved baseline revisions for the project must be made known to all of the project's stakeholders.** It must be confirmed that the project's stakeholders acknowledge the existence of these revised baselines; else, **the project could get itself into another troubled situation with some of the stakeholders working to one set of previous baselines while others work to another set of revised baselines.**

14 GET WITH THE PROGRAM, AND EXECUTE THE RECOVERY PLAN

Finally, with the sign-off of the Baseline Change Request (i.e. those authorized revised baselines), it is time for the Project Implementation Team to get back to working at full steam and recommence the Executing Phase in earnest with respect to the "all new and approved" baselines.

There should be an independent **mini kick-off meeting or stand-up meeting** held with the Project Implementation Team and the Project Working Group (and if need be with the Project Steering Committee) to inform them of:

1) the project's change of direction,

2) what the new baselines are, and

3) how they will be affected by these changes.

 **Watch out for project stakeholders that try to get a "scrapped" Scope component reintroduced after the Project Recovery Plan and/or Baseline Change Request has been signed-off and authorized.**

To allow these de-scoped components back in would constitute scope creep and can result in the project getting into trouble again.

---

15 ACTIVELY WATCH OUT FOR THE REOCCURRENCE OF THE PROBLEM

Continue to monitor the project and specifically for those causes that derailed the project the first time, as these problems may arise again; see [Chapter 12].

Reviewer's Comment ... Adjust course, but stay on target.

In my experience, unless you have hit a wall at full speed, everyone still wants the project to be a success, so the objective is to make it work. Therefore, stay on target, even if the course has to be "adjusted".

Don't get me wrong ... if the project has been totalled, then trying to bring it back to life is just a waste of effort, as it should be put down if it is lingering in pain. However, the trick is knowing which ones to save and those that need to die.

War Story ... Good news is we rescued the project, the bad news

I once inherited a project that had run aground. The executive made it clear that this project could not be allowed to flounder, as there was a multi-million dollar investment in its outcome. Hence, my role was to get the project off the beach and back on course, which we (the project team) managed to do. However, the executive who had presided over the project when it crashed wanted it back on schedule as well. Thus, as he vigorously insisted, we ramped the project up to full speed, but were never able to recover the lost time, and by trying to do so, the budget was blown.

To use a racing analogy; if the project crashes into a barrier then forget about winning the race, as the best that can be hoped for is to finish, and by finishing then maybe just maybe pick-up a minor placing.

War Story ... The solution, found by "out of the box" thinking.

There was this small company that had a major problem with finding security cleared electrical sub-contractors to do these irregular bespoke installations of their product. The newly hired project manager's "out of the box thinking" solution was to go to the navy base and "just sit in the car park", note down the logos on the tradesman's shirts going into & out of the base, then simply telephone those identified companies until he found a firm that could & would do these installations on the company's behalf.

JUST BECAUSE THE FINISH LINE HAS
BEEN CROSSED DOES NOT NECESSARILY
MEAN THAT THE RACE IS OVER ...
AS THERE IS STILL THE
AFTER RACE SCRUTINEERING.

17. CLOSING Phase

17.1. Introduction

WELL, THAT IS IT FOLKS ...
YOU CAN STOP RIGHT THERE.

PLEASE SPEND THE REST OF TODAY WRAPPING UP ...
DOCUMENT WHAT YOU CAN, THEN ARCHIVE IT AWAY ...
THERE ARE NEW HORIZONS STARTING TOMORROW.

OH, BY THE WAY ... IT IS WITH REGRET THAT
MR BUNNY-MAN HAS DECIDED TO LEAVE US TO PURSUE
OTHER OPPORTUNITIES AND HENCE HE HAS
DEPARTED IMMEDIATELY.

This is a pathetic way to end.

After you have spent days, weeks, months reading this far, and now you find out that someone has decided to pull the plug and end it all so abruptly.   *... Oh, how rude.*

I kinda feel ripped-off.

Well, it is surprising the number of projects that end exactly this way ... suddenly,

as though shot-down in some dark alleyway by senior management gangsters who then unceremoniously bury the evidence out of sight. Alternatively, the project slowly fades away as person after person is requisitioned off to work on other activities. What an inglorious death, but alas one that occurs all too often for SDLC projects in the real-world.

Example Case: Do not worry, two thirds of projects around here get canned.

Once upon a time, there was a prominent telecommunications development organization, and a young bright-eyed & bushy-tailed software engineer was employed to work on one of their latest development projects.

A year into the project, the organization's executive management decided "overnight" to cancel this project, because they had acquired a smaller firm who happened to be building a similar product (and they were further advanced in their development).

This young engineer felt devastated, as if personally responsible for the project's failure.

> *"If only I had worked harder, then the project would have succeeded."*

To which a battle-worn senior engineer put it into perspective with an off-the-cuff remark...

> *"Don't worry, two thirds of projects around here get canned"* ...
> *"it is seen as a stern management decision"* ...
> *"so don't take it personally, it definitely ain't your fault".*

You honestly have to wonder, how could so many of these projects be destine to fail?

How could project after project end so tragically?

More importantly, how could so many projects be allowed to start, to only end in

FAILURE?   *... Doesn't anyone give a damn?*

Why do projects end?

A project can be closed out because:

1) **The project successfully achieved its objectives**; i.e. the project's deliverables were handed over to the customer with the agreed [Scope], by the agreed [Time], within the agreed [Cost] budget, and has the agreed level of [Quality] to satisfy the customer.

2) **The project failed to achieve its objectives** or at the current rate of progress, the project would take too long or be too costly to eventually achieve its objectives**.**

3) **The project never succeeded or failed because it is no longer required**;
   i.e. the objectives are no longer feasible given the current situation and the prevailing circumstances. *For example; a decline in the economy, a change in the targeted market segment, or a change in the business's direction.*

Just as with the project's other phases, the **Closing Phase needs to be approached with a structured mentality**, and **not** just **muddled through** in an **ad-hoc** manner ... as though the project was a never-ending story.

I would recommend that ... irrespective of whether the project is going to end as a deemed success or as a failure, **the Closing Phase should be considered as its own project and not thought of as the wind down of the current project**.

This is because, it is very easy to rush to get this phase over & done with so as to move on to the next paying job; rather than ensuring that all of this current project's loose ends are tied off satisfactorily.

17.2. Overview

As illustrated below in [Figure 139], the Closing Phase is concerned with the darker shaded area of the Project Management Process model.

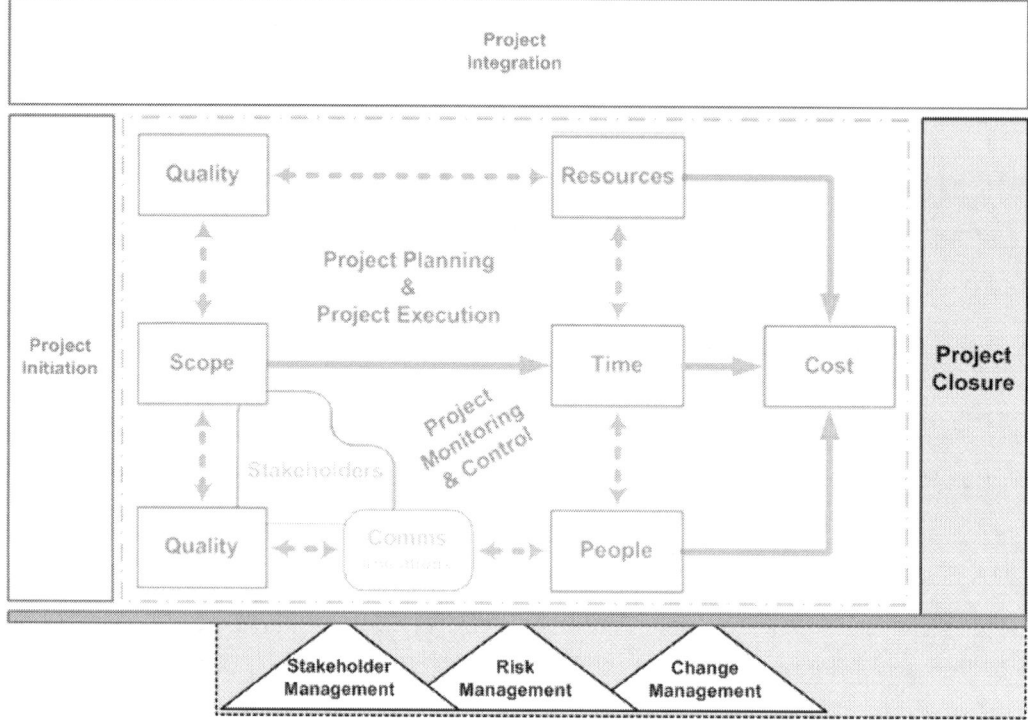

Figure 139: The Closing Phase of the Project Management Process model.

NOTICE how in [Figure 139] the Closing Phase still involves aspects of Stakeholder Management, Risk Management, and Change Management. This is because, just as first impressions are important to establishing the relationship between the project's stakeholders, the impression that these stakeholders obtain during the Closing Phase is essential to the project being deemed a success or a failure. *(As well as forming the foundation for any future relationship).* Therefore, the **project's Closure Phase requires just as much management attention as did the Initiation Phase and the Planning Phase.**

17.3. Purpose

The purpose of the **Closing Phase** is to perform those activities necessary to **officially conclude the project or to culminate the current milestone release**.

That is, the Closing Phase endeavours to:

1) **Wrap up the** project's **Executing Phase** (in entirety or just for the current release).

2) Ensure that the project's **[People] and [Resources] are released** to other activities.

3) Ensure that, in a logical sequence, all of the **project stakeholders** are **communicated with**; to inform them that the project is ending, stating why the project is ending, and from their perspectives when is the project to be concluded.

4) Ensure that the **specified & agreed deliverables are handed over** to the appropriate project stakeholders, and **list any outstanding deliverables**. This also includes the availability of the associated **Objective Quality Evidence (OQE)**.

5) Ensure that these **deliverables are officially accepted** by the relevant project stakeholders, and **list any deviations from the contracted deliverables**.

6) Ensure that the **administrative functions are closed out**; i.e. to finalize the financial accounts, legal obligations, quality evidence, project reports & records, project documentation, intellectual property, and archiving the project's artefacts.

Alternatively, the project Closing Phase is to answer the following questions:

1. Did **WE GET IT RIGHT**?

2. Does **EVERYONE AGREE** it is **ALL DONE**?

3. Let's say **GOOD BYE**... as friends.

4. What did **WE LEARN** from this?

17.4.   The Big Questions

17.4.1.   Question: "Did WE GET IT RIGHT?"

As stated back in [Section 1.3] and listed below, there are four traditional determinates for judging a project as a success or a failure. Accordingly, to be able to determine "Did WE GET IT RIGHT" the following questions need to be answered:

1) **Do all of the project's deliverables conform to the agreed [Scope Baseline]** as was defined in the approved Detailed Specifications (and those approved Change Requests) that were signed-off by the representatives of both the customer and performing organizations?

2) **Have these deliverables been provided within the agreed [Time Baseline]?**
    That is, before or after the milestone due-dates, as was laid out in the signed-off master project schedule / milestone list.

3) **Have these deliverables been produced within the approved [Cost Baseline]?**
    That is, for more or less than the amount specified in the authorized budget.

4) **Do these deliverables and artefacts conform to the agreed [Quality Baseline]** as prescribed in the signed-off Acceptance Criteria?

5) Have these deliverables and their associated artefacts been handed over to the relevant recipients?

Given that, the determination of whether the project is considered a success or a failure comes down to an examination of these project constraints & variables, then wouldn't it be advantageous to treat the Closing Phase as a work-through of each one of these project parameters?

Where, the examination of these project constraints & variables could be undertaken in much the same way as was done for the Planning Phase, back in [Chapter 4].

That is, as a walk-through of the implementation of the entire project to ascertain whether, what was planned to be done was in fact what was actually done.

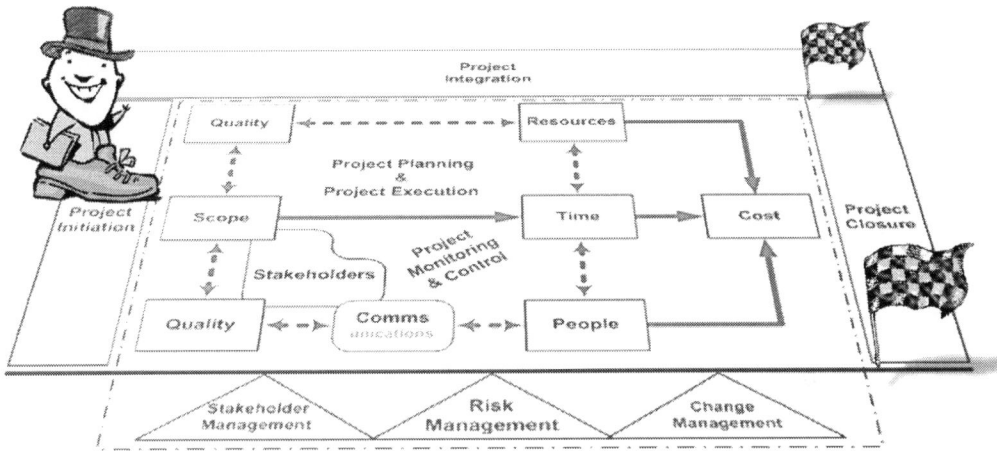

Hmm, you could consider the Closing Phase as a project unto itself.

17.4.1.1. [Scope] Verification & [Quality] Deliverables

The first of these project constraints & variables to be examined is the [Scope] and the [Quality]. More specifically, the first step towards the project's official conclusion is the **verification that the agreed [Scope] has been covered, and that the project deliverables & associated project artefacts have been supplied with the agreed level of [Quality].**

Hmm, interesting agreed [Scope] and agreed level of [Quality].

Yes, that is correct, the **Closing Phase is a confirmation that what was "agreed to" is in fact what was delivered.**

$$\underset{\text{SATISFACTION}}{\text{CUSTOMER}} = \text{QUALITY} + \text{SCOPE}$$

From [Section 1.3], the **combination of [Scope] and [Quality]** can be thought of as "**customer satisfaction**".

Shouldn't the objectives of the Closing Phase be to ensure that the deliverables are fit for use, and to confirm that the customer's needs have been met?

NO ... the **satisfying** of **the customer's needs (and wants), or** more specifically the **confirmation that the deliverables are "fit-for-use" are NOT the objectives of the Closing Phase** ... nor were these the objectives of the project.

Recall back in [Section 8.1] the warning that;

> "**Quality is NOT about** producing project deliverables that **precisely meet the customer's needs & wants.**" ... rather,

> "**Quality is delivering exactly what was agreed to** between the performing organization and the customer's representative(s) as recorded in the signed-off / approved Detailed Specifications."

Also, recall back in [Section 7.1] the warning that;

> "Be wary of this demarcation between [Scope] and [Quality], as there has been many a disagreement between the customer organization and the performing organization due to a deliverable conforming to the specifications yet it did not conform to the purpose for which the customer had intended."

PAUSE HERE... I would like to refocus this current line of thought onto similar 'big' questions that were covered during the Initiating Phase, see [Section 3.3].

Recall that, part of the purpose of the Initiating Phase was to answer;

"*What do THEY WANT?*" and "*What will WE GIVE them?*"

Well, it is **the Closing Phase where the project's primary-core stakeholders find out exactly how effectively the project manager and the project team have understood and interpreted what the customer was wanting**.

"SIMILAR" IS NOT CLOSE ENOUGH TO THE "SAME", AS IT WILL ONLY END IN FLAMES AND BLAME.

This current step in the Closing Phase is questioning, "Did WE GET IT RIGHT" and in essence, did the performing organization cover the agreed [Scope] and deliver to the agreed level of [Quality]. **Because the Closing Phase is when those differences in understanding & interpretation will definitely manifest themselves as disagreements between the representatives of the customer organization and the performing organization.**

IF WE ARE THINKING THE "SAME", THERE SHOULD BE NO PAIN.

Requirements Traceability Matrix (RTM)

An effective way of **record**ing **the coverage of the [Scope Baseline] that was delivered** is to tick-off each agreed component of [Scope] as it is incorporated into the project's deliverables. The **Requirements Traceability Matrix (RTM)** is the perfect vehicle for this purpose, as colour coding can be used to emphasize **what functionality has & has not been included** in each release. For more information on the RTM, refer [Section 12.1.1].

BR ID No.	Title	Short Description	Category	Overall Priority	Release Priority	Scenario	FS ID No.	Release	TS ID No.	Comments	Status
BR001	channel receiver	receive differentiated signals	System Input	5 (high)	5 (high)		FS001	RES_01	TS011		DONE
BR002	video display	display moving images	Video output	5	5		FS002	RES-01	TS012		DONE
BR003	play audio	play audio sound	Audio output	5	5		FS003	RES-01	TS013		DONE
BR004	channel selector	switch between channels	User input	5	3		FS014	RES-03	TS031		TBD
BR005	volume changer	volume change up & down	User input	3	4		FS004	RES-03	TS032		WIP
BR006	brightness change	brightness change up & down	User input	5	2		FS005	RES-03	TS033		WIP
BR007	contrast changer	contrast change up & down	User input	5	1		FS006	RES-03	TS034		WIP
BR008	Antenna	external antenna connection	System Input	5	3 (medium)		FS015	RES_01	TS010		DONE
BR010	loud speaker	front panel speakers	Audio output	5	4		FS016	RES_02	TS020		DONE
BR011	black & white dis	black & white display	Video output	5	5		FS017	RES_01	TS010		DONE
BR012	coloured display	colour display	Video output	4	5		FS018	RES_04	TS040		TBD
BR013	hand-held remote	user hand-held controls	User input				NA	NA	NA		OOS
BR014	on screen menu	user menus on screen	Video output				NA	NA	NA		OOS
BR015	auxilary port	auxilary input	System Input				NA	NA	NA		OOS
BR015	auxilary port	auxilary output	System Output				NA	NA	NA		OOS

Figure 101: "Candy-striped" Requirements Traceability Matrix (RTM).

Based on the examination of the RTM, it should be a relatively simple matter to determine what functionality has been implemented, what functionality is yet to be implemented, and what functionality was agreed to have been out-of-scope for the current milestone release.

Hence, with this clear understanding of what is in, what is out, and what is still lying about, then it should be possible to come to a mutual agreement between the representatives of both the performing and customer organizations on:

- **Whether or not to proceed** to the next stage with the functionality that currently is,

- **How to handle any concessions** for functionality that was not as agreed, and

- **Should compensation be forthcoming** to any party that was affected by changes to the included functionality?

Testing Coverage Matrix (TCM)

While the Requirements Traceability Matrix (RTM) is very useful at highlighting the scope coverage, the RTM is not usually designed to also indicate whether that functionality was implemented correctly or not. A **Testing Coverage Matrix (TCM)** is usually crafted specially for the purpose of **recording the PASS : FAIL [Quality] of the functionality that was delivered**.

As an aside, different industries will use various names for this "testing centric matrix"; *e.g. Testing Traceability Matrix (TTM), Verification Cross Reference Matrix (VCRM), or Requirements Traceability Verification Matrix (RTVM).* Irrespective of whatever this Testing Coverage Matrix is called, its **purpose is to indicate that functionality that has been verified as operating in accordance with the agreed Acceptance Criteria and/or the approved Detailed Specifications, that functionality that has been found to be not conforming, and that functionality that has not yet been tested.**

This Testing Coverage Matrix would contain;

1) **references to test cases and test scenarios** (in the Acceptance Test Plans) against which each agreed component of the [Scope] would be verified,

2) **cross-references to** the relevant points and **sections of the approved Detailed Specifications** (possibly with a reference to the corresponding entry in the Requirements Traceability Matrix),

3) **PASS : FAIL indicator** of the functionality tested, and

4) potentially **references to the associated Defect Reports**, Concessions, and Waivers.

 **DO NOT use the Testing Coverage Matrix to document the reasons why a test did not pass**, as this is the purpose of each individual Defect Report.

Relationship of the RTM to the TCM

As depicted below in [Figure 140] that was derived from [Figure 79] in [Section 8.1].

❖ **Requirements Traceability Matrix (RTM) records whether the individual requirements contained within the approved Detailed Specifications have been included in the project deliverables**.

❖ **Testing Coverage Matrix (TCM) records the PASS : FAIL confirmation of each requirement when compared with the agreed Acceptance Criteria**.

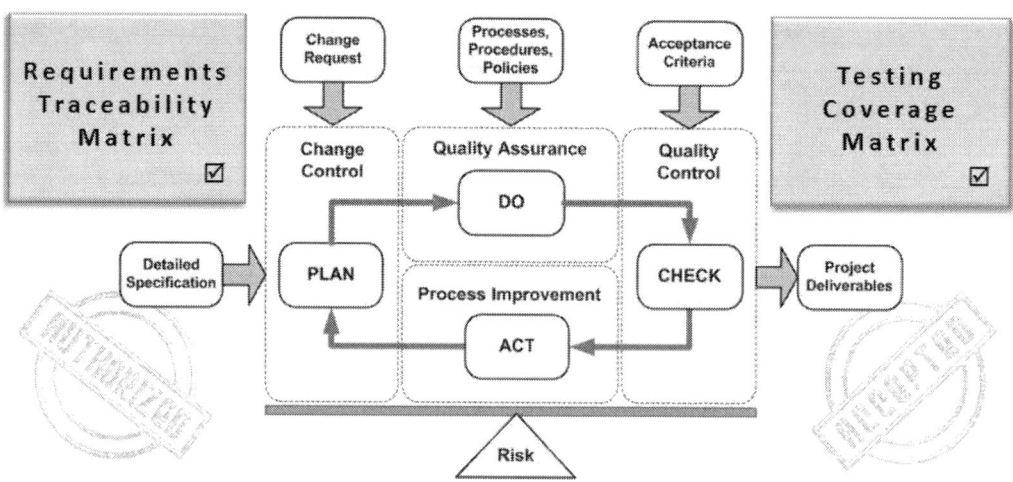

Figure 140: Relationship of the Requirements Traceability Matrix (RTM) to the Testing Coverage Matrix (TCM) to the "PLAN-DO-CHECK-ACT" model of Quality Assurance, Quality Control, Process Improvement, and Change Control.

In essence the RTM and TCM are the records of the inputs and outputs to the Quality Assurance Processes and the Quality Control Procedures used to determine the "correctness" of the project deliverables that were agreed to. This "correctness" is subsequently determined via the undertaking of a sequence of acceptance tests.

Acceptance Test Suites

Recall back in [Section 13.1] where the "Functional" Acceptance Test (FAT), "Satisfaction" Acceptance Test (SAT), and the "Usability" Acceptance Test (UAT) formed the final sequence of quality control checks. As reproduced below in [Figure 127], it is the suite of these **acceptance tests** that will **confirm whether the approved [Scope Baseline] and the agreed Acceptance Criteria have been met**; i.e. **"Scope Verification'**.

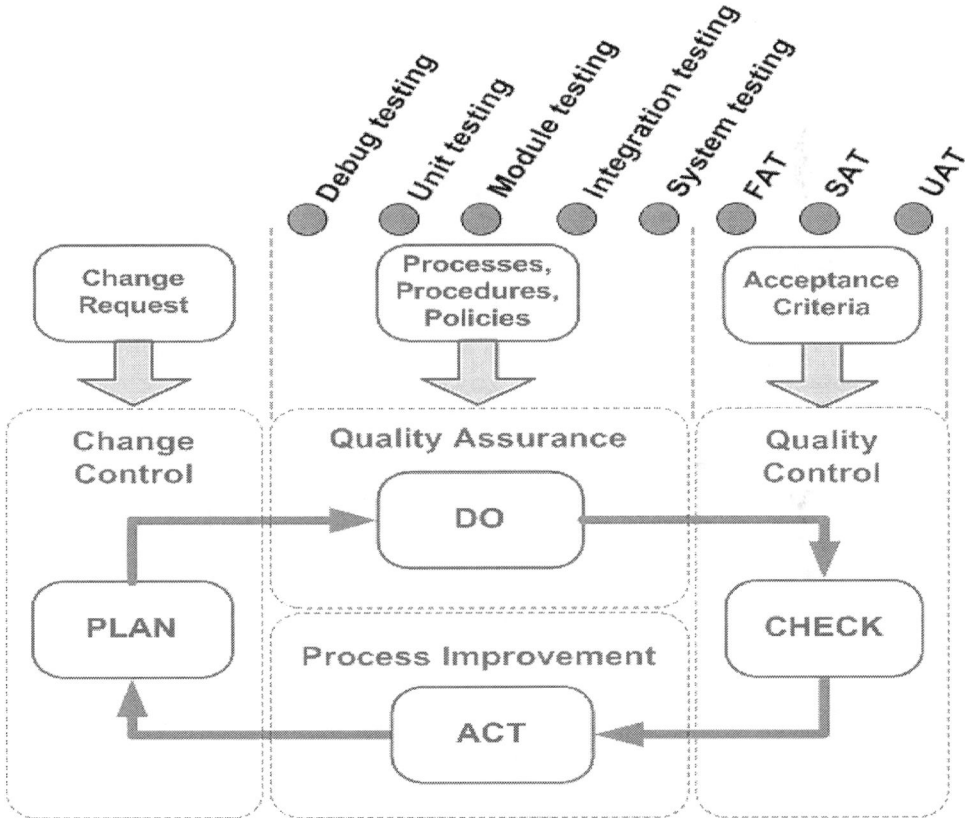

Figure 127: Relationship of testing to Quality Assurance and Quality Control.

NOTE ☐ These **acceptance tests DO NOT confirm that what the customer "meant" has in fact been "satisfied"**; rather, these acceptance tests **only confirm that the deliverables do or don't conform to what was agreed to in the signed-off Detailed Specifications (and NOT against the Customer Requirements).**

NOTE   ☐ During these acceptance tests, it could be found that **a deliverable does in fact confirm with the agreed specifications yet it does not correspond with what the customer subsequently "wanted"**. When faced with such a situation, **a Change Request must be raised and not the generation of a Defect Report**.

[Scope] & [Quality] Concluding Thought

This step in the Closing Phase should have verified that the agreed [Scope Baseline] has been included in the deliverables, and that the delivered [Scope] is of such [Quality] that it does comply with that functionality defined in the approved Detailed Specifications and those authorized Change Requests.

However, if that **functionality is not as specified then a Defect Report should be raised, but if the functionality is not as the customer intended then a Change Request should be generated**.

The differentiators of what constitutes a defect versus a change needs to be established prior to proceeding into the Acceptance Testing, as a significant amount of [Time] could be consumed sorting out the piles of Defect Reports and Change Requests.

 For more information on Change Management, Change Requests, and Defect Reports, please refer to [Section 10.2] and [Section 15.4].

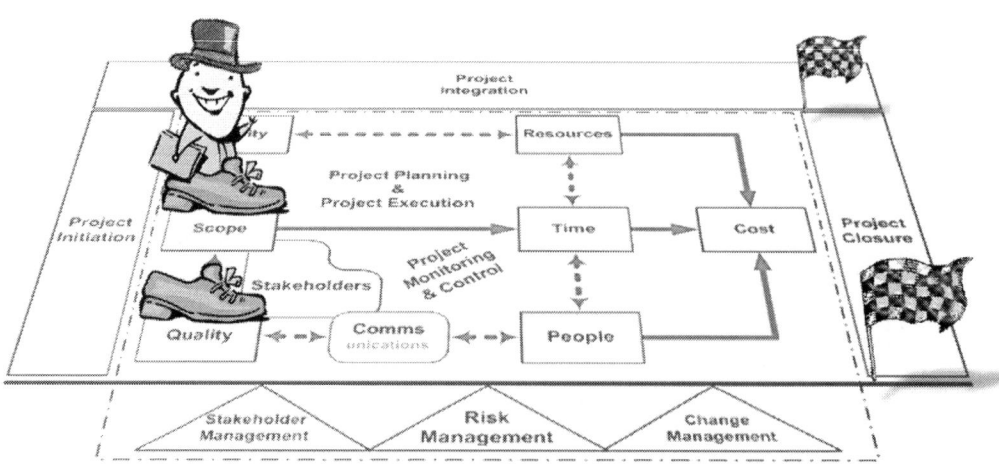

17.4.1.2. [People] and Communications

With the project or the milestone release coming to an end, then the involved [People] and [Resource] providers will have to be communicated with, so as to notify them of whether their services will be required any more, and also to inform them of what is expected from them during the conclusion of their participation.

Additionally, the project's primary-core stakeholders, secondary-strategic stakeholders, (and environmental stakeholders if they are affected by the outcomes) should also be informed of the project's (or the release's) concluding status.

I would recommend that ... the remaining members of the project team should be gathered together for a kickoff meeting specifically targetted at the Closing Phase of the project / release.

During this 'closure kickoff' meeting the project manager should outline the following:

- What will occur for the remainder of the project / release?
- Who will be responsible for what closing activities?
- What is the expected duration for the Closure Phase?

The project manager should also encourage the project team members to:

1) Come an talk if they have any concerns about the Closing Phase and/or what they will be doing after their project participation has ended.

2) Encourage them to remain focused on completing the job at hand as effectively and efficiently as possible.

Let's not lose concentration ... not let the game slip away in the final minutes, be beaten by the bell.

Or as the saying goes, "it aint over till the fat lady sings". And, I don't hear her warming up yet.

Third Party [People]

If the services of a **Third Party** (i.e. **external to both the performing and customer organizations**) will no longer be required then:

1) Any **timesheets and financial claims** (for services rendered / goods supplied) that the Third Party has outstanding will need to be **submitted for approval** by an appropriate representative of the performing organization (or in special cases by the customer organization).

The project manager should actively pursue the timely submission of these outstanding timesheets and any associated expense claims / invoices to ensure that these are logged and processed well in advance of the project's financial accounts being closed out. Thereby, guaranteeing that there are **no residual claims against the project by the time that the final invoice is submitted to the customer organization.**

Outstanding residual claims DO NOT induce a very good impression of the project manager's competence.

Especially after the project's account codes have been closed out.

Any residual claims, especially those discovered after the customer has already been invoiced can result in some very uncomfortable, "Please Explain" questioning of the project manager by the performing organization's senior management and possibly

by an irate representative of the customer organization. This displeasure is because, the performing organization may have already closed the project's billing / Cost Account Codes, and the customer's representative may not have any accessible / approved funds leftover from the now-closed project (and/or the previous financial year) to pay for any such outstanding claims.

2) Any **outstanding financial obligations** between the Third Party, the performing organization, and/or the customer organization will have to be **paid-out**.

Similarly, any **unresolved financial claims** (including liquidated damages) will have to be **settled**. While the project manager may not be in a position of authority to resolve these issues, the project manager should (when possible) notify the involved parties of the situation, and the project manager should (when possible) try to facilitate the resolution of these issues.

I would recommend that ... the performing organization "politely enquire" of the Third Party (i.e., the primary contractor) as to whether they have been fulfilling their financial obligations to the subcontractors that they have engaged to work on the project.

This show of concern is because, you don't want to end up in the situation of having the project's progress come to a grinding halt, or held to ransom by disgruntled subcontractors who have not been paid by the Third Party for their services / supplies.

Clear the books and resolve all outstanding claims by 3rd parties, as you do not truly have control over how they will react and how far they are prepared to go to recover amounts owed; i.e. litigation, unionized black-listing, social network slandering, media assassination., and destruction / sabotage of their part of the deliverables.

3) Any **outstanding performance bonuses** for the Third Party will have to be **judged and awarded** to the relevant recipient(s).

While the project manager may not be in a position of authority to distribute performance bonuses, the project manager should (when possible) notify the involved parties of the situation, and the project manager should (when possible) inform the judger (of such performance bonuses) of whether in the project manager's opinion the Third Party is worthy of such a bonus.

4) Any **existing contract agreements** between the Third Party, the performing organization, and/or the customer organization **will have to be concluded or updated accordingly**. All of the parties involved with that contract agreement will have to be notified of the pertinent changes that affect them specifically.

5) Any **sourced materials** (i.e. documentation, equipment, information, and data) that were provided to the Third Party by either the performing organization or the customer organization will have **to be returned or responsibly disposed of**.

   The project manager should confirm that such items have been dealt with accordingly. For more details about the processing of such items, then please refer to the [Resources] part in [Section 17.4.1.3].

6) Any **warrantees** (*e.g. hardware warrantees*), **licenses and certificates** (*e.g. export licenses, end-user certificates*) that the Third Party is obligated to provide will have to be turned over to appropriate representatives of either the performing organization or the customer organization. Similarly, any **software license keys**, **activation codes**, and the **related paperwork** will also have to be **turned over** to the appropriate recipient.

   The project manager should maintain a list (and where possible copies) of the transactional records for all of these items. The project manager should also confirm when each of these items have been handed over by the Third Party, and when in turn these have been handed over to the relevant final recipients.

   That is, the project manager needs to **know what items are and are no longer within the boundaries of the performing organization's responsibility of delivery**.

7) Any **documentation, schematics, drawings, and handover notes** that the Third Party is required **to produce** will have to be of an **acceptable level of quality & presentation to be usable by the intended audience**.

   The project manager should confirm that such usable documentation has been produced and handed over to an appropriate representative of the recipient organization. *... Not just accept a verbal promise that it will be done.*

8) **Intellectual Property** (*such as source code, schematics, algorithms, artwork, copyrights, and patents*) and Intellectual Property Rights (i.e. Foreground IP and Background IP) that the Third Party is **contractually obligated to provide will have to be turned over** to the appropriate representative of the performing organization or the customer organization.

Alternatively, if specified in the contract agreements, **it may be necessary for the intellectual property to be held in mutual trust by an independent escrow agent** (just in case the licenser of that intellectual property goes out of business).

The intellectual property that is received will have to be recorded in an **Intellectual Property Register (IPR)**, securely held / archived using suitable facilities and means of retrieval; see [Chapter 18] on Project Integration & Information Management.

That is, this intellectual property will have to be catalogued and stored in a usable form so that it can be accessed without much difficulty by those authorized persons who require its use. Yet, not allowing admittance to those persons who do not have access rights to such information and materials.

During the life of the project, the project manager should also keep track of what is happening with this intellectual property and who currently has access to it.

9) Is there **anything else** that is **specific to this project**, to the **industry / market-segment** that the performing organization and/or the customer organization are engaged?

10) Is there **anything else prescribed by national, state, or statutory authorities** that must be taken into consideration?

You had better figure out what exactly is required to be handed over now, as there may not be the opportunity, or that 3ʳᵈ Party may no longer exist, to get missing stuff from at some point in the future.

War Story ... Resurrection of the long since dead.

During my career, it is surprising how many times the situation has arisen that after a project has been laid to rest and the project team members have long since moved onto other endeavours (or left for other organizations), ... then all of a sudden there is a critical need to recall information, retrieve documentation artefacts, and/or access intellectual property for that archaic project. This information is urgently required to; rectify a problem encountered by the customer, or it is needed to aid another project, or it is required to resurrect that long dead project.

Though, what usually happens is that this required information is discovered as having been lost, not complete, or was never obtained from its original source (i.e. from the third party). Subsequently, this lost information has to be either:

- Accepted as lost and the company has to wear the consequences; e.g. issues with the customer, or missed business opportunities.

- The lost information has to be reproduced, which introduces its own built-in [Time] & [Cost], and takes [People] & [Resources] away from the servicing of new and existing business.

First Party and Second Party

The following points are for both the First Party and the Second Party.

The **First Party includes permanent employees of the performing organization as well as perpetual contractors** whom the performing organization **engages on a continual or semi-continual basis** from one project to the next (as "fixtures" of the organization).

The **Second Party includes temporary contractors and loaned members of the customer organization** who come into the performing organization and **participate** as members of the project team **for the duration of the project** or for a significant portion of the project.

I would recommend that ... even though from an accounting perspective these perpetual contractors would be treated differently to those permanent employees, **from the project team's perspective each perpetual contractor's participation should be thought of as the same as that for permanent staff.**
That is, think of the perpetual contractor as having an "alternate payment lifestyle".

This blurring of the line between perpetual contractor and permanent employee is because, often the majority of the performing organization's permanent employees won't know or even realize that the perpetual contractor is not a permanent employee.

> *For example; I know of some contractors who have been with their companies longer than most of the permanent staff.*

To treat a perpetual contractor with indifference and to give them an inglorious exit could have a significant effect on the morale of the project team and the other permanent staff.

Hence, it is often beneficial to give these perpetual contractors similar promotional and educational opportunities as with the permanent employees. The only difference being that, the permanent employees should be considered first for new opportunities and if retrenchments are to occur then the perpetual contractors should be the first candidates considered for being let go.

If a First Party / Second Party's services will no longer be required then:

1) The performing organization's **senior management and other project managers** will need to be **informed in advance that the First Party / Second Party will soon be available** for other projects and activities.

2) The **First Party / Second Party** will need to be **notified of when their services will no longer be required** on this project.

When & where possible, the First Party / Second Party should be informed of what other projects or activities that they **will be moving onto, or** at least to **whom** they will need **to communicate with** in order **to find out what they will be doing next.**

 **DO NOT leave the First Party / Second Party persons hanging in limbo wondering what is next for them** once their involvement with the project has ended. Because their fear of the unknown could result in a decrease in their morale, a reduction in their work effort (as a covert end-date delaying tactic), and/or a subsequent slide in their attention to detail.

Consequently, their change in characteristics could have a negative effect on other members of staff, which could contagiously result in several of them looking for alternate work opportunities outside of the performing organization. ... Before, according to scuttlebutt, they are all given the "boot out the door".

Yep, it is really nice to hear by the back-channels (hearsay) that your services will no longer be required once the project ends. ... Especially given that because of their promises and reassurances, you left a relatively safe job for that particular "once in a lifetime can't be missed" opportunity.

---

3) If the departing First Party / Second Party are required to generate project documentation then they should be **made clearly aware of what** they are **expected to produce, to what standard, by when**, and **to whom it is to be delivered**.

4) If the departing First Party / Second Party is required to provide **hand-over notes** for their successor in the project team then they should be made clearly aware of **what is required** of them, **to what standard, by when**, and **to whom it is intended** (even if no specific recipient has been identified). Where possible the project manager should **arrange for a face-to-face handover between the predecessor and their successor**.   *… Not a 5-15 minute "dump it in your lap" mental download.*

5) Any **successor** person should be **informed of their new role & responsibilities, when** it is expected that they will be **taking over that role**, and **to whom** they should **approach for more information** about their new role.

Downtime between projects

With some of these First Party persons no longer required for the project, the last thing that the project manager wants for the project's bottom-line is to have these persons malingering about the project, and submitting timesheet entries against the project's job codes for non-productive activities; i.e. project [Time] consumed with "**bench sitting**".

As you will read in the following part on the closure of [Time] & [Cost] … and as was described previously in [Section 12.2.3] … those persons who are still booking their [Time] to the project job code but are no longer doing productive work on the project, then these "bench sitting" [People] will have a significant impact on the [Time Baseline] and [Cost Baseline] determination of whether the project is judged as a success or failure.

I would recommend that … an "administration" (aka "bench sitting") job code should be provided specifically for those persons currently in a work activity loll between projects.

Example Case: Burnout, a break, and the drive to get onto the next project.

Once upon a time, there were two organizations that had the same regional employment market and alike ethnic cultural backgrounds. Yet even though both organizations and their [People] were always busy, these two performing organizations handled the changeover between major projects in very different ways.

With one of the organizations, when the project team members were no longer required (and with that team having worked long and hard on that project) then these [People] would be instantly thrust into another project that had some soon to be approaching deliverable milestone (or worse the project was already behind schedule). Hence, these implementers were continually under pressure to deliver. Consequently, there were multiple cases of team members experiencing **"burnout"**; i.e. **lack of drive and energy to work on the next endeavour**, and growing amounts of unseasonal sick leave.

Burn Out, guaranteed to limit productivity & restrain creativity.

With the second organization, when the project team members were no longer required then these [People] would have a few days of reduced work duties prior to commencing their involvement with their next project. This "twiddling their thumbs" time meant that these implementers became a little bit bored. As a result, these [People] were highly motivated to get on with the next project and burnout was evidently not an issue.

In retrospect, there may not have been some brilliant covert scheme on behalf of the second organization's senior management to deliberately **"rest" individuals prior to their next project**. This may have in fact been an indication of a lack of preplanning and resource management. However, this short break between major projects did happen to have a very **therapeutic effect of re-energizing the [People]**.

Don't underestimate the positive effect that Time-In-Lieu can have on people's morale and the feeling of having one's effort being appreciated.

[People] and Communications A Concluding Thought

This step in the Closing Phase should respectfully & humanly deal with those [People] who have participated in the project, those who remain, those who are leaving, and those who have joined the project specifically for its conclusion.

NOTE ❏ During the project's Closing Phase, new stakeholders and previously lower priority stakeholders will become more involved and important.

> *For example; the quality assurance & control group, the finance & accounts department, and even statutory authorities.*

War Story ... *A past relationship influencing the new.*

During my professional career, it is surprising how many times that I have encountered someone from a previous job; either as a work colleague, customer, or as a supplier. Sometimes I have been in the "position of power", and other times they were in that position.

The interesting thing is that, theirs and my initial reaction towards each other at the re-meeting was partially influenced by how that previous relationship ended. ... If you or they were a "Jerk" last time, then possibly the new relationship won't be starting with the best foot forward. Therefore, it is better to end the current relationship on positive terms.

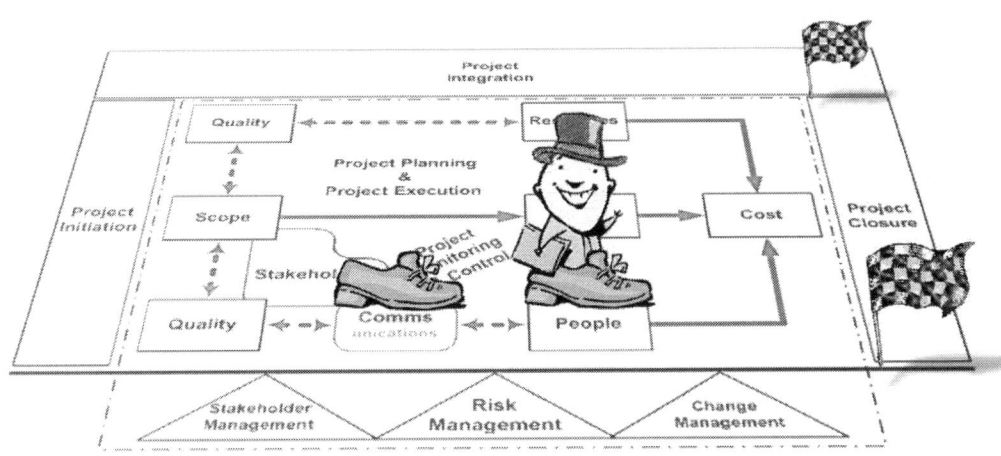

17.4.1.3. [Resources] Return

Back in [Section 14.2] for the practical monitoring & control of [Resources] there was a clear distinction made between the inanimate [Resources] and the [People] involved with the project. Thus, with the project or the milestone release coming to an end, then there needs to be a focus on the closure activities related to those inanimate [Resources] and the procurement completion process.

In [Section 14.2.1.1], these [Resource] items aka sourced materials (of documentation, information, data, equipment, components, etc.) were procured by begging, borrowing, negotiation, and/or formalized procurement from First Party persons internal to the performing organization, from the Second Party customer organization, and/or from external Third Parties.

If the providing Party and/or the [Resource] items that are being furnished by that particular party will no longer be required for the remainder of the project, then these "residual" [Resource] items will need to be either:

- returned to the provider possibly with some form of **usage payment.** ... *At least, a "thank you" note and verbally stated.*

- will not be returned to the provider and hence the provider will have to be **compensate**d appropriately.

Handling the Dispatch & Delivery of "Residual" [Resources]

Please refer to [Section 14.2.2] and [Section 14.2.3] for information relating specifically to the packaging, dispatch, and shipping of these no longer required "residual" [Resource] items.

Though, **when returning [Resource] items, these need to be handed over to an appropriate and authorized representative of the recipient's organization.**

At each stage during the transfer, these items need to be received by someone who is capable, willing, and authorized to handle such items.

That is, these items cannot be turned over to just anyone; because, for **traceability reasons there needs to be a clear "Line Of Custody" from beginning to end, else who will be "held accountable"** if these items were to go missing or were to be received in a damaged or unusable state.

 I would recommend that ... whenever a physical item is sent to another party, the project manager should email that recipient party to notify them that the particular item is on its way. The project manager should also detail; when the item is expected to arrive, how the item is expected to get there, and to whom the item is addressed. Because the last thing the project manager needs is for the item to go missing (and worse then sometime after the project's accounts are closed, a claim is raised against the project for the disappeared or damaged item).

 I would recommend that ... a short time after the item is anticipated to have been received by the recipient, the project manager should email that recipient to enquire as to whether the item has in fact arrived. It is also advisable for the project manager to keep a watchful eye on the freight/courier's tracking information; especially for big dollar items, items of strategic importance, or items of a time critical nature.

 I would recommend that ... the project manager email and not solely telephone the recipient (or at least make sure that the results of the conversation are reiterated in a summation email).
The reason for this email is to provide traceability evidence if something unforeseen happens to that item.

 I would also recommend that ... whenever a physical item is sent out by any member of the project team then some form of centralized **Dispatch Record** should register the facts of this transfer.
The purpose of this Dispatch Record is to enable the determination of "who to ask" and "what to ask" if the item happens to go missing.

Lessons Learnt: Traceability Records and Dispatched Consignments.

For traceability purposes, it is essential to maintain records of what happened to these items. A centralized **Dispatch Record** should be kept for each consignment of items that leaves the performing organization and/or the project team.

This **Dispatch Record** should at least capture the following details:

1) A list of **exactly what items were sent** in the consignment? ... *Serial Numbers.*

2) To whom was the consignment of items **addressed to**?

3) **Who organized** and **who authorized** the sending of this consignment of items, and of course their **signatures** and the **dates** when they signed?

4) On what **date** was the consignment of **items sent / collected** by the courier?

5) What was the **Consignment Tracking Number** provided by the courier?

6) Who **was the courier**, and what was the **method of transfer**ring the item to its destination; *e.g. by land, air, overnight express, by hand.*

The physical consignment of items should have some form of **"To:"** and **"From:"** **identification** attached; *e.g. a label on the exterior of the package.*

Clearly indicate 'TO' who is the intended recipient, and 'FROM' whom was the item sent.

Just as you would have a "To:" and a "From:" field completed for an email.

I would recommend that ... a **Cover Letter** accompany the item(s) if any additional explanatory information is required about why the item(s) has been sent, to whom is the final intended recipient of the item(s), and/or for what purpose the item is intended.

Yep, such dispatch records got me out of a potentially expensive mess when a big dollar item went missing. Fortunately, our records proved that the item had been delivered and received at their warehouse, but they "misplaced" it. ... How do you misplace a computer rack the size of a large family refrigerator? ... Not like, it can be used as a paperweight.

Handling the Return and Receipt of "Residual" [Resources]

For traceability purposes, when delivered items are received, the following information should at least be recorded:

1) **What was received** in this consignment of items **and were these items in an acceptable condition?**

2) **Who received** the consignment and **when were** these items **delivered?**

3) **From whom** were the consignment of **items sent?**

Due to the quantity of items coming into and exiting the project, as well as to and from the various project participants, **it is important for the project team** (and the project manager) **to record the movements, last known location, transfers, and the custodian of each of these items**.

Therefore, the project's records need to capture such information as:

1) **What items does the performing organization need to return to the customer organization and/or any third parties?**

   The details of these items (*e.g. serial numbers*) should be maintained in a **Third Party Property Register (TPP)** that is either established generically for the performing organization or created specifically for the project.

   NOTE   ☐ By the time, **that the project is closed out the Third Party Property Register should be empty of those items that related specifically to the project**. This register should also have recorded all of the transitions for items into and out of the project / performing organization (even for temporary "boomerang" transitions).

   Beware carrying too greater value of Third Party Property (for too long) because such items would need to be covered by the performing organization's contents insurance. If there was a fire / disaster could the presence of these "TPP" items exceed the policy's financial coverage. ... Then, who pays?

2) **What items need to be returned to the performing organization by the customer organization and/or by third parties?**

The details of these items would be retained in an **Asset (Tracking) Register**.
Established generically for the performing organization or created specifically for the project.

Details will also have to be kept on:

(1) These items will need to be **returned to where**?

(2) These items will need to be **addressed to whom**?

(3) These items will need to be **returned by when**?

(4) These items were **returned by whom**?

(5) **Who** is/was **responsible** for these items?

QUESTION: If [Resource] items go missing to never be seen again, then who is liable?

Is the performing organization responsible for the cost of these missing items, is it the last person or organization to be seen utilizing the item of concern, is it just bad luck for the provider of that missing item, or will it be considered as a consumable to be invoiced to the customer?

That is, who pays whom if an item is lost or damaged?

WARNING: The project manager is often the one held responsible and accountable for overseeing the return of [Resources] and the closure of the associated procurement arrangements.

So keep a watchful eye on the movement and transfer of items,
ELSE an unreasonable amount of project [Time] could be consumed
and [People] involved with the search for an item that long ago left the
possession of the performing organization.

Handling Never To Be Returned "Residual" [Resources]

If these [Resource] items are no longer going to be used on the project and these items are not going to be returned then these items will need to be **decommissioned**, **scrapped**, or **disposed of "in a responsible manner"**. Where, "in a responsible manner" is concerned with; the environmental impact of "dumping" these items, and/or the possibility of inappropriate access to classified or confidential information that should not fall into the wrong hands; i.e. be obtained by competitors or be accessed by the general public.

Alternatively, will "ownership" of these items be transferred from one party to another? If the item is to be transferred into or out of the performing organization then will it be the project manager's responsibility to organize the transfer of the asset into/off the **Asset Tracking Register** (and also to notify the appropriate administrative persons about this transfer)? Additionally, will this asset transfer involve some type of financial transaction or simply be a "swap" of assets between the parties?

Accountability For [Resources]

When these [Resource] items are being transferred between the project's participants then the performing organization's asset / property registers and accounting ledgers will have to be reconciled accordingly; aka "balancing the books".

That is:

- **Received invoices** verified as being for the correct items and monetary amounts. Then the invoice will have to be paid within the payment period specified in the Terms And Conditions on the supplier's quotation or invoice.

- **Leased and hired items** will have to be returned, any residual payments paid out, and any security deposits recovered.

- Any **services rendered** will have to be paid for.

- The owner of any **lost or damaged** item will have to be compensated.

- Any items **returned to stores** (i.e. the organization's stock inventory) will have to be **credited / reimbursed back to the project's account** or to the paying organization's account.

- Any **decommissioned / scrapped / disposed** of items (*e.g. development prototypes*) will have to be **written-off against the project's Cost Account Code**.

 I would recommend that ... the project manager strive to ensure that these [Resource] items are dealt with well in advance of the project's closure. The reason for doing this is because, work-life can turn into a real administrative nightmare when there are outstanding claims for [Resource] items; especially claims lodged after the project has been declared as concluded and the project's Cost Account Codes have been closed.

Hence, the project manager will need to **maintain up-to-date records** for [Resource] items that have been received, returned, transferred, decommissioned, and disposed of. That is, those [Resource] items that the project manager is being held (directly and indirectly) responsible for, these all have to be **accurately tracked and accounted for**.

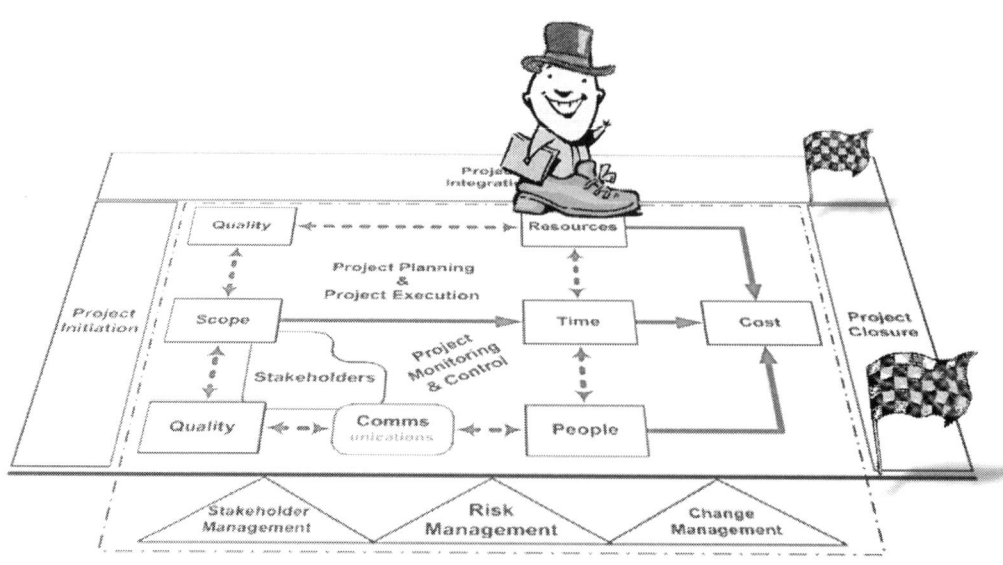

17.4.1.4. [Time] & [Cost] ... Book Closure

This section will focus on the closure activities related to the [Time] & [Cost] project variables; specifically the closure of tracking and recording of both [Time] & [Cost].

Simplified Cost To Date Formula

For many an SDLC project, a significant portion of the project's [Cost] is due to the accumulated monetary value of the [Time] that the [People] have spent working on the project. Recall back in [Section 7.3], the simplified 'Cost To Date' equation;

$$COST_{TO\ DATE} = (PEOPLE_{RATES} \quad X \quad TIME_{USED})$$
$$+$$
$$(RESOURCE_{UNITS} \quad X \quad AMOUNT_{USED})$$
$$+$$
$$(FIXED\ COSTS_{TO\ DATE})$$

Also, recall from [Chapter 4] for the Planning Phase, that the project's [Scope] was broken down into work-packages via the Work Breakdown Structure (WBS). These work-packages were then turned into tasks with estimated [Time] durations in the project's schedule, and then [People] & [Resources] were assigned to each of these scheduled tasks.

Where each of these [People] & [Resources] happened to have known monetary values based on their hourly cost rates, the quantities used, and/or some fixed set of costs.

Relationship to the Budget At Completion

As reproduced below in [Figure 47], the final link in the chain was to connect these individual WBS work-packages to the project's budget; thereby enabling the calculation of the expected accumulated cost for each of these work-packages, and hence deriving the project's "**Budget At Completion**".

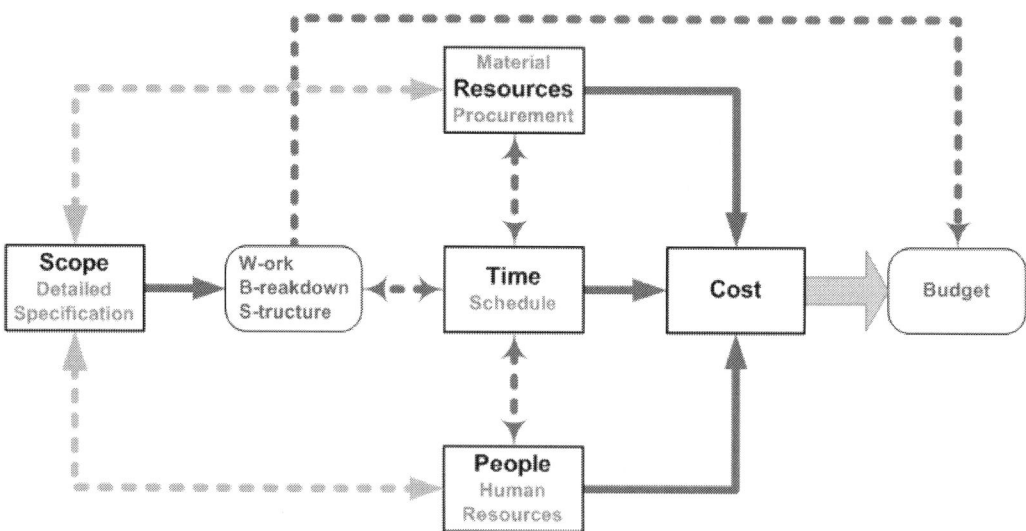

Figure 47: The Planning Process Model to determine the project's cost & budget.

Relationship of Actuals Versus Planned

With the relationships between [Scope], work-packages, [Time], [People], [Resources], [Cost], and the budget having been established for the project, then the next thing was to bind the "Planned" project variables & constraints with their "Actual" counterparts by associating **[Time]** with the **Timesheet System** and by assigning **[Cost]** codes in the **Cost Accounting System**. With the "Planned" and the "Actual" having been bound together from at least the beginning of the Executing Phase then at defined intervals (*e.g., daily, weekly, fortnightly, or monthly*) the [People] working on the project would use the Timesheet System to enter numerical values against the corresponding job codes that represented the work-packages that they (supposedly) had been working on.

At another defined interval (*e.g. weekly, fortnightly, or monthly*) the performing organization's administrative personnel would (send out friendly reminder notices that all staff are requested to submit their timesheets by the End-Of-Business on a specific day, after which they would) extract from the timesheet system the accumulated hours recorded against each job code. This extracted data would then be imported into the Accounting System to be mapped to the corresponding Cost Account Codes.

This consolidated data would then be used to compare the project's "Actual" [Cost] and the "Actual" [Time] progress against the project's "Planned" baselines; possibly using Earned Value Performance Measures as described previously in [Section 12.3.1].

Now that the project or the milestone release is coming to an end, then each of those elements that form this simplified "Cost To Date" equation will need to be closed out / deactivated.

Closing the gate before the Exit Us

The previous parts of this chapter have dealt with the:

- the closure of the [Resources] and reconciling the associated accounting records,

- the relevant [People] were told that their services would no longer be required, and

- these [People] were asked to submit their final timesheets, invoices, and expense claim forms.

With each of these [People] & [Resources] associated hours, expenses, and claims having now been received and processed then **as part of the Closing Phase, the related time reporting Job Codes and Cost Account Codes need to be disabled**.

NOTE　❑ The reason for disabling access to these Job Codes and Cost Account Codes as soon as feasibly possible is to prevent spurious "bogus" hours, expenses, and claims being tacked onto the project. With these codes having been closed out then the persons submitting any residual hours, expenses, or claims will have to **justify why** these should be **added onto the project's bottom line**.

Recall back in [Section 1.3] that [Scope], [Time], [Cost], and [Quality] were stated as the traditional determinates used to judge whether the project is deemed a success or failure.

Well, if these Job Codes and Cost Account Codes were to remain open longer than necessary then inexplicably additional [Time] & [Cost] claims are often drawn to the project like rubbish dumped on a vacant block of suburban land.

> *For example; the workload for a member of the project team has reduced significantly since their involvement in the project has practically ended. However, this individual finds himself or herself in need of some job code to justify what they have been doing for the last few hours / days, but they are hesitant to be honest and put surfing the net, social networking, or just hiding away in the corner. Hence, they decide to put this "bench-sitting" time down against some reasonable sounding project job code that they still have access to.*

> *Similarly, "miscellaneous" [Resource] items can be attached to one of the project's open Cost Account Codes because the attaching person has no other codes accessible to use. Moreover, they may think, "no one will notice it among all of those other items".*

 I would recommend that ... across all projects and development groups that a generic job code be established specifically for "self-learning" / "bench-sitting". This job code should be used for the situation when a person has not yet been assigned new/additional work to do. Thereby, this non-billable time will not be incorrectly attached to the project's [Time] & [Cost] and subsequently will mitigate against inadvertent inflation of the project's ~~success~~ failure determinates.

This also, makes it apparent to senior management when they need to reassign this available staff to new taskings such as other projects.

 I would also recommend that ... the project manager inform those persons transitioning from the project, that it is okay to use this "self-learning" job code until instructed otherwise.

[Time] & [Cost] Book Closure Concluding Thought

With all of the project's timesheets having been submitted & processed, and all of the project's bills & invoices having been paid then the project's [Time] Job Codes and [Cost] Account Codes will need to be disabled as soon as feasibly & reasonably possible, so that no extraneous additions can be booked against the closed project.

This may also require the **closure of any external financial accounts** related specifically to this project, and the subsequent distribution of the closing balances of such accounts to the authorized representatives of the relevant parties (i.e. to the First Party performing organization, to the Second Party customer organization, and to the appropriate Third Party organizations).

IF YOU DON'T CLOSE IT,

THEN SOMEONE WILL KEEP BOOKING TIME

AND CHARGING COSTS TO IT. ...

SO BETTER, CLOSE IT A.S.A.P. OR ELSE.

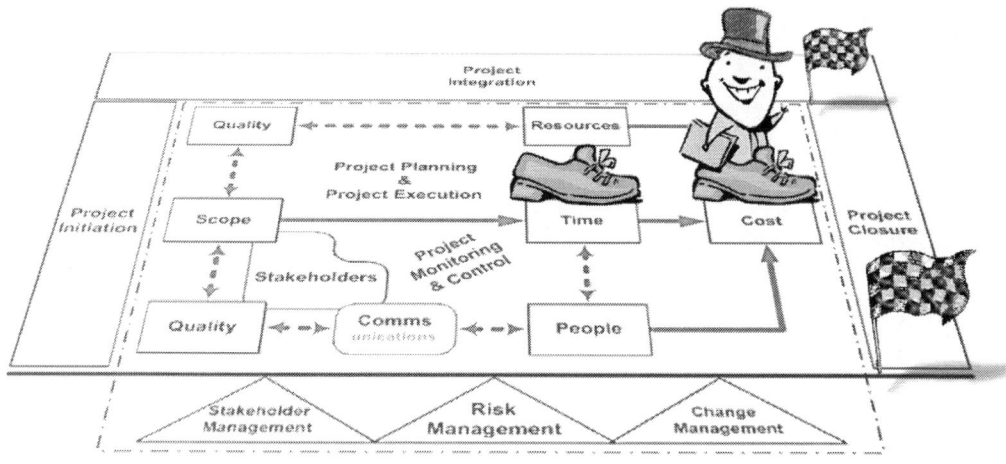

17.4.2. Question: "Does EVERYONE AGREE it is ALL DONE?"

To answer this question of, "does EVERYONE AGREE it is ALL DONE" then the following sub-questions need to be replied to with a "YES".

1) Have the **activities** associated with this project been **closed out**?

2) Have **all of the project constraints been settled "satisfactorily"**?

3) If non-conforming project deliverables were produced then have **Concessions been generated, reviewed, and approved** for these non-conformances?

4) Have the **project deliverables been transferred** to the appropriate recipients, and has the **acceptance** of each deliverable been **signed-off** by an authorized representative of the recipient organization?

   NOTE   ❒ The recipient may not only be the customer organization but also statutory authorities, escrow agents, third party service providers ... etc.

5) Have the required **project artefacts & documentation been handed over** to the authorized & appropriate representative of the recipient organization?

6) Has the **Objective Quality Evidence (OQE) been produced, cumulated, and signed-off** by the relevant quality authorities?

7) Has the **project's Intellectual Property** (Foreground IP and Background IP) been **accumulated, filtered, archived, and indexed for future reference**? ... *And, has this information been successfully verified as retrievable?*

8) Have the project's externally **procured resources** (*e.g. third party and customer furnished equipment & information*) been **returned** to their authorized & approved recipients? ... *And, has this fact been recorded as such?*

9) Have these resources' associated **financial accounting** been finalized and closed out?

10) Have the project / milestone release's **Job Codes & Cost Account Codes** been **disabled**, and are the **associated records & reports up-to-date**?

11) Have all of the **contractual obligations** been fulfilled?

12) Are there any **outstanding issues** that need to be resolved, have these been noted in the Risks & Issues Register, and have these been notified to the appropriate project stakeholders for resolution?

13) Has the project / milestone release been confirmed as having satisfactorily complied with the **Exit Criteria** as detailed in the Project Charter?

14) Has a **Post Completion Review** been **held** and has a **Project Closure Report** been **generated** for this project / milestone release? See [Section 17.4.2.2].

15) Has the **Project Closure Report** been **signed-off** (and dated) by representatives from both the performing and customer organizations **to mark the formal closure of the project / milestone release**?

16) Has a celebratory function / luncheon been held to thank the Project Team for their efforts on this now closed project?

17) *Has a "Lessons Learnt" session been held with the Project Team?*

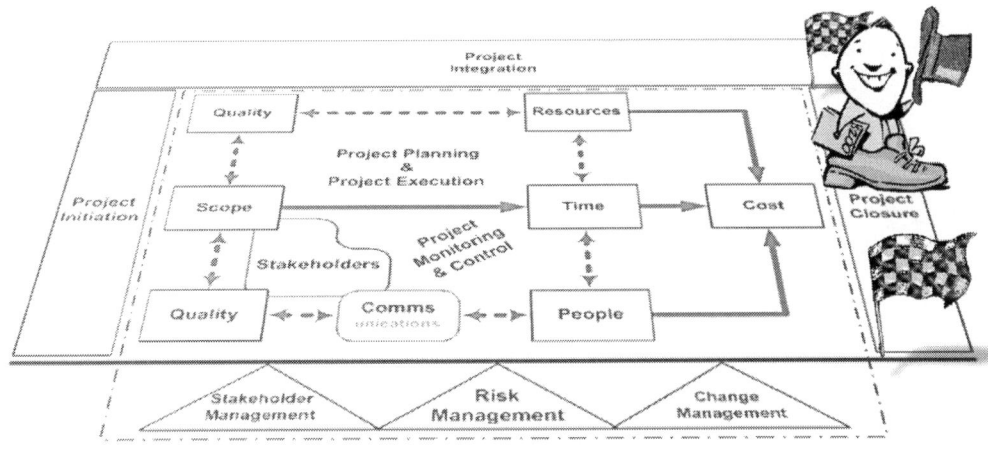

17.4.2.1. Post Implementation Review

The purpose of the **Post Implementation Review (PIR)** is to confirm that what was supposed to have been implemented has in fact been done.

Alternatively, the Post Implementation Review could be thought of as **the sanity check of the entry criteria that must be satisfied in order** for the project **to be allowed to proceed to the Closing Phase**; i.e. checklist review of the "readiness" indicators. Subsequently, the outcome of this review should be an informed & responsible decision to proceed, rework, or halt the project / milestone release.

This Post Implementation Review would usually involve; the project manager, selected members of the project team, the representatives of the customer organization, possibly representatives of the performing organization's senior management (*e.g. the quality authority, technical authority, the business group manager, manufacturing, legal, IT & network administration, other internal entities as need be*), and maybe representatives of any other concerned stakeholders (such as strategic third party vendors & suppliers).

NOTE   ❑ The Post Implementation Review is **one of the sequence of Phase Completion Reviews** that should be held at the conclusion of each phase in the project life cycle, see [Section 2.2.1] on "gates".

Requirement Traceability Matrix (RTM)

During (or prior to) the Post Implementation Review, the **Requirement Traceability Matrix** (RTM) [Section 12.1.1] and [Section 17.4.1.1] would be examined and updated to establish:

- **What functionality** has been **included**?

- **What functionality** has **not** been **included**?

- **What functionality** was **agreed** to have been **in-scope & out-of-scope** for this current release?

If there are **any discrepancies** found between what has "actually" been included and what was agreed / baselined to have been included then;

1) A **concession** would be generated, reviewed, and **approved** for this missing or re-scoped functionality so as to be accepted as having been left out of the project / milestone release. **OR**

2) The Project Implementation Team would be required to **go back and rectify** this **missing or re-scoped functionality**, before any possible permission to advance would be granted.

NOTE ❐ The appraisal of the RTM is often done as a sit-down or teleconference meeting between the project manager (lead members of the Project Working Group / Project Implementation Team) and the customer's representative.

Testing Coverage Matrix (TCM)

In addition to the Requirements Traceability Matrix, during (or prior to) this Post Implementation Review, the **Testing Coverage Matrix** (TCM) [Section 17.4.1.1] would be examined and updated to **establish the "correctness" of the functionality** that has been implemented; i.e. the PASS : FAIL operational status. If there are found to be any non-conformances (i.e. outstanding defects, or departures from the specified functionality) then;

1) A **waiver / deviation** would be generated, reviewed, and **approved** to permit the acceptance of this non-conforming functionality. Thereby, the associated deliverable may be accepted as ready for inclusion in the Closing Phase. **OR**

2) The Project Implementation Team would be required to **go back and rectify** this **non-conforming functionality** before any possible permission would be granted for project advancement.

Relationship of the RTM to the TCM

The Requirement Traceability Matrix (**RTM**) lists **what functionality has been included**, the Testing Coverage Matrix (**TCM**) lists **what functionality has been verified as working**. The RTM and the TCM can also provide:

- Identification of changes to the [Scope Baseline] and notes any outstanding issues,

- lists outstanding change proposals and notes the impact of such change proposals on the [Scope Baseline], and

- lists outstanding waivers & deviations and notes the impact of such waivers & deviations on the [Scope Baseline].

Objective Quality Evidence (OQE)

As part of the Post Implementation Review, prior to allowing the project to progress to the final stages, there may be other check-listed questions that need to be positively answered (or at least the presence of concessions pertaining to these):

1) **Status of the Test Procedures**

   - Have the **test procedures been documented**, so that each specific test can be **accurately reproduced**?

   - Have these **test procedures been reviewed & approved internally** by the performing organization; i.e. by a relevant "technical authority" (*e.g. the testing manager / team-leader*), the "quality authority" (*e.g. the quality manager*), and the project manager?

   - Have these **test procedures been reviewed & agreed to externally** by the representatives of the customer organization?

   - Are the documented **test procedures under version control** so that the appropriate revision of the test procedures can be retrieved for each corresponding release of the deliverable?

   - Have these **test procedures been "Dry Run" tested** to verify the correctness and appropriateness of each test? If procedural imperfections or inaccuracies have been found then have these mistakes/omissions been rectified, reviewed, approved, and have the test procedures version been up-issued?

   - Have the results of the **Dry Run testing been recorded** to form part of the Objective Quality Evidence?

2) **Status of the tests conducted**

   - Was the **test undertaken by a pertinent person** / group; *e.g. a certified test house, or persons in the performing organization who have appropriate testing job roles?*

- Have the **results** of the testing been **recorded** in some form of **Inspection & Test Record (ITR)** and/or an **Acceptance Test Record (ATR)**?

- Has the **Testing Coverage Matrix (TCM)** been updated to indicate the **PASS : FAIL status** of each particular test?

That is, have all of the relevant **Test & Traceability Records** been accumulated as OQE in preparation for any possible future scrutineering of the testing that was conducted?

3) **Status of the test environment**

- Has a **test environment been established** to verify the correctness of the deliverable?

- Have **test harnesses & simulators been validated** as functioning correctly to replicate real-world situations (as close as is feasibly possible)?

- Has the configuration of this **test environment been documented** for any possible future need to **reproduce** that specific test environment?

- Has the **test equipment been calibrated** and where applicable certified as having been properly calibrated by an approved **calibration organization**?
  Does the performing organization have records detailing when the test equipment was last calibrated, and what is the **period of validity of such calibration**?

 For certain industries (*e.g. defence, telecoms, biomedical*), it is an imperative that such test equipment (and the test environments and harnesses) be calibrated, validated, and certified before being used for recorded testing. The use of un-calibrated, un-validated, and/or with expired period of validity can mean that those tests conducted using such test equipment & environments are excluded / rejected as being acceptable Objective Quality Evidence irrespective of whether those tests were successful and conducted properly.

---

4) **Status of the product baseline**

- Has the **Manufacturing Data Packages (MDP)**, the **Build Of Materials (BOM)**, and the **Factory Release Test (FRT)** been **reviewed & approved** prior to full production commencing?

  > MDP – the manufacturing instructions, BOM – the list of components necessary to build an individual deliverable unit, FRT – the instructions on how to verify the correctness of each manufactured deliverable unit.

  That is, once production starts can it be assured that each subsequent unit produced will be exactly the same as the "First Of Type" unit that received signed-off acceptance. Similarly, will all of the units that follow be identical to all of the units that came before?

 **Beware of conducting acceptance tests on a deliverable that does not exactly conform to the same specifications as that of the final deliverable**; i.e. testing on a pre-production unit / prototype.

Acceptance tests must be conducted on a representative sample of the final deliverable units, and not some one-off bespoke unit.

---

5) **Status of the readiness to support acceptance testing**

- Has the availability of all of the necessary [Resources], [People], and test witnesses been arranged? That is, have these [People] & [Resource] providers been informed of when they need to turn-up to participate in the acceptance testing?

  Often acceptance testing is conducted by an independent **Verification, Validation & Test (VV&T)** team / department / individual who tests multiple projects' deliverables via a conveyor-belt-like work-stream approach. Consequently, if the project does not have its deliverables ready for testing when the assigned test window is available then the project may have to wait until another window-of-opportunity arises in the VV&T work-stream.

Implementers' Objective Quality Evidence

Other Objective Quality Evidence (OQE) that maybe requested for inspection during the Post Implementation Review relates specifically to the Executing Phase.

1) **Defect Tracking** of **Defect Reports** both CLOSED (i.e. fixed and cancelled) as well as OPEN (i.e. not-fixed and deferred).

2) **Issue Tracking** of matters that require further investigation and issues that were investigated.

   For example; peculiarities between the operation in the development environment and the test environment; i.e. not a defect par say but unexpected or unexplained behaviour.

3) **System Test Results** as well as the **test cases and test scenarios** that were used to confirm the correct operations of functional modules of the deliverable.

4) **Integration Test Results** for the outcomes of merging the deliverable components together as a deliverable system.

5) **Dry-Run Test Results** for those practice tests conducted to establish confidence that the deliverable will successfully PASS the pending **Acceptance Tests**.

6) **Documentation** and **Release Notes** (*e.g. Software Version Description – SVD*) pertaining to the deliverable.

NOTE   ❐ The whole point with implementers' Objective Quality Evidence is not to burden the implementers with bureaucracy, but rather de-burden the implementers from future quality inspections / investigations into the work that they have performed. Hence, **the implementers need to provide sufficient evidence of their due-diligence with design-develop-test when undertaking their assigned tasks**.

Concluding Thoughts

The **purpose of the Post Implementation Review is to obtain approval from the various primary-core stakeholders** (and secondary-strategic stakeholders) **to commence the closure of the project / milestone release**.

QUESTION: If it was your personal money and your family name on the line, would you honestly be happy to receive the deliverables that are about to come out of this project / milestone release?

NOTE ☐ The formality and the amount of details involved with the Post Implementation Review is very much dependent on; the type of industry that the project is being conducted in, the scale of the project, the strategic importance of the project, the mandatory quality standards & guidelines, and the expectations of the representatives of the customer organization (and the senior management of the performing organization).

17.4.2.2. Post Completion Review

The finish line has been crossed; it is all over but for the cheering of the admiring crowds. But wait, the race is **not quite finished yet**; as there is still the **post-race scrutineering** of the driver, the inspection of the vehicle, the selective comparison against the technical regulations, and the hearing of any outstanding protests that have been lodged.

Oh, to only lose now on a technical infringement or worse to be caught cheating after the race has been run.

The **purpose of the Post Completion Review (PCR) is to reach a consensus among the various primary-core stakeholders that the project** / milestone release has been **completed satisfactorily in accordance with the agreed Acceptance Criteria**.

Additionally, the Post Completion Review could also be an orchestrated judgement of how the project performed against its objectives. That is, the Post Completion Review could be used to adjudicate on the [Scope], [Time], [Cost], and [Quality] determinates of the project's success or failure; see [Section 1.3].

However, by the time that the Post Completion Review is held, the various representatives of the performing and customer organizations would already have a good idea as to whether they consider the project to be a success or a failure. Hence, these stakeholders usually just want to make sure that their organization is getting what is due to them, so that they can move onto the next endeavour.

 I would recommend that ... the analysis of the project's determinates of Success : Failure be done during the regular project inspections (i.e. during the progress reviews and milestone reviews), and not left solely to The End during the Post Completion Review. Because, by that time, it could be too late to effectively & efficiently correct any discovered problems.

The Post Completion Review would be held after the project or milestone deliverables have been handed over to the appropriate recipients and the deliverables have been signed-off ... possibly these deliverables have been operating in a steady state for some period of "judgement time".

Usually the Post Completion Review would involve the project manager, representatives of the customer organization, and representatives of the performing organization's senior management.

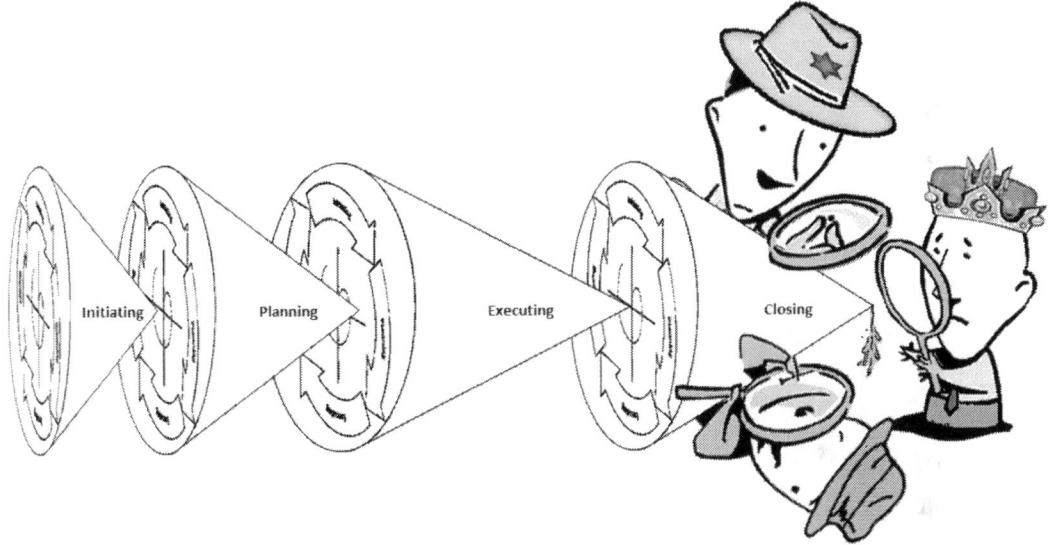

Once the Post Completion Review is completed then the Project Closure Report (i.e. Closeout Report) **would be created to record the outcomes of the review,** then be examined for correctness of the facts before being **signed-off by representatives from both the performing and customer organizations,** and thereby **sealing the project's closure**.

However, depending on the type of project, and the expectations & perspectives of the project's primary-core stakeholders then the performing organization senior management may require an internal closeout report that is independent of the external closeout report that was handed over to the representative of the customer organization.

External – Project Closure Report

This external Project Closure Report would contain a summation of the details that are of relevant importance to the customer organization. Such details as:

1) **Degree of Compliance with the Agreed Deliverables**

- List of / reference to those **deliverables** that were **provided** to the authorized and approved recipients.

  Noting when these deliverables were handed over and to whom these deliverables were handed over to.

- List of / reference to any **deliverables** that **changed** (significantly) **in specifications**.

  Noting if these changes were authorized, by whom, and when these changes were approved; i.e. Change Request Number.

- List of / reference to any **outstanding deliverables**.

  Noting if these omissions were authorized, by whom, and when these omissions were approved. Provide details or references to how these outstanding deliverables will or will not be resolved; i.e. Concessions.

- List of / reference to any **compensation** that is **to be rendered** for changed specifications and/or outstanding deliverables (*e.g. contract deviations and liquidated damages*).

  Noting the form of compensation, to whom the compensation is due, and by when compensation is due.

2) **Summation of the Change Requests and Defect Reports**

- List of / reference to any **Change Requests** that were **implemented**.

  Noting if these Change Requests were authorized, by whom, and when these Change Requests were approved.

- List of / reference to any **outstanding Change Requests** that were not included within the deliverables. Provide details or references to how these outstanding Change Requests will or will not be handled.
  If need be, initiate the Change Control Process (CCP) to ensure that these Change Requests receive the attention that these deserve.

- List of / reference to any **outstanding defects** that have not been closed out (i.e. not fixed or not cancelled). Provide details or references to how these outstanding **Defect Reports** will or will not be resolved.

3) **Objective Quality Evidence**

- Check that the **Acceptance Test documentation** (i.e. FAT, SAT, UAT) have been **signed-off** by representatives of the performing organization (and where appropriate by a representative of the customer organization).

- Check that any other **Objective Quality Evidence** (*e.g. waivers, concessions, declarations of conformity, certificate of conformance, third party certificates, licenses, reports, and audits*) has all been **certified** by the appropriate authorizing signatories.

4) **Book Keeping**

- List of / reference to any **outstanding financial claims** (*e.g. unpaid invoices*) that remain open. Provide details or references to how these outstanding financial claims will or will not be handled.

- List of / reference to any **residual resources** (*e.g. equipment, materials, and data*) that needs to be returned, transferred, decommissioned, or disposed of. Provide details or references to how these residual resources will or will not be handled.

- List of / reference to any **Intellectual Property Rights** (Foreground IP and Background IP) that have and have not been transferred, and/or put into escrow. Provide details or references to how this residual IP will or will not be handled.

- List of / reference to any **financial accounts** that have to be **reconciled**, funds distributed, and accounts closed. Provide details or references to how these outstanding financial accounts will or will not be handled.

5) **Risks & Issues**

- List of / reference to any **outstanding risks & issues** (i.e. problems) that remain unresolved. Provide details or references to how these outstanding risks & issues will or will not be handled.

  Should these outstanding risks & issues be incorporated into the performing organization and/or customer organization's lists of **Problem Reports**?

The Post Completion Review or more specifically the Project Closure Report marks THE END of the Closure Phase for the project's life ... in much the same way that the Project Charter marked THE START of the project's life in the Initiating Phase; see [Section 3.3.2] and [Section 3.4.4].

Given that the Project Closure Report will be archived away as a significant project artefact for historical reference, it is only appropriate that the Project Closure Report contain some context information about the project in the same way as the Project Charter did.

I would recommend that ... the following sections of the Project Charter be selectively plagiarized and redressed for the introductory context section of the Project Closure Report.

Examine the 'overview' (including the 'background', 'goals', 'benefits', 'desired outcomes', 'measurable & tangible objectives'), 'assumptions', 'constraints', 'scope boundaries', references to the requirements, list of major deliverables, and the milestones.

That is, summarize whether the customer's documented & agreed needs, wants, concerns, expectations, and perceptions have been satisfied.

This does not need to be verbose, just sufficient to provide a contextual grounding for a future reader.

Internal – Project Closure Report

This internal Project Closure Report would contain summation details that are of relevant importance to the senior management of the performing organization.

 **This internal Project Closure Report may contain highly sensitive** and **confidential information** that may prove detrimental to the performing organization if it was to fall into the wrong hands; such as the representatives of the customer organization, competitors, or subcontractors; i.e. **FOR INTERNAL USE ONLY.**

The internal Project Closure Report could contain such information as:

1) **Did the completed project deliver the functionality that was agreed to [Scope]?**

   That is, is there missing or additional functionality known or unknown to the customer organization?

2) **Was the project delivered by the agreed milestone dates [Time]?**

   That is, those internal target dates prior to the inclusion of buffering.

3) **Was the project completed within the allocated budget [Cost]?**

   That is, was the project a profit or loss making endeavour?

4) **Are there any outstanding unresolved issues with the deliverables that need to be dealt with [Quality] and <Risks>?**

   That is, are there any problems with the deliverables (known or unknown to the customer organization) that need to be resolved at the most opportune moment?

5) **Are there any outstanding unresolved issues with the [People] & [Resources] involved with the project that need to be dealt with?**

   That is, are there any outstanding claims against the performing organization related to this project's undertaking?

6) **Was the customer's representative(s) satisfied with the project's deliverables** as presented or were their expectations & perceptions different to that which they signed-off?

7) **Did the completed project result in the desired outcomes for the customer?**

8) **Did the completed project provide the perceived benefits for the customer?**

9) **Does the customer's representative consider the project to be a success or a failure?**

10) **Did the completed project result in the desired outcomes for the performing organization?**

11) **Did the completed project provide the perceived benefits for the performing organization?**

12) **Does the performing organization's senior management consider the project to be a success or a failure?**

 I would recommend that ... this Internal Project Closure Report address "the Big Questions" that were asked during the Initiating Phase; see [Section 3.3].

Internal – Self Assessment

Now for some "honest" self-assessment and self-reflection: ... i.e. "Lessons Learnt".

1) **What were the notable mistakes made** during each phase of this project's lifecycle?

2) **What could be done better** next time to ensure that any future projects (like this one just completed) are subsequently a success?

3) **What were the project's major failings** and how could these be approached differently next time?

4) **What were the positives** from each phase of the project lifecycle?

5) **What were the negatives** from each phase of the project lifecycle?

6) **What were the project's significant achievements**?

7) **Who were the project's standout performers, what did they do that worked so well, and why did it work so well**?

8) **Who did not perform as well as expected, and why was this lack of performance so**? *This ain't meant to be a witch-hunt, just looking for actual reasons.*

9) **What could be improved next time to guarantee success**?

And finally, some "thought provoking" opinionated questions:

10) In hindsight, **did this project meet the performing organization's Selection Criteria for acceptable projects**?

11) In hindsight, **did this project undertaking make sense by providing value to the performing organization** and more specifically its owners / shareholders?

All of these are very important questions that needed to be answered ... but alas, these questions are all too often not asked, let alone answered (honestly) by anyone of relevant authority in the managerial hierarchy.

17.4.3. Question: "Let's say GOOD BYE … as friends"

With the project / milestone release ending, then it is time to say goodbye to those departing project participants.   *… Just as you would with an old friend.*

 I would recommend that … relationships with all of the project's participants (i.e. Third Parties, Second Parties, and First Parties) should be ended in as amicable manner as possible. This is because, it may be of strategic value in the future to be able to successfully ~~leverage off~~ rekindle these past relationships. That is, it is a lot easier to reconnect if the previous relationship ended on good terms, rather than overcoming vows of never working together again.

To end the relationship amicably then those departing project participants should be:

1) **Privately** and **publically thanked** for their efforts and their contributions.

2) **Rewards** and **recognitions** distributed to those participants whose project performances have earned such gratitude (and potential financial bonuses).

3) For a relatively long duration project and/or for a project that required extra-ordinary efforts from the participants then a **celebratory** lunch is a good way to **sociably conclude** the project.

4) Ensure that **final payments are made within a reasonable amount of time** after each participant's departure (i.e. immediately when possible, but not several weeks or months later). *… The longer that it takes to pay them, then the greater will be the disintegration of the working relationship.*

5) Try to **resolve any outstanding grievances or underlying conflicts** with any of the departing project participants (and with those participants that remain).

 **DO NOT burn down relationship bridges** with any of the project participants, because in the future you may find yourself or the organization at an impasse due to a damaged former relationship.

Example Case: Burning relationship bridges is not a wise business strategy.

Once upon a time in "the land of hopes and schemes"... there was a shrewd businessman whose commercial enterprise was the construction of bridges throughout the kingdom. Because of the geographic disbursed nature of the land, the businessman would contract each bridge building project to one of the local builders. Each of these primary contractors would in turn hire their own subcontractors as the project's needs dictated.

This contracting arrangement turned out to be highly productive and reasonably cost effective; but then, one day the businessman fell ill and was off sick for a few weeks and thus he was delayed in paying each of his primary contractors on the various projects.

To the businessman's surprise he discovered that these delayed payments had resulted in a beneficial side-effect of earning him more monthly interest on his bank account balance then was usual.

"*Hmmm*", the businessman thought, "*what if I delayed these payments on a continual basis?*" Henceforth, the businessman decided that his new "profit driven" business strategy would be to delay all payments to his contractors.

Month after month the additional bank interest grew, so month after month the businessman extended out the duration before he would pay. A little bit more each time until he found the sweet spot of where the contractors would come begging for their invoices to be paid so that they could in turn pay their own workers. With the contractors having to beg, the businessman felt empowered, and he did so enjoy rubbing it in with his new signature saying of, "*it's nothing personal, it is only business*".

Then one sunny day with profits booming, the businessman was offered his lifelong goal of the prestigious contract to construct the royal bridge into the capital city.

His name would be known throughout the land, he would be a man of men,
he would be richer than rich ... he could only smile with glee.

Once upon a time in "the land of hopes and schemes"... there was a shrewd businessman whose life-long goal was slipping from his grasp as contracted builder after builder turned down his proposition to come work on the prestigious project to construct the royal bridge into the capital city.

His name was backhand spoken throughout the land, he was a man that no other man wanted to be, he was on the verge of bankruptcy... he could only lament his past "profit driven" business strategies.

Then one day when things were at their worst, a smiling builder came forth and offered his services to the businessman.

The businessman smiled with glee; finally, a hope to revive his dream.

"*I will pay you handsomely*", said the businessman.

The builder replied, "*Not so fast, as I'll be the one setting the conditions, and you'll pay me half up front and the remainder within one month of completion at this fee*".

"*What !!!*", gulped the businessman as he read the rate of quadruple the standard fee.

With a smirk the builder replied, "*It's nothing personal, it is only business*".

Mutually Beneficial Relationships

For the purposes of **sustaining productive working arrangements** into the future, it is essential to maintain **mutually beneficial relationships** with all of the project's participants. This can be achieved by, **building each relationship on the principle of minimizing any additional burden that could be incurred by the other party** (excluding that which each party has agreed to accept and be compensated for).

That is, **DO NOT build relationships based solely on one's own self-interest and sole advancement**.

> **"YOU MUST NEVER TRY TO MAKE ALL THE MONEY THAT'S IN A DEAL. LET THE OTHER FELLOW MAKE SOME MONEY TOO, BECAUSE IF YOU HAVE A REPUTATION FOR ALWAYS MAKING ALL THE MONEY, YOU WON'T HAVE MANY DEALS."**
>
> J PAUL GETTY

To put it another way; the **relationship**'s future prospects **will be short lived if** one or more of the parties involved feels that they are being **taken unfair advantage** of, and they perceive that they are expected to take on a **disproportionate amount of the burden / risk** when compared to their Return On Investment; i.e. a misalignment between their risk versus reward.

> *For example; a primary contractor who pushes the majority of the project's risks down onto the subcontractors, but the primary contractor still manages to take home a disproportionately large part of the proceeds. Eventually some very disgruntled subcontractors will leave to pursue more **equitable opportunities**. Worse, the customer could find out exactly what has been going on and decide that for all future endeavours it would be better to cut out the **non-contributing** primary contractor (aka **middleman**) and instead switch to a more equitable primary contractor or work directly with the subcontractors.*

If the other party (*e.g. the subcontractor*) feels that they have previously not been receiving a fair & equitable deal then when it comes time for the next deal or extension of the current deal, that party (having realized their own importance to the project's success) could become a lot more difficult to negotiate with and wanting greater remuneration. That is, the negotiation could turn into a protracted debate with inflated pricing and/or scope minimization as they (*the subcontractor*) tries to recoup some of that which they feel has previously been "stolen" from them.

 I would recommend that ... in some situations of where the other party (*e.g. the subcontractor*) has suffered documented extra costs that were beyond their control (i.e. **force majeure**) then it may be in the long-term best interests of the performing organization (and even the customer organization) to ensure that the other party is compensated sufficiently, so that the other party at least covers the costs of their participation on the project.

That is, the performing organization and/or customer organization make up the difference out of their own pocket, thereby ensuring that the other party at least "**breaks even**".

This "break-even" compensation may be advisable for self-interested strategic reasons such as:

- The subcontractor is the only one who really understands how the system/application is held together. Hence it is not in the organization's short-term interest for the subcontractor to go broke (i.e. out-of-business) and therefore to be unable to support the system/application during its critical warrantee run-in period.

- It may be in the organization's long-term best interest to throw some temporary work in the sub-contractors direction, knowing that there is a more important / profitable endeavour on the horizon that would significantly benefit from this sub-contractor's participation.

This is, not being charitable on the performing organization's part, but rather being strategically savvy for the success of future endeavours.

17.4.4.   Question: "What did WE LEARN from this?"

Ever wondered, why some projects end in failure, while others are successful?

Think of the last ~~"failed"~~ unsuccessful project that you were involved with.

In your "honest opinion", why did that project ~~fail~~ not succeed?

1) Was it due to **mismanagement of the traditional project determinants** of
   [Scope], [Time], [Cost], and [Quality]?

2) Was it due to **restraints imposed** on the availability of [People] and/or [Resources]?

3) Was it due to **unforeseen technical difficulties**?

4) Was is due to **lack of understanding** of what the customer intended, and/or
   lack of relevant knowledge?

So, did real thought go into determining why that project (and other projects) ended so

badly, or conversely, why some projects were considered successful?

Was there some form of post-mortem to determine why the project ended the way it did?

That is, has there been some investment made into **"learning the lessons from the past"**,

or will the next project's success or failure be simply left to chance?

LEARNING FROM YOUR MISTAKES IS SMART;
LEARNING FROM THE MISTAKES OF OTHERS
IS WISE.

Umm, **there was no concerted effort to learn from that project or other projects; why is this so?**

I would put this lack of effort to learn from the past, down to a deficiency of commitment from the project's ex-team members and ex-stakeholders to participate in a meaningful analysis & review of that project.

While there may be superficial support for conducting such analysis & review, its undertaking often gets minimal effort or it gets "put onto the backburner" because there is some other time critical thing that needs to be taken care of first, ... then something else requires immediate attention, ... until, *"I would like to help you with that, but my time would be better spent on my current project, rather than going over that project we finished months ago"*.

 I would recommend that ... some form of project analysis & review be conducted as part of the project closure process and NOT as a tack-on that is held weeks or months after the project has ended.

 I would also recommend that ... as a way to overcome this problem of lack of meaningful participation, it is important to emphasise the benefits to their new project of considering the useful ideas and good practices that they could take from this old project and thereby help improve the management & implementation of their current project. That is, emphasise that the closed project's analysis & review is "forward thinking" rather than "backwards focused".

This project closure analysis & review can be undertaken via a few different forms:

1) Project Audit

2) Project Inspection

3) Project Autopsy

4) Project Reminiscing

THOSE WHO DO NOT LEARN FROM THE PAST ARE BOUND TO REPEAT IT.

Each of these listed project closure analysis & review types is based on differing proportions of "quantitative" and "qualitative" information exchange.

- ❖ **Quantitative Project Closure Analysis & Review** – is concerned with the **analytical comparison of the "planned" and the "expected" values against the "actual" values.**

  That is, a **consideration of the cold hard facts about the project**; *e.g. earned value performance measures, delivery dates for project milestones, profit & loss, return on investment, scope coverage, numbers of defects raised & closed, the severity & priority of these defects, agile burn down velocity, risks & issues closures, durations of risks & issues resolution, people & resource utilization rates, inventory levels, ... etc.*

  Quantitative project closure analysis & review is about putting numerical values to attributes & aspects that describe the project. Thereby enabling a systematic & logical judgement of the project as either a success or a failure. Consequently, providing a rational & unequivocal PASS : FAIL verdict on the performance & competence of the project team and the project's management.

- ❖ **Qualitative Project Closure Analysis & Review** – is concerned with **anecdotal evidence and accounts of experiences that descriptively compare the "expected" outcomes against the "actual" outcomes.**

  That is, the participants provide their **thoughts & opinions** on what helped to make the project a success, and what hindered or pushed the project towards failure.

  Qualitative project closure analysis & review is about understanding the interrelationships that defined the project's outcomes. Thereby, enabling the participants to take away; personal memories, comprehensions, and reference notes on what could work well for them on their next project and what would probably not work well for them in the future.

Which is better to use, "qualitative" or "quantitative" project closure analysis & review?

The type of project closure analysis & review that should be utilized per project is dependent on what are the intended uses of the findings and outcomes of such an analysis & review ... and, who will participate in this review.

Let's be honest, in 3 – 6 – 12 months' time after the project is completed, are any of the participants (let alone anyone else in the performing or customer organizations) going to get much value from these quantitative metrics and bullet pointed findings?

In the real-world, the answer is "NO", unless it is to produce some analytical report for senior management to ponder. For the majority of the time, this historical documentation (*e.g. the project reports and administrative records*) won't be looked at by the members of the project team nor by other projects. Most probably, the only occasion when such information will be accessed is when plagiarising portions of its contents to help accelerate the creation of similar quantitative analysis & review materials for some other project.

Quantitative analysis & review is useful for establishing Quality Control policies & procedures, where the resultant numbers can be correlated & compared in some form of process feedback loop.

Qualitative analysis & review is useful for establishing Quality Assurance guidelines, where the description of how things were done and should have been done can change the thinking of those persons involved with the Process Improvement loop.

While **quantitative project closure analysis & reviews** (i.e. Project Audits and Inspections) serve quality and administrative purposes, this **SHOULD NOT supersede the need to undertake qualitative project closure analysis & reviews** (i.e. Project Autopsies and Reminiscing).

17.4.4.1. Project Audit

A **project audit is a formalized objective appraisal of the project's previous processes & procedures.** This audit will **systematically examine whether pre-establish sets of Audit Criteria** (up to the time of the audit) **have or have not been complied with by the project's participants**.

This project audit would be **undertaken by** an accredited organization / individual external to both the performing and customer organizations; i.e. an **independent third party**.

This **auditing party would be contracted to determine whether the specified project processes & procedures** (such as recognised industrial quality standards & guidelines) **were followed, and whether managerial due-diligence has been exercised.**

Though, the resultant findings have to be based on a pre-approved set of Audit Criteria; i.e. something tangible to pass judgement against.

This **auditing party should be impartial in their appraisal, and they should have no invested interest in whether the project is a success or a failure.** Secondly, this auditing party may not even know or understand the subject matter (or the underlining technology) on which the project is based.

The **outcome of this project audit will be an Audit Report containing the auditing party's observations** on the project's "apparent conformance" to those specified project processes & procedures, identification of revealed non-conformances & irregularities, notes about these discovered non-conformances, and maybe ~~recommendations~~ *suggestions* on quality improvements.

Due to potential legal ramifications, the auditor probably will not make "recommendations", rather they may only provide "suggestions".

NOTE ❑ The **project audit can only ever provide a reasonable assessment** as to whether the specified processes & procedures were followed, and pass comment on the extent of project's management due-diligence.

Hence, think of the project audit as a stranger coming into the performing organization and examining whether, "the i's have been dotted and the t's have been crossed". Though, not necessarily confirming that the sentences make a lot of sense in the overall scheme of things.

The auditing party would **confirm evidentiary conformance with the pre-approved Audit Criteria** by:

1) **Visual inspection of the project artefacts and associated Objective Quality Evidence (OQE)** such as; the various project management plans, the acceptance test reports, the risks & issues register, ... concessions, certificates, licences, manufacturing data, BOMs, ITRs, ATRs, COCs, DECLs, RTM, TCM, BCR, WBS, ... etc.

   That is, examining the ingredients of the acronym soup.

2) **Interviews with members of the project team**, though this is usually limited to the project's manager and the person who is functioning as the performing organization's quality authority.

NOTE ❑ What the auditing party would not be interested in is the Earned Value Performance Measures and KPIs; see [Section 7.3.1] and [Section 12.3.1]. However, this type of metrics would be of concern to the Project Inspection.

Project Audits is a High Altitude Inspection

Due to constraints imposed on the audit (such as time limitations, and the classification and confidential nature of project materials), often the project audit is limited to (the analogy of) a high altitude inspection of the project's terrain, followed by a focused zoom-in on those areas that standout as strange, unusual, or obvious targets for further examination. Hence, the auditing party may only be able to verify:

1) whether the project artefacts do in fact exist (such as those documents laid out in the project's documentation hierarchy in [Figure 142]),

2) that some form of document management and version control is being utilized,

3) that some form of content / configuration management has been applied to controlling what is going into the deliverables,

4) that the project documentation selected for closer scrutineering does appear to contain meaningful information, and

5) that there are appropriate cross-references between related documents in the project's documentation hierarchy.

Due to the amount of cross checking and glossary examination, it is important that the project manager prepare for this audit. That is, to ensure that all of the project artefacts and documentation are completed, up-to-date, and that these are easily accessible when requested. Consequently, **it is essential that the project's planning documentation and the Objective Quality Evidence (OQE) is accumulated, catalogued, and given a once-over prior to inspection**; see [Section 17.4.2.1].

 The **project audit cannot provide assurances that the required processes & procedures will be followed in the future**, nor that project management will henceforth exercise due-diligence.

The need for these "assurances" is the reason for Project Inspections.

17.4.4.2. Project Inspection

A **project inspection is a semi-formal assessment of the project's current performance**; i.e. an examination of the current phase in the project's life cycle (the Initiating, Planning, Executing, Monitoring & Control, or Closing Phase).

While the Project Audit maybe a one-off event, **Project Inspections should be performed multiple times throughout the life of the project**, either:

- **Routinely** at specific intervals (*e.g. monthly, quarterly*).

  These "Routine" Project Inspections should be an anticipated part of the project's regular activities. As such, these inspections should be short duration events that are conducted in as unobtrusive manner as possible, so as not to greatly disrupt the current work activities of the Project Implementation Team.

- **Triggered** at the conclusion of a project phase (i.e. a Phase Completion Review 'Gate' such as a Post Implementation Review [Section 17.4.2.1]), the culmination of specific project activities, or in association with a major project milestone (i.e. as a Post Completion Review [Section 17.4.2.2]). Such an inspection will likely be of a longer duration event that requires the preparation of the project manager and can also involve various members of the project team, as well as including the senior management of the performing organization, and potentially the representatives from the customer organization.

- **Arbitrarily** at the **discretion** of the project's stakeholders. The "Discretionary" Project Inspection can be called for at any time by the senior management of the performing organization and potentially by the representatives from the customer organization. This inspection can be of whatever duration is deemed necessary to address the major concerns of the project's primary stakeholders, and can involve as many members of the project team (and other "situational experts") as required to satisfy the concerns of these primary-core stakeholders.

While the project manager may not be able to schedule in advance for the Discretionary Project Inspection, the project manager should schedule for the occurrence of the Routine and Triggered Project Inspections, and prepare in advance when each one of these inspections are due to occur.

What is the difference between Project Inspection and a Project Audit?

❖ **Project Audit** is an **objective appraisal of the** project's **previous activities**, and hence is an **assessment of** the conformance proficiency of the **"then" project participants** when **compared to** a pre-specified **set of Audit Criteria**.

That is, a **Project Audit is "historically" focused** so as to learn what changes need to be made to how things are done in the future.

❖ **Project Inspection** is an **assessment of** the project's **current progress**, and hence is a **judgement on** the competency & capabilities of the **"now" project participants** when **compared to a set of tangible measures**.

That is, a **Project Inspection is "today" focused** so as to learn what changes need to be made to how things are currently being done.

Other differences between a Project Inspection and a Project Audit are:

● Unlike the Project Audit, the Project Inspection is not conducted by an independent third party. Rather, the **Project Inspection is carried out by** either; a **First Party** (internal to the performing organization), **or a Second Party** (internal to the customer organization). This **inspecting party probably has an invested interest in whether the project will be a success or a failure**, and this inspector may know & understand the subject matter and/or underlining technology on which the project is based.

● Unlike the Project Audit, the **Project Inspection would be interested in the Earned Value Performance Measures**, as well as KPI **Traffic Light Indicators** used to draw attention to that performance area of immediate concern. See [Section 12.3].

The inspecting party would determine; whether the project is currently progressing as per the approved baselines, whether established processes & procedures are currently being followed, and whether the project's management are currently exercising appropriate due-diligence?

That is, the Project Inspection would investigate:

1) **How is the project progressing when compared to the [Scope Baseline]?**

   What is the coverage of the **Requirements Traceability Matrix (RTM)** and the **Work Breakdown Structure (WBS)**? See [Section 12.1].

2) **How is the project progressing when compared to the [Time Baseline]?**

   On the **project schedule (Gantt chart)**, are the progress lines on the task bars in front of or behind the current date line? See [Section 12.2.1].

   Alternately for an agile based project, is the actual burn down line tracking close to that of the ideal burn down line on the **Burn Down Chart**, or is the actual burn down line swinging away from the ideal? Are the story cards traversing the **Task Board** at a regular velocity, or are too many cards being moved back as needing "To Do"? See [Section 12.2.2].

3) **How is the project progressing when compared to the [Cost Baseline]?**

   Are the **Earned Value Performance Measures** of the Actual Costs (AC) and the Earned Value (EV) tracking closely to the Planned Value (PV) S-Curve? See [Section 12.3.1] and [Section 12.3.2].

   Is the trend for the **Cost Performance Index (CPI)** and the **Schedule Performance Index (SPI)** staying between the upper and lower comfort lines of the Performance Index Window and converging towards a consistency of ONE? See [Section 12.3.3].

   Is the project motoring along nicely, or does the project lack traction as it stationarily spins its wheels with lots of smoke and noise, but not much real progress to show?

4) **What is the current state of the Key Performance Indicators (KPIs)** that are being used to represent the current state of the project? See [Section 15.5].

5) **Is the project satisfying the [Quality] expectations?**

   - Are the opened **Defect Reports** exceeding those being closed? [Section 13.1.3].
   - Is the **Rates Of Failure** for the deliverable units on the rise or fall? [Section 13.1.4].
   - What are the reports from 'Verification, Validation & Testing' (VV&T) team noting about the **Acceptance Tests**? [Section 13.1.1].
   - What is the PASS : FAIL coverage of the **Testing Coverage Matrix (TCM)**? [Section 17.4.1.1].

6) **How rigidly have the Quality Assurance and the Quality Control processes & procedures been complied with, and has the Objective Quality Evidence been produced that would satisfy the scrutiny of a project audit?**

   Are the members of the project team abiding by the project's designated quality standards & guidelines, and is the project team adhering to the performing organization's established best practices?

   Or is the project just a free-for-all of activity and half-baked documentation?

7) **How effectively & efficiently has the project utilised its allocation of [People] and [Resources]?**

   Or, has the project been sucking in everything around it - like some swirling black hole in the centre of the office?

   Has the project's material components been piling up to the ceiling?

   Have project team members been overworked & irritable, or are they obviously bored & restless?

8) **Have the prescribed project management processes & procedures been followed?**

Has the project manager been using sound project management principles, and has the project been operating in accordance with the methodologies & strategies laid out in the Project Management Plan?

Or is the project manager just making it up on a wing and a prayer?

9) **How capable is the current project management** at supervising and exerting control over the conduct of the project?

Does the project manager know what is going on with the project, and does the project manager know what they are doing?

Or is the project manager just a waste of space … an oxygen thief?

10) **What is the current state of the project** and does this equate to that which was previously described in the **Project Status Reports**?

Have the project's primary-core stakeholders been presented with a realistic representation of the project's current predicament?

Or is the regular Project Status Report better utilized as toilet-paper?

11) **What is the status of project's risks & issues** and has the Risks & Issues Register been kept up-to-date?

Are there any previously unidentified or unreported concerns that could threaten the success of the project?

Or is the project said to be coming along swimmingly, only to be flooded with problems closer to the next major milestone-release?

12) **Is the project ready to advance** to the next milestone or to the next phase?

All is fine so let's move along, or should a do-over be commenced?

Consequently, the outcome of each type of project inspection should be a better understanding by the project's primary-core stakeholders as to the true state of the project, so that they can decide how best to proceed responsibly with this project.

Should they be considering whether this particular project needs to be managed differently, re-baselined, and ultimately deciding on the fate of this project ... or should they instead be focusing on another project that requires more of their attention?

In conclusion, the purpose of the Project Inspection is to:

1) determine whether there are any things that need improving on the project,

2) to notice if there are any inefficiencies with how the project is being conducted,

3) to detect any oversights with the project's management, and

4) to suggest ways to rectify any such imperfections.

Hence, think of the **Project Inspection** as the quality & management cops **checking that the project "has all of its ducks in a row"**.

PROJECT INSPECTION...
WHEN OTHERS FIND OUT WHETHER THE PROJECT'S MANAGEMENT HAS ALL OF ITS DUCKS IN A ROW.

Watch out for project inspectors that try to **retrospectively enforce quality processes & procedures that were not specified / required** (or not standard operational procedure) at the time of the project's Planning and Execution. That is, imposing the retrofitting of quality processes & procedures that were **not agreed to be abided by prior to and during the project's undertaking**.

Retrofitting Quality Standards & Guidelines

Due to the occurrence of the Project Inspection there may be a recommendation made to retrofit quality standards & guidelines to the project. Depending on where the project is up to in its life cycle will influence whether this recommendation is a worthwhile proposal or a pending bureaucratic disaster.

If the project is nearing completion (or has finished) then the retrofitting of quality standards & guidelines is not a good idea, because such retrofitting will result in a significant increase in the [Time], [Cost], [People], and [Resources] burden on the project and on the performing organization. ... Unless of course, the lack of these particular quality standards & guidelines will mean that the performing organization risks having the project deliverables being rejected, or the performing organization risks losing / not obtaining quality accreditation that is essential for the ongoing operation of the business.

I would recommend that ... the project manager should "push back" (actively oppose) the undertaking of any activities that won't directly contribute to the successful conclusion of the project. Meaning that, conformity to quality standards & guidelines for bureaucracy sake is not an acceptable resolution.

Hence, it maybe necessary to compromise on a subset of the quality standards & guidelines that will be applied to the project, so as to ensure that the project's agreed [Scope], is delivered on [Time], within [Cost], yet still satisfying the [Quality] expectations.

Furthermore, due to the effort involved with the retrofitting of such quality standards & guidelines, this can result in very disgruntled and resentful project team members who have to do what they consider to be "non-productive" "bureaucratic" work.

These persons' resultant "annoyance" can fester into a perception that those particular quality standards & guidelines will be a burden to any future projects that they happen to be involved with. *... Oh, go away you QuACK!!!*

17.4.4.3. Project Autopsy

Right from the get-go, the **Project Autopsy review must be "conducted in the spirit of learning", and not as a search for someone to blame, shame, and flame for the project's failures**. If there is, the slightest perception that the Project Autopsy will turn into a "blame-game" or "witch-hunt" then the potential attendees of such a review will be averse to participating, knowing that they could end up being the one who is "cut into".

DO NOT turn the Project Autopsy review meeting into a search for someone to blame for the project's failures.

DO NOT turn the review meeting **into a point scoring exercise** of overstating some groups / individuals to the detriment of others.

DO NOT turn the review meeting **into a systematic & analytical appraisal** of the project's [Scope], [Time], [Cost], [Quality], [People], and [Resources] metrics.

DO NOT place the **emphasis on Earned Value Performance Measures**.

In such a negative environment, the reviews participants may decide that it is in their own self-interest to "save face" and either; hide those problems that they knew / know about (i.e. sweep it under the carpet, or hide it in the closet), be reluctant to reveal anything that is less than perfect or redress the facts (i.e. "sugar-coating"), or worse redirect the focus and eventual blame onto someone else's domain (i.e. "to throw them under the bus").

While the title of Project Autopsy does sound like it would be a forensic review of the project's outcomes, this was actually the purpose of the various Project Inspections and the Project Audit ... when actions could have been taken to affect the "then pending" outcome of the project / milestone release.

Consequently, **the Project Autopsy should be used as purely a "learning experience"**.

"REALITY IS ... NO MATTER HOW SKILLED THE PROJECT MANAGER AND THE PROJECT TEAM ... SOMETIMES BAD THINGS DO HAPPEN.

THE CAUSE OF THE PROJECT'S FAILURE MAYBE WAY OUTSIDE THEIR CONTROL. ... THE IMPORTANT THING IS TO, FIGURE OUT WHAT WENT WRONG, WHY IT WENT WRONG. ... AND, HOPEFULLY LEARN FROM THIS, SO AS TO MINIMALIZE ITS POTENTIAL OCCURRENCE IN THE FUTURE.

BUT DEFINITELY NOT TO ASSIGN BLAME, NOR TO REPRIMAND THOSE PERCEIVED AS THE WRONG DOERS."

The **Project Autopsy is carried out after** the project / milestone release has concluded and **the associated emotions have subsided to a level where an open & honest conversation can be held** to discuss:

1) **What went well** on the project?

   - What processes & procedures contributed to the project's successful outcome?

   - What activities worked to the benefit of each participant?

2) **What didn't go so well** on the project?

   - What processes & procedures hindered success or contributed to the project's unsuccessful outcomes?

   - What activities & attitudes should not be encouraged in the future?

3) **What** in each participant's opinion **were the mistakes made** with the project?

   - What should be done and what changes should be made to ensure that such mistakes do not occur in the future?

4) **Were there any participants who could have been more effectively utilized** in a different way? Alternatively, were there participants who were utilized in an optimal manner?

5) **What** resultant **good ideas** should henceforth be incorporated into the performing organization's suite of **best practices**?

6) **What was the feedback received from the primary-core stakeholders?**

7) **What was the feedback from the end-user customer?**

 I would recommend that ... discreet independent review meetings be held with the Project Steering Committee, the Project Working Group, and the Project Implementation Team. In this way, the previously listed points can be discussed in a peer based environment, devoid of the watchful eye and influences of the hierarchy above.

Notice how the previously listed points for the Project Autopsy were all **based around "descriptive exploration" instead of an analytical analysis**; this is because:

- Often by the time that the Project Autopsy review occurs then those previous project participants would generally be more concerned with their current (new) project or the next milestone release rather than reliving that which has now passed.

- It is going to be difficult enough obtaining commitment from the participants to attend such a review without facing an end-of-term grading session.

- In a month, six months, a year, or more time these participants would not retain much benefit from an analytical review, but the benefit would be retained from remembering, "Next time I need to ...".

NOTE ☐ It is **highly beneficial to have** this **Project Autopsy review at the conclusion of each major milestone release**, as well as at the conclusion of the entire project. With an agile implementation, this self-assessment would involve the Project Implementation Team getting together at the end of each sprint for a round-table discussion; see [Section 7.4].
For a waterfall implementation, this discussion could be at the conclusion of each Phase Completion (Gate); see [Section 3.6].

Hence, **the Project Autopsy needs to provide benefit to each participant so as to help them ensure that their next project is a successful endeavour.**

17.4.4.4. Project Reminiscing

While that project is now just a distant memory, it is surprising, how many times throughout each project participant's career; office chitchat will contain some reminiscing about the events related to that now past project.

"When I was a developer, we ..."

While on the surface, such project reminiscing may not provide much quantitative information as with the other forms of project closure analysis & review; however, **this reminiscing does contain the essence of the major lessons to be learnt from these past projects** (given that these lessons are still being recalled today).

"That project was such a disaster, because ..."

17.5.   Outputs

17.5.1.   Output: Documents & Deliverables

Now that the big questions have been answered, the next step is to ensure that these answers have been documented. The resultant documentation from the Closing Phase is highlighted below in [Figure 141], and the inter-relationship of this documentation to other documents produced during the project's life is illustrated in [Figure 142].

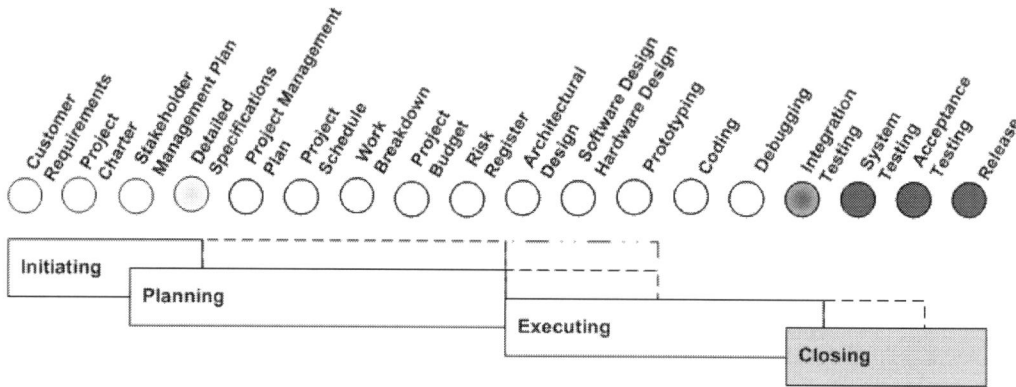

Figure 141: Document output of the Closing Phase.

Though, [Figure 141] and [Figure 142] does not contain all of the possible Closing Phase output documentation, as some of the required documentation will be dependent on the industry/market segment that the performing and customer organizations operate in.

Some of this documentation will be a mandatory deliverable that is stipulated in the project's requirements, while other documentation will be necessary for compliance with relevant industrial quality standards & guidelines.

The trick is to provide the necessary documentation to a level of quality that will satisfy the authorizing project stakeholders.

NOTE   ❑ Several of the outputs of the Closing Phase serve to complement / "bookend" the outputs from the Initiating Phase and the Planning Phase.

17.5.2. Output: PMBOK-SDLC Map

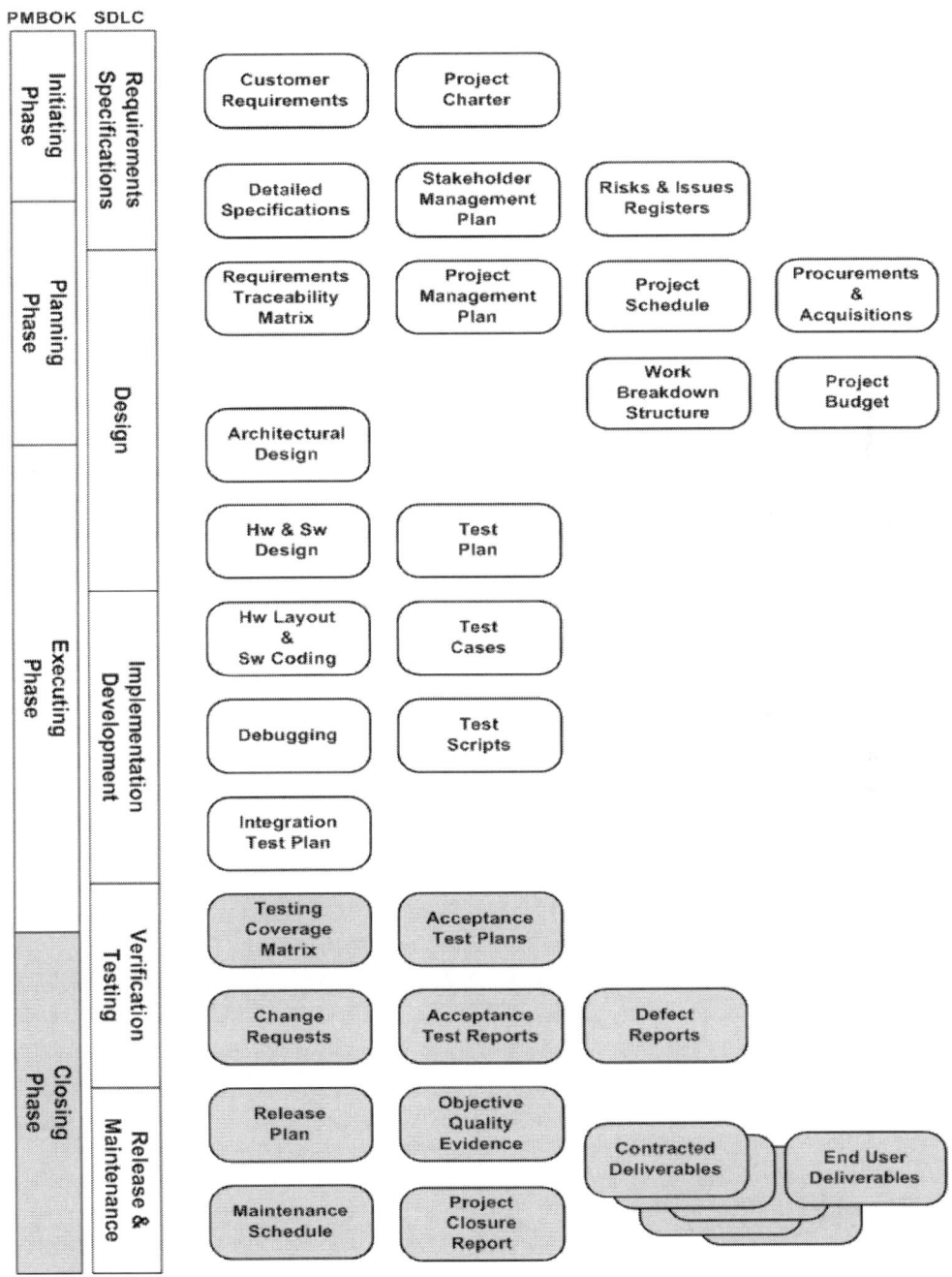

Figure 142: Relationship of PMBOK - SDLC and the documents & deliverables produced for the Closing Phase.

Table 2: Cross-Reference of the Closing Phase outputs to Initiating and Planning.

Initiating Phase	Planning Phase	Closing Phase
Project Charter	Project Management Plan	Project Closure Report
Customer Requirements	Detailed Specifications Baseline Change Requests	Acceptance Test Reports, Baseline Change Authorization
Detailed Specifications (features & functionality)	Requirements Traceability Matrix	Testing Coverage Matrix, Defect Reports
Customer Requirements	Detailed Specifications (Acceptance Criteria)	Objective Quality Evidence

17.5.3.   Output: Project Closure Report

The major output document of the Closing Phase is
the **Project Closure Report** because this particular
document **denotes THE END of the project / milestone
release**. When for all intents, the project / milestone
release is finished then the Project Closure Report
would be signed-off (and dated) by Representatives
from both the performing and customer organizations.

NOTE   ☐ It is essential that the **Project Closure Report be signed-off in a timely
manner, as these signatures signify that the project has concluded**.
Subsequently, any outstanding invoices / claims against the project by involved
parties should be concluded accordingly.

     For more information on the Project Closure Report; see [Section 17.4.2.2]

17.5.4.   Output: Objective Quality Evidence

❖ **Objective Quality Evidence (OQE)** is the **accumulated "Artefacts Of Proof"** that **the project's deliverables do satisfy the mutually agreed Acceptance Criteria**, and hence these deliverables should thus be accepted by the customer's representative.

The Objective Quality Evidence is the "Goldilocks" deliverable of the Closing Phase; i.e. not too little, not too much, but just enough to make the relevant project stakeholders happy to "tick-off" the acceptance of that deliverable.

NOTE   ❑ For some organizations, and/or projects then a minimal quantity of OQE is required. However for other industries, market-segments, organizations, and/or projects then a sizeable amount of OQE will have to be produced and presented ("in triplicate") for the deliverables to be deemed acceptable.

The trick is to provide the amount of OQE that will satisfy the relevant project stakeholders ... nothing more nothing less ... just proof of satisfaction.

 For more information on Objective Quality Evidence; please see [Section 13.1.2] and [Section 17.4.2.1].

17.5.5.   Output: [Baseline] Change Requests

The **Change Requests details what were the requested alterations to the project's [Baselines]**; predominately [Scope] changes that request modification to the features & functionality, and/or to the stipulated deliverables. The repository of these Change Requests (i.e. CMS – Change Management System) will need to provide traceability of those changes that were and were not agreed to be implemented, as well as those changes that were and were not authorized to be implemented in specific project releases.

 For more information on Change Requests; see [Section 10.2] & [Section 13.1.2].

17.5.6. Output: Defect Reports

The **Defect Reports details what non-conformances in functionality & features were detected during the acceptance testing**. The repository of these Defect Reports (i.e. the Defect Tracking System) will need to provide traceability of those defects that were and were not closed, in which release of the deliverable was the defect found, in which release was the defect rectified, as well as identifying those defects that were ruled-out as being Change Requests.

 For more information on Defect Reports; see [Section 10.2], [Section 13.1.2], and [Section 13.1.3].

 I would recommend that ... those internally discovered defects be kept separate from those defects found by the customer's representatives, because a high number of reported defects (irrespective of their relevance to the project deliverables) will impact on the perceived quality of the deliverables.

17.5.7. Output: Acceptance Test Plans and Test Reports

The **Acceptance Test Plans details how specific features & functionality (specified in the approved Detailed Specifications) will be tested to verify & validate conformity**.

The **Acceptance Test Reports details the outcomes & findings of each of those tests that were and were not conducted during the execution of the acceptance testing** (using the approved Acceptance Test Plans as guide to those tests to be performed). It is not unusual for these Acceptance Test Reports to be handwritten records on the Acceptance Test Plan. The important thing is for there to be a traceable record of; what tests were and were not conducted, what was the PASS : FAIL status of each of these tests, and what were the documented findings & conclusions of each of these tests.

 For more information on acceptance testing; [Section 13.1] & [Section 17.4.1.1].

17.5.8.   Output: Testing Coverage Matrix

The **Testing Coverage Matrix (TCM) details what planned tests** (i.e. as defined in the Acceptance Test Plans) **were and were not undertaken during the execution of the acceptance testing, and what was the confirmed PASS : FAIL status for each of these conducted tests**. That is, the Testing Coverage Matrix presents the confirmed coverage of those features & functionality that were defined in the approved Detailed Specifications; as illustrated in [Figure 140] in [Section 17.4.1.1].

Ideally, the Testing Coverage Matrix that is presented to the customer's representative should **generate a sense of confidence that the pending deliverables will be as per the agreed Acceptance Criteria**. As demonstrated by the candy-stripped colour coded spreadsheet illustrated previously in [Figure 101].

For more information on the Testing Coverage Matrix; see [Section 17.4.1.1] and [Section 17.4.2.1].

17.5.9.   Output: Release Plans and Release Notes

The **Release Plans details when and how specific milestone releases of the deliverables will be provided to the customer organization.**

The **Release Notes details any specific information about the milestone release's deliverables that the customer organization / end-user should know.** These Release Notes would include such things as; what features & functionality were included (*e.g. a Change Log*), what features& functionality were not included (*e.g. things that were left out that may have been expected to have been included*), specific configurations & instructions that will be required for the delivered features & functionality to work properly, and any restrictions / constraints on the operation of the features & functionality provided.

That is, what does the customer or end user need to know so that they will not turn in any spurious Defect Reports or make unjustified warrantee claims.

17.5.10. Output: Maintenance Schedules and Warrantee Agreements

The **Maintenance Schedule is** analogous to **the regular periodic maintenance plan** for your new car, so that the manufacturer's warrantee won't be null & void. The performing organization and/or third party vendors & suppliers may impose similar maintenance and Warrantee Terms & Conditions on the deliverables provided to the customer organization.

17.6. Things to Watch Out for

When undertaking the Closing Phase there are a few things that have to be watched out for, these are:

- **Know when to call it quits ...**

  That is, **DO NOT allow the project to linger on any longer than it needs to**. There can be a tendency for some projects to continue almost indefinitely well after the project deliverables have been turned over to the customer's representative. **This situation often occurs when there is no clear distinction made between the stages of; producing the deliverables, warrantee support of those deliverables, and the ongoing maintenance of the deliverables**.

  Recall back in [Section 1.2];

  > *"A project is a limited duration unique endeavour that produces a one-off set of deliverables that are not brought about by continually ongoing repetitive operations; i.e. not Business As Usual (BAU)."*

  Hence, **during the Initiating Phase and the Planning Phase, a clearly defined set of Exit Criteria should have been agreed to (and signed off)** by the representatives from both the performing and customer organizations.

These Exit Criteria would be detailed in the Project Charter sections of the 'Project Acceptance Criteria', the 'Tangible Measures', and the 'Project Closure & Termination'; see [Section 3.4.4].

 **Beware that the project does not evolve into an ongoing program of work.** It would be best to close out the current project and then start up a new job / Cost Account Code for the BAU activities.

 I would recommend that ... a large project be broken up into smaller milestone projects that have their own unique job / Cost Account Codes (and subsequent payment milestones).

For example a multiple staged project of;
(1) development and delivery of the First-Of-Type unit,
(2) delivery of the production run of units,
(3) warrantee support of the production units,
(4) field support & maintenance of the production units.

- **Know when tweaking and defect fixes transcends what was intended ...**

Sometimes it can be difficult to clearly identify when the concluding activities of tweaking the performance & operational characteristics of the project deliverables and defect fixes crosses over to being an extension of what the customer intended when the project was conceived and not what was agreed & authorized to be implemented. That is, **scope creep disguised as defect rectification**.

 **Beware when project team members (in good faith) go beyond the bounds of the project's scope and give more than what was intended.** That is, keep an eye on what is a defect, and what is a Change Request in disguise. As the mistaken implementation of Change Requests as defect rectifications can easily result in the project haemorrhaging in [Time] & [Cost].

- **Know when and when not to give away work for "free" ...**

With many a project there will be a grey area of what is and is not within the [Scope] of the project. While the approved Detailed Specifications may have been written to describe the deliverables (as best as possible), there will always be some scenario / situation that the Detailed Specifications did not cover in enough detail to clarify the deliverable's features & functionality.

Hence, **the representatives of both the performing and customer organizations will need to "compromise" on whether such "shade to grey" work is or is not within the scope of the project**.

- **Know when not to get between an irate customer and a spoilt brat ...**

We have all witnessed the situation (most probably just after Christmas) when a shop-assistant is bailed up by an irate customer (parent) because their "little darling" of a child (read "spoilt-brat prima-donna") didn't get the exact model / version of the present that was wanted. It is as though it is the shop-assistant's failing that they didn't understand exactly what was wanted, even though the customer / parent requested the wrong thing in the first place.

Well it may or may not come as a surprise that, in the corporate / business world similar redirection of blame for incorrectly specified end-user deliverables can be passed from the customer organization (representatives) to the performing organization (representatives). This shift in responsibility & accountability is based on the premise that the performing organization (as the "Subject Matter Expert") should have known what was intended and not what was asked for.

 I would recommend that ... the performing organization fall-in behind the protective shield of the approved Detailed Specifications and the various documented communications exchanges (e.g. email summations).

 **BEWARE** the impact of insufficient end-user training on project success.

- **DO NOT underestimate the impact of insufficient end-user training ...**

Insufficient end-user training can have a significant impact on the perception of the project's deliverables and subsequently the project as a failure. **Without appropriate end-user training then these end-users can very easily misunderstand, misinterpret, and not know what the deliverables are and are not capable of. Consequently, they could decree the deliverable as "broken" when in fact they are not using the deliverables as it was specified, designed, developed, and agreed as acceptable.**

Hence, due to a lack of relevant & appropriate training these end-users' expectations may not be satisfied, thereby prejudicing their perceptions of [Quality] deliverables.

17.7. Phase Completion Check List

1) Has the **point of completion** for the project **been identified?**	✓
2) Has the **point of completion** for the project **been arrived at?**	✓
3) Has it been understood what the customer wanted & needed to be delivered, and when it was to be delivered? **Hence, will they be satisfied when the box of deliverables are handed over?**	✓
4) Has it been understood what were the desired outcomes, and what the customer perceived as the benefits from undertaking this project? Hence, **will the customer be satisfied when the actual outcomes and the actual benefits are known?**	✓

5) Has it been understood what were the requirements? Hence, **will the customer be satisfied with what they will be getting?**	✓
6) Has it been understood what was the scale of the project, the scope of the project, and the project's measurable & tangible objectives? Hence, **will the customer be satisfied when the deliverables arrive** on their doorstep **and the invoice** for the deliverables **is presented to them?**	✓
7) Have all the required **Objective Quality Evidence and project artefacts been produced** and are these ready for inspection?	✓
8) Have all of the **project's participants and project stakeholders been communicated with** to inform them of what is happening with the conclusion of the project?	✓
9) Have all of the **project's people & resources been released** from service?	✓
10) Have all of the **project's time recording and Cost Account Codes been** completed, up-to-date, and subsequently **deactivated?**	✓
11) Have the **project's risks & issues been laid to rest** or are there outstanding problems that could still cause the project some pain?	✓
12) Does everyone agree that the project is over and has the **Project Closure Report been signed-off?**	✓

Reviewer's Comment ... Success is opinionated.

Just because a project has not complied with the stated baselines and milestones does not necessarily mean that in the primary stakeholders' opinions the project was a failure.

What they deem as a successful project is greatly influenced by each stakeholder's perspective on what could have been achieved given the situation and the prevailing circumstances, and the project participants' actions & attitudes.

War Story ... Obituary of a start-up company.

Well today, I just witnessed the demise of my employer's start-up company, and there was nothing that I as the senior project manager, nor the General Manager, nor the functional managers, nor the general staff could do to prevent this. ... When, the company's Executive had not revealed the true state of things that were occurring behind the scenes.

Lessons Learnt:

1. Cash flow, cash flow, cash flow ... without a continuous cash flow stream, an organization will starve to death, irrespective of how hardworking & efficient the development teams and the staff. A self-generating cash flow stream must be established ASAP. And, not be dependent solely on investment capital, nor the periodic injection of government grants, nor the timely return of taxation concessions.

2. Even with a sizable amount of funds to start with, it is essential to dispense this diligently on what is needed today and tomorrow. And, not excessive capital outlays on distant future plans.

3. Five-year objectives are nice dreams, but realistic business plans for the upcoming 6 and 12 months periods must be in place (and known by operational management) for the business to have a chance to survive.

4. A start-up company must have only one realistic target at a time. And, not dissipate the workforce onto independent product lines. That is, "establishment of a beachhead before trying to invade the mainland".

5. Trust the implementers & the field commanders to do their jobs. And, not second guess their capabilities, as multiple paper degrees & academic theory does not make up for real-world experiences & understandings. And, learn to quickly accept the realities of the current situation and quickly adapt to the prevailing circumstances, else expect to die.

6. People, people, people ... without its people, an organization will end. And, when their respect & trust is lost, it's nay impossible to rebuild again.

INFORMATION PRODUCED BY THE PROJECT IS ONLY AS BENEFICIAL AS THAT WHICH CAN BE ARCHIVED, RETRIEVED, DISSEMINATED, AND READILY REUSED.

INFORMATION NOT RECORDED AND STORED IS KNOWLEDGE THAT SOMEONE ELSE WILL EVENTUALLY HAVE TO SPEND A LOT OF TIME RE-FIGURING OUT.

18. Project Integration and Information Management

18.1. Introduction

If you are reading this book in the physical form then you would probably agree that it is telephone book thick, and hopefully you will consider that it contains a lot of useful information.

Similarly, if you printed out all of the documentation & information related to and produced during the course of your current project then it to would probably be enough to fill a telephone book or few, and for some projects, there would be enough material produced to fill a suite of encyclopaedias.

Thus, the question is ... **How to handle the copious amount of information that will be associated with each and every project?** ... And, **how to intelligently search & readily access this information?**

Like this book, many a project started out simply as a light bulb of inspiration in a dark corner of someone's mind, but as time went on, and as others got involved then the project information grew until there was just too much stuff & goings-on to be kept track of inside one person's solitary head.

Hence, this chapter will touch on the integration of the various project management processes & procedures, as well as raise awareness of the need to manage all of the information that will be accumulated over the life of the project.

18.2. Project Integration Management

It is strange to realize that this book on the practical application of project management has so far not addressed the topic of Project Integration Management (given that it is covered in the PMBOK® Guide). However, there is a simple explanation for why this topic has yet to be covered.

Just as most people don't need to remind themselves that they have to breathe, and that their heart needs to pump blood ... Project **Integration Management is the underlying synergy by which the project is managed on a daily basis**.

As illustrated below in [Figure 143], **Project Integration** can be thought of as **the capstone for the Project Management Process** model.

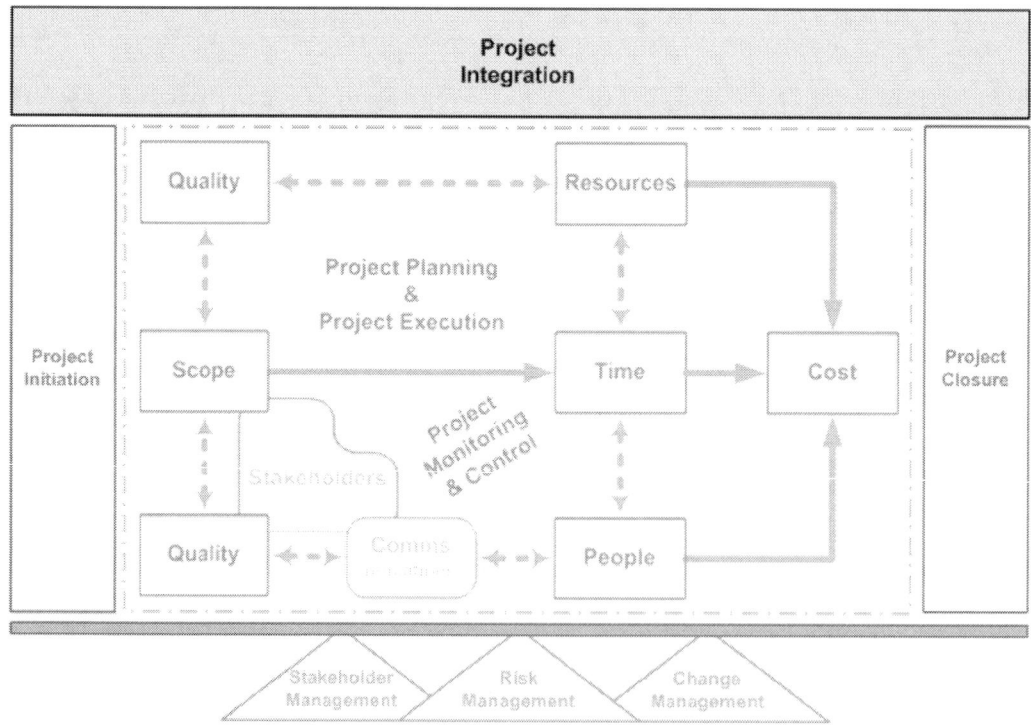

Figure 143: Relationship of Project Integration Management to the Project Management Process model.

From a "practical" perspective, **Project Integration Management is the daily intermixing of the various project management models** that have been detailed throughout this book.

Please, refer to the PMBOK® inspired "Project Management Process" model, the "Project Variable Star" model, the "PLAN – DO – CHECK – ACT" model, the "R.I.S.C. Management" model, and the "Four Re's" model of Project Rescue & Project Recovery.

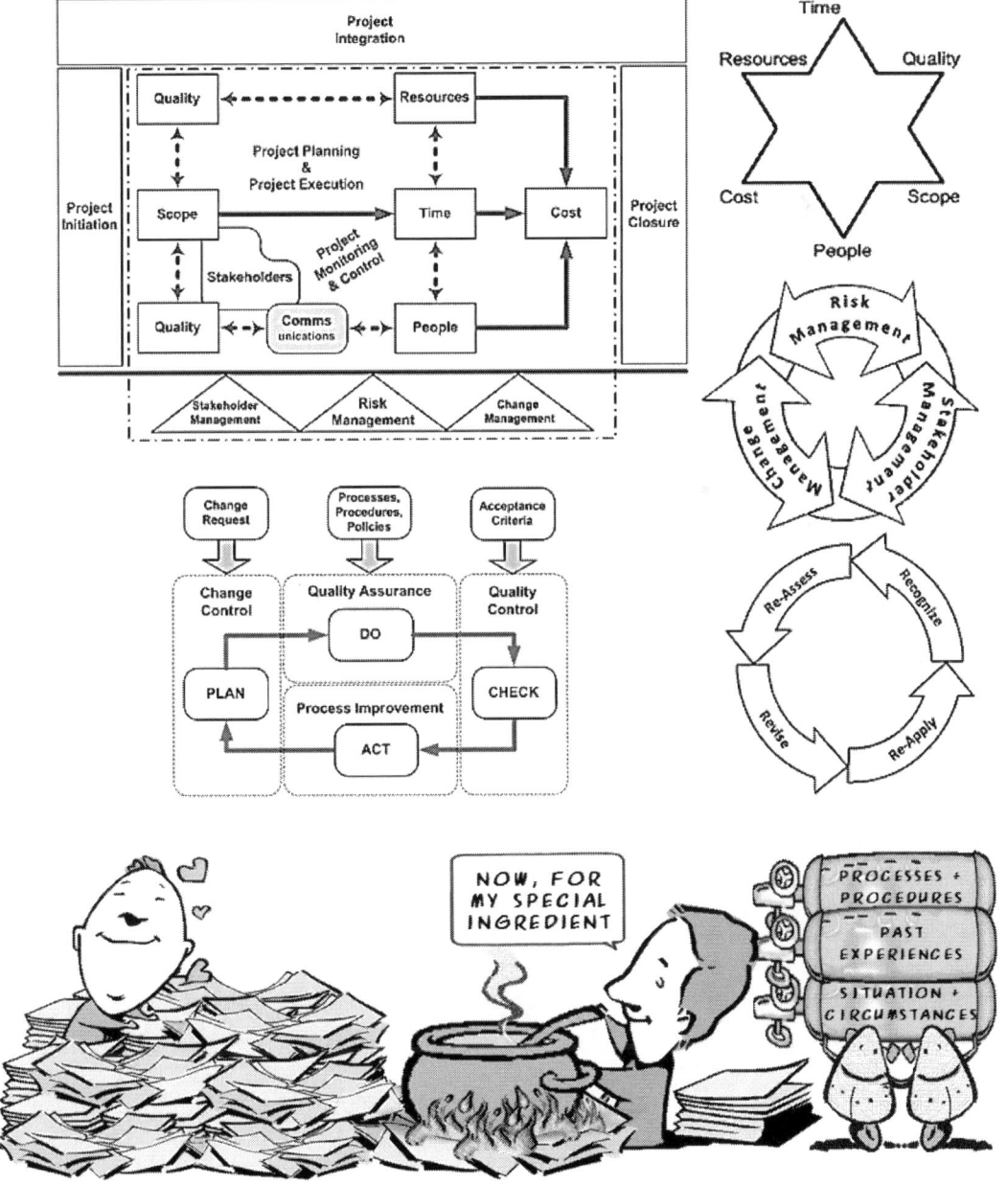

The PMBOK® Guide describes Project Integration Management as consisting of:

1. **Develop Project Charter** – the process of getting the project formally authorized, and determining / documenting what are the project's core-primary stakeholders' needs, wants, and expectations; i.e. the Initiating Phase.

2. **Develop Project Management Plan** – the process of establishing the rules for managing the project, and determining / documenting the plans for achieving the project's objectives; i.e. the Planning Phase.

3. **Direct & Manage Project Execution** – the process of getting the planned project work done; i.e. the Executing Phase.

4. **Monitor & Control Project Work** – the process of ensuring that the work done will achieve the project's stated objectives, and to make course corrections when necessary; i.e. the Monitoring & Control Phase.

5. **Perform Integrated Change Control** – the process of reviewing & approving all of the changes made to the project, specifically those changes made to the project's deliverables; i.e. R.I.S.C. Management.

6. **Close Project** – the process of completing the project work and formally concluding the project; i.e. the Closing Phase.

Reviewer's Comment ... Evolution of project integration.

Over the course of your career, you will be exposed to an assortment of projects, work places, organizations, industries, and personalities ... subsequently, your personal perspective will change on what project management techniques (in hindsight) do and don't work.

Hence, based on your lessons learnt, you will evolve over time to integrate an assortment of various project management processes & procedures so as to better adapt to the current situation and the prevailing circumstances that now confront your project.

> **AUTHOR'S NOTE**
>
> While you are looking through the PMBOK® Guide, you will also encounter the **Project Management Information System (PMIS).** Where the PMIS is an information system used to accumulate, incorporate, and disseminate the inputs & outputs of the Project Management Process.
>
> However, given that the project will have so much information being generated, as well as information coming & going; the project will need a system that can handle more than just the project management information. That is, the project will require some form of **Project Information Management System (PIMS).**

18.3. Project Information Management System (PIMS)

Over the life of the project, there will be a large amount of data, documentation, and artefacts (aka "stuff") that will be generated & accumulated due to the various processes & procedures utilized during the project's Initiating, Planning, Executing, Closing, Monitoring & Control Phases (as well as for any Project Rescue & Project Recovery that happens to be undertaken).

All of this "stuff" will be distributed amongst different project participants, and often this "stuff" will end up residing in various "private stashes". Some of these locations may be known to many of the project's participants, while other locations will be known to just a few, and other locations will be forgotten about in entirety.

However, with the use of such a **"Randomly Distributed Information System"**, some of these project participants will not have a clear understanding of what "stuff" does & does not exist, nor will they know where to find the latest versions of particular "stuff", let alone where they should look for this "stuff" at some point in the near or distant future.

Though, for the project participants to have any realistic chance of maintaining and controlling this "stuff" (aka project "information"), then there needs to be some way to accumulate this "information" into a **"Structured Searchable Centralized Repository"**.

Via this "Structured Searchable Centralized Repository" (i.e. the **Project Information Management System)**, this "information" could then be readily utilized; i.e. accumulated, incorporated, disseminated to the relevant project stakeholders, and thereby ensuring that the most up-to-date versions are made communally accessible during the entire life of the project.

Additionally, some of this "information" may need to be accessed long after the project has concluded, when possibly all of the project's participants have moved onto other things. This is when such a **Project Information Management System would provide significant retrospective benefit to the performing organization's knowledge base**.

PROJECT DATA, DOCUMENTATION, AND ARTEFACTS THAT ARE "SAFELY LOCKED AWAY" (OUT OF SIGHT) IN "PRIVATE STASHES" WILL SOON BECOME OUT OF MIND, AND ARE THUS PRACTICALLY USELESS FOR FUTURE ENDEAVOURS.

PIMS: Structure and Architecture

A **Project Information Management System (PIMS)** could be as simple as a group of associated directories / folders on a **network file share**. Alternatively, the PIMS could be composed of a **Document Management System (DMS)** and/or a **Content Management System (CMS)**; i.e. a collection of **Electronic Document Management System (EDMS)**. The PIMS could also contain **semi-autonomous or automated business process workflows**. This PIMS will come in very handy when the project and the performing organization happen to be quality audited; see [Section 17.4.4.1].

NOTE ❐ The PIMS does not necessarily have to be one system; the PIMS may be fabricated from a collection of independent or interoperable systems dedicated to specific purposes.

PIMS: Features

Note: this part won't be detailing a specific PIMS product to be used.

Irrespective of the size of the performing organization and/or the size of the project, the **PIMS employed will need to be at least capable of storing, maintaining, preserving, and retrieving the project's information**.

Additionally, it would be highly beneficial if the PIMS could:

- **Archive and then selectively retrieve different versions of the individual entries** of stored information; i.e. **version control**.

- Allow the **segmentation of the information into discrete containers of related materials**; *e.g. the siloing of information related to specific groupings.*

- Allow the **segregation of the information to guarantee the security, confidentiality, and functional usability** of the information contained within.

- **Control who has access** (read-write, read-only, none) to specific containers of information, as well as **recording who accessed** which particular information.

- **Control the simultaneous access** to information by multiple users (i.e. check-in / check-out), and allow the users to **collaborative**ly create & manipulate such information.

- **Identify / notify** the users of the **new and latest versions** of information so that obsolete / out-dated information is not "unintentionally" used.

- Manual or automatic **assignment of metadata** (i.e. descriptive indexing attributes) to individual entries of information so that the PIMS **retained information can be organized, navigated, filtered, and searched**.

- Allow the information to be **accessed via different mediums** (such as by file exploring and web browsing) as well as **via other services & systems**.

- A **scalable solution** that allows more and more information to be added to the PIMS, and optimally **allow the PIMS capabilities to be expanded upon** (via additional software features and hardware) as the projects' and the organization's needs dictate.

Often small to medium size organizations (with small to medium size projects) will think that they can forgo the establishment of some form of PIMS because the PIMS is not a project deliverable.

Thereby, mistakenly believing that this decision will save [Time] and restraint the [Cost] burden to the project.

Unfortunately, **the non-existence of an appropriate form of PIMS does not save on [Time] & [Cost], because the non-presence of relevant forms of PIMS often results in nightmares of missing information, use of the wrong versions of information, and rework to correct those mistakes due to the lack of an effective PIMS.**

PIMS: Establishment

When establishing the Project Information Management System (PIMS) that is to be utilized by the project, a few questions need to be answered during the selection process:

1) **What types of information are required to be stored** in the PIMS?

    For example; DOCs & PDFs, spreadsheets, diagrams, schematics, source code.

2) **How long** after the project has been completed, **will the information be kept**?

    • **Are there statutory requirements on how long a period does this information need to be retained?**

3) **How will** the project's **information be retained, archived, and preserved**?

4) **What computer operating systems** / platforms need to be supported?

5) **What other features** will this particular PIMS be required / expected to have?

    For example; integration with MS-Project & MS-Outlook, KPI reporting, workflows... calendars, to-do lists, document imaging ... etc.

6) **How will the cost of the PIMS be charged against the project**?

That is, as a **direct cost** to the project, **or** will this cost be incorporated into the business's ongoing **operational overhead** because this PIMS will also be used by other projects and personnel within the performing organization.

7) **Who is going to setup the PIMS** and **who will pay the cost of this setup**?

- Is this going to be a job for the I.T. department, an external contractor, a service hosting organization, or is a member of the project team going to have to do this?

- Has the Setter-Upper's [Time] & [Cost] been **marked against the project, or** is this considered as **zero-cost & zero-time** for the project?

  *For example; this establishment activity is considered as **"Business As Usual"** for the I.T. personnel in the same way as rolling out a workstation for a new employee.*

8) **Will the PIMS require dedicated [Resources]** specifically for this project, **or** will the PIMS utilize a **pool of resources**; *e.g. MS SharePoint, a server farm, cloud services... etc.*

9) **Where will the PIMS be located?**

That is, hosted internally or externally to the performing organization?

*For example; on a local server which is administered by the performing organization's I.T. personnel, on a server at the customer's organization, or provided by an external third party vendor as an internet accessible **Software As A Service (SAAS)**.*

10) **Who will administer the PIMS as a whole system and who will administer that portion related specifically to this project**; i.e. who will control access to what, assign read-write/read-only privileges, impose usage quotas, and police the "Terms Of Use"?

11) **Who will maintain the PIMS**; i.e. who with oversee the system's performance levels, install additional hardware, and upgrade the PIMS software?

12) **What will be the user layout of the PIMS** related to this project (i.e. the folder / directory structure / website map)?

The PIMS structure could be based on any variant or combination of the following (plus any other possible alternative arrangements):

- Broken up along the **demarcation lines** of the project lifecycle phases or SDLC stages as depicted in [Figure 26] and [Figure 31].

- Based on the **grouping of people** working on the project; i.e. by departments, teams, organizations.

- Based on the **layering or slicing used to modularize** the project, see [Section 2.3]; e.g. *Graphical User Interface (GUI), database, input sensors, output lines, real time kernel, hardware drivers, Application Programming Interface (API) ... etc.*

- Based on the **release milestones** or multiple collections of releases.

As previously stated, the PIMS can be much more than a collection of directories / folders on a file server ... but alas, **the PIMS at the project's disposal is very much dependent on those facilities made available either by the performing organization or by the customer organization. Consequently, the project's participants often have to make do with the tools that they are given and/or can hobble together.**

Reviewer's Comment ... The bounds of the project manager.

In my experience ... it is not unusual for the project manager to end up being the person who has to; choose which PIMS is to be used for the project, then decide on the layout of information segmentation, administer the segregation of the information, and generally act as the local administrator for that portion of the PIMS that is being utilized by the project. And all the while, trying to stay within the [Time], [Cost], [People], [Resource] limitations imposed on the project.

Oh, and while the PIMS is "technically" outside the domain of the project manager, when things go wrong and stuff is lost, it is often the project manager who is disproportionately held accountable for the occurrence of the problem and subsequently for the problem's resolution.

18.4. Project Information Segmentation & Segregation

As the project's information starts to be accumulated into the various sub-systems that compose the Project Information Management System (PIMS), then this information will need to be segmented & segregated.

❖ **Information Segmentation** – is dividing / partitioning of the information into **distinct silos of related content**; *e.g. how the PIMS subfolders are arranged.*

❖ **Information Segregation** – is **enforcing the boundaries between the various information silos**; *e.g. who has access to each particular PIMS subfolder.*

Such information segmentation & segregation enables different facilities & functionality to be provided to the specific individuals and user groups who "need to utilize" a particular information silo. Additionally, such information segmentation & segregation enables different Terms Of Use and utilization processes & procedures to be imposed on each particular silo; i.e. how a certain silo may be accessed, used, and monitored & controlled.

The Importance of Information Segmentation & Segregation

During the project's life, it is not in the performing organization's best interest to have just anybody trawling through the various project information.

- Only the members of the Project Steering Committee will "need" and "want" to have **access to confidential and sensitive information** such as, the contractual agreements between the customer and performing organizations. Similarly, the project's budget / cost calculations and the associated profit margins is information that the performing organization's senior management definitely do not want to have known by the customer's representatives, let alone have this information fall into a competitor's hands, nor be viewed by nosey employees & contractors.

- The members of the Project Working Group will need **read-write access** to the project's management information, but do not want just anyone "tinkering" at the project's controls. Yet, other parties will require **read-only access** to such project management information such as, project schedules, Detailed Specifications, acceptance test results.

- The performing organization does not want just anyone sifting through; the software source code, the hardware designs, and schematics ... as who knows what intellectual property could slip out a side door. Yet, the members of the Project Implementation **Team will need the freedom** to be able to build their various interim iterations of the deliverables (and its sub-components) **without the restraints imposed by draconian supervision & oversight**.

- The performing organization **does not want the customer's representative to obtain project documentation** or witness the project deliverables **until these are in a form that the performing organization intends to have viewed**. Given that, the perception of the performing organization can be negatively impacted upon if the customer's representative were to read a draft document before it has been sanitized to remove any incorrect information and censored any political inconveniences contained within (which implementers often overlook because they don't see the bigger picture).

  In addition, the customer representative's faith in the quality of the deliverable could be greatly diminished if they happened to read an internal Defect Report detailing the spectacular crash of an interim prototype prior to that defect's rectification.

ACCESS TO INFORMATION HAS TO BE ON A "NEED TO KNOW" BASIS AND NOT ON A "WANT TO KNOW" BASIS.

It is not unusual, that it falls to the project manager to decide who has access to what information for the project.

Early on in the project's life, the beginning of the segmentation & segregation of the project information needs to occur (i.e. **with clear lines of demarcation, and no-go boundaries**) based on the confidentiality of the information, the sensitivity of the information, and the purpose & usability of the information.

> *For example; contractual information, planning information, information furnished provided by the customer, information provided to the customer, information furnished from third parties, information sent to third parties, intellectual property (foreground and background), architectural & technical designs, software source code, hardware designs & schematics, development environments, test environments, objective quality evidence ... etc.*

Information Segmentation & Segregation makes life easier

While it is very important to monitor & control who has access and privilege to each particular silo of information, it is also **important that each information silo's authorized users are easily able to effectively & efficiently utilise the contained information** that they require to undertake their assigned project duties.

> *For example; the software development team need to check-in / checkout / merge code with the minimal of fuss without having to jump through multiple authorization hoops.*

> *Just give me the Frickin information that I need to do my job!!!*

 **If the Terms Of Use and the stipulated utilization processes & procedures impose too great a burden** and present barriers that retard those authorized users from doing their assigned project activities, **then eventually these users will establish alternative arrangements and ways to enable them to get on with what they need to do** (*such as creating their own uncontrolled private stashes of information*).

Restrictive Information Segmentation & Segregation

NOTE ☐ **Restrictive segmentation & segregation of the project information can inadvertently result in the covert creation of "Randomly Distributed Information Systems"** that the PIMS was originally intended to eliminate.

Additionally, such **restrictive utilization processes & procedures can be demoralizing and demotivating to the members of the project team.**

Do I really have to go through this crap every day just so that I can do my job? There has to be a better way of doing this!

In frustration, they will find an (uncontrolled) alternative way to get on with their job ... unfortunately this will be at the expense of Quality Assurance and Quality Control that the segmentation & segregation practices were supposed to benefit.

 I would recommend that ... serious thought be given to who exactly needs to access what information, and how such information will be utilized, as the Terms Of Use and the utilization processes & procedures may need to be adjusted accordingly in order to enable those particular authorized users to get on with doing their jobs.

 I would recommend that ... the project manager should escalate the issue to senior management when misguided corporate policies & procedures are riding rough-shot over the project team's ability to effectively & efficiently achieve their assigned project activities.

 Rules and regulations for the sake of rules and regulations is a sure-fire way to restrain the successful implementation of the project.

THERE'S MORE TO PROJECT MANAGEMENT
THAN JUST KNOWING ALL OF THE THEORY
AND USING ALL OF THE CORRECT JARGON.
AT THE END OF THE DAY; YOUR SUCCESS
OR FAILURE AS A PROJECT MANAGER WILL
DEPEND SOLELY ON YOUR ABILITY TO ...
ADAPT & IMPROVISE TO THE PROJECT'S
CURRENT SITUATION & THE PREVAILING
CIRCUMSTANCES, AND THE UTILIZATION OF
THAT WHICH IS AVAILABLE TO YOU.

THESE ARE THE TOOLS BY WHICH YOU WILL GUIDE THE PROJECT TO SUCCESS.

[1st Edition]

As I sit here at the back of the bus, marooned in a sea of red taillights, I come to the realization that, *"this is it, THE END"*, the final chapter in this epic journey.

It has taken over two years to get this far, but alas here we are with only a publisher's imposed constraint on the total number of physical pages that are permissible.

[2nd Edition]

Well, it is coming up on two year since the 1st Edition of this textbook was published, and during that time, some things have changed, while others have remained the same.

I published a sci-fi novel, "HarliQuinn Dreams and the Celestial Queen".

I invested a lot of time & effort into a rocket-ship ride of a start-up company, only to have it fall from the sky due to their surprised major cash flow crisis.

To then dust off from that demise, and begin again.

As for those who contributed to this book ... the (global) economic climate has had its highs and its lows; for some industries it has been full steam ahead, while for other industries it has been the living end, and for those in between it has been a period of continued stagflation.

Many real-world example cases have been suggested and some personally experienced, and selected ones have been adapted for incorporation into this latest edition of the textbook.

Thank you to those who contributed ... given your work situation and circumstances at the time of writing.

Moreover, thanks to you and those others who have read my book.

Cheers.

19. THE END ... project management, is it the right choice for you?

A View From A Bridge

Well you have decided to take that next big step with your career and move on from being a project implementer to becoming a project manager.

In this final chapter I will cover some things that you may want to give serious thought to before you take that next step; because, you could (in retrospect) find yourself pondering whether in fact you've chosen to step off a bridge rather than advancing your career.

That is, this chapter will question whether project management is currently the right career move for you, your family, your lifestyle, and ... also, to question whether project management is in fact going to advance you in the direction you really want to go.

TO BE OR NOT TO BE...
NOW THAT IS, THE QUESTION

DOER LAND

PM LAND

Why did you decide that becoming a project manager was the next step for you?

Was it simply because you have progressed as far as you can go in the technical hierarchy (at your current employer's organization), and thus moving across into project management is the next step required for you to advance any further with your career?

Is the change to project management something that is expected of you by; your employer, your industry, your peers, your family, or just what seems natural to do?

...

So, **is this something that you really want to do for yourself; just because, being an implementer "no longer feels right" for you ... or is this what you are expected to do?**

Are you willing to accept that not everything you have learnt and done so far in your career may be applicable to enable you to succeed as a project manager?

Just because you **have good technical skills and possess an abundance of experience with project implementation does not necessarily mean that you will be a competent project manager**. These skills & experiences could in fact cause you to FAIL ... BIG TIME.

GOOD TECHNICAL SKILLS + EXPERIENCE ≠ COMPETENT PROJECT MANAGER

This conundrum is because some of the techniques & aptitudes that helped you to achieve your previous technical implementation successes could actually work against your future long-term success as a project manager.

Please refer to [Section 11.1].

Are you going to be comfortable with saying, "We made this" rather than "I made that"?

As an implementer, you could have been a mono-thought individual who methodically worked their way through the technical problems to come up with the best possible solutions. But now, **as the project manager you will no longer be the one who works through these technical problems; rather, other people are going to work on these types of problems, while you strive for an overall resolution.**

That is, **your role as project manager will no longer be 'TO DO' but rather to make sure 'IT GETS DONE'.** So do you think that you can be; a communicator, a listener, a negotiator, an influencer, a conflict resolver, a problem solver, a planner, an organizer, a delegator, a motivator, a leader, a team builder, a multi-tasker, … but not an implementer?

Please refer to [Section 5.1] and [Section 11.3].

Are you going to be comfortable with giving others the implementation reigns while you remain hands-off?

As an implementer you may have really *"known your stuff"*, but as the project's manager you will occasionally find yourself in the uncomfortable situation where the implementers assigned to work on your project team *"don't know shit"* as your technical competency exceeds theirs.

That is, your project's implementers could be producing work and deliverables nowhere near the efficiency & effectiveness that you personally are capable of delivering, let alone to your personal quality standards. Under these circumstances **are you the type who would yell, *"Get out of the way you fool"*, *"If you want something done right, you"*.**

Way to go … you just cheesed off that developer and all the other members of the project team, with your evident lack of faith in them.

Are you going to be comfortable with trusting your implementers with the freedom to do their work in a manner that they deem appropriate?

As the project manager, do you think that, to ensure the success of the project, that you would try to control and organize every aspect of the tasks that you have assigned to your implementers? Hence, **would your management style be based on Theory X where you feel that you need to micromanage everything that your implementers do?**

Alternatively, do you think you would trust them enough to give them the opportunity and the latitude to undertake their assigned tasks as they deem necessary, without your continual oversight; i.e. **will your management style incorporate Theory Y?**

IMHO, many a project manager doesn't really want to be Theory X, but they're afraid that if they don't manage that way then their people will only stuff-up. Alas, this manager doesn't understand that coaching, mentoring, and freedom (to occasionally fail) can help their people to establish their own problem solving and solution creation methods.

Please refer to [Section 11.3] and [Section 14.1.2].

"WHY MUST I BE SURROUNDED BY FRICKEN IDIOTS?"

AUSTIN POWERS 1997

Are you prepared to go from the top of the technical pile to being at the bottom of the management heap?

By this stage in your career, you are most probably one of the senior project implementers at your employer's organization. **As an implementer you are respected for your skills, local knowledge, expertise, and your past experiences**. Your technical opinion counts as it is sought by even the most senior of staff, and you know all of the ins & outs of executing a project at your employer's organization.

However ... **once you move across into project management, the GAME RESETS**.

You will be starting again at the bottom rung of the ladder, as a mere junior project manager. Senior management are not going to be seeking your opinion or advice on management issues, and if you are consulted about some future project, it may only be to confer with you as the awe-inspiring implementer that you once were. This situation is even worse if you had to jump-ship to a new employer to get the opportunity to move into project management, as your years of technical experience will probably count for almost nothing, as those who may have sought such advice will not know of your past exploits simply because you are new to their organization.

So, **are you honestly prepared for all of that hard technical work to be put aside, and to then start the long climb back up the "earn respect mountain"?**

Are you really prepared to spend all of that time to gain the relevant experience as a manager, and toss aside some or most of that cred which you gained as an implementer?

Are you prepared to forget some of what you have already learnt (if not most of what you thought you knew), and start the learning process again; like a mature age student going back to finish high school or university?

And ... are you prepared for the possibility of being subordinate to someone a lot less knowledgeable than you, just because they started earlier on the climb up the management ladder at your current employer's organization?

Are you prepared to be first in line for retrenchment if things go financially wrong?

As a senior implementer, the thought of your employer even contemplating making you redundant was "inconceivable". You had way too much relevant experience on their projects, products, and operations; plus, you possessed copious amounts of (mystical) proprietary intellectual knowledge that your employer strategically could not afford to lose. Even during the worst of a financial crisis, your employer would have to have been on the verge of bankruptcy before they would contemplate letting you go ... as they (and you both) knew that you were exactly what they required to rebuild their crippled business, let alone enable them to survive the current predicament.

When you were a junior implementer then you were probably too naive to imagine that such a thing as retrenchment / redundancy could "happen to you". Well, **by moving across into project management, you have effectively repositioned yourself as a junior again.** Hence, **as a junior project manager, you are no longer considered as essential personnel for the crisis survival of the organization.**

Oh yeah, ever noticed that when the proverbial "shit hits the financial fan" it is the lower ranks of management and the junior implementers who are the first ones out onto the street.

This is because, when *"push comes to shove"*, it is the junior project managers and the middle project managers that the senior management at the organization choose to shoot first, rather than themselves. Senior management know that under the worst financial conditions, they can burden themselves with also doing your project management role, but even they realize that they cannot do the implementers roles.

However, sometimes they don't even know who are the implementers essential to their organization's survival.

Are you prepared to carry the destiny of your employer on your shoulders, and to hold the fate of your colleagues' livelihood in your hands?

Depending on the size of the organization that you are working for (and how far up the management hierarchy you go), you could very well find yourself in the unenviable position of managing the project or operations that is going to either financially make or break your employer's business.

You will be carrying a lot of responsibility on your shoulders. How will you feel if it all goes wrong (irrespective of whether it was or was not beyond your control) **and your employer's survival necessitates that some of your colleagues, your team are let go?**

War Story ... Sorry, but I would choose to shoot you first.

A project manager work-friend of mine once told me at a backyard BBQ that if he had to make the decision; then I would be the first person that he would have chosen to retrench. Before I could say, "you Bastard", he added, "you're single with limited commitments and I am very confident that you would have a new job in no time at all" (which was true at that time and point in my career), "but some of these older guys ... well I just couldn't bear that weighing on my mind every time I tried to go to sleep".

Jump forward a few years and I knew exactly how he felt ... Sitting there surreally in a coffee shop sipping on caffelatte with the two heads of the business, discussing the pros & cons of which of the staff members have to be let go. While strategic worth to the organization was a major concern, that individual person's home life circumstances did surprisingly come into the selection equation. I suppose this is partly why younger employees (who don't have such a high financial & family burden) and new employees (who we don't really know yet) are often the prime candidates for the first round of redundancies.

Your Morals Vs. Others Lack Of Business Ethics?

As an implementer, life was a pretty simple arrangement, where you were required to turn some specifications – requirements – ideas into something real, that did something that someone somewhere wanted and needed. Thus, at the end of the day you could honestly and proudly say, "*I made this*".

Now, at some point in your project management career (especially when the economy is bad and your employer is not doing so well), **you could be requested to do something(s) that are at odds with your sense of business ethics and your personal morals**.

> *For example; you are (verbally) requested by senior management to invoice the customer for a portion of the project work that is yet to be done. ... Though, your boss has reassured you that it will be done first thing next week / month. ... BUT, it needs to be invoiced now, so as to improve the organization's financial results for this month / quarter / End Of Financial Year's reporting.*

You could raise your concerns, highlight your hesitation to do as requested, or you could even humbly decline to do this request. However, **would your "*high morals*" stance** be marking you as "*not a team player*", and hence could you **possibly be limiting your career's future advancement at that organization (under that management's reign)?**
And, what would happen if you decided to escalate this issue, to the bosses above them?

Sometimes the situation is not so black & white, as to what is ethically right or wrong.

> *For example; there are two unrelated but simultaneous projects for the same customer. Both of these projects have been "sold" to the customer as being 1000 hours of effort each. However, during the implementation, one project is a lot easier than expected and only required 750 hours of effort, whereas the other project was harder than expected and required 1250 hours to complete. Your senior management tell you, as the project manager, "to just", shift hours in the timesheet system from the over-hours project to the under-hours project, and then invoice the customer as though each project was in factuality a 1000 hour job.*

In that previous example case; *"no overall harm was done"*, as the customer got what was promised, and the performing organization *"broke even"* on both of these projects.

Now, let's *"bend"* the relationship a little bit more.

> *For example; there is a fixed price project for the customer, and at the same time, that customer also has an ongoing contract with the performing organization to provide a certain number of maintenance hours per year for their suite of products. However, during the project's implementation, it goes way over the hours of effort and will be a loss maker for the performing organization. Your senior management tell you, as the project manager, "to just", shift hours in the timesheet system from the over-hours project to the maintenance contract's job code. ... Given that, the customer is not using up all of those maintenance hours, and the project's product output would just use up those maintenance hours when it's released, so might as well just use those maintenance hours now during the product's development.*

So, **HOW FAR IS YOUR SENSE OF REAL-WORLD BUSINESS ETHICS PREPARED TO BEND?**

Before you decide on what is right & wrong, reconsider those previous examples and the following examples, and set these during a major (global) financial crisis for your employer.

> *For example; there are two unrelated but simultaneous projects for two unrelated customers. One project is going really well and is not consuming all of the planned hours of effort, but the other project has real problems and is burning through its hours of effort. Your senior management tell you, as the project manager, "to just", shift hours in the timesheet system from the over-hours project to the under-hours project, and then invoice each customer as per the timesheet factualities.*

> *For example; there are two unrelated but simultaneous projects for two unrelated customers, but there is a common body of work to be developed. Your senior management tell you, as the project manager, "to just", book both customers for the full amount of the common work. In essence, invoice double dipping.*

HOW BENDABLE IS YOUR ETHICS, *"just to"* retain your job during such an economic crisis?

Just how far up the management hierarchy, do you want to go?

Have you mapped out a 5 to 10 year plan?

Is your plan; a few years as a project manager, then on to becoming a program manager, then the head of a department or business unit, then general manager, then managing director, then chief executive, then the sky is the limit?

Well this is all up to you, and to be honest, I should not and cannot try to advise you on your decision because it is exactly that ... your decision, not mine.

However, please take note that **the higher up the management hierarchy you go then the more of a "project accountant" and "politician" you will have to become**.

You will be more and more involved with budgets, profits & loss, revenues & expenses, returns on investments, share prices, organizational politics, and interactions with heads of other organizations and with the company's shareholders.

Over time, as you rise **higher up through the management ranks**, you will be less and less managing the implementations, **less and less involved with** understanding the **technical aspect** of what your organization is building.

Therefore, if you still want to be able to say, "*I / we made this*" then most probably there will be a limit to how far you will want to advance in your employer's organization before you honestly feel uncomfortable with what you are doing. To remedy this you could venture off into business development and product roadmaps, but here to, you may feel that the products are becoming more and more black boxes to you.

You may occasionally come and reminisce with a project team about your past glory days as an implementer, but while they will be polite and listen to your war stories ... in the back of your mind, you will be wondering if they are thinking, "Gee that is so yesterday, old man." ... As even more grey hairs, join the others as a memento of your latest battle campaign.

Just how much time are you prepared to commit?

Are you generally looking for a 9 to 5, Monday to Friday job with maybe the occasional overtime required?

Are you prepared to do extra work on the weekends and at night when you get home?

Are you prepared to be the first into work and the last one to leave?

Well, depending on how new you are to project management (at your current employer's organization), how many projects you are responsible for and have on the go at any one time, or how far up the management hierarchy you plan to go, you could find that it is necessary for you to continually work a lot more (unpaid) hours than you did as an implementer.

Therefore, **will your decisions about becoming a project manager or a higher management rank mean that in addition to your day job, you will also have a night job of catching up on that work which you didn't get done during the day?**

WHY DO WE GO TO WORK IN THE DARK, YET WORK INSIDE WHILE IT IS SUNNY, THEN GO HOME IN THE DARK?

Just how much time are your spouse / partner and your family prepared to commit?

Your decision to move into project management (and eventually to move higher up in the management hierarchy) **doesn't just affect yourself, as it will also affect your spouse / partner, your family, and your friends**.

You may now have a higher paying job (though with the trade-off of requiring more of your time); but, just how long is it going to be before the other members of your home life start to question what this "promotion" really entails for them and not just for you?

Your spouse / partner, your children, and even your friends are not going to be prepared to put their lives on permanent hold while you complete another "terribly important" project.

Therefore, unless you are planning to have no family, no friends, and no social life then **you will need to decide (with your personal-life stakeholders):**

1) What is your "level of commitment" to your work and to your home life?
2) What are the potential consequences for your personal-life stakeholders?
3) What is your plan for balancing your work & home lives?

War Story … A winner at work, yet a Loser at home.

One night, after another long day at the office, are you going to walk in the front door of your home to find; someone's bags are packed, divorce papers are sitting on the dinner table next to another cold cooked meal, you don't even recognize your son & daughter, you can't recall the last time you heard more than a grunt out of them, and your once favourite tabby cat looks at you as if to say, "who you, why you here, we don't need you, this my place, go away you LOSER".

EVERYONE HAS TWO PARTS TO THEIR LIFE ...
WORK AND HOME.

WHEN ONE OF THESE IMPOSES TOO MUCH
ON THE OTHER THEN THERE WILL BE
NEGATIVE CONSEQUENCES THAT
SOMEONE WILL EVENTUALLY HAVE TO PAY.

Most likely, YOU.

AUTHOR'S NOTE

Unfortunately, the physical size of this book has reached the publisher's limitation on the maximum number of allowable pages. Hence, niceties such as a traditional index have been left out. To find the specific topic that you are looking for, please look at the grid index, or download the eBook version and do a word search.

20. References & Index

References

- PMI (2008/13), "A Guide to the Project Management Body of Knowledge (PMBOK® Guide) 4th/5th Editions", Project Management Institute, PA, USA.

- www.Wikipedia.org, the Free Encyclopaedia. Wikimedia Foundation, Inc.

- Belbin, Meredith. (1981), "Management Teams", London, UK.

- Clarkson, M. (1999), "Principle of Stakeholder Management", Toronto, Canada.

- Cleland, D. (1998), "Stakeholder Management", New York, USA.

- DeMarco ,Tom & Lister, Timothy (1999), "Peopleware, Productive Projects and Teams" 2nd edition.

- Deming, W. Edwards (1986), "Out of the Crisis", Cambridge.

- Gardiner, P.D. (2005), "Project Management: a Strategic Planning Approach", New York, USA

- Herzberg, F., Mausner, B. & Snyderman, B.B. (1959), "The Motivation to Work", New York, USA.

- Kniberg, Henrik (2007), "Scrum and XP from the Trenches", USA.

- Kniberg, Henrik & Skarin, Mattias (2010), "Kanban and Scrum, making the most of both", USA.

- Lewin, K.; Lippitt, R.; White, R.K. (1939). "Patterns of aggressive behavior in experimentally created social climates", USA.

- Maslow, Abraham (1954), "Motivation and Personality", New York, USA.

- McGregor, Douglas (1960), "The Human Side of Enterprise", USA.

- Shewhart, Walter Andrew (1939), "Statistical Method from the Viewpoint of Quality Control", New York, USA.

- Tuckman, B. W. & Jensen, M. A. (1977), "Stages of small-group development revisited", USA.

- *Sources unknown ... from those dark crevasses of my mind, from wherever I recalled this stuff ... while imprisoned on this motionless bus stuck in another meaningless traffic jam ... ARRRRRRRRRRR !!!*

Index - Grid

Key Word	Theory	Initiating Phase	Planning Phase	Executing Phase	Monitoring & Control	Closing Phase
Acceptance Testing					13.1	17.4.1, 17.5.7
Agile - SDLC	2.2				6.4.4, 6.4.5, 7.4, 12.2.2	
Budgets			4.3.3, 4.4.5			
Business Analysis		3.3.1			6.4	
Change Control					8.1, 8.1.3	17.4.1, 17.4.2
Change Management	2.5		4.3.3		8.1, 10.2, 15.1, 15.4	
Change Requests					10.2.1	17.4.1, 17.5.5
Communications Management			4.3.4		10.3.4	17.4.1
Completion Checklist		3.7	4.8	5.9		17.7
Completion Review		3.6	4.7	5.8		17.4.2, 17.4.4
Cost - Project Variable	1	3.3.3	4.3.3		7.3, 12.3	17.4.1
Customer Requirements		3.4.4	4.3.1			
Defect Reports					10.2.2, 13.1.3	17.5.6
Defect Resolution					13.1.4	
Detailed Specifications		3.4.6	4.3.1, 4.4.4			
EVPM - Earned Value Performance					7.3.1, 9.1.2, 12.3	
Iterative - SDLC	2.2, 2.4				6.4.4, 6.4.5	

Key Word	Theory	Initiating Phase	Planning Phase	Executing Phase	Monitoring & Control	Closing Phase
KPI - Key Performance Indicators					12.3.3, 12.3.4, 15.5	
OQE - Objective Quality Evidence					13.1.5	17.4.2, 17.5.4
Packaging & Dispatch					14.2.2, 14.2.3	17.4.1
People - Project Variable	1		4.3.2, 4.3.3		9.1, 14.1	17.4.1
Plan Do Check Act		2.6.1		5.3	6.2.2, 8.1	
PMBOK	2.8	3.1, 3.4	4.1, 4.4	5.2, 5.5	6.2	17.2, 17.5
Process Improvement					8.1, 8.1.3	
Project Charter		3.4.4				
Project Management Plan			4.4.3			
Project Manager - humanities	19				11.1, 11.2, 11.3, 11.5	
Project Recovery					16.4	
Project Rescue					16.3	
Project Team - humanities					11.4, 11.5, 14.1	
Quality - Project Variable	1		4.3.4		8.1, 13	17.4.1
Quality Assurance			4.3.4		8.1, 8.1.1, 13.1.1	17.4.4
Quality Control			4.3.4		8.1, 8.1.2, 13.1.1	17.4.4
Resources - Project Variable	1		4.3.2, 4.3.3		9.1, 14.2	17.4.1
Risk Management	1.6	3.4.5	4.3.5		10.1, 15.1, 15.2	

Key Word	Theory	Initiating Phase	Planning Phase	Executing Phase	Monitoring & Control	Closing Phase
Risk Matrix					10.1.1, 15.2.1	
Risks & Issues Register		3.4.5	4.4.6		15.2.1	
RTM - Requirement Traceability Matrix			4.4.5		7.1, 12.1.1	17.4.1, 17.4.2
Schedule			4.3.1, 4.3.2, 4.3.3, 4.4.5		7.2.1, 12.2.1, 12.3.1	
Scope - Project Variable	1	3.3.3	4.3.1		7.1, 12.1	17.4.1
SDLC - Life Cycle	2				6.3, 6.4.4, 6.4.5	
Stakeholder Matrix					10.3.3, 10.3.4	
Stakeholders Management	2.4	3.4.7	4.3.4, 4.4.7		10.3, 15.1, 15.3	17.4.1, 17.4.3
Testing Coverage Matrix					13.1.2	17.4.1, 17.4.2, 17.5.8
Time - Project Variable	1	3.3.3	4.3.1, 4.3.2, 4.3.3		7.2, 12.2	17.4.1
Waterfall – SDLC	2.2, 2.4				6.4.4, 6.4.5	
Work Breakdown Structure			4.3.1, 4.4.5		7.1, 12.1.2	

Printed in Great Britain
by Amazon